Methods in Molecular Biology

Series Editor
John M. Walker
School of Life and Medical Sciences
University of Hertfordshire
Hatfield, Hertfordshire, UK

For further volumes:
http://www.springer.com/series/7651

For over 35 years, biological scientists have come to rely on the research protocols and methodologies in the critically acclaimed *Methods in Molecular Biology* series. The series was the first to introduce the step-by-step protocols approach that has become the standard in all biomedical protocol publishing. Each protocol is provided in readily-reproducible step-by-step fashion, opening with an introductory overview, a list of the materials and reagents needed to complete the experiment, and followed by a detailed procedure that is supported with a helpful notes section offering tips and tricks of the trade as well as troubleshooting advice. These hallmark features were introduced by series editor Dr. John Walker and constitute the key ingredient in each and every volume of the *Methods in Molecular Biology* series. Tested and trusted, comprehensive and reliable, all protocols from the series are indexed in PubMed.

Protein-Protein Interactions

Methods and Protocols

Edited by

Shahid Mukhtar

Department of Biology, University of Alabama at Birmingham, Birmingham, AL, USA

Humana Press

Editor
Shahid Mukhtar
Department of Biology
University of Alabama at Birmingham
Birmingham, AL, USA

ISSN 1064-3745 ISSN 1940-6029 (electronic)
Methods in Molecular Biology
ISBN 978-1-0716-3326-7 ISBN 978-1-0716-3327-4 (eBook)
https://doi.org/10.1007/978-1-0716-3327-4

This Humana imprint is published by the registered company Springer Science+Business Media, LLC, part of Springer Nature.
The registered company address is: 1 New York Plaza, New York, NY 10004, U.S.A.

Preface

This book provides a comprehensive overview of classic and cutting-edge methods and techniques in mapping protein-protein interactions. The chapters will include a variety of in vitro and in vivo experimental methods covering cell biology, biochemistry, and biophysics. In addition, the book will also focus on in silico methods including sequence-, structure-, and phylogenetic profile-based approaches as well as gene expression and machine learning methods. The book is equally useful as a graduate textbook or as standard laboratory protocols while entertaining the specialized audience.

Birmingham, AL, USA *Shahid Mukhtar*

Contents

Contributors

AFTAB AHMAD • *Department of Anesthesiology and Perioperative Medicine, Birmingham, AL, USA*

MOPELOLA O. AKINLAJA • *Department of Biochemistry and Molecular Biology, University of British Columbia, Vancouver, BC, Canada; Michael Smith Laboratories, Vancouver, BC, Canada*

LOUAI ALIDRISSI • *BESE Division, Plant Cell and Developmental Biology, King Abdullah University of Science and Technology, Thuwal, Kingdom of Saudi Arabia*

FIN AMIN • *Department of Electrical and Computer Engineering, North Carolina State University, Raleigh, NC, USA*

PEPIJN BAIS • *BESE Division, Plant Cell and Developmental Biology, King Abdullah University of Science and Technology, Thuwal, Kingdom of Saudi Arabia*

PADAM S. BHATT • *Department of Biology, Texas State University, San Marcos, TX, USA*

MICHAEL R. BLATT • *Laboratory of Plant Physiology and Biophysics, School of Molecular Biosciences, University of Glasgow, Glasgow, UK*

IKRAM BLILOU • *BESE Division, Plant Cell and Developmental Biology, King Abdullah University of Science and Technology, Thuwal, Kingdom of Saudi Arabia*

LISA VAN DEN BROECK • *Department of Plant and Microbial Biology, North Carolina State University, Raleigh, NC, USA*

BENOÎT CASTANDET • *Institute of Plant Sciences Paris-Saclay (IPS2), Université Paris-Saclay, Université Paris-Cité, CNRS, INRAE, Université Evry, Gif sur Yvette, France*

QUEENIE W. T. CHAN • *Michael Smith Laboratories, Vancouver, BC, Canada*

SIXUE CHEN • *Department of Biology, University of Mississippi, Oxford, MS, USA*

SAIFUL M. CHOWDHURY • *Department of Chemistry and Biochemistry, University of Texas, Arlington, TX, USA*

CHARLES CHRISTOFFER • *Department of Computer Science, Purdue University, West Lafayette, IN, USA*

SHAREL CORNELIUS • *Department of Chemistry and Biochemistry, University of Texas, Arlington, TX, USA*

JOREN DE RYCK • *Department of Plant Biotechnology and Bioinformatics, Ghent University, Ghent, Belgium; Center for Plant Systems Biology, Ghent, Belgium; iRIP Unit, Laboratory of Microbiology, Department of Biochemistry and Microbiology, Ghent University, Ghent, Belgium*

ÉTIENNE DELANNOY • *Institute of Plant Sciences Paris-Saclay (IPS2), Université Paris-Saclay, Université Paris-Cité, CNRS, INRAE, Université Evry, Gif sur Yvette, France*

GUNNAR DITTMAR • *Department of Infection and Immunity, Luxembourg Institute of Health, Strassen, Luxembourg; Department of Life Sciences and Medicine, University of Luxembourg, Esch-sur-Alzette, Luxembourg*

DANISH DIWAN • *Department of Biology, University of Alabama, Birmingham, AL, USA*

ARMIN DJAMEI • *Department of Plant Pathology, Institute of Crop Science and Resource Conservation (INRES), University of Bonn, Bonn, Germany*

SIDDHARTHA DUTTA • *Department of Microbiology and Biotechnology, Sister Nivedita University, Kolkata, West Bengal, India*

J. Mitch Elmore • *USDA-Agricultural Research Service, Cereal Disease Laboratory, St. Paul, MN, USA; USDA-Agricultural Research Service, Corn Insects and Crop Genetics Research, Ames, IA, USA; Department of Plant Pathology, Entomology and Microbiology, Iowa State University, Ames, IA, USA*

Ali Zeeshan Fakhar • *Department of Biology at University of Alabama, Birmingham, AL, USA*

Leonard J. Foster • *Department of Biochemistry and Molecular Biology, University of British Columbia, Vancouver, BC, Canada; Michael Smith Laboratories, Vancouver, BC, Canada*

Prabu Gnanasekaran • *Department of Plant Pathology, Washington State University, Pullman, WA, USA*

Sofie Goormachtig • *Department of Plant Biotechnology and Bioinformatics, Ghent University, Ghent, Belgium; Center for Plant Systems Biology, Ghent, Belgium*

Alain Goossens • *Department of Plant Biotechnology and Bioinformatics, Ghent University, Ghent, Belgium; Center for Plant Systems Biology, Ghent, Belgium*

M. Michael Gromiha • *Department of Biotechnology, Bhupat and Jyoti Mehta School of Biosciences, Indian Institute of Technology Madras, Chennai, India; Department of Biological Sciences, Purdue University, West Lafayette, IN, USA*

Lore Gryffroy • *Department of Plant Biotechnology and Bioinformatics, Ghent University, Ghent, Belgium; Center for Plant Systems Biology, Ghent, Belgium; iRIP Unit, Laboratory of Microbiology, Department of Biochemistry and Microbiology, Ghent University, Ghent, Belgium*

Kannan Harini • *Department of Biotechnology, Bhupat and Jyoti Mehta School of Biosciences, Indian Institute of Technology Madras, Chennai, India; Department of Biological Sciences, Purdue University, West Lafayette, IN, USA*

Mehadi Hasan • *Department of Biology, University of Alabama at Birmingham, Birmingham, AL, USA*

SeyedMohsen Hosseini • *Department of Computer Science, University of Western Ontario, London, ON, Canada*

Ahmad Hassan Ibrahim • *Bioinformatics Graduate Program, Graduate School of Natural and Applied Sciences, Muğla Sıtkı Koçman University, Muğla, Turkey*

Lucian Ilie • *Department of Computer Science, University of Western Ontario, London, ON, Canada*

Veronique Jonckheere • *iRIP Unit, Laboratory of Microbiology, Department of Biochemistry and Microbiology, Ghent University, Ghent, Belgium*

Mattson Jones • *Department of Infection and Immunity, Luxembourg Institute of Health, Strassen, Luxembourg*

Abu Hena Mostafa Kamal • *Department of Molecular and Cellular Biology, Baylor College of Medicine, Houston, TX, USA*

Hong-Gu Kang • *Department of Biology, Texas State University, San Marcos, TX, USA*

Onur Can Karabulut • *Bioinformatics Graduate Program, Graduate School of Natural and Applied Sciences, Muğla Sıtkı Koçman University, Muğla, Turkey*

Rucha Karnik • *Laboratory of Plant Physiology and Biophysics, School of Molecular Biosciences, University of Glasgow, Glasgow, UK*

Betül Asiye Karpuzcu • *Bioinformatics Graduate Program, Graduate School of Natural and Applied Sciences, Muğla Sıtkı Koçman University, Muğla, Turkey*

Salla Keskitalo • *Institute of Biotechnology, HiLIFE Helsinki Institute of Life Science, University of Helsinki, Helsinki, Finland*

Mamoona Khan • *Department of Plant Pathology, Institute of Crop Science and Resource Conservation (INRES), University of Bonn, Bonn, Germany*
Daisuke Kihara • *Department of Biological Sciences, Purdue University, West Lafayette, IN, USA; Department of Computer Science, Purdue University, West Lafayette, IN, USA*
Sung-Il Kim • *Department of Biochemistry and Biophysics, Texas A&M University, College Station, TX, USA*
Nilesh Kumar • *Department of Biology, University of Alabama at Birmingham, Birmingham, AL, USA*
Pey Yee Lee • *UKM Medical Molecular Biology Institute (UMBI), Universiti Kebangsaan Malaysia, Kuala Lumpur, Malaysia*
Qing X. Li • *Department of Molecular Biosciences and Bioengineering, University of Hawaii at Manoa, Honolulu, HI, USA*
Zhuoran Lin • *State Key Laboratory of Crop Stress Biology for Arid Areas and College of Life Sciences, and Institute of Future Agriculture, Northwest Agriculture & Forestry University, Yangling, Shaanxi, China*
Jinbao Liu • *Department of Biology at University of Alabama, Birmingham, AL, USA*
Kun-Hsaing Liu • *Department of Molecular Biology and Centre for Computational and Integrative Biology, Massachusetts General Hospital, and Department of Genetics, Harvard Medical School, Boston, MA, USA; State Key Laboratory of Crop Stress Biology for Arid Areas and College of Life Sciences, and Institute of Future Agriculture, Northwest Agriculture & Forestry University, Yangling, Shaanxi, China*
Shaoqun Liu • *College of Horticulture, South China Agricultural University, Guangzhou, China*
Xiaonan Liu • *Institute of Biotechnology, HiLIFE Helsinki Institute of Life Science, University of Helsinki, Helsinki, Finland; Department of Physiology, Faculty of Medical Sciences in Katowice, Medical University of Silesia, Katowice, Poland*
Teck Yew Low • *UKM Medical Molecular Biology Institute (UMBI), Universiti Kebangsaan Malaysia, Kuala Lumpur, Malaysia*
Claire Lurin • *Institute of Plant Sciences Paris-Saclay (IPS2), Université Paris-Saclay, Université Paris-Cité, CNRS, INRAE, Université Evry, Gif sur Yvette, France*
Imani Madison • *Department of Plant and Microbial Biology, North Carolina State University, Raleigh, NC, USA*
Aqsa Majeed • *Department of Biology, University of Alabama at Birmingham, Birmingham, AL, USA*
Patricia Martínez-Botía • *Instituto de Investigación Sanitaria del Principado de Asturias (ISPA), Oviedo, Spain*
Sergio Martin-Ramirez • *Wageningen University & Research Laboratory of Biochemistry, Wageningen, The Netherlands*
Ashima Mehta • *Department of Chemistry and Biochemistry, University of Texas, Arlington, TX, USA*
Binoop Mohan • *Department of Biology, University of Alabama, Birmingham, AL, USA*
Dario Monachello • *Institute of Plant Sciences Paris-Saclay (IPS2), Université Paris-Saclay, Université Paris-Cité, CNRS, INRAE, Université Evry, Gif sur Yvette, France*
Shahid Mukhtar • *Department of Biology, University of Alabama at Birmingham, Birmingham, AL, USA*
Vijaykumar Yogesh Muley • *Independent Researcher, Jijamata Nagar, Hingoli, India; Instituto de Neurobiología, Universidad Nacional Autónoma de México, Querétaro, Mé xico*

JI CHUL NAM • *Department of Molecular Biosciences, Institute for Cellular & Molecular Biology, The University of Texas at Austin, Austin, TX, USA*
KAROLINA M. PAJEROWSKA-MUKHTAR • *Department of Biology, University of Alabama, Birmingham, AL, USA*
HANU R. PAPPU • *Department of Plant Pathology, Washington State University, Pullman, WA, USA*
DANIEL PEREZ-HERNANDEZ • *Department of Infection and Immunity, Luxembourg Institute of Health, Strassen, Luxembourg*
KARI SALOKAS • *Institute of Biotechnology, HiLIFE Helsinki Institute of Life Science, University of Helsinki, Helsinki, Finland*
JEN SHEEN • *Department of Molecular Biology and Centre for Computational and Integrative Biology, Massachusetts General Hospital, and Department of Genetics, Harvard Medical School, Boston, MA, USA*
ELWIRA SMAKOWSKA-LUZAN • *Wageningen University & Research Laboratory of Biochemistry, Wageningen, The Netherlands*
MATTHEW D. SMITH • *Department of Biology, Wilfrid Laurier University, Waterloo, ON, Canada*
KUNCHENG SONG • *Bioinformatics Research Center, North Carolina State University, Raleigh, NC, USA*
ROSANGELA SOZZANI • *Department of Plant and Microbial Biology, North Carolina State University, Raleigh, NC, USA*
R. GREG STACEY • *Michael Smith Laboratories, Vancouver, BC, Canada*
JENTE STOUTHAMER • *Wageningen University & Research Laboratory of Biochemistry, Wageningen, The Netherlands*
BINMEI SUN • *College of Horticulture, South China Agricultural University, Guangzhou, China*
BARIŞ ETHEM SÜZEK • *Bioinformatics Graduate Program, Graduate School of Natural and Applied Sciences, Muğla Sıtkı Koçman University, Muğla, Turkey; Department of Computer Engineering, Faculty of Engineering, Muğla Sıtkı Koçman University, Muğla, Turkey*
DONI THINGUJAM • *Department of Biology, University of Alabama, Birmingham, AL, USA*
ERDEM TÜRK • *Bioinformatics Graduate Program, Graduate School of Natural and Applied Sciences, Muğla Sıtkı Koçman University, Muğla, Turkey; Department of Computer Engineering, Faculty of Engineering, Muğla Sıtkı Koçman University, Muğla, Turkey*
PETRA VAN DAMME • *iRIP Unit, Laboratory of Microbiology, Department of Biochemistry and Microbiology, Ghent University, Ghent, Belgium*
MARKKU VARJOSALO • *Institute of Biotechnology, HiLIFE Helsinki Institute of Life Science, University of Helsinki, Helsinki, Finland*
VALERIA VELÁSQUEZ-ZAPATA • *Department of Plant Pathology, Entomology and Microbiology, Iowa State University, Ames, IA, USA; Program in Bioinformatics & Computational Biology, Iowa State University, Ames, IA, USA*
VALERIA VELÁSQUEZ-ZAPATA • *Program in Bioinformatics & Computational Biology, Iowa State University, Ames, IA, USA; Department of Plant Pathology, Entomology and Microbiology, Iowa State University, Ames, IA, USA*
ROGER P. WISE • *USDA-Agricultural Research Service, Corn Insects and Crop Genetics Research, Ames, IA, USA; Department of Plant Pathology, Entomology and Microbiology, Iowa State University, Ames, IA, USA; Program in Bioinformatics & Computational Biology, Iowa State University, Ames, IA, USA*

JIANMEI XU • *College of Horticulture, South China Agricultural University, Guangzhou, China*
XIAODI YANG • *State Key Laboratory of Agrobiotechnology, College of Biological Sciences, China Agricultural University, Beijing, China*
RUIQIANG YE • *Department of Molecular Biology and Centre for Computational and Integrative Biology, Massachusetts General Hospital, and Department of Genetics, Harvard Medical School, Boston, MA, USA; CAS Center for Excellence in Molecular Plant Sciences, Chinese Academy of Sciences, Shanghai, China*
ZIDING ZHANG • *State Key Laboratory of Agrobiotechnology, College of Biological Sciences, China Agricultural University, Beijing, China*
JINGYAN ZHENG • *State Key Laboratory of Agrobiotechnology, College of Biological Sciences, China Agricultural University, Beijing, China*

Chapter 1

Yeast Two-Hybrid Technique to Identify Protein–Protein Interactions

Prabu Gnanasekaran and Hanu R. Pappu

Abstract

Protein–protein interactions are specific and direct physical contact between two or more proteins, and the interaction involves hydrogen bonding, electrostatic forces, and hydrophobic forces. Majority of biological processes in the living cell are executed by proteins, and any particular protein function is regulated by numerous other proteins. Thus, knowledge of protein–protein interaction is necessary to understand the biological processes. In this chapter, we explain the widely used yeast two-hybrid assay to identify the protein-interacting partners.

Key words Y2H, Yeast, GAL4, Bait, Prey, Protein–protein interaction

1 Introduction

Yeast two-hybrid (Y2H) is a well-established genetic approach used to identify the novel interacting protein partners [1]. Yeast strain Y2HGold contains distinct reporter (*AUR1-C, HIS3, ADE2*, and *MEL1*) genes each under the control of GAL4-responsive promoter and these strains are auxotroph for the reporter gene product [2]. To identify the protein-interacting partner, the protein of interest is fused to the DNA-binding domain (BD) of the GAL4 transcription factor. Library cells contain plasmids that are coding for prey protein fused to the activation domain (AD) of the GAL4 transcription factor [3, 4]. If there is an interaction between bait and prey protein, this interaction recruits the activation domain of GAL4 and activates the expression of the reporter gene [5]. Thus, the yeast cell transformed with constructs required for GAL4-based protein interaction could grow on the screening media that lack reporter gene product (Fig. 1). For a pairwise Y2H screen, proteins of interest, protein X and protein Y are fused with the binding domain (BD) and activation domain (AD) of the GAL4

Shahid Mukhtar (ed.), *Protein-Protein Interactions: Methods and Protocols*, Methods in Molecular Biology, vol. 2690, https://doi.org/10.1007/978-1-0716-3327-4_1,

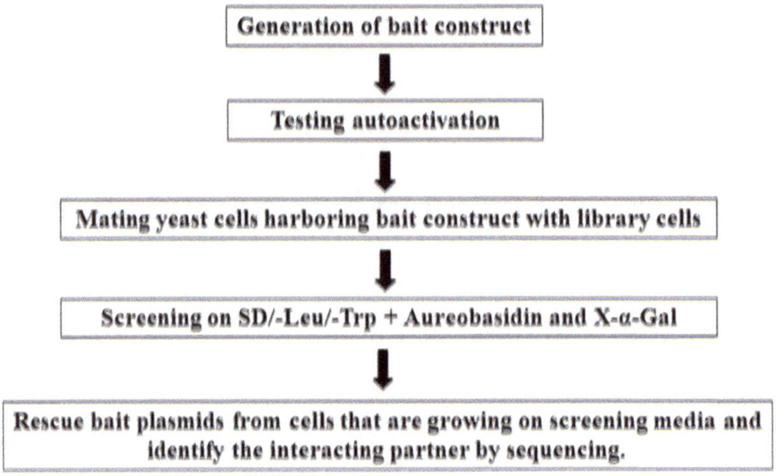

Fig. 1 Steps in Y2H library screening

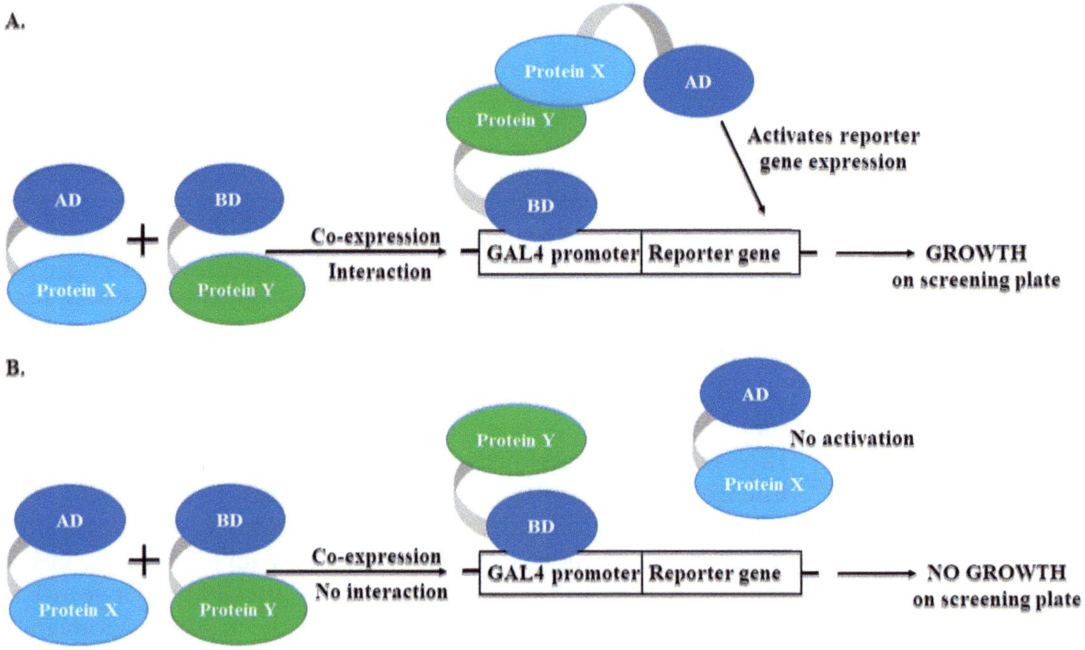

Fig. 2 Schematic representation of principle of Y2H assay

transcription factor [6]. If there is an interaction between AD-X and BD-Y fusion proteins, this interaction recruits the activation domain of GAL4 and activates the expression of the reporter gene. Thus, the yeast cell transformed with constructs required for GAL4-based protein interaction could grow on the screening media that lack reporter gene product (Fig. 2).

If there is no interaction between AD-X and BD-Y fusion proteins, then there is no expression of the reporter gene. Therefore, the yeast cells harboring these constructs could not grow on screening media that lack reporter gene product (Fig. 2).

2 Materials

2.1 Yeast Two-Hybrid Library Construction

1. Total mRNA isolation kits.
2. cDNA synthesis from RNA using smart technology.
3. Linearized pGADT7-Rec cloning vector.
4. Yeast Y187 strain.

2.2 Constructs for Pairwise Y2H

1. pGADT7 (has coding sequence of GAL4 AD, and Trp1).
2. pGBKT7 (has coding sequence of GAL4 BD, and Leu2).
3. pGADT7-X: Generate this construct by cloning the open reading frame (ORF) coding for the protein of interest X in frame with the GAL4 AD in the pGADT7 vector.
4. pGBKT7-Y: Generate this construct by cloning the ORF coding for the protein of interest Y in frame with the GAL4 BD in the pGBKT7 vector.
5. Positive control plasmid combinations:
   (a) pGADT7-T_{Ag} and pGBKT7-P_{53} is a well-recognized positive control combination for Y2H assays.
6. Negative control plasmid combinations:
   (a) pGADT7 empty vector + PGBKT7 empty vector.
   (b) pGADT7-X + PGBKT7 empty vector.
   (c) pGADT7 empty vector + PGBKT7-Y.
7. Y2H test plasmid combination:
   (a) pGADT7-X + PGBKT7-Y.

2.3 Yeast Transformation

1. Y2HGold yeast strain: Y2HGold yeast strain contains *ADE2*, *HIS3*, *MEL1*, and *AUR1-C* reporter genes that are under the control of distinct GAL4-responsive promoter.
2. Y2H constructs are pGADT7 (contains coding sequence of GAL4 AD, and Trp1), pGBKT7 (contains coding sequence of GAL4 BD, and Leu2), and pGBKT7-Bait protein construct (*see* **Note 1**).
3. Positive control plasmid combinations include pGADT7-T_{Ag} and pGBKT7-P_{53}.
4. Negative control plasmid combinations are pGADT7 empty vector + PGBKT7 empty vector.

5. Test plasmid combinations are pGADT7 empty vector + PGBKT7-Bait.
6. Yeast transformation carrier DNA: Salmon sperm DNA (5 mg/mL).
7. 10× Tris-Ethylenediaminetetraacetic acid (TE) buffer: Dissolve 1.576 g of Tris-HCl to approximately 80 mL of distilled water and adjust to pH 7.4. Then, add 2 mL of 0.5 M Ethylenediaminetetraacetic acid (EDTA) (pH 8) and make up the volume to 100 mL with distilled water. Autoclave it and store it at room temperature.
8. 10× lithium acetate solution: Dissolve 10.2 g of lithium acetate in 100 mL of distilled water and adjust to pH 7.5. Autoclave it and store it at room temperature.
9. 50% polyethylene glycol 3350 (PEG) solution: Freshly prepare the 50% PEG (w/v) solution in distilled water and autoclave it.
10. 1× TE/lithium acetate solution: To prepare this solution, add 1 mL of 10× TE buffer, 1 mL of 10× lithium acetate, and 8 mL of autoclaved distilled water.
11. 1× TE/lithium acetate solution/PEG solution: To prepare this solution, add 2 mL of 10× TE buffer, 2 mL of 10× lithium acetate, 2 mL of 50% PEG, and 14 mL of 50% PEG solution.
12. Dimethyl sulfoxide (DMSO).
13. Yeast peptone dextrose adenine (YPDA) broth.
14. YPDA agar plates.
15. SD/−Leu agar plates: This medium is used for selecting cells that have prey (pGADT7) plasmid.
16. SD/−Trp agar plates: This medium is used for selecting cells that have bait (pGBKT7) plasmid.
17. SD/−Leu/−Trp agar plates (2DO): This medium is used for selecting cells that have bait and prey plasmids.
18. SD/−Leu/−Trp/-His (3DO): This medium is used for selecting cells that have bait and prey plasmid and in addition, to test the interaction. The activation of expression of the GAL4-responsive gene, *HIS3*, during interaction allows the yeast cells carrying interacting protein plasmids to grow on media lacking histidine.
19. SD/−Leu/−Trp/-His +3-aminotriazole (3DO + 3AT): This medium is used for screening the interacting cells with high stringency. 3AT acts as a competitive inhibitor of *HIS3* gene product, and it inhibits if there is any *HIS3* gene expression in the absence of interaction, due to leaky gene expression.
20. SD/−Leu/−Trp + Aureobasidin: This medium is used for selecting cells that have bait and prey plasmid and in addition,

to study the activation of the GAL4-responsive gene, AUR1-C. Expression of AUR1-C gene product provides resistance against Aureobasidin antibiotic.

2.4 Y2H Library Screening

1. 2× YPDA broth.
2. Screening plates (SD/−Leu/−Trp containing Aureobasidin 60 ng/mL, and X-α-Gal).

3 Methods

3.1 Yeast Two-Hybrid Library Construction

1. Isolate total mRNA and generate cDNA library with adapter for homologous recombination using kits and following manufacturer's instruction.
2. Transform the linearized pGADT7-Rec vector and cDNA library into yeast strain Y187 and plate them on SD/−Leu medium. SD/−Leu medium is used for selecting cells that have prey (library) plasmids.
3. Incubate the plates at 30 °C for 4–5 days.
4. Pool all the transformants in the SD/−Leu broth, and aliquot, and store at −80 °C.
5. Use 1 mL of this aliquot for each library screening.

3.2 Yeast Transformation

1. For primary culture, inoculate one or two colonies of Y2H yeast strain in 50 mL of YPDA broth and grow them overnight at 30 °C and shaking at 200 rpm.
2. For secondary culture, inoculate 300 mL of YPDA broth to 0.15 optical density (OD) using the overnight grown primary culture and grow at 30 °C and shake at 200 rpm till the OD reaches approximately 0.3 OD.
3. Harvest the cells by centrifuging at room temperature and 3000 rpm for 5 min and discard the supernatant.
4. Wash by resuspending the cell pellet in sterile distilled water and centrifuging at room temperature and 3000 rpm for 5 min, and discard the supernatant.
5. Resuspend the cells in 1.5 mL of 1× TE/lithium acetate solution and aliquot 100 μL of this into 1.5 mL microcentrifuge tubes.
6. Add 100 ng of the desired plasmid DNA and 50 μg of heat-denatured and snap-chilled carrier DNA (salmon sperm DNA) and mix gently by pipetting.
7. Add 600 μL of 1× TE/lithium acetate solution/PEG solution and mix well by vortexing briefly and incubate at 200 rpm shaking and 30 °C for 1 h.

8. Add 70 μL of DMSO and mix immediately.
9. Incubate the tube in 42 °C water bath for 15 min and snap chill on ice for 10 min.
10. Centrifuge at room temperature and 13,000 rpm for 1 min and discard the supernatant.
11. Resuspend the pellet in 500 μL of 1× TE buffer, plate 100 μL on the appropriate selection plate, and incubate at 30 °C for 2–3 days.

3.3 Testing Autoactivation

1. Transform pGBKT7 (empty), pGBKT7-P_{53}, and pGBKT7-bait constructs individually into yeast strain Y187 and plate it on SD/−Trp. Similarly, transform pGADT7 (empty) and pGADT7-T_{Ag} constructs individually into yeast strain Y2HGold and plate it on SD/−Leu. Incubate plates at 30 °C for 2–3 days till the transformant yeast colonies grow up to 1–2 mm.
2. For mating, mix the transformed yeast cells in 0.5 mL of YPDA broth in the following combinations as pGBKT7 + pGADT7 (negative control), pGBKT7-P_{53} + pGADT7-T_{Ag} (positive control), and pGBKT7-Bait + pGADT7 (test).
3. Incubate the yeast cells for mating at 30 °C and shake at 200 rpm overnight.
4. Plate 100 μL of 1/10, 1/100, and 1/1000 of each combination onto SD/−Leu/−Trp and SD/−Leu/−Trp/+Aureobasidin agar plates and incubate at 30 °C for 2–3 days (*see* **Note 2**).

3.4 Yeast Two-Hybrid Library Screening

1. Inoculate yeast transformant harboring bait construct in 50 mL of YPDA broth and grow them overnight at 30 °C and shaking at 200 rpm.
2. Harvest the cells by centrifuging at room temperature and 3000 rpm for 10 min and discard the supernatant.
3. Resuspend the cells in 50 mL of 2× YPDA broth and add 1 mL of Y2H library aliquot after thawing at 37 °C.
4. Incubate the yeast cells for mating at 30 °C and shaking at 200 rpm for 20–24 h.
5. Harvest the cells by centrifuging at room temperature and 3000 rpm for 10 min and discard the supernatant. Resuspend the cells in 10 mL 1× TE buffer.
6. Plate the resuspended cells on screening plates (SD/−Leu/−Trp containing Aureobasidin 60 ng/mL and X-α-Gal) and incubate at 30 °C for 5–7 days.
7. Rescue the prey plasmid from the yeast colonies growing on the screening plate and sequence them to identify the putative interacting partner of the bait protein (*see* **Note 3**).

3.5 Pairwise Y2H Screening

Transform the yeast cells with the test, positive, and negative control combination plasmids, plate 100 μL on the 2DO selection plate, and incubate at 30 °C for 2–3 days.

1. Streak the yeast transformant with test, positive, and negative control combination plasmids on 2DO, 3DO, and 3DO + 3AT plates and incubate at 30 °C for 3–6 days (*see* **Note 4**).
2. Positive control: The yeast cells transformed with the positive control plasmid combination should grow on 2DO, 3DO, and 3DO + 3AT plates.
3. Negative control: The yeast cells transformed with the negative control plasmid combination should grow on 2DO but not on 3DO, and 3DO + 3AT plates.

4 Notes

1. Generate this construct by cloning the ORF coding for bait protein in frame with the GAL4 AD in the pGADT7 vector.
2. Yeast cells transformed with all combinations (positive, negative, and test) should be able to grow in the 2DO plates. Positive control combination must grow in the 2DO + Aureobasidin agar plates. Negative control combination should not grow in the 2DO + Aureobasidin agar plates. If there is no growth of test combination transformed cells on 2DO + Aureobasidin agar plates, this result suggests that the tested bait protein does not autoactivate the GAL4-responsive reporter gene and this bait construct could be used for library screening (Fig. 1).
3. Interacting partner identified by Y2H library screening has to be tested by other independent protein interaction techniques such as pairwise Y2H assay, Glutathione S-tranferase (GST) pulldown assay, and bimolecular fluorescence complementation assay.
4. If the yeast cell transformed with test plasmid combination (pGADT7-X + PGBKT7-Y) grow on 2DO, 3DO, and 3DO + 3AT plates, then there is an interaction between protein X and Y. If the yeast cell transformed with test plasmid combination (pGADT7-X + PGBKT7-Y) grows only on 2DO plates and not on 3DO, and 3DO + 3AT plates, then there is no interaction between protein X and Y.

References

1. Chien CT, Bartel PL, Sternglanz R et al (1991) The two-hybrid system: a method to identify and clone genes for proteins that interact with a protein of interest. Proc Natl Acad Sci U S A 88(21):9578–9582
2. Giniger E, Varnum SM, Ptashne M (1985) Specific DNA binding of GAL4, a positive regulatory protein of yeast. Cell 40(4):767–774

3. Fields S, Song O (1989) A novel genetic system to detect protein-protein interactions. Nature 340(6230):245–246
4. Ruden DM, Ma J, Li Y et al (1991) Generating yeast transcriptional activators containing no yeast protein sequences. Nature 350(6315): 250–252
5. Hannon GJ, Demetrick D, Beach D (1993) Isolation of the Rb-related p130 through its interaction with CDK2 and cyclins. Genes Dev 7(12a):2378–2391
6. Gnanasekaran P, Ponnusamy K, Chakraborty S (2019) A geminivirus betasatellite encoded betaC1 protein interacts with PsbP and subverts PsbP-mediated antiviral defence in plants. Mol Plant Pathol 20(7):943–960

Chapter 2

Cytotrap: An Innovative Approach for Protein–Protein Interaction Studies for Cytoplasmic Proteins

Binoop Mohan, Doni Thingujam, and Karolina M. Pajerowska-Mukhtar

Abstract

Protein–protein interaction mapping has gained immense importance in understanding protein functions in diverse biological pathways. There are various in vivo and in vitro techniques associated with the protein–protein interaction studies but generally, the focus is confined to understanding the protein interaction in the nucleus of the cell, and thus it limits the availability to explore protein interactions that are happening in the cytoplasm of the cell. Since posttranslational modification is a crucial step in signaling pathways and cellular protein interactions harnessing the cytoplasmic protein and evaluating the interaction in the cytoplasm, this protocol will provide more information about studying these types of protein interactions. Cytotrap is a type of yeast-two-hybrid system that differs in its ability to anchor along the membrane, thus directing the protein of interest to anchor along the membrane through the myristoylation signaling unit. The vector containing the target protein contains the myristoylation unit, called the prey, and the bait unit contains the protein of interest as a fusion with the hSos protein. In an event of interaction between the target and the protein of interest, the hSos protein unit will be localized to the membrane and the GDP/GTP exchange unit will trigger the activation of the Ras pathway that leads to the survival of the temperature-sensitive yeast strain at a higher temperature.

Key words Cytotrap, Cytoplasmic protein, Protein, protein interaction, Yeast-two-hybrid (Y2H) system

1 Introduction

Protein–protein interactions (PPIs) are critical to maintaining the rhythm of life [1]. Most of the vital biological activities are regulated and properly executed through the complex networks involved in PPI [2–11]. Understanding the PPI kinetics will be crucial in determining the function of the protein in diverse biological systems making this process a crucial step in systems biology [12–20]. Protein is a complex structure and its activities are highly significant in regulating an overall process; an estimate of the activity of a protein can be clearly understood through its interaction mechanism [21]. The study of PPI has found its

Shahid Mukhtar (ed.), *Protein-Protein Interactions: Methods and Protocols*, Methods in Molecular Biology, vol. 2690,
https://doi.org/10.1007/978-1-0716-3327-4_2,

applications in understanding different research perspectives such as biochemical [12], molecular-based dynamics [22], and also in the interpretation of various signal transduction pathways [23]. Eventually, there are a couple of methods widely administered in the study of PPI to gain more insight into the level of interaction to provide a clear indication of the characteristic property of the target protein [24]. Among the in vivo and in vitro methods that are widely used in the study of PPI, the most commonly used are yeast-two-hybrid system (Y2H) [25], co-immunoprecipitation [26], biotin-based proximity labeling [27], affinity purification [28], protein microarrays, protein-fragment complementation assay [29], light scattering [30], as well as other computational-based [31] protein interaction tools. In this chapter, we will focus on an in vivo technique termed "Cytotrap" [32], which has similarities with the conventional Y2H system but works differently to understand protein interaction along the cytoplasm instead of the nucleus (as is the case in a conventional Y2H system.

In vivo techniques associated with the PPI will help to highlight the function of the desired protein of interest and the most used technique is the Y2H system. This particular technique is widely exploited in understanding the role of protein in a protein–protein complex system [33] as well as in protein–DNA interaction systems [34]. The basic mechanism through which this system functions is that it comprises two different sequences, the reporter sequence and the activation sequence, and the activation of the reporter sequence will be triggered by the binding of a transcription factor *GAL4* [35] onto the activation sequence. The transcription factor is a combination of two individual fragment units: the DNA binding domain that can bind with the activation sequence, and the activation domain that will trigger the activation of the transcription process. The Y2H system is designed in a way that the activation domain and the DNA binding domain are present within the prey and bait, respectively, and activation of the reporter sequence will only happen if the prey and bait will interact with each other [36]. While the conventional Y2H system is an excellent system for PPI it limits our focus to the proteins that are generally localized into the nucleus and will not be sufficient to understand the characteristic features of membrane-associated target proteins [37]. Specifically, the classical Y2H system exhibits limitations when it comes to integral membrane proteins, cytosol-associated proteins, and proteins that are localized along the subcellular compartments [38].

Cytotrap [32] works with the same principles as the Y2H system but has a different perspective when it comes to identifying a protein–protein interaction. Apart from the already existing system, Cytotrap is well advanced in its mechanism to investigate the possible protein interactions that are occurring in the cytoplasm and along the membrane structure; thus it opens up wider opportunities for examining a protein for its interaction. Cytotrap uses yeast as a model system to study the interaction and uses the human protein hSos that will be fused with a bait protein. The second protein is the prey protein that will be expressed as a fusion along with the myristoylation signal, which helps the protein complex to anchor along the membrane of the yeast cell. The yeast strain used in this experiment is a temperature-sensitive yeast that will not be able to survive at higher temperatures during the incubation process; this sensitive strain is used to validate the interaction of two sets of proteins. In this mechanism, if the proteins are interacting with each other, then the hSos protein will be recruited toward the cell membrane. The recruitment of hSos protein will help in the activation of the Ras pathway signaling; this process is crucial for the yeast cell to survive at higher incubation temperature. Cytotrap system has a complex architecture, and it mainly comprises two distinct vector units generally referred to as the prey and the bait. The pSos and pMyr are the two vector units that are used in the Cytotrap system; the bait protein includes the protein of interest, and, at the DNA level, it is ligated along the cloning site of pSos to produce a fusion complex comprising of the protein of interest and the hSos (Fig. 1). pMyr is the prey vector that has specific cloning sites for incorporating the desired target protein and this vector will express the target protein as a fusion complex containing the myristoylation signal unit that targets the complex to adhere along the membrane region. The protein fusions that are generated for this system will interact inside the yeast cytoplasm, which in turn will activate the Ras signaling pathway that favors the yeast cell growth. The yeast strain used will be sensitive to the higher incubation temperature but following the interaction with the protein of interest, the Ras pathway will be activated, which promotes the growth of temperature-sensitive yeast cells to thrive at higher temperatures.

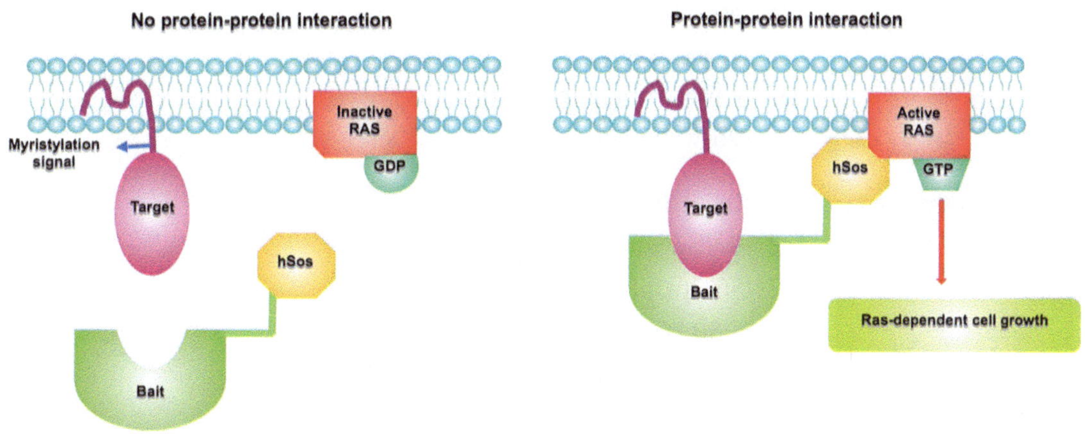

Fig. 1 Mechanism of Cytotrap protein–protein interaction screening system

2 Materials

2.1 Reagents and Supply Items

1. Cryogenic tubes 2 mL
2. Sterile inoculation loop
3. Petri plates Fisher Scientific
4. Parafilm
5. 15 mL centrifuge tubes
6. Glycerol Extra pure
7. Cryo marker pen
8. Cyto Trap vector kit
9. PMyr vector Agilent
10. pSos vector Agilent
11. cdc25H yeast strain
12. Chloroform
13. Phenol
14. Ammonium acetate
15. 250 mL conical flask
16. One-liter conical flask
17. 50 mL falcon tube
18. Salmon sperm DNA
19. LiSorb
20. Polyethylene glycol–lithium acetate solution
21. Dimethylsulfoxide (DMSO)
22. 1.5 mL Eppendorf tubes
23. Yeast transformation kit
24. Beta-mercaptoethanol
25. Sorbitol
26. Glass beads
27. Chloramphenicol antibiotic
28. 96-well plate

2.2 Specialist Equipment

1. Incubator
2. Refrigerator
3. Spectrophotometer
4. −80 °C freezer
5. Thermocycler
6. Centrifuge

7. Vortex machine
8. Shaking incubator
9. Water bath

2.3 The Composition of All Buffers, Media, and Solutions

Synthetic media (SD) glucose-leucine and uracil (LU) 500 mL: Yeast nitrogen base lacking amino acids 0.85 g; Ammonium sulfate: 2.5 g; Dextrose: 10 g; Agar: 7.5 g; Amino acid supplement -Leucine/-Uracil/-Histidine/−Tryptophan: 0.3 g; Histidine: 0.01 g; and Tryptophan: 0.025 g. Bring up the volume to 500 mL using a measuring cylinder and pour it onto a one-liter conical flask and sterilize by autoclaving at 121 °C for 20 min and add the filter-sterilized amino acids to the conical flask after the media has reached 50 °C.

SD galactose-LU 500 mL: Yeast nitrogen base lacking amino acids 0.85 g; Ammonium sulfate: 2.5 g; Galactose: 10 g; Raffinose: 5 g; Agar: 7.5 g; Amino acid supplement -Leucine/-Uracil/-Histidine/−Tryptophan: 0.3 g; Histidine: 0.01 g; and Tryptophan: 0.025 g. Bring up the volume to 500 mL using a measuring cylinder and pour it onto a one-liter conical flask and sterilize by autoclaving at 121 °C for 20 min and add the filter-sterilized amino acids to the conical flask after the media has reached 50 °C.

Luria-Bertani medium (LB) Agar 500 mL (Ampicillin + chloramphenicol): LB agar: 12.5 g; and Agar: 7.5 g. Bring up the final volume to 500 mL using distilled water and adjust the final pH to 7.0 using 1 N NaOH; sterilize by autoclaving at 121 °C for 20 min for preparing antibiotic-specific media; to prepare Ampicillin 100 mg/mL and chloramphenicol 100 mg/mL filter sterilize and add 0.5 mL and 0.15 mL of the solution into the media before pouring.

Cell lysis solution: Yeast 500 mL: Triton x-100: 10 mL; Sodium dodecyl sulfate: 5 g; Sodium Chloride: 2.9 g; 0.5 M Tris-HCl: 10 mL; and 1 M ethylenediaminetetraacetic acid (EDTA): 0.5 mL. Bring up the final volume to 500 mL using sterile distilled water.

Yeast peptone adenine sulfate dextrose (YPAD) Agar/Broth 500 mL: Yeast Extract: 5 g; Protease peptone: 10 g; Dextrose: 10 g; and Agar optional for solid media 7.5 g. Bring up the final volume to 500 mL using distilled water and adjust the final pH to 5.8; sterilize by autoclaving at 121 °C for 20 min.

Ligation mix: Vector DNA: 1.0 μL; Insert DNA: 1.0 μL; rATP10mM: 1.0 μL; Ligase buffer: 1.0 μL; T4 DNA ligase enzyme: 0.5 μL; Nuclease-free water: 5.5 μL.

3 Methods

3.1 Processing the Yeast Strain and Glycerol Stock Preparation Process (Fig. 2)

1. Take the yeast glycerol stock from the freezer and keep it on ice for thawing (*see* **Note 1**).
2. For streaking, use a sterile space and a previously prepared YPAD plate.
3. Transfer a loop full of yeast culture from the vial and evenly streak it onto the YPAD agar plate by following the quadrant streaking method.
4. The agar plate containing the culture can be incubated at 25 °C. Wait for approximately four days for the culture to form individual colonies (*see* **Note 2**).
5. Plates containing the individual colonies can be protected from drying by taping them with a covering film (Parafilm) and can be stored in the refrigerator at 4 °C for up to one week.
6. The YPAD broth can be used in the preparation of yeast culture for making glycerol stock for long-term storage.
7. Take a 15 mL sterile tube and add 4 mL of YPAD, scrap out a single colony from the YPAD agar plate, and add it into the sterile tube; cover the lid and incubate the culture at room temperature at 25 °C until the OD_{600} of the culture has reached 0.8.
8. Use 100% glycerol solution in a sterile cryo vial and add an equal volume of culture into the vial and mix it evenly.
9. Prepare an individual stock of culture following step 8; label the vial using cryo marker; tighten the screw cap and store it at −80 °C to be used for further experiments.
10. Screen the yeast culture by growing the culture in a YPAD agar plate and incubate it at two different temperatures, one at room temperature of 25 °C and the other at 37 °C. Observe growth in each plate for four days and confirm that the culture is not showing growth at 37 °C (*see* **Note 3**).

3.2 Construction of the Bait–hSos Protein Fusion and Target Plasmid

1. pSos vector needs to be restriction digested with the respective restriction enzyme by selecting the unique site from the multiple cloning site region of the desired vector fragment (*see* **Note 4**).
2. For preparing the target plasmid, the specific DNA of the protein of interest can be inserted into the pMyr vector by selecting the unique site from the multiple cloning site region of the desired vector fragment.
3. The cloning process for both the bait and the target plasmid can be carried out as follows:

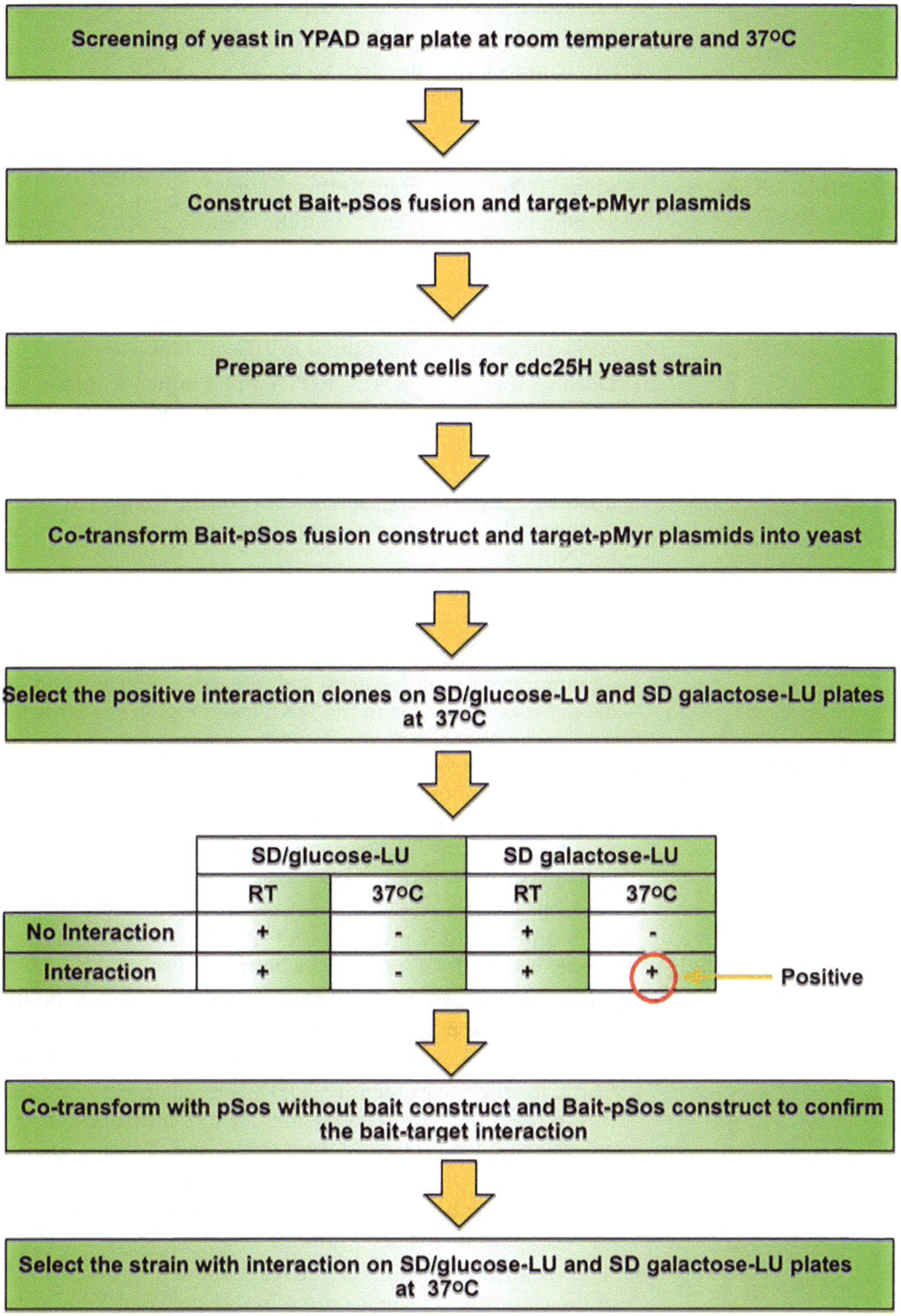

Fig. 2 Flowchart of Cytotrap screening procedure

4. Confirm the concentration of the DNA vector to be 4 μg, and digest it in a final working volume of 40 μL.
5. Extraction of the digested vector DNA can be done using phenol-chloroform at an equal proportion till it reaches a clear state.
6. The process can be further repeated to form a clear solution by adding chloroform alone to the reaction.
7. Separate the aqueous layer from the reaction above and add an equal amount of 4 M ammonium acetate.
8. Add 320 μL of absolute alcohol and centrifuge the sample to pellet out the vector DNA.
9. Purify the sample by washing it in 70% ethanol.
10. Dissolve the pellet in Tris EDTA (TE) buffer to make the vector DNA concentration equal to the insert DNA concentration.
11. The ligation for the vector DNA and the insert DNA can be done by fixing the appropriate ratio of the samples as mentioned in the ligation mix and incubating at 12 °C overnight (*see* **Note 5**).
12. The ligation mix is used for transformation in *Escherichia coli* competent cells.
13. Transformed screened colonies will be verified to check the orientation for bait protein and the pSos domain and to check for any mutation in the DNA insert region (*see* **Note 6**).

3.3 Competent Cell Preparation for cdc25H Yeast Strain

1. Start by streaking the cdc25H yeast strain using a sterile loop from the previously prepared glycerol stock onto a freshly prepared YPAD agar plate and place it for incubation at room temperature of 25 °C until single colonies are formed; it might take approximately four days to grow (*see* **Note 7**).
2. From the individual colonies, pick around five colonies and mix them onto a freshly prepared 750 μL YPAD and mix it thoroughly using a 1 mL pipette (*see* **Note 8**).
3. Prepare 40 mL of YPAD in a 250 mL conical flask and add the culture from step 2 into the media and incubate by shaking at 200 rpm for 16 h at room temperature of 25 °C (*see* **Note 9**).
4. Check for the optical density (OD) to reach above 1 and if it fails to reach the required OD after 19 h of incubation, try to proceed with the procedure from **step 2** and continue again.
5. Dilute the culture into a one-liter flask to make the OD at 0.2 and further incubate the culture for 3 h at room temperature of 25 °C by shaking it at 200 rpm.
6. After incubation, the OD should reach above 0.7. Take 100 μL of the culture and plate it onto a freshly prepared YPAD agar

plate and incubate the culture for approximately 6 days at 37 °C. Count the number of colonies; if the number of colonies exceeds 30, discard the entire culture and try this for other batches of cultures as well.

7. Transfer the culture into a sterile 50 mL conical tube and centrifuge the culture at 4000 rpm for 10 min. Remove the media and resuspend it in 40 mL sterile distilled water; mix it gently and centrifuge and pellet out the culture.
8. Mix the pelleted sample with 50 mL of LiSorb and keep for incubation at room temperature of 25 °C for 30 min.
9. After incubation, add 400 μL salmon sperm DNA of 20 mg/mL and place it in boiling water, then add 600 μL LiSorb to the boiling mixture and cool it to 25 °C.
10. Pellet out the mixture from **step 8** and add 300 μL LiSorb.
11. To the mixture from **step 9**, add the 300 μL reaction mixture from step 10 and mix the entire reaction with 600 μL of salmon sperm DNA.
12. To this reaction mix from step 11, add 5.4 mL of polyethylene glycol–lithium acetate solution and 530 μL of DMSO solution. Mix it evenly and transfer 500 μL of the competent cells into sterile 1.5 mL tubes; for greater efficiency, use the sample directly for transformation.

3.4 Transformation of the Plasmids into Yeast and Identification of the Interaction System

1. Use both pSos Bait and pMyr with 300 ng/mL concentration of the plasmids and mix it with 100 μL of the yeast competent cells.
2. To the tube containing the plasmid and the yeast cell, add 2 μL of 1.4 M beta-mercaptoethanol; mix it evenly and incubate for 30 min with gentle mixing in between intervals.
3. Place the tubes in a water bath for a heat shock treatment set at 42 °C for approximately 25 min.
4. Transfer the tube to ice and pellet out the sample by centrifugation at maximum rpm.
5. Add 0.5 mL of 1 M sorbitol into the pellet and gently mix it to spread evenly.
6. Plate the reaction mixture onto the respective agar plates containing synthetic glucose media without uracil and leucine and incubate the plates at 25 °C for up to six days (*see* **Note 10**).
7. The plates with individual colonies are further selected to be incubated at 37 °C to test the protein–protein interaction.
8. For plating, individual colonies from the plates are mixed with 30 μL sterile water.

9. Using 2 μL from the reaction, the samples will be further spotted on the two separate plates containing synthetic dextrose/glucose/-LU and synthetic dextrose galactose/-LU.
10. One plate from the two sets of plates will be incubated at 25 °C and the other plate will be incubated at 37 °C.

3.5 Confirmation of the Interaction Between Bait Protein and Target Protein Using a Co-transformation Approach

1. Scrap out the positive colony and mix it into 2.5 mL SD/glucose media; continue to incubate the sample at 22–25 °C until it reaches an OD_{600} above 1.
2. Centrifuge the sample at maximum rpm and remove the media and suspend the pellet in 0.5 mL of the lysis buffer.
3. Add 350 μL of phenol-chloroform reagent into the tube and place 100 μL of the glass beads into it; vortex it completely for two minutes.
4. Centrifuge the reaction mix at maximum rpm for two minutes and pipette out the aqueous layer to a fresh Eppendorf tube.
5. Using chilled 100% ethanol, precipitate the DNA and centrifuge the sample to pellet the DNA and discard the excess ethanol.
6. Wash the DNA with 70% ethanol two consecutive times, centrifuge the sample at maximum rpm and discard the excess ethanol.
7. Dry the excess ethanol by keeping it open under a flow hood.
8. After the excess ethanol is removed, dilute the pellet using 30 μL of distilled water.
9. Using the DNA, transform the plasmid into an *E. coli* cell, and being the pMyr plasmid it can be screened by plating it onto an LB with chloramphenicol.
10. The transformed colonies can be used for plasmid isolation and they can be digested with restriction enzyme(s) for further steps.
11. Do the co-transformation of the pSos plasmid with the pMyr plasmid.
12. Once the transformation is completed, use the transformed yeast culture for plating it onto SD/glucose and incubate it at room temperature.
13. The resulting colonies can be further scratched out and plated onto two different sets of plates containing SD/glucose-LU and SD/galactose-LU.
14. Use the two sets of plates to be incubated at two incubation conditions: one set of each plate at 25 °C and the other set at 37 °C.

15. Based on the growth pattern, the strain with interaction will show no growth in glucose media but will show growth in galactose media when it is being incubated at 37 °C (*see* **Note 11**).

3.6 Confirmation of the Interaction Between Bait Protein and Target Protein Using Yeast Mating System

1. The positive clone is used in this experiment where the colony is dissolved onto 4 mL of SD/glucose-U and grown at 22–25 °C.
2. Into freshly prepared SD/glucose-U agar plate, spread 300 μL of the culture and incubate it at 25 °C for four days.
3. From the individual colony, transfer around 35 colonies into SD/glucose-U and -L. The resulting colonies that show growth in -U plates and that fail to grow in -L are those that have lost the pSos plasmids and further, these colonies are used to mate with the yeast strain that incorporates the plasmid containing pSos.
4. Add 30 μL of sterile water to the wells of a 96-well plate. Pick cdc25H alpha and cdc25H, a strain that is needed to be mated to check the interaction.
5. After the mating process of the two yeast strains using the standard process, plate 3 μL of the reaction mixture from step 4 into a YPAD agar plate. Keep the plates for incubation at 22–25 °C.
6. Prepare another 96-well plate, add 30 μL of sterile water into each well and add the remaining culture into the 96-well plate.
7. Using 2 μL from the reaction, the samples will be further spotted on the two separate plates containing SD/glucose-LU and SD galactose-LU.
8. One plate from the two sets of plates will be incubated at 25 °C and the other plate will be incubated at 37 °C.
9. Based on the growth pattern, the strain with interaction will show no growth in glucose media but will show growth in galactose media when it is being incubated at 37 °C.

4 Notes

1. The yeast strain is sensitive to temperature and chances of reverting will be higher if the number of generations increased; hence, carefully handle the strain through the Y2H protocol and reduce the number of generations.
2. Take special care to grow the culture at 22–25 °C; higher incubation temperature can revert the mutant strain.

3. During incubation, if colonies appear at a higher temperature, discard the entire yeast culture and start the culture from a fresh stock culture.
4. Dephosphorylating the vector using alkaline phosphatase will help remove fragments and reduce the background.
5. The ratio of the insert and vector can be ideal in the ratio of 1:1.
6. Co-transform the bait plasmid with the negative control plasmid and confirm that it is not interacting by growing the transformed culture at a higher temperature.
7. Be prepared with the required reagents and the specialized media before proceeding through the competent cell preparation process.
8. Competent cell preparation needs to be carried out using two of the mating types of yeast cells, i.e., a and α.
9. Prepare multiple sets of yeast cultures to confirm that the strains have not reverted their temperature-sensitive property and discard the cultures that can grow at a higher temperature.
10. Use sterilized glass beads for the spreading process to maintain the even spreading of the yeast inoculum.
11. The confirmation can be made from the selection media: the interaction can be positive if the mutants are growing at 37 °C in the galactose media and there are no positive colonies in the glucose media.

Acknowledgments

This research was funded by the National Science Foundation (IOS-2038872).

References

1. Kumar N, Mishra B, Mehmood A et al (2020) Integrative network biology framework elucidates molecular mechanisms of SARS-CoV-2 pathogenesis. Iscience 23(9):101526
2. Ahmed H, Howton TC, Sun Y et al (2018) Network biology discovers pathogen contact points in host protein-protein interactomes. Nat Commun 9(1):2312. https://doi.org/10.1038/s41467-018-04632-8
3. Arabidopsis Interactome Mapping C (2011) Evidence for network evolution in an Arabidopsis interactome map. Science 333(6042):601–607. https://doi.org/10.1126/science.1203877
4. Garbutt CC, Bangalore PV, Kannar P et al (2014) Getting to the edge: protein dynamical networks as a new frontier in plant-microbe interactions. Front Plant Sci 5:312. https://doi.org/10.3389/fpls.2014.00312
5. Gonzalez-Fuente M, Carrere S, Monachello D et al (2020) Effector K, a comprehensive resource to mine for Ralstonia, Xanthomonas, and other published effector interactors in the Arabidopsis proteome. Mol Plant Pathol 21(10):1257–1270. https://doi.org/10.1111/mpp.12965
6. Klopffleisch K, Phan N, Augustin K et al (2011) Arabidopsis G-protein interactome reveals connections to cell wall carbohydrates and morphogenesis. Mol Syst Biol 7:532. https://doi.org/10.1038/msb.2011.66

7. Kumar N, Mishra B, Mehmood A et al (2020) Integrative network biology framework elucidates molecular mechanisms of SARS-CoV-2 pathogenesis. iScience 23(9):101526. https://doi.org/10.1016/j.isci.2020.101526
8. Kumar N, Mishra B, Mukhtar MS (2022) A pipeline of integrating transcriptome and interactome to elucidate central nodes in host-pathogens interactions. STAR Protoc 3(3): 101608. https://doi.org/10.1016/j.xpro.2022.101608
9. Lopez J, Mukhtar MS (2017) Mapping protein-protein interaction using high-throughput yeast 2-hybrid. Methods Mol Biol 1610:217–230. https://doi.org/10.1007/978-1-4939-7003-2_14
10. McCormack ME, Lopez JA, Crocker TH et al (2016) Making the right connections: network biology and plant immune system dynamics. Current Plant Biology 5:2–12
11. Mishra B, Kumar N, Mukhtar MS (2019) Systems biology and machine learning in plant-pathogen interactions. Mol Plant-Microbe Interact 32(1):45–55. https://doi.org/10.1094/MPMI-08-18-0221-FI
12. Mishra B, Kumar N, Mukhtar MS (2019) Systems biology and machine learning in plant–pathogen interactions. Mol Plant-Microbe Interact 32(1):45–55
13. Mishra B, Kumar N, Mukhtar MS (2021) Network biology to uncover functional and structural properties of the plant immune system. Curr Opin Plant Biol 62:102057. https://doi.org/10.1016/j.pbi.2021.102057
14. Mishra B, Kumar N, Shahid Mukhtar M (2022) A rice protein interaction network reveals high centrality nodes and candidate pathogen effector targets. Comput Struct Biotechnol J 20:2001–2012. https://doi.org/10.1016/j.csbj.2022.04.027
15. Mishra B, Sun Y, Ahmed H et al (2017) Global temporal dynamic landscape of pathogen-mediated subversion of Arabidopsis innate immunity. Sci Rep 7(1):7849. https://doi.org/10.1038/s41598-017-08073-z
16. Mishra B, Sun Y, Howton TC et al (2018) Dynamic modeling of transcriptional gene regulatory network uncovers distinct pathways during the onset of Arabidopsis leaf senescence. NPJ Syst Biol Appl 4:35. https://doi.org/10.1038/s41540-018-0071-2
17. Mott GA, Smakowska-Luzan E, Pasha A et al (2019) Map of physical interactions between extracellular domains of Arabidopsis leucine-rich repeat receptor kinases. Sci Data 6: 190025. https://doi.org/10.1038/sdata.2019.25
18. Mukhtar MS, Carvunis AR, Dreze M et al (2011) Independently evolved virulence effectors converge onto hubs in a plant immune system network. Science 333(6042):596–601. https://doi.org/10.1126/science.1203659
19. Smakowska-Luzan E, Mott GA, Parys K et al (2018) An extracellular network of Arabidopsis leucine-rich repeat receptor kinases. Nature 553(7688):342–346. https://doi.org/10.1038/nature25184
20. Wessling R, Epple P, Altmann S et al (2014) Convergent targeting of a common host protein-network by pathogen effectors from three kingdoms of life. Cell Host Microbe 16(3):364–375. https://doi.org/10.1016/j.chom.2014.08.004
21. Kumar N, Mishra B, Mukhtar MS (2022) A pipeline of integrating transcriptome and interactome to elucidate central nodes in host-pathogens interactions. STAR protocols 3(3): 101608
22. Mishra B, Kumar N, Mukhtar MS (2022) A rice protein interaction network reveals high centrality nodes and candidate pathogen effector targets. Comput Struct Biotechnol J 20: 2001–2012
23. Pawson T, Nash P (2000) Protein–protein interactions define specificity in signal transduction. Genes Dev 14(9):1027–1047
24. Titeca K, Lemmens I, Tavernier J et al (2019) Discovering cellular protein-protein interactions: technological strategies and opportunities. Mass Spectrom Rev 38(1):79–111
25. Fields S, O-k S (1989) A novel genetic system to detect protein–protein interactions. Nature 340(6230):245–246
26. Ryu JY, Kim J, Shon MJ et al (2019) Profiling protein–protein interactions of single cancer cells with in situ lysis and co-immunoprecipitation. Lab Chip 19(11): 1922–1928
27. Terracciano R, Preianò M, Fregola A et al (2021) Mapping the SARS-CoV-2–host protein–protein interactome by affinity purification mass spectrometry and proximity-dependent biotin labeling: a rational and straightforward route to discover host-directed anti-SARS-CoV-2 therapeutics. Int J Mol Sci 22(2):532
28. Morris JH, Knudsen GM, Verschueren E et al (2014) Affinity purification–mass spectrometry and network analysis to understand protein-protein interactions. Nat Protoc 9(11): 2539–2554
29. Remy I, Ghaddar G, Michnick SW (2007) Using the β-lactamase protein-fragment complementation assay to probe dynamic protein–

protein interactions. Nat Protoc 2(9): 2302–2306

30. Mishra K, Chakrabarti A, Das PK (2017) Protein–protein interaction probed by label-free second harmonic light scattering: hemoglobin adsorption on spectrin surface as a case study. J Phys Chem B 121(33):7797–7802
31. Skrabanek L, Saini HK, Bader GD et al (2008) Computational prediction of protein–protein interactions. Mol Biotechnol 38(1):1–17
32. Fields S The CytoTrap™ Two-Hybrid System. Cytotrap Brochure (leibniz-fli.de)
33. Lopez J, Mukhtar MS (2017) Mapping protein-protein interaction using high-throughput yeast 2-hybrid. In: Plant genomics. Springer, pp 217–230
34. Jinju G, Liangliang S, Rufang W et al (2022) Construction of yeast two-hybrid library and screening of McRPF interacting proteins in the fruit of bitter gourd. Journal of Nuclear Agricultural Sciences 36(7):1293
35. Osman A (2004) Yeast two-hybrid assay for studying protein-protein interactions. In: Parasite genomics protocols. Springer, pp 403–422
36. Chen Q, Wei T (2022) Membrane and nuclear yeast two-hybrid systems. In: Plant virology. Springer, pp 93–104
37. Causier B, Davies B (2002) Analysing protein-protein interactions with the yeast two-hybrid system. Plant Mol Biol 50(6):855–870
38. Brückner A, Polge C, Lentze N et al (2009) Yeast two-hybrid, a powerful tool for systems biology. Int J Mol Sci 10(6):2763–2788

Chapter 3

Analyzing Protein–Protein Interactions Using the Split-Ubiquitin System

Rucha Karnik and Michael R. Blatt

Abstract

The split-ubiquitin technology was developed over 20 years ago as an alternative to Gal4-based, yeast-two-hybrid methods to identify interacting protein partners. With the introduction of mating-based methods for split-ubiquitin screens, the approach has gained broad popularity because of its exceptionally high transformation efficiency, utility in working with full-length membrane proteins, and positive selection with little interference from spurious interactions. Recent advances now extend these split-ubiquitin methods to the analysis of interactions between otherwise soluble proteins and tripartite protein interactions.

Key words Yeast mating-based split-ubiquitin screening, Membrane protein, protein interactions, Tripartite protein interaction assays, Semi-quantitative interaction analysis

1 Introduction

Researching eukaryotic cellular processes that rely on protein–protein interactions requires robust techniques that can examine protein binding in vivo and in vitro. A wide variety of biophysical and biochemical methods are available to study protein–protein interactions using native or heterologous expression systems such as bacteria, yeast, insect, and mammalian cell lines for protein expression [1–4]. Of these, the affinity or antibody-based co-immunoprecipitation and pulldown assays are popular for resolving multiprotein immune complexes using target-specific antibodies in samples such as cell lysates [3, 4]. Once isolated, the protein complexes can be analyzed using proteomics to identify new binding partners, determine binding affinities, and measure the kinetics for binding reactions [5–7]. Such biochemical assays are technically challenging and expensive, and they generally require optimization for each set of proteins to obtain reliable results [3].

Shahid Mukhtar (ed.), *Protein-Protein Interactions: Methods and Protocols*, Methods in Molecular Biology, vol. 2690,
https://doi.org/10.1007/978-1-0716-3327-4_3,

Other methods designed primarily to assess binary protein–protein interactions offer a more modular approach that circumvents the need for extensive optimization steps and can be rapid and cost-effective in high-throughput analyses. Examples include imaging-based bimolecular fluorescence complementation (BiFC), Förster resonance energy transfer (FRET), and split-luciferase methods, all of which are useful in vivo for semi-quantitative analysis of protein binding using suitable controls as reference [1, 8–10]. These assays provide the advantage that analysis can be carried out in native or near-native tissues if they can be transformed, and several of these methods are suitable for detecting single-molecule interactions. Specialized imaging equipment and expertise in its operation are required, however, and the analysis can be time-consuming [11, 12].

Modular yeast-based assays that track bait and prey protein binding in vitro (Xing et al., 2016) have gained much popularity because they are quick, easy to perform, and inexpensive compared to the biochemical and imaging-based assays [13]. These assays commonly depend on the bait–prey protein complex to pass through the nuclear pore and activate reporter gene expression in the yeast cells, either to facilitate cell growth on selective media or to produce a color reaction. The conventional yeast-two-hybrid (Y2H) assay, for example, can be easily adapted for high-throughput analysis of bait protein interactors using matrix or library screens [1, 13, 14], but the outcomes are commonly subject to a high level of false-positive interactions. As an alternative, the yeast mating-based split-ubiquitin system (mbSUS) technology is gaining increasing attention, as it circumvents many of the constraints of Y2H assays, and it provides a high degree of reliability with reduced false-positive signals [15–22]. The mbSUS approach enables the user to detect binary and tertiary protein–protein interactions and has become a method of choice when full-length proteins are to be examined, especially integral membrane proteins. In this chapter, we outline the mbSUS assay, including the recent enhancements that provide added advantages in tests for binary and tertiary protein–protein interactions, both of membrane-bound and soluble proteins.

1.1 Yeast mbSUS Reporter System

The mbSUS assay offers positive selection and relies on yeast growth on selective media as a reporter for protein–protein interactions [15]. Growth is enabled by the reassembly of two halves of the ubiquitin protein, reconstituting ubiquitin activity that leads to cleavage of a transactivator. The transactivator then translocates to the nucleus and activates reporter gene expression. This assay is highly reliable when used with large, membrane-localized bait proteins that cannot mislocalize to the nucleus [15, 23]. The bait vector is designed to express a protein fusion with a C-terminal tag comprising the C-terminal half of ubiquitin (Cub) fused to a

ProteinA-LexA-VP16 (PLV) transactivator (Y-Cub). Prey vectors are designed to express the N-terminal half of Ubiquitin (Nub) fused to proteins at their N-terminus (Nub-X) [24]. Bait and prey proteins are expressed in THY.AP4 and THY.AP5 yeast cells, respectively. The yeast strains are mated to jointly express bait and prey for interaction; the mated yeast cells are selected for growth on thiamine and uracil to ensure assays in the diploid yeast. Binding between the split-ubiquitin-tagged bait and prey proteins reconstitutes ubiquitin activity, cleaving the PLV transactivator to enable its entry to the nucleus. In turn, the transactivator activates the expression of LexA-driven genes for the synthesis of adenine and histidine essential for yeast growth on selective media lacking these compounds [15, 16, 24–26]. Thus, yeast growth on the mbSUS selection media provides a visual readout for protein interaction.

User control over PLV transactivator location is a key factor for reliable results in mbSUS assays. In principle, Y-Cub fusions can activate reporter gene expression if they can pass to the nucleus. So it is vital that the bait protein is anchored and is unable to relocate to the nucleus. Anchoring in the classical mbSUS is achieved through attachment of the bait fusion with an integral membrane protein [16, 23, 27]; enhancements to mbSUS methods noted below relax this restriction for work with otherwise soluble bait proteins without undermining its reliability.

mbSUS assays incorporate a mutation of ^{13}Isoleucine to Glycine (NubG) in the N-terminal half of the wild-type Ubiquitin (NubI) to prevent spontaneous reassembly [15]. NubG expressed on its own with the bait is normally included as a negative control in the assays, while NubI serves as a positive experimental control. For added reliability and as a test for interaction specificity, the bait (Y-Cub) construct is placed under the control of the yeast met25 promoter that allows for bait suppression in the presence of submillimolar concentrations of methionine. Adding methionine to the selection medium allows the user to test for the strength of interactions by suppressing bait protein expression levels, thus further reducing the likelihood of false positives [15, 16, 25].

In general, the mbSUS approach offers several advantages over the conventional Y2H methods. In addition to methionine-based testing for the strength of interaction, mating-based recombination greatly enhances transformation efficiencies. mbSUS assays have proven effective across a range of experimental designs [15, 16, 25]. For example, they are ideal for examining binding between specific protein domains using domain swap chimeras, and they are proven equally effective in the screening of single-site mutations to define interaction motifs [17, 21, 27–30].

One consideration in undertaking mbSUS assays is the need to ensure the correct orientation of the putative interactors. Normally, both bait and prey ubiquitin moieties fused with the PLV transactivator must be situated within the cytosol, so if cleaved, the LexA-VP16 PLV transactivator can reach the nucleus to trigger selective yeast growth. Topological considerations dictate that the bait protein is anchored N-terminally from the Cub-PLV fusion and Nub moieties are normally situated at the N-terminus of the prey protein in standard mbSUS assays [15, 16, 25].

In a recent enhancement, we developed a system for bait anchoring using a hydrophobic signal sequence from a glycosylphosphatidylinositol (GPI) protein, the GPI peptide-anchored split-ubiquitin (GPS) assay, that introduces a synthetic membrane anchor and is highly effective in working with soluble and low molecular weight bait proteins [25]. This anchor ensures bait localization to the plasma membrane in yeast cells, enabling highly reliable analysis of binary interactions between otherwise soluble baits with prey proteins [17, 31].

1.2 mbSUS Assays for Analysis of Tripartite Interactions

mbSUS methods can be extended for analysis of higher-order protein complexes, much like the yeast-3-hybrid assay [32, 33], and have been employed to study interactions between three proteins [27, 34]. An early modification of mbSUS assays [35] allowed testing for interactions with a "bridging" protein, but this modification relied on an additional transformation step to introduce a third plasmid, thereby reducing transformation efficiencies. These assays were limited also in testing how this third protein influenced bait–prey binding, as this "bridging" protein was expressed constitutively. Overcoming these limitations, the Tri-SUS system extends testing for tripartite interactions using a bait vector and a bicistronic vector that co-expresses the prey and a third "partner" protein under the control of met25 promoter [26]. Thus, like the classic mbSUS, the Tri-SUS allows the user to test whether the "partner" protein facilitates or outcompetes "bait" and "prey" interactions through comparisons of growth as a function of methionine concentration to suppress partner protein expression [26]. Together with the mbSUS assay, and its enhanced forms in the GPS and Tri-SUS systems, these methods allow for testing of a wide range of interactions and are tractable for use with both soluble and membrane-anchored baits with high confidence (Figs. 1 and 2). A complete list of vectors is provided in Table 1.

2 Materials

2.1 Yeast SUS Assays Chemicals List

Adenine hemisulfate salt (Sigma-Aldrich, A3159)

L-Methionine (Sigma-Aldrich, M5308)

L-Tryptophan (Merckmillipore, 1.08374.0010)

L-Leucine (Sigma-Aldrich, L8912)

L-Histidine (Merckmillipore, 1.04351.0025)

Uracil (Sigma-Aldrich, U1128)

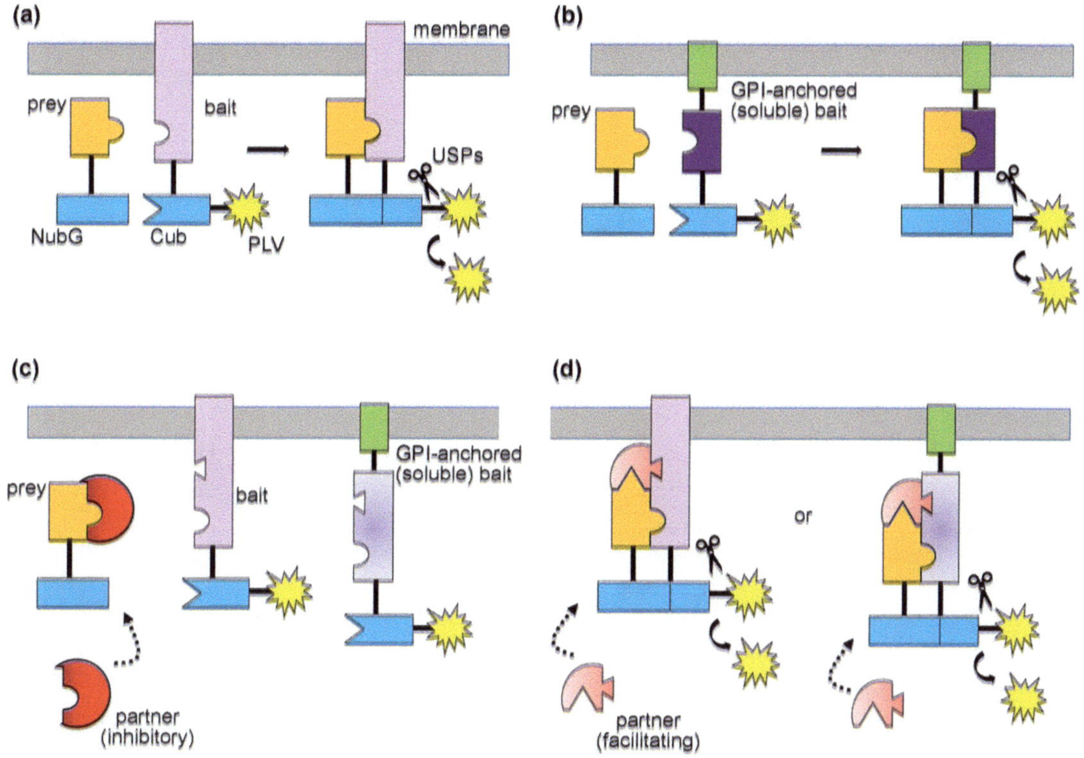

Fig. 1 Schematic representation of classic mbSUS, GPS, and Tri-SUS mbSUS assays. (**a**) Classical mbSUS assay schematic showing prey protein fused to N-terminal ubiquitin (NubG) and bait protein tagged with C-terminal ubiquitin (Cub) fused with the transcriptional activator ProteinA-lexA-VP16 (PLV) complex. Interactions between the bait and prey proteins reconstitute functional ubiquitin, which is cleaved to release the PLV activator. While the bait protein must be anchored to the membrane, the prey may be soluble or membrane-bound. (**b**) Yeast GPS system expresses N-terminus GPI membrane anchor fused with the Bait-Cub-PLV, enabling use in mbSUS assays for soluble baits. (**c**, **d**) Tri-SUS assay offers both classical and GPI-fused bait-Cub-PLV fusion vectors. The prey-partner vector co-expresses the NubG-prey fusion and a methionine-repressible partner protein, allowing to test if the partner inhibits (**c**) or facilitates (**d**) the binding between bait and prey proteins. The mbSUS vector sets (Fig. 2) are available for the expression of bait, prey, and/or partner and bridge proteins and are Gateway compatible. The pMetYC-Dest vector is used for bait expression and the pNX35-Dest vector is used to express prey proteins in the regular mbSUS [16]. The mbSUS GPS assays use the pMetExg2YC-Dest vector for bait [25]. The Tri-SUS system includes ptdh3YC-Dest and ptdh3Exg2YC-Dest vectors for bait expression and pPrey-PartnerSUS-2in1-Dest vector for prey and partner protein expressions [26]. Standard Gateway cloning approaches are used to prepare expression vectors following the manufacturer's instructions. Vector maps, and sequences for the extended mbSUS systems, are available for download from https:/psrg.org.uk. A complete list of vectors is provided in Table 1

Salmon sperm (ss)-DNA (Sigma-Aldrich, D7656)

Ammonium persulfate (Fisher Scientific, 10020020)

Potassium phosphate monobasic (Sigma-Aldrich, P9791)

Yeast nitrogen base (YNB) without ammonium persulfate, without potassium phosphate(MP Biomedicals, 114029622)

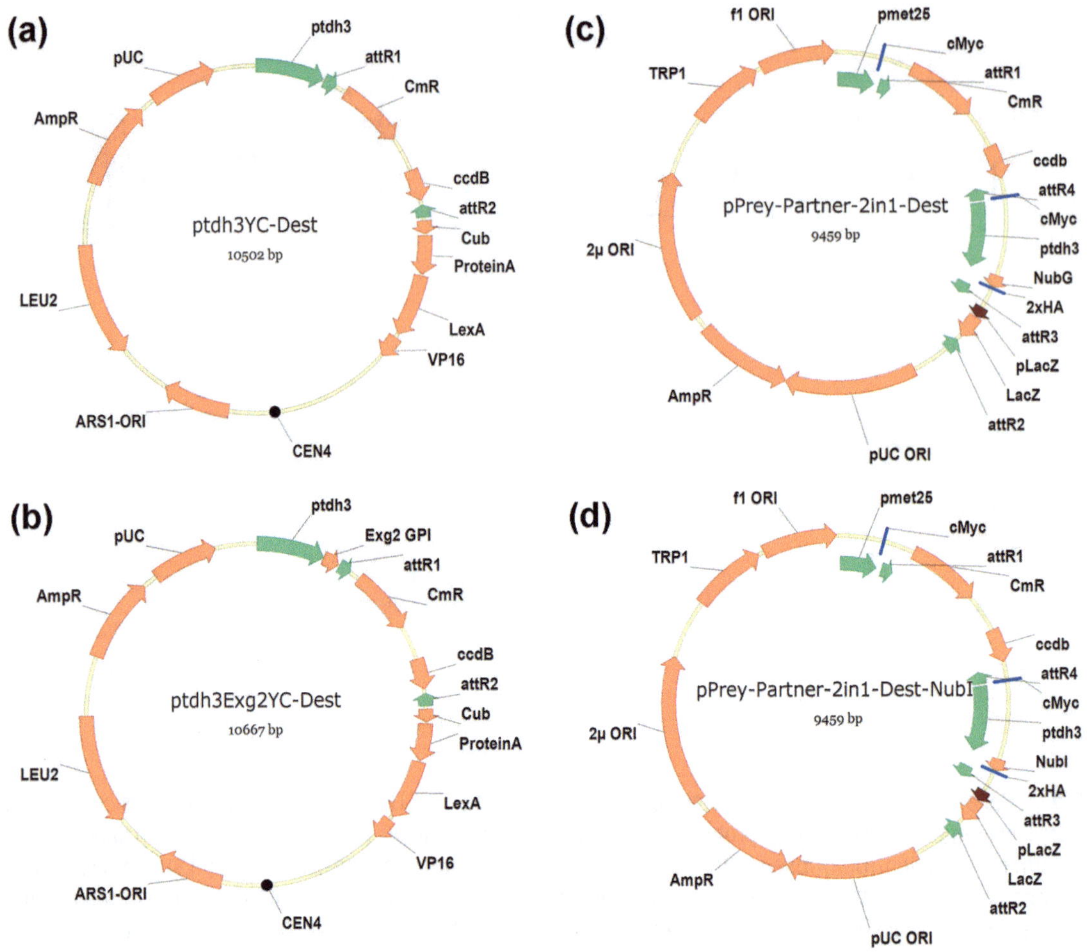

Fig. 2 Gateway destination vectors (expression constructs) for Tri-SUS protein–protein interaction assay in yeast. Schematic maps showing key features incorporated in mbSUS vectors. (**a**) ptdh3YC-Dest vector transformed into yeast strain THY.AP4 for Tri-SUS assay, expressing bait proteins fused with C-terminal Cub, protein A, LexA, and VP16 tags under constitutive tdh3 promoter. (**b**) ptdh3Exg2YC-Dest vector transformed into yeast strain THY.AP4 for Tri-SUS assay, expressing N-terminal membrane-anchored bait proteins fused with C-terminal Cub, protein A, LexA, and VP16 tags under constitutive tdh3 promoter. (**c**) pPrey-Partner-2in1-Dest bicistronic vector transformed into yeast strain THY.AP5 for Tri-SUS assay, expressing prey protein with N-terminal NubG and haemaglutinin (HA) tags under tdh3 constitutive promoter and partner proteins with N-terminal cMyc tag under met25 inducible promoter. The met25 promoter activity is suppressed by methionine in a selective medium. The vector was used to express NubG and partner protein as a negative control. (**d**) pPrey-Partner-2in1-Dest-NubI for expressing NubI and partner protein as positive control. Features of the vectors are: the origin of replicons (pUC ORI in bacteria, 2 µ ORI, ARS1-ORI, and CEN4 in yeast), promoter (ptdh3, pmet25, and pLacZ), Gateway attachments (attR1, attR2, attR3, and attR4), selection markers in bacteria (AmpR: Ampicillin; CmR: Chloramphenicol; ccdB: type II toxin; and LacZ: X-Gal), selection marker in yeast (LEU2: leucine; TRP1: tryptophan), antigens (ProteinA, VP16, HA, and cMyc), transcriptional repressor (LexA), ubiquitin (Cub, C-terminal part of ubiquitin), membrane GPI anchor (Exg2), and ubiquitin (NubG, N-terminal part of ubiquitin with ^{13}Isoleucine to Glycine; or NubI, N-terminal part of ubiquitin)

Table 1
mbSUS Gateway destination vectors list

Vector	Gateway attachments	Selection (bacteria/yeast)	Note
pNX35-Dest	attP1, attP2	Ampicillin/Tryptophan	mbSUS, Prey
pMetYC-Dest	attP1, attP2	Ampicillin/Leucine	mbSUS, Bait
pMetExg2YC-Dest	attP1, attP2	Ampicillin/Leucine	mbSUS, Bait with a membrane anchor
ptdh3YC-Dest	attR1, attR2	Ampicillin/Leucine	Tri-SUS, Bait
ptdh3Exg2YC-Dest	attP1, attP2	Ampicillin/Leucine	Tri-SUS, Bait with a membrane anchor
pPrey-PartnerSUS-2in1-Dest	attR1/R4, attR3/R2	Ampicillin/Tryptophan	Tri-SUS, Prey, and Partner

Table 2
Yeast strains used in mbSUS, GPS, and Tri-SUS systems

Yeast strain	Genotype	Function	Reference
THY.AP4	*MATa; ade2-, his3-, leu2-, trp1-, ura3-; lexA::ADE2, lexA::HIS3, lexA::lacZ*	Bait carrying in vector pMetYC-Dest, pMetExg2-Dest, or ptdh3Exg2YC-Dest	[36]
THY.AP5	*MATα; URA3; ade2-, his3-, leu2-, trp1-*	Prey carrying in vector pNX35-Dest, pPrey-Partner-2in1-Dest, or NubI	[36]

Dropout complete supplement mixture (CSM) (MP Biomedicals, 114560222)

Lithium acetate dihydrate (Sigma-Aldrich, L4158)

Agar No. 1 (Sigma-Aldrich, LP0011)

2.2 Antibodies

Anti-HA High Affinity (Roche, 11867423001, 1:10,000 dilution)

Anti-VP16 tag (Abcam, ab4808, 1:10,000 dilution)

Anti-Myc tag (Chromotex, 9E1, 1:5000 dilution)

Rabbit Anti-Rat IgG H & L horseradish peroxidase (HRP) conjugated (Abcam, ab6734, 1:10,000 dilution)

Goat Anti-Rabbit IgG H & L horseradish peroxidase(HRP) conjugated (Abcam, ab6721, 1:20,000 dilution)

2.3 mbSUS Yeast Strains

Yeast strains for use in mbSUS assays are listed in Table 2. Briefly, the THY.AP4 and THY.AP5 yeast strains are used for all mbSUS assays [36]. Both these strains are unable to grow without a supply of adenine (A), histidine (H), tryptophan (T), and leucine (L), and THY.AP4 yeast additionally requires uracil (U) supplements in their growth media. All mbSUS approaches rely on the mating of transformed THY.AP4 and THY.AP5 yeast to co-express bait, prey, and/or partner proteins. The THY.AP4 cells are transformed with the bait expression vector that restores the synthesis of leucine (L), and THY.AP5 yeast cells are transformed with a prey expression vector that restores the expression of tryptophan (T). Therefore, the mated yeast cells are selected for growth on a complete supplement mixture (CSM) supplemented medium lacking tryptophan, leucine, and uracil (CSM_{-TLUM} medium). Additionally, the THY. AP4 strain carries the reporter genes to synthesize adenine (A) and histidine (H) under the control of an activatable LexA promoter. Binding between the bait and prey proteins releases PLV transactivator for the LexA promoter such that mated yeast can grow on $CSM_{-AHTLUM}$ medium. mbSUS assays normally are run together with matings of THY.AP4 yeast with NubG and NubI in THY.AP5 yeast to provide negative and positive controls, respectively [15, 16]. In the Tri-SUS assay, the expression of soluble NubG and its partner proteins serves as negative control; either the prey Gateway cassette is left empty, or a non-interacting "prey" protein is expressed. pPrey-PartnerSUS-Dest-NubI is used as the positive control. VP16-tagged bait and HA-tagged prey protein expression are verified by immunoblot using anti-VP16 and anti-HA antibodies, respectively, to ensure that lack of growth is not due to the absence of either bait or prey proteins. *Saccharomyces cerevisiae* yeast transformations are carried out using standard LiAc/ ss-DNA/polyethylene glycol (PEG) methods for competent yeast cells [16].

2.4 The Reagents and Media for Yeast Growth and Transformation.

Yeast Extract-Peptone-Dextrose (YPD) medium: 2% w/v peptone, 2% w/v glucose, 1% w/v yeast extract, (pH 6.0); add 20 g L-1 agar No.1 to prepare agar plates. Store at room temperature (*see* **Note 1**).

1M/0.1M LiAc: Dissolve lithium acetate in distilled water and adjust the pH to 7.5 with 1M HCl; dilute 10-fold with distilled water to get 0.1M LiAc. Store at room temperature (*see* **Note 2**).

ss-DNA: Dissolve 10 mg mL^{-1} Salmon sperm DNA in 0.1M LiAc and heat at 95 °C for 5 min followed by immediate cooling in ice before usage. Store at –20°C (*see* **Note 2**).

PEG solution: For 50 mL, dissolve 25 g PEG3350 and adjust the final volume to 50 mL with distilled water to a final concentration of 50% w/v. Store at room temperature (*see* **Note 2**).

2.5 Yeast Selection Media for mbSUS, GPS, and Tri-SUS Systems

CSM-$_{AHTLUM}$ medium: 0.7 g L^{-1} yeast nitrogen base (YNB) powder w/o amino acids, w/o ammonium sulfate, w/o potassium phosphate, 1 g L^{-1} potassium phosphate monobasic, 5 g L^{-1} ammonium sulfate, 20 g L^{-1} glucose, and 0.55 g L^{-1} CSM dropout powder; adjust pH to 6.0 with KOH; add 20 g L^{-1} agar No.1 to prepare agar plates. Store at 4 °C (*see* **Note 1**).

CSM-$_{TUM}$ medium: Selective medium for prey strains in THY.AP5 with adenine, histidine, and leucine (*see* **Note 1**).

CSM-$_{LM}$ medium: Selective medium for baits in THY.AP4 with adenine, histidine, tryptophane, and uracil (*see* **Note 1**).

CSM-$_{TLUM}$ medium: Yeast mating verification medium with adenine and histidine (*see* **Note 1**).

CSM-$_{AHTLUM}$ medium ± methionine (Met): Yeast growth assay medium with additional methionine to suppress bait expression (*see* **Note 1**).

2.6 The Amino Acid and Nucleobase Stock Solutions for mbSUS, GPS, and Tri-SUS Systems (see Note 3)

Adenine (Ade, A): 4 mg mL^{-1} stock solution

Uracil (Ura, U): 4 mg mL–1 stock solution

Leucine (Leu, L): 20 mg mL^{-1} stock solution

Tryptophan (Trp, T): 4 mg mL^{-1} stock solution

Histidine (His, H): 4 mg mL^{-1} stock solution

Tryptophan (Trp, T): 15 mg mL^{-1} stock solution, 0.1M

3 Methods

3.1 Protocol for Yeast Transformation and Selection for mbSUS, GPS, and Tri-SUS Assays (See Note 4)

1. Inoculate yeast strains THY.AP4 and THY.AP5 separately into 5 mL of Yeast Extract-Peptone-Dextrose (YPD) liquid medium in round bottom falcon tubes and grow at 30 °C, 180 rpm overnight.
2. Inoculate the overnight culture into 45 mL of YPD in a 150 mL flask and incubate at 30 °C, 180 rpm for 3 h until the culture reaches 0.6–0.8 optical density (OD_{600}).
3. Harvest cells by centrifugation at 2000 g, 4 °C for 10 min in 50 mL centrifuge tubes.
4. Decant the supernatant and resuspend the pellets with 20 mL of distilled water; repeat **step 3**.
5. Decant the supernatant and resuspend the cells with 1.5 mL 0.1 M LiAc (lithium acetate) and transfer to 2.0 mL microcentrifuge tubes.
6. Pellet by centrifugation at 2000 g, 4 °C for 30 s and discard the supernatant.

7. Resuspend the pellet with an appropriate volume of 0.1 M LiAc (20 μL for each transformation) and incubate at room temperature for 30 min. Competent cells are ready for a transformation.
8. Heat Salmon sperm (ss) DNA (10 mg mL^{-1}) at 95 °C for 5 min and place immediately on ice.
9. Mix 10 μL of heat-denatured ss-DNA with 5 μL of plasmids (150 ng mL^{-1}) in 0.2 ml polymerase chain reaction (PCR) tubes.
10. Prepare master mixture in ratio 1:7:2 for 1 M LiAc, 50% w/v PEG3350, and competent cells from **step 7**. For each transformation, use 100 μL of the master mixture.
11. Mix the master mixture with the DNA mixture from **step 9**.
12. Incubate the mixtures at 30 °C for 30 min in the thermal cycler.
13. Heat shock the mixture by incubation at 43 °C for 30 min in a thermal cycler/water bath (Note: temperature should be accurate at 43 ± 0.3 °C).
14. Harvest the cells by centrifugation at 2000 g for 1 min, and then aspirate the supernatant by pipetting.
15. Resuspend the pellets in 100 μL of distilled water and repeat **step 14**.
16. Resuspend the cells in 50 μL of distilled water and plate on the appropriate selective medium: THY.AP4 carrying bait vector on CSM_{-LM} agar plate, THY.AP5 carrying prey vector on CSM_{-TUM} agar plate.
17. Incubate at 30°C for 48–72 h.

3.2 The Steps for the Protocol for Yeast Growth Assay (See Note 4 and Fig. 3)

1. Collect 10–15 transformed yeast colonies, and inoculate them into 2 mL of liquid selection medium (CSM_{-LM} for THY.AP4 and CSM_{-TUM} for THY.AP5) in round bottom culture tubes for overnight growth at 180 rpm and 30 °C.
2. Harvest the yeast cell pellets in a 2.0 mL microcentrifuge by centrifugation at 2000 g for 1 min.
3. Resuspend pellets in YPD liquid medium. For each mating combination, 10 μL of cell suspension is required for yeast expressing bait and prey each in mbSUS and GPS systems or bait and prey with partner each in Tri-SUS system.
4. Mix by pipetting 10 μL of yeast suspension (bait) with 10 μL of yeast suspension (prey with/out a partner) in 96-well plates or 0.2 ml PCR tubes.
5. Drop 5 μL of mixed yeast on YPD agar plates and incubate at 30 °C overnight (16–20 h).

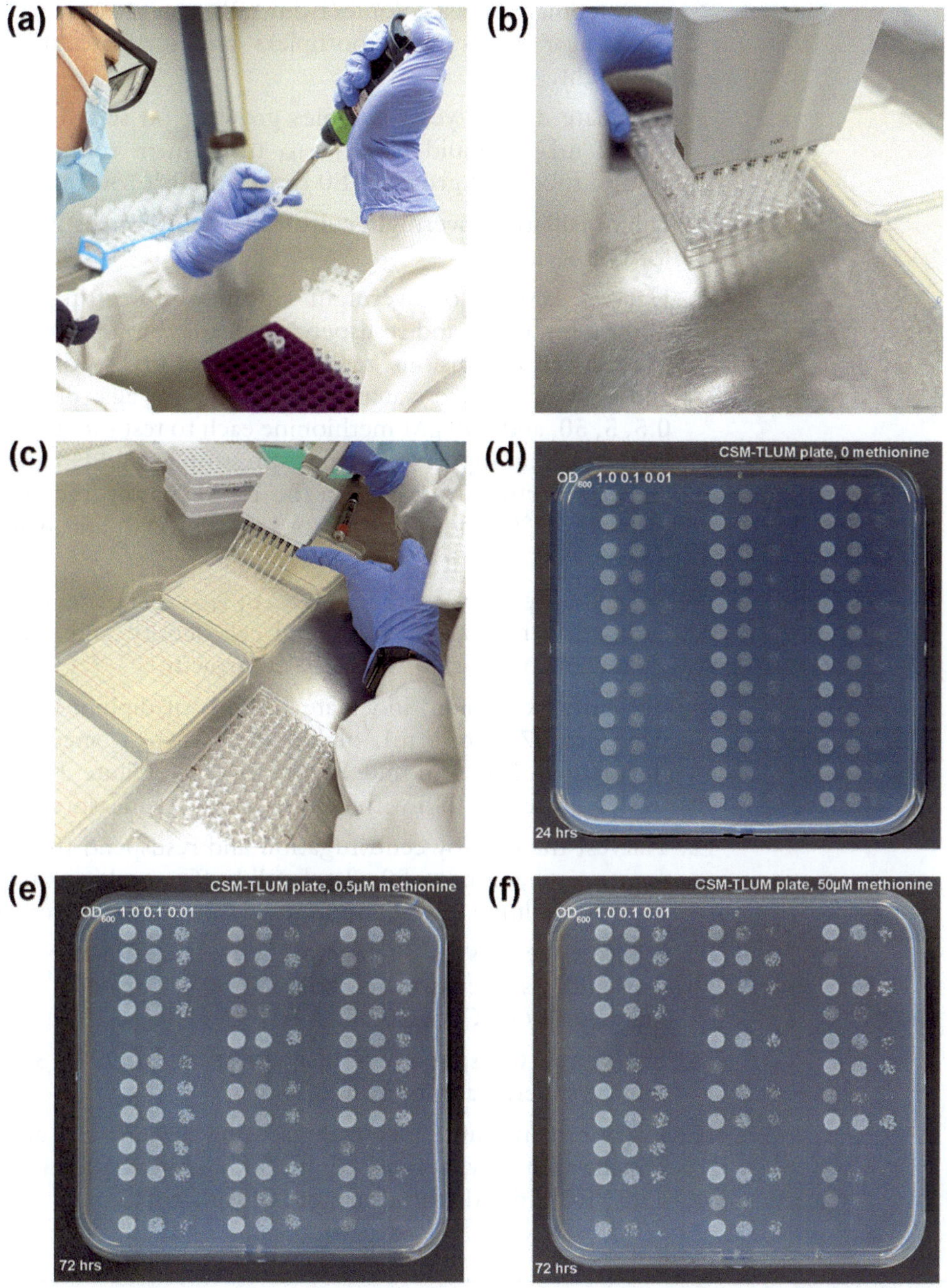

Fig. 3 Images showing steps for yeast handling in mbSUS, GPS, and Tri-SUS assays. Mated yeast pellets are suspended in water at OD_{600} 1.0 (**a**), serial dilutions are carried out (**b**), and yeast suspensions are dropped on selective media plates (**c**). Yeast mating efficiency is verified on CSM_{-TLUM} plate containing no methionine (**d**). Yeast growth is detected after incubation at 30°C for 72 h on $CSM_{-AHTLUM}$ agar plates with increasing concentrations of methionine, here 0.5 μM methionine (**e**) and 50 μM methionine (**f**), to suppress bait expression. Yeast growth in the presence of higher methionine indicates a strong interaction between the bait and prey protein

6. Transfer mated yeast colonies from YPD agar plates onto CSM-LTUM agar plates with toothpicks and incubate at 30 °C overnight.
7. Inoculate diploid yeast colonies growing on selection media into 2 mL of liquid CSM-LTUM medium in round bottom culture tubes and grow at 180 rpm and 30 °C overnight.
8. Dilute 100 μL of overnight cultures 10-fold with distilled water for OD_{600} assay.
9. Harvest 100 μL of overnight cultures by centrifugation at 2000 g for 1 min and resuspend in distilled water. Drop 5 μL of yeast cell suspensions, serially diluted at OD_{600} of 1.0, 0.1, and 0.01 in distilled water on CSM-AHLTUM agar plates, with 0.5, 5, 50, and 500 μM methionine each to test the strength of interaction on suppressing bait expression in mbSUS and GPS assays or partner expression in the Tri-SUS assay. Incubate the plates at 30 °C, and scan to record yeast growth for 48 and/or 72 h.
10. Drop the diploid yeast cells on CSM-LTUM agar plates (control) to confirm mating efficiency and cell density, and scan after 24 h incubation at 30 °C to record growth.
11. To verify expression, inoculate 1.8 mL of diploid yeast cells from **step 7** into 1 mL of liquid CSM-LTUM medium and incubate at 180 rpm and 30 °C for 2 h. Dilute 100 μL of fresh cultures 10-fold with distilled water to measure OD_{600}.
12. Harvest the cells by centrifugation and resuspend them in an appropriate amount of 2× Laemmli buffer reaching OD_{600} at 50–100 (volume must be more than 100 μL for sonication).
13. Sonicate the cells for 30 s (amplitude setting 20% for 10 s and pause for 5 s; keep samples in ice) to release proteins and then incubate at 60 °C for 30 min.
14. Vortex for 10 s, spin down the lysate at full speed for 30 s, and use the supernatant for immunoblots.
15. Carry out immunoblots analysis using an anti-HA antibody (Roche) for prey, anti-VP16 antibody (Abcam) for bait, and anti-Myc antibody for a partner (Chromotex).

4 Notes

1. Autoclave sterilization method needs to be adapted.
2. Sterilize by filtration through a 0.22 μM filter.
3. For auxotrophic selection, amino acid powders are dissolved into warm water (40 °C) and sterilized by filtration and stored at −20 °C. Avoid prolonged exposure to light. Methionine

(5 mL of stock solution per liter of medium) should be added after medium autoclaving, to prevent disintegration.

4. All steps of the mbSUS assays are carried out in sterile conditions

Acknowledgments

We thank Lingfeng Xia for helping with the figures and for proofreading the text. Funding from the UK Biotechnology and Biological Sciences Research Council and the Royal Society of London is acknowledged.

References

1. Xing S, Wallmeroth N, Berendzen KW, Grefen C (2016) Techniques for the analysis of protein-protein interactions in vivo. *Plant Physiol* 171:727–758
2. Vikis HG, Guan KL (2004) Glutathione-S-transferase-fusion based assays for studying protein-protein interactions. *Methods Mol Biol* 261:175–186
3. Louche A, Salcedo SP, Bigot S (2017) Protein-protein interactions: pull-down assays. *Methods Mol Biol* 1615:247–255
4. Tang Z, Takahashi Y (2018) Analysis of protein-protein interaction by co-IP in human cells. *Methods Mol Biol* 1794:289–296
5. Maccarrone G, Bonfiglio JJ, Silberstein S, Turck CW, Martins-de-Souza D (2017) Characterization of a protein interactome by co-immunoprecipitation and shotgun mass spectrometry. *Methods Mol Biol* 1546:223–234
6. Jamge S, Angenent GC, Bemer M (2018) Identification of in planta protein-protein interactions using IP-MS. *Methods Mol Biol* 1675:315–329
7. Fujiwara M, Uemura T, Ebine K et al (2014) Interactomics of Qa-SNARE in arabidopsis thaliana. *Plant and Cell Physiology* 55:781–789
8. Grefen C, Blatt MR (2012) A 2in1 cloning system enables ratiometric bimolecular fluorescence complementation (rBiFC). *Biotechniques* 53:311–314
9. Walter M, Chaban C, Schutze K et al (2004) Visualization of protein interactions in living plant cells using bimolecular fluorescence complementation. *Plant Journal* 40:428–438
10. Mehlhorn DG, Wallmeroth N, Berendzen KW, Grefen C (2018) 2in1 vectors improve in planta BiFC and FRET analyses. *Methods Mol Biol* 1691:139–158
11. Lerner E, Barth A, Hendrix J et al (2021) FRET-based dynamic structural biology: challenges, perspectives and an appeal for open-science practices. *Elife* 10
12. Hecker A, Wallmeroth N, Peter S, Blatt MR, Harter K, Grefen C (2015) Binary 2in1 vectors improve in planta (co-) localisation and dynamic protein interaction studies. *Plant Physiology* 168:776–787
13. McAlister-Henn L, Gibson N, Panisko E (1999) Applications of the yeast two-hybrid system. *Methods* 19:330–337
14. Buckholz RG, Simmons CA, Stuart JM, Weiner MP (1999) Automation of yeast two-hybrid screening. *J. Mol. Microbiol. Biotechnol.* 1:135–140
15. Grefen C, Obrdlik P, Harter K (2009) The determination of protein-protein interactions by the mating-based split-ubiquitin system (mbSUS). *Methods Mol Biol* 479:217–233
16. Horaruang W, Zhang B (2017) Mating based split-ubiquitin assay for detection of protein interactions. *Bio Protoc* 7:e2258
17. Baena G, Xia L, Waghmare S, Karnik RA (2022) SNARE SYP132 mediates divergent traffic of plasma membrane H+-ATPase AHA1 and antimicrobial PR1 during pathogenesis. *Plant Physiol*
18. Karnik R, Zhang B, Waghmare S et al (2015) Binding of SEC11 indicates its role in SNARE recycling after vesicle fusion and identifies two pathways for vesicular traffic to the plasma membrane. *Plant Cell* 27:675–694
19. Zhang B, Karnik R, Alvim J, Donald N, Blatt MR (2019) Dual sites for SEC11 on the SNARE SYP121 implicate a binding exchange during secretory traffic. *Plant Physiol* 180:228–239

20. Zhang B, Karnik R, Waghmare S, Donald N, Blatt MR (2017) VAMP721 conformations unmask an extended motif for K+ channel binding and gating control. *Plant Physiol* 173: 536–551
21. Zhang B, Karnik R, Wang Y et al (2015) The arabidopsis R-SNARE VAMP721 interacts with KAT1 and KC1 K^+ channels to moderate K^+ current at the plasma membrane. *Plant Cell* 27:1697–1717
22. Waghmare S, Lefoulon C, Zhang B et al (2019) K(+) channel-SEC11 binding exchange regulates SNARE assembly for secretory traffic. *Plant Physiol* 181:1096–1113
23. Grefen C, Lalonde S, Obrdlik P (2007) Split-ubiquitin system for identifying protein-protein interactions in membrane and full-length proteins. *Current Protocols Neurosci.* 5: 5–27
24. Grefen C, Obrdlik P, Harter K (2009) The determination of protein-protein interactions byt the mating-based split-ubiquitin system (mbSUS). *Methods in Molecular Biology* 479: 1–17
25. Zhang B, Karnik R, Donald N, Blatt MR (2018) A GPI signal peptide-anchored split-ubiquitin (GPS) system for detecting soluble bait protein interactions at the membrane. *Plant Physiol* 178:13–17
26. Zhang B, Xia L, Zhang Y, Wang H, Karnik R (2021) Tri-SUS: a yeast split-ubiquitin assay to examine protein interactions governed by a third binding partner. *Plant Physiol* 185:285–289
27. Honsbein A, Sokolovski S, Grefen C et al (2009) A tripartite SNARE-K+ channel complex mediates in channel-dependent K+ nutrition in Arabidopsis. *Plant Cell* 21:2859–2877
28. Grefen C, Karnik R, Larson E et al (2015) A vesicle-trafficking protein commandeers Kv channel voltage sensors for voltage-dependent secretion. *Nat Plants* 1:15108
29. Grefen C, Chen ZH, Honsbein A et al (2010) A novel motif essential for SNARE interaction with the K^+ channel KC1 and channel gating in arabidopsis. *Plant Cell* 22:3076–3092
30. Hachez C, Laloux T, Reinhardt H et al (2014) Arabidopsis SNAREs SYP61 and SYP121 coordinate the trafficking of plasma membrane aquaporin PIP2;7 to modulate the cell membrane water permeability. *Plant Cell* 26:3132–3147
31. Xia L, Mar Marques-Bueno M, Bruce CG, Karnik R (2019) Unusual roles of secretory SNARE SYP132 in plasma membrane H(+)-ATPase traffic and vegetative plant growth. *Plant Physiol* 180:837–858
32. Zhang J, Lautar S (1996) A yeast three-hybrid method to clone ternary protein complex components. *Analytical Biochemistry* 242:68–72
33. Tirode F, Malaguti C, Romero F et al (1997) A conditionally expressed third partner stabilizes or prevents the formation of a transcriptional activator in a three-hybrid system. *J Biol Chem* 272:22995–22999
34. Grefen C, Blatt MR (2012) Do calcineurin B-like proteins interact independently of the serine threonine kinase CIPK23 with the K^+ channel AKT1? lessons learned from a menage a trois. *Plant Physiology* 159:915–919
35. Grefen C (2014) The split-ubiquitin system for the analysis of three-component interactions. *Methods Mol Biol* 1062:659–678
36. Obrdlik P, El Bakkoury M, Hamacher T et al (2004) K + channel interactions detected by a genetic system optimized for systematic studies of membrane protein interactions. *P Natl Acad Sci USA* 101:12242–12247

Chapter 4

Detection of Protein–Protein Interactions Utilizing the Split-Ubiquitin Membrane-Based Yeast Two-Hybrid System

Siddhartha Dutta and Matthew D. Smith

Abstract

Identifying the interactors of a protein is a key step in understanding its possible cellular function(s). Among the various methods that can be used to study protein-protein interactions (PPIs), the yeast two-hybrid (Y2H) assay is one of the most standardized, sensitive, and cost-effective in vivo methods available. The most commonly used GAL4-based Y2H system utilizes the yeast transcription factor GAL4 to detect interactions between soluble proteins. By virtue of involving a transcription factor, the protein–protein interactions occur in the nucleus. The split-ubiquitin Y2H system offers an alternative to the traditional GAL4-based Y2H system and takes advantage of the reconstitution of split-ubiquitin in the cytosol to identify interactions between two proteins. Moreover, new membranous and soluble interacting partner(s) can be identified by screening a target protein against proteins produced from a cDNA library using this system.

Key words Protein–protein interaction, Yeast two-hybrid, Split-ubiquitin, Interactors, library screening

1 Introduction

Proteins predominantly operate in complexes rather than as isolated species for performing their functions in vivo [1–3]. Thus, identifying and characterizing protein–protein interactions (PPIs) is crucial for unravelling the cellular functions of candidate proteins. Due to several factors, including affinity and stability, interactions between proteins may be permanent or transient. To date, multiple in vivo and in vitro approaches, such as affinity chromatography, yeast two-hybrid (Y2H), co-immunoprecipitation, protein-fragment complementation, and protein microarray have been extensively employed to investigate PPIs [1, 4, 5]. Amongst the various methods employed for PPI studies, Y2H systems have been one of the most commonly used approaches due to the advantages of sensitivity, cost-effectiveness, and detecting interacting partners in a cellular environment under controlled laboratory conditions.

The Y2H system is a technique used to identify PPIs in vivo in the yeast, *Saccharomyces cerevisiae.* Originally described by Stanley Fields in 1989 [6], the principle relies on the transcriptional activation/initiation of a reporter gene(s) for the identification of PPIs. The basic idea involves splitting a transcription factor (TF) into two separate functional domains, the DNA binding domain (DBD) and the transcription activation domain (AD). The DBD is fused to one of the proteins of interest (generally called bait) and the AD is fused to the other protein of interest (called prey). The bait and prey fusion proteins are then co-expressed in the yeast host. Interaction of the two proteins leads to reconstitution of the functional TF, thereby driving the expression of selectable yeast marker and reporter genes that are under the control of a promoter that is controlled by the TF. Expression of the selectable yeast marker genes and reporter proteins, respectively, allows the yeast cells to grow on minimal media and develop a characteristic color when using a colorimetric substrate. The classic Y2H system is based on the GAL4 transcription factor, which consists of an N-terminal DNA binding domain (GAL4DBD) and a C-terminal activation domain (GAL4AD). In the GAL4-based system, GAL4DBD is linked to the bait protein and GAL4AD is linked to the prey. If the bait and prey interact with one another through protein–protein interaction, a functional GAL4 TF will be reconstituted from the two domains and can trigger transcription of any gene that is under the control of the GAL4 promoter. The selectable marker genes used in this system include the *HIS3* and *ADE2* genes, and the *Lac Z* gene from *Escherichia coli* is used as the reporter. The *HIS3* gene encodes imidazole glycerol-phosphate dehydratase which catalyzes the sixth step of the histidine biosynthesis pathway, whereas the *ADE2* genes encode AIR carboxylase which is involved in adenine biosynthesis [7, 8]. Thus, host yeast cells can only grow in the absence of histidine and adenine when these two genes are expressed upon the interaction between bait and prey, and these cells also develop color in a ß-galactosidase assay. Although the Y2H assay was initially used for studying only pairwise interactions involving two proteins of interest, over time it has gained acceptance for large-scale library screening to identify novel prey proteins that interact with the bait [9–11]. A prerequisite for the GAL4-based Y2H system is that the reconstituted TF generated due to interaction between bait and prey must be localized to the nucleus and the interactions must persist in the nucleus to activate the reporter genes. Thus, target proteins that are membrane-associated and/or require cytosolic posttranslational modifications for interactions are not likely to be identified through the GAL4-based Y2H system.

In this chapter, we present protocols based on an alternate Y2H system, the split-ubiquitin system, for studying the interactions of integral membrane proteins and their interacting partner(s) (either soluble or integral membrane proteins) [12, 13]. Originally developed in 1994, the split-ubiquitin system is based on the molecular function of ubiquitin in ubiquitin-mediated proteasomal degradation [14–17]. The ubiquitin-mediated proteasomal degradation process involves tagging the target protein with a chain of ubiquitin molecules (poly-ubiquitin tagging), followed by its proteasomal degradation. To recycle the ubiquitin, a deubiquitinating enzyme, ubiquitin-specific-processing protease (UBP), recognizes the folded ubiquitin and cleaves it from the target protein, thus releasing monomeric ubiquitin in the cytosol. Nils Johnsson and Alexander Varshavsky in 1994 demonstrated that the yeast ubiquitin can be split into N-terminal (Nub) and C-terminal (Cub) halves [15]. When expressed separately in yeast, they exhibit partial conformation and hence are not recognized by UBPs. On co-expressing the Nub and Cub in the same yeast cell, they interact to form an intact, folded ubiquitin, referred to as "split-ubiquitin," that could be recognized and cleaved by UBPs. The inherent high affinity of wild-type Nub (termed NubI) with Cub was shown to be abolished by replacing the isoleucine present at position 3 with a glycine (hereinafter called NubG). When co-expressing NubG with Cub, efficient reconstitution of "split-ubiquitin" and activation of UBPs occurs only when NubG and Cub are forced to come in close proximity through the interactions of the two proteins that they are fused with. In the split-ubiquitin membrane-based Y2H system, the bait protein is fused to Cub, which is in turn attached to an artificial transcription factor, LexA-VP16 (hereafter referred to as B-Cub-LexA-VP16). The B-Cub-LexA-VP16 fusion protein is targeted to the endoplasmic reticulum, where it resides assuming that the bait is a membrane protein. NubG is fused to either terminus of the prey protein (NubG-P or P-NubG). The two fusion proteins are co-expressed in the same cells, and if the bait and the prey interact with each other, the B-Cub-LexA-VP16 and P-NubG/NubG-P co-expression leads to the formation of split-ubiquitin and activation of UBPs, and therefore release of the artificial transcription factor, LexA-VP16 into the cytosol. The released LexA-VP16 re-localizes to the nucleus where it binds to the *LexA* operators situated upstream of the selectable yeast genes *HIS3* and *ADE2* and activate their transcription thereby allowing the yeast cell to grow on minimal media lacking histidine and adenine. The LexA-VP16 also activates the transcription of the *Lac Z* gene, the reporter gene that is also under the control of the *LexA* operator, encoding the reporter enzyme ß-galactosidase, whose activity can be detected using a standard X-Gal test (Fig. 1).

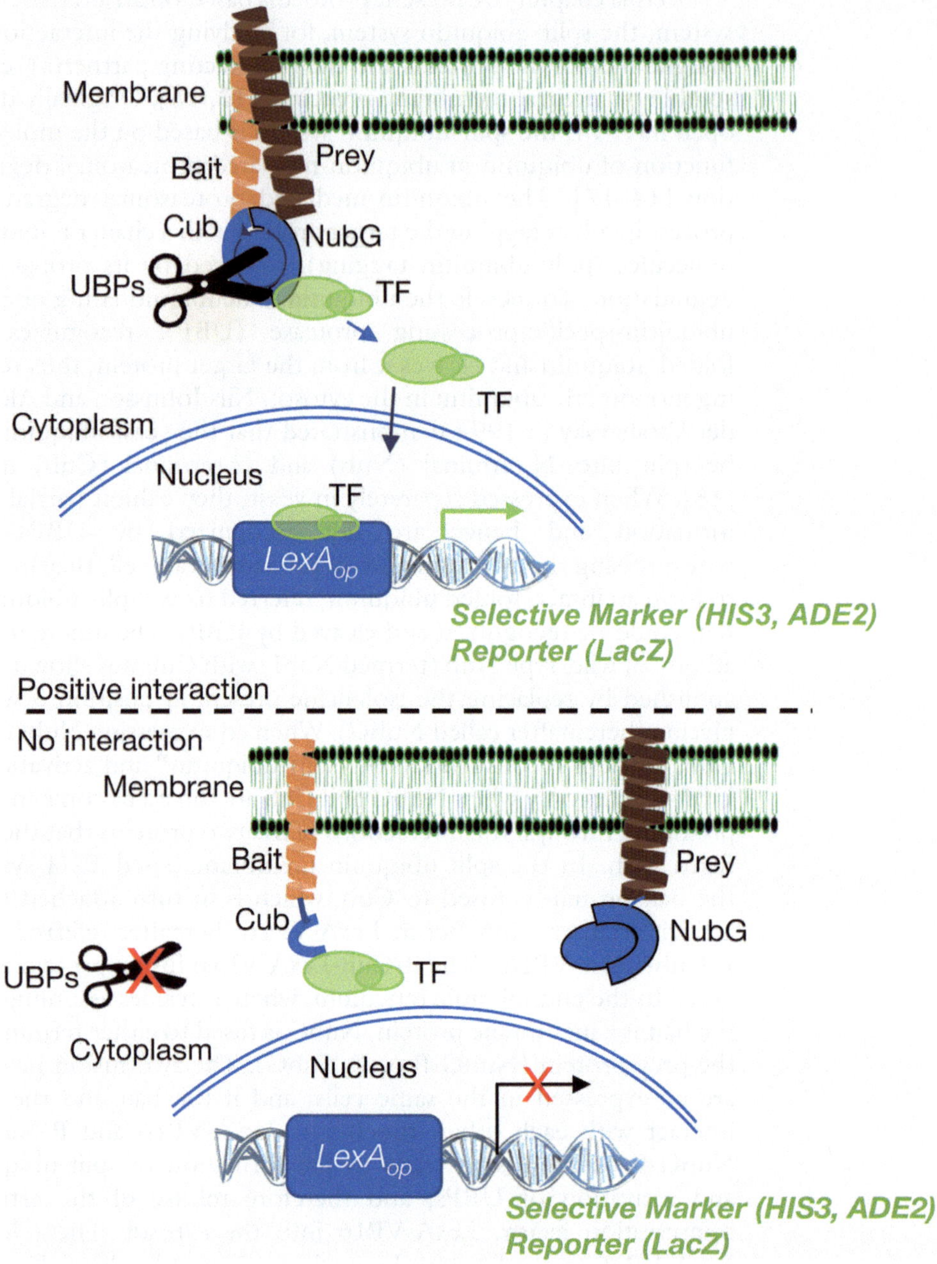

Fig. 1 The split-ubiquitin membrane-based yeast two-hybrid system. The bait protein is fused to the Cub domain and the artificial transcription factor (TF) LexA-VP16 to generate the fusion bait protein. The fusion is targeted to the endoplasmic reticulum membrane using the STE2 leader peptide. The prey protein is fused to the NubG domain. Upper Panel: Interaction between the bait and prey proteins results in the formation of "split-ubiquitin." The split-ubiquitin is recognized by the UBPs that cleave the artificial TF, thereby liberating the TF into the cytosol. The free TF is targeted to the nucleus, where it interacts with the LexA operator, thus switching on the selectable marker and reporter genes (HIS3, ADE2, LacZ). Interaction of the bait and prey proteins, therefore, results in the growth of the yeast cells on the SD-LWHA selective medium and gives the colonies blue coloration in the presence of the ß-galactosidase substrate X-Gal. Lower Panel: If there is no interaction between bait and prey, the "split-ubiquitin" is not generated, resulting in the absence of free cytosolic TF. Thus, no activation of the selectable markers and reporter genes occurs in the yeast cells

2 Materials

1. *Saccharomyces cerevisiae* reporter strain NMY51: *MATa his3Δ200 trp1–901 leu2–3, 112 ade2 LYS2::(lexAop)4-HIS3 ura3::(lexAop)8-lacZ ade2::(lexAop)8-ADE2 GAL4.*
2. pBT-3 STE bait plasmid: Encodes Cub fused to LexA-VP16 DNA binding domain (Cub-LexA-VP16 ORF) with *LEU2* auxotrophic marker (Dualsystems Biotech, Zurich, Switzerland). A multiple cloning site (MCS) meant for insertion of cDNA/gene/fragment sequence of interest is present upstream of the Cub-LexA-VP16 ORF, thereby resulting in a bait vector encoding a fusion protein consisting of the bait protein upstream of the C–terminal Cub-LexA-VP16 fusion partner (i.e., Cub-LexA-VP16 will be fused to the C-terminal end of the bait protein). To ensure targeting of the heterologous fusion protein to the endoplasmic reticulum, a yeast-specific *STE2* leader sequence is inserted immediately upstream of the MCS. The STE2-Bait-Cub-LexA-VP16 fusion cassette is driven by a weak promoter, p*CYC1*, that provides a low-level expression of the fusion protein. The pBT3-STE plasmid contains the CEN/ARA origin of replication that allows only 1–2 copies of the bait plasmid to be maintained per yeast cell. A bacterial selective marker gene conferring resistance to kanamycin is also present to enable cloning in *E. coli* (Fig. 2a).
3. Bait construct: The bait construct to express Cub-LexA-VP16 fused to the protein of interest can be obtained by inserting the coding sequence of the bait protein at the MCS in-frame with the upstream STE2 leader and downstream Cub-LexA-VP16 fusion. The fragment encoding the bait protein should lack the initiating ATG codon and last stop codon in order to facilitate successful translation of the STE2-Bait-Cub-LexA-VP16 fusion cassette in the yeast.
4. pPR3-N prey plasmid: NubG activation domain fusion vector with *TRP1* auxotrophic marker (Dualsystems Biotech, Zurich, Switzerland). The MCS meant for prey/library insertion is present downstream of the NubG coding sequence thereby resulting in a prey vector encoding fusion proteins consisting of the prey protein downstream of the N–terminal NubG partner. An HA epitope tag sequence is present upstream of the MCS. The pPR3-N contains the 2 μ origin of replication origin that allows a high copy number bait plasmid (10–30 copies per yeast cell) to be maintained. A bacterial selective marker conferring resistance to ampicillin is included to facilitate cloning in *E. coli* (Fig. 2b).

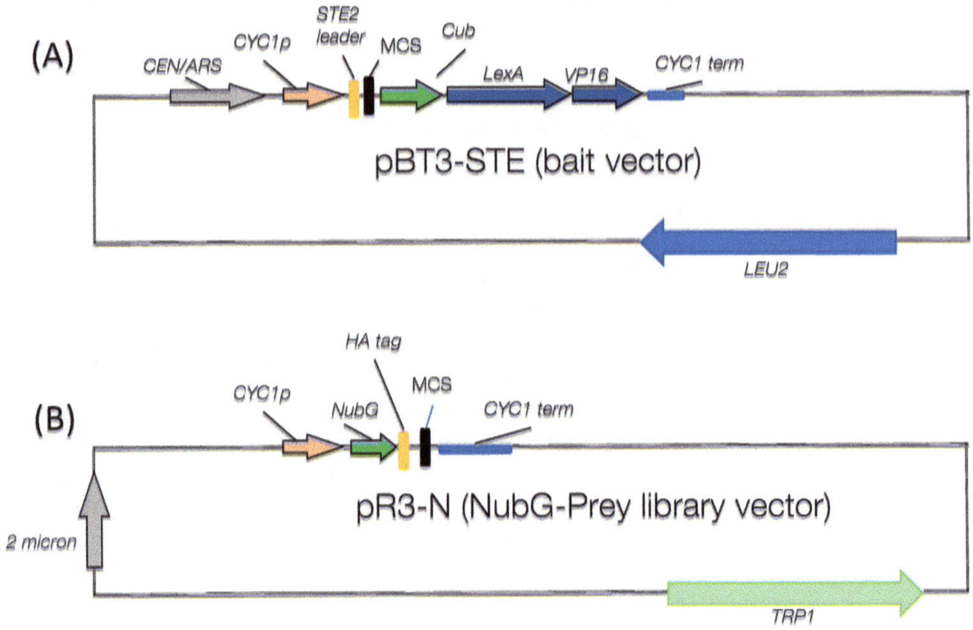

Fig. 2 The split-ubiquitin membrane-based yeast two-hybrid system bait (**a**) and prey (**b**) vectors. (**a**) The bait vector pBT3-STE (Dualsystems Biotech, Zurich, Switzerland) consists of the Cub-LexA-VP16 cassette downstream of the MCS. The coding sequence for a STE2 leader sequence is present upstream of the MCS. (**b**) The prey vector pPR3-N (Dualsystems Biotech, Zurich, Switzerland) contains the NubG sequence present upstream of the MCS. Both bait and prey plasmids contain the CYC1 promoter and CYC1 terminator that drive the expression of bait and prey fusion proteins in yeast cells

5. Y2H cDNA library: A NubG-HA-Prey library is constructed by cloning the cDNA inserts into the unique SfiI restriction site of the MCS of the pPR3-N plasmid. The cDNA library is constructed by following a standard protocol according to Sambrook and Russell (2001) [18]. The mRNA is isolated from the cell/tissue of interest and first strand cDNA is synthesized by reverse transcription using oligo-dT primers followed by double-stranded cDNA (ds cDNA) synthesis by PCR amplification. The recognition sequence for the restriction enzyme SfiI is introduced into the ends of the ds cDNA through oligonucleotide linkers such that the cDNA fragments can be inserted into the MCS of the prey plasmid. The cDNA is size fractionated using a CHROMA SPIN TM-400 column (Takara Bio USA), ligated into the pPR3-N prey vector, and transformed into *E. coli* cells. The cDNA library is amplified and stored in −80 °C for later use.
6. Yeast extract Peptone Adenine Dextrose (YPAD) medium: YPAD medium was obtained in powder form from Dualsystems Biotech. The medium contains peptone (20 g/L), yeast extract (10 g/L), adenine hemisulfate (20 mg/L), and dextrose (20 g/L) in optimal proportions for the growth of

Table 1
The stock and final concentrations of the amino acid solutions used to prepare the yeast growth medium

Amino acid	Stock solution	Final concentration
Adenine (A, Ade)	2 g/L	10 mg/L
Histidine (H, His)	10 g/L	20 mg/L
Leucine (L. Leu)	10 g/L	100 mg/L
Tryptophan (W, Trp)	10 g/L	20 mg/L

Saccharomyces cerevisiae. The entire content of a 1× YPAD pouch is dissolved in 1 L sterile water and autoclaved. For 2× YPAD medium, two pouches of 1× YPAD are dissolved in 1 L sterile water and autoclaved. For agar plates, 20 g of agar is added to 1 L of liquid YPAD medium before autoclaving. Autoclaved YPAD medium can be stored at 4 °C.

7. Amino acid stock solution: Concentrations of leucine (Leu, L), tryptophan (Trp, W), histidine (His, H), and adenine (Ade, A) stock solutions are listed in Table 1. For the preparation of stock solutions, 200 mg amino acid powder was dissolved in 20 mL (for L, W, and H) or 100 mL (for A) of sterile water. The stock solutions were filter sterilized using a 0.2 μm filter and stored at 4 °C.
8. 3-Aminotriazole (3-AT) stock solution: For the preparation of 1 M 3-AT stock solution, 4.2 g powdered 3-AT was dissolved in 50 mL of sterile water. The stock solution was filter sterilized using a 0.2 μm filter and stored at −20 °C.
9. Synthetic Dropout (SD) medium: The SD-LWHA medium was obtained in powder form from Dualsystems Biotech. The medium contains yeast nitrogen base and dextrose in optimal proportion for the growth of *Saccharomyces cerevisiae* and lacks leucine, tryptophan, histidine, and adenine amino acids. The SD medium is prepared by adding the supplementary amino acid(s) that is/are absent from the medium leaving out the appropriate amino acid from the dropout medium. Thus, the SD medium contains all the ingredients required for optimum growth of yeast but lacks the appropriate amino acid: no leucine (SD-L), no tryptophan (SD-W), no leucine/tryptophan (SD-LW), no leucine/tryptophan/histidine (SD-LWH), no leucine/tryptophan/histidine/adenine (SD-LWHA) or no leucine/tryptophan/histidine/adenine with added 3-AT (SD-LWHA/3-AT). For agar plates, 20 g/L of agar is added to a liquid medium before autoclaving. For SD-LWHA/3-AT, the appropriate final concentration of 3-AT is obtained by adding the required amount of 3-AT from the 1 M 3-AT stock solution.

10. Single-stranded carrier DNA: 2.0 mg/mL of Salmon sperm DNA. Prior to use, the Single-stranded carrier DNA should be denatured by boiling for 5 min and cooled on ice.
11. Sterile 50% PEG 4000 (Polyethylene glycol).
12. Sterile 1 M LiOAc.
13. Sterile 10X TE pH 7.5: 100 mM Tris HCl, 10 mM EDTA.
14. PEG/LiOAc master mix: For transformation of the bait construct into NMY51 (protocol described in Subheading 3.1.1), mix 180 μL of 1 M LiOAc, 1.2 mL of 50% PEG and 125 μL of single-stranded carrier DNA in a sterile 50 mL falcon tube. For bait functional assay (protocol described in Subheading 3.1.3), mix 360 μL of 1 M LiOAc, 2.4 mL of 50% PEG and 250 μL of single-stranded carrier DNA in a sterile 50 mL falcon tube. For standardization of screening stringency (protocol described in Subheading 3.2), mix 1.2 mL of 1 M LiOAc, 1.2 mL of 10 × TE pH 7.5, and 9.6 mL of 50% PEG in a sterile 50 mL falcon tube. For library transformation (protocol described in Subheading 3.3.1), mix 1.5 mL of 1 M LiOAc, 1.5 mL of 10 × TE pH 7.5, and 12 mL of 50% PEG in a sterile 50 mL falcon tube. Prepare the master mix fresh just before use (*see* **Note 1**).
15. LiOAc/TE master mix: For standardization of screening stringency (protocol described in Subheading 3.2), mix 0.88 mL of 1 M LiOAc, 0.88 mL of 10 × TE pH 7.5, and 6.24 mL of sterile water in a sterile 50 mL falcon tube. For library transformation (protocol described in Subheading 3.3.1), mix 1.1 mL of 1 M LiOAc, 1.1 mL of 10 × TE pH 7.5, and 7.8 mL of sterile water in a sterile 50 mL falcon tube. Prepare the master mix fresh just before use in the methods.
16. Dimethylsulfoxide (DMSO).
17. Sterile 0.9% NaCl.
18. 500 μm of Acid-washed glass beads.
19. Z buffer: 0.06 M $Na_2HPO_4.7H_2O$, 0.01 M $NaH_2PO_4.H_2O$, 0.01 M KCl, 0.001 M $MgSO_4$, 0.05 M β-mercaptoethanol (BME), adjust to pH 7.
20. X-gal (5-Bromo-4-chloro-ß-D-galactopyranoside): 20 mg/mL dissolved in N,N-dimethylformamide).
21. Z buffer/X-gal solution (freshly prepared): To make 100 mL of Z buffer, add 0.27 mL of β-mercaptoethanol and 1.67 mL of X-gal stock solution.
22. Whatman No.5 filter paper.
23. Mouse monoclonal anti-LexA antibody.
24. 50% TCA.

25. Sequencing primers: For the prey clones in pPR3-N vector, the set of primers used are,

$$FP-pPR3-N: 5'\ GTCGAAAATTCAAGACAAGG\ 3'$$

$$RP-pPR3-N: 5'\ TTTCTGCACAATATTTCAAGC\ 3'$$

3 Methods

The screening of a cDNA library using a split-ubiquitin membrane-based Y2H system can be performed by a large-scale transformation of yeast cells expressing the bait protein with the prey (library) plasmids. The overall library screening procedure is outlined in Fig. 3.

3.1 Generation of NMY51 Yeast Strain Expressing the Functional Bait Protein

3.1.1 Transformation of the Bait Construct into NMY51

1. Inoculate several colonies of the NMY51 from a fresh YPAD master plate into 50 mL YPAD liquid medium in a sterile 250 mL flask. Incubate at 30 °C with shaking at 200 rpm overnight (*see* **Note 2**).
2. Use the overnight culture (hereinafter referred to as primary culture) to inoculate 50 mL of fresh YPAD medium (hereinafter referred to as secondary culture) such that the OD_{546} of the secondary culture is ~0.15. Incubate the secondary culture at 30 °C with shaking at 200 rpm until the OD_{546} reaches ~0.6 (approximately 3–5 h) (*see* **Note 2**).
3. Cells from the culture are pelleted in a sterile 50 mL falcon tube by centrifuging at 3000× g for 5 min at room temperature. Discard the supernatant and resuspend the cell pellet in 2.5 mL sterile water.
4. Transfer 1.5 μg of bait construct plasmid and empty bait plasmid into two independent sterile 1.5 mL microcentrifuge tubes.
5. Add 300 μL of freshly prepared PEG/LiOAc master mix to each tube and vortex briefly.
6. Transfer 100 μL of resuspended NMY51 cells from **step 3** to each tube. Vortex each tube for 1 min to mix all the components.
7. Incubate for 45 min in a 42 °C water bath.
8. Collect the cells by centrifugation at room temperature for 5 min at 1000× g.
9. Resuspend each pellet in 100 μL of 0.9% NaCl.
10. Spread the entire volume of resuspended cells from each tube onto two separate 100 mm SD-L selective plates. Incubate the plates at 30 °C until the colonies are ~2 mm in diameter (~3–4 days).

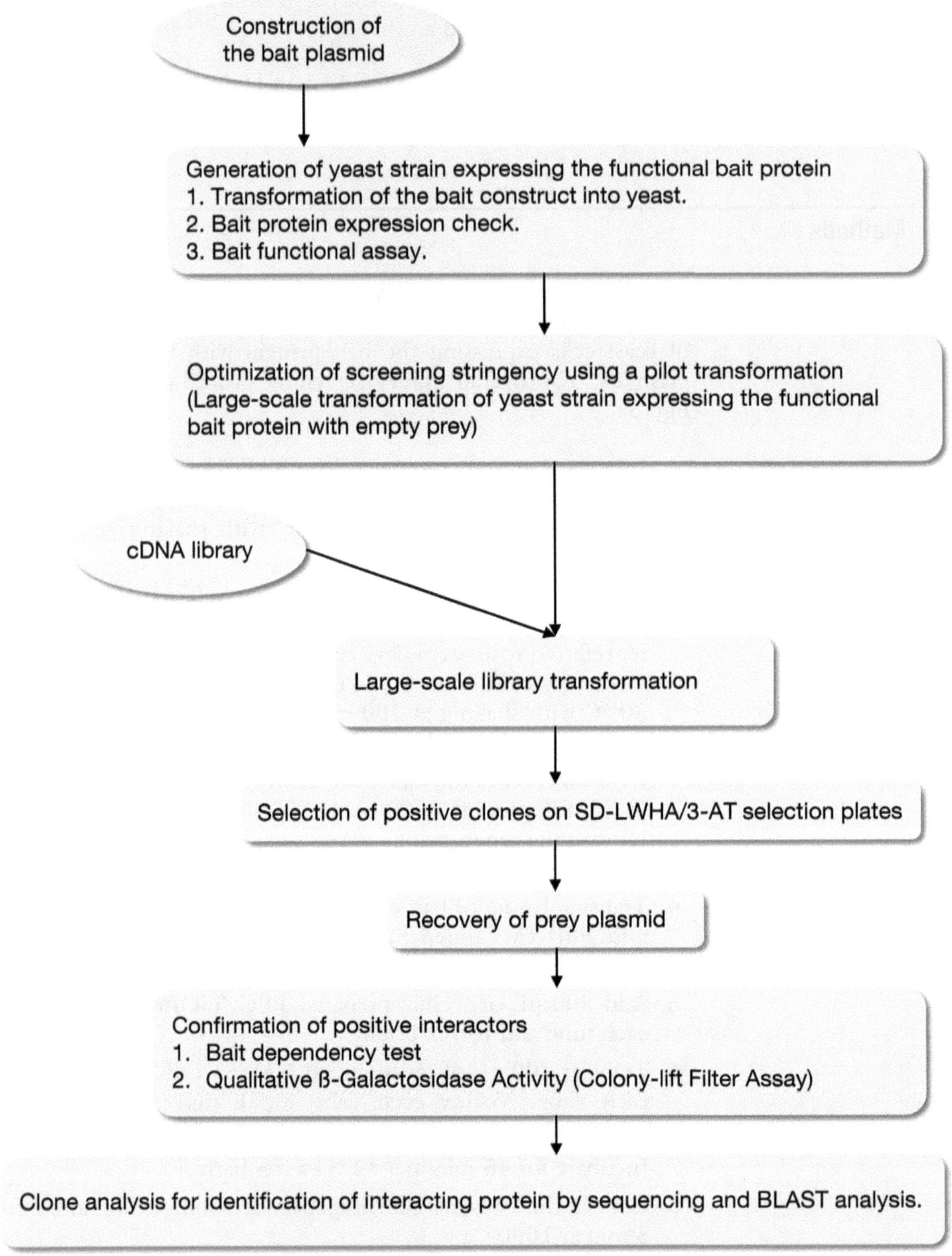

Fig. 3 General flowchart outlining the steps used for split-ubiquitin membrane-based yeast two-hybrid library screening

3.1.2 Bait Protein Expression Check by Western Blot

1. Inoculate several colonies of the NMY51 transformed with bait construct into 10 mL SD-L liquid medium in a sterile 50 mL falcon tube. Incubate at 30 °C with shaking at 200 rpm overnight.
2. Use the overnight culture (primary culture) to inoculate 50 mL of fresh YPAD medium (termed, secondary culture) such that the OD_{546} of the secondary culture is ~0.15. Incubate the secondary culture at 30 °C with shaking at 200 rpm until the OD_{546} reaches ~0.6 (approximately 3–5 h).
3. Transfer 10 mL of the secondary culture to a sterile 15 mL falcon tube and centrifuge at 1000× g for 5 min at room temperature to collect the cells in a pellet. Discard the supernatant and wash the pellet in 2.5 mL of 1 mM EDTA.
4. Resuspend the pellet in 200 μL of 2 M NaOH. Carefully transfer the resuspended cells to a 2 mL Eppendorf tube. Incubate on ice for 10 min.
5. Add 200 μL of 50% TCA and mix well. Incubate on ice for 2 h.
6. Centrifuge at 4 °C for 20 min at 13,000× g. Discard the supernatant and resuspend the pellet in 200 μL of ice-cold acetone.
7. Centrifuge at 4 °C for 20 minutes at 13,000x g. Discard the supernatant and resuspend the pellet in 200 μL of 5% SDS by pipetting.
8. Incubate at 37 °C for 15 min with gentle shaking.
9. Centrifuge the extract for 5 min at 13,000× g.
10. Transfer the extract to a new Eppendorf tube. Proceed immediately with the SDS-PAGE and Western Blot analysis using standard methods. To detect the bait, probe the blot with the monoclonal LexA antibody (raised against the LexA protein present in the Cub-LexA-VP16 cassette). If the bait protein is properly expressed, a band corresponding to the molecular weight of the bait protein plus 38 KDa (corresponding to Cub-LexA-VP16) will be detected on the blot. Extract from NMY51 yeast cells transformed with an empty bait vector can be used as a negative control (*see* **Note 3**).

3.1.3 Bait Functional Assay

After verifying the expression of the bait fusion protein in yeast, a co-transformation of yeast with the bait construct and control prey construct consisting of non-interacting prey protein fused to NubI, is performed to check the functional status of the bait protein. Co-expression of the functional bait-Cub-LexA-VP16 and non-interacting prey fused to NubI (P_{non}-NubI) should result in reconstitution of "split-ubiquitin" due to a strong inherent affinity between Cub and wild-type Nub (NubI) proteins, thus switching on the expression of the selectable marker (*HIS3* and *ADE2)* and reporter (*LacZI*) genes.

1. Inoculate several colonies of the NMY51 from a fresh YPAD master plate into 50 mL YPAD liquid medium in a sterile 250 mL flask. Incubate at 30 °C with shaking at 200 rpm overnight.
2. Use the overnight culture (primary culture) to inoculate 50 mL of fresh YPAD medium (termed, secondary culture) such that the OD_{546} of the secondary culture is ~0.15. Incubate the secondary culture at 30 °C with shaking at 200 rpm until the OD_{546} reaches ~0.6 (approximately 3–5 h).
3. The culture is centrifuged in a sterile 50 mL falcon tube at 2500× g for 5 min at room temperature to collect the cells in a pellet. Discard the supernatant and resuspend the pellet in 2.5 mL sterile water.
4. Set up two co-transformation reactions. Reaction 1 corresponds to the co-transformation of cells with bait construct and non-interacting prey protein fused to NubI (P_{non}-NubI) and Reaction 2 corresponds to the co-transformation of cells with bait construct and empty prey vector (pPR3-N). Use 1.5 μg of each of the plasmid for the co-transformation reactions 1 and 2.
5. Add 300 μL of freshly prepared PEG/LiOAc master mix to each tube and vortex briefly.
6. Transfer 100 μL of resuspended NMY51 cells from **step 3** to each tube. Vortex each tube for 1 min to mix all the components.
7. Incubate for 45 min in a 42 °C water bath.
8. Collect the cells by centrifugation at room temperature for 5 min at 1000× g.
9. Resuspend each pellet in 150 μL of 0.9% NaCl.
10. Spread 50 μL of the resuspended cells from each Reaction tube onto two separate 100 mm SD-LW and SD-LWHA selective plates.
11. Incubate the plates at 30 °C until the colonies are ~2 mm in diameter (~3–4 days for SD-LW and ~ 4–6 days for SD-LWHA).
12. Count the total number of colonies (co-transformants) in all the plates for Reaction 1 and Reaction 2 (from **step 11**). Calculate the "% growth under selection" for each Reaction by using the following equation:

$$\begin{aligned}&\%\text{Growth under selection}\\&\quad=(\text{no.of colonies on SD}-\text{LWHA plate})\\&\quad\times 100/\text{no.of colonies on SD}-\text{LW}\end{aligned}$$

If the bait is expressed and functional, Reaction 1 should yield a % Growth under a selection value between 10% and 100% depending on the expression and stability of the bait protein. Reaction 2 should yield no significant growth on the SD-LWHA as the bait does not interact with NubG fused nonsense-peptide expressed from the pPR3-N control prey (*see* **Note 4**).

3.2 Standardization of Screening Stringency Using a Pilot Transformation

As the *HIS3* gene tends to have a leaky expression in yeast strains, it is necessary to optimize the basic screening stringency by conducting a pilot with the bait construct and an empty prey plasmid. 3-aminotriazole (3-AT) is an inhibitor of the *HIS3* gene product. Thus, the addition of 3-AT to the selective plate increases the stringency of *HIS3* selection and significantly reduces the chances of false positive colonies appearing when screening a Y2H library. Therefore, a pilot screening is performed before the large-scale transformation thereby enabling titration of the optimum concentration of 3-AT needed for a specific bait.

1. Streak an SD-L plate with NMY51 expressing the bait protein and incubate the plate at 30 °C until colonies appear (~3–4 days).
2. Inoculate one colony of the NMY51 expressing the bait protein from the SD-L master plate into a 10 mL SD-L medium in a sterile 50 mL falcon tube. Incubate at 30 °C with shaking at 200 rpm for 8 h.
3. Transfer the entire volume (10 mL) of the starter culture to 100 mL of SD-L in a 250-mL flask and incubate at 30 °C with shaking at 200 rpm overnight.
4. Measure the OD_{546} of the overnight culture using water as blank.
5. Pellet down an amount of culture corresponding to 22.5 OD_{546} units in a sterile 50-mL falcon tube by centrifuging at 1000× g for 5 min at room temperature.
6. Resuspend the cells in 10 mL of prewarmed (30 °C) 2 × YPAD medium, and transfer the entire volume to a 1 L sterile flask. Rinse the falcon tube with 40 mL of prewarmed (30 °C) 2 × YPAD medium and transfer the entire volume of medium to the 1 L flask.
7. To the above 1 L flask add 150 mL of prewarmed (30 °C) 2 × YPAD medium. Measure the OD_{546} of the culture to ensure that the OD_{546} is ~0.15.
8. Incubate the culture at 30 °C with shaking at 200 rpm until the OD_{546} reaches ~0.6 (4–5 h).
9. Collect 150 mL culture from **step 8** in three 50-mL sterile falcon tubes, and centrifuge at room temperature for 5 min at 1000× g.

10. Discard the supernatant and resuspend each pellet in 30 mL of sterile water by vortexing. Pellet down the cells by centrifuging at room temperature for 5 min at 1000× g.
11. Discard the supernatant and resuspend each pellet in 1 mL of freshly prepared LiOAc/TE master mix by vortexing. Pellet down the cells by centrifuging at room temperature for 5 min at 1000× g.
12. Discard the supernatant and resuspend each pellet in 600 μL of freshly prepared LiOAc/TE master mix by vortexing. The yeast cells are now ready to be transformed.
13. Prepare the library transformation reaction by adding 10 μg of library plasmid, 600 μL of competent cells (from **step 12**), 100 μL of single-stranded carrier DNA, and 2.5 mL of PEG/LiOAc master mix in a sterile 50-mL falcon tube. Set up three separate transformation reactions (*see* **Note 1**).
14. Vortex the library transformation reaction tubes to thoroughly mix all the components. Incubate the tubes for 45 min at 30 °C. Mix the components every 15 min by gently tapping the tubes.
15. At the end of the incubation, add 160 μL of DMSO to each tube. Mix the components by gently tapping the tubes.
16. Incubate the tubes for 20 min at 42 °C.
17. Collect the cells by centrifugation at room temperature for 5 min at 1000× g.
18. Recovery step: Discard the supernatant and resuspend each pellet in 3 mL of prewarmed (30 °C) 2 × YPAD medium. Combine the cell suspensions from all three tubes into a sterile 50-mL falcon tube. Incubate the culture medium at 30 °C with slow shaking at 150 rpm for 1.5 h.
19. Centrifuge the cells at room temperature for 5 min at 1000× g. Discard the supernatant and resuspend the pellet in 3.6 mL of sterile 0.9% NaCl.
20. To determine transformation efficiency prepare 1:100, 1:1000, and 1:10000 serial dilutions in sterile 0.9% NaCl and plate 100 μL of each dilution on an SD-LW plate. Incubate the plates at 30 °C until the colonies appear (~3–4 days).
21. Spread the remaining resuspended cells (300 μL resuspended cells per plate) on the 150 mm SD-LWHA selective plates supplemented with 0, 1, 2.5, 5, 7.5, and 10 mM 3-AT plates. Incubate the plate at 30 °C (~4–6 days, *see* **Note 5**).
22. Count the total number of colonies (co-transformants) in SD-LW plates (**step 20**). Calculate the "total number of co-transformants" and "transformation efficiency" by using the following equations:

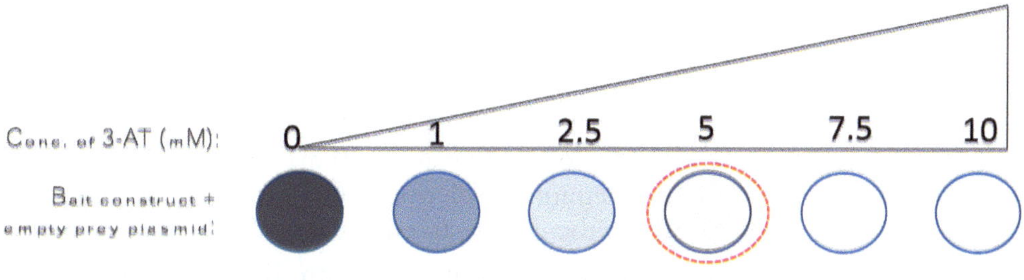

Fig. 4 Determination of optimal 3-AT concentration for library transformation by pilot screening using the bait construct and empty prey plasmid. The co-transformed yeast cells are plated on a series of SD-LWHA selective plates containing increasing concentrations of 3-AT. The very first concentration at which no colonies appear after 3–4 days (or 1–2 colonies, indicating a very low background) is optimal for library screening for that specific bait protein of interest (indicated by red dotted circle). In the representative image, the deep blue to white color gradients of the plates is a quantitative representation of the number of colonies on the 3-AT gradient. Deep blue: higher number of background colonies, white: no background colonies

$$\text{Total number of transformants} = \text{no.of co} - \text{transformants on SD} - \text{LW} \times \text{dilution factor} \times 10 \times 3.6$$

$$\text{Transformation efficiency (clones/g DNA)} = \text{Total no.of transformants}/21\ \mu\text{g}$$

The total number of transformants should be higher than 8 × 10^5 cfu for a meaningful interpretation of the result. Choose the selective stringency (lowest conc. of 3-AT) where no colonies grow. This is the concentration of 3-AT that should be used in SD-LWHA plates for library screening (Fig. 4) (*see* **Note 5**).

3.3 Library Screening and Confirmation of Interaction

For library screening, the *Arabidopsis* cDNA library fused to the C-terminus of NubG (NubG-prey) is introduced into the yeast strain expressing the bait protein.

3.3.1 Library Transformation

1. Streak an SD-L plate with NMY51 expressing the bait protein and incubate the plate at 30 °C until colonies appear (~3–4 days).
2. Inoculate one colony of NMY51 cells expressing the bait protein from the SD-L master plate into a 10 mL SD-L medium in a sterile 50-mL falcon tube. Incubate at 30 °C with shaking at 200 rpm for 8 h.
3. Transfer the entire volume (10 mL) of the starter culture to 100 mL of SD-L in a 250-mL flask and incubate at 30 °C with shaking at 200 rpm overnight.
4. Measure the OD_{546} of the overnight culture by using water as blank.
5. Pellet down an amount of culture corresponding to 30 OD_{546} units in a sterile 50-ml falcon tube by centrifuging at 1000× g for 5 min at room temperature.

6. Resuspend the cells in 10 mL of prewarmed (30 °C) 2 × YPAD medium and transfer the entire volume to a sterile 1 L flask. Rinse the falcon tube with 40 mL of prewarmed (30 °C) 2 × YPAD medium and transfer the entire medium to the 1 L flask.
7. To the above 1 L flask add 150 mL of prewarmed (30 °C) 2 × YPAD medium (200 mL total final volume). Measure the OD_{546} of the culture to ensure that the OD_{546} is ~0.15.
8. Incubate the culture at 30 °C with shaking at 200 rpm until the OD_{546} reaches ~0.6 (4–5 h).
9. Divide the 200 mL culture from step 8 into four 50-mL sterile falcon tubes and centrifuge at room temperature for 5 minutes at 1000 × g.
10. Wash 1: Discard the supernatant and resuspend each pellet in 30 mL of sterile water by vortexing. Pellet down the cells by centrifuging at room temperature for 5 min at 1000 × g.
11. Wash 2: Discard the supernatant and resuspend each pellet in 1 mL of freshly prepared LiOAc/TE master mix by vortexing. Pellet down the cells by centrifuging at room temperature for 5 min at 1000× g.
12. Discard the supernatant and resuspend each pellet in 600 μL of freshly prepared LiOAc/TE master mix by vortexing. The yeast cells are now ready to be transformed.
13. Prepare the library transformation reaction by adding 10 μg of library plasmid, 600 μL of competent cells (from **step 12**), 100 μL of single-stranded carrier DNA, and 2.5 mL of PEG/LiOAc master mix in a sterile 50-mL falcon tube. Set up four separate library transformation reactions. (see **Note 1**).
14. Vortex the library transformation reaction tubes to thoroughly mix all the components. Incubate the tubes for 45 min at 30 ° C. Mix the components every 15 min by gently tapping the tubes.
15. At the end of the incubation, add 160 μL of DMSO to each tube. Mix the components by gently tapping the tubes.
16. Incubate the tubes for 20 min at 42 °C.
17. Collect the cells by centrifugation at room temperature for 5 min at 1000× g.
18. Recovery step: Discard the supernatant and resuspend each pellet in 3 mL of prewarmed (30 °C) 2 × YPAD medium. Pool all the cell suspensions from all four tubes into a sterile 50-mL falcon tube. Incubate the culture medium at 30 °C with slow shaking at 150 rpm for 1.5 h.
19. Centrifuge the cells at room temperature for 5 min at 1000× g. Discard the supernatant and resuspend the pellet in 4.8 mL of sterile 0.9% NaCl.

20. For determination of transformation efficiency prepare 1:100, 1:1000, and 1:10000 serial dilutions in sterile 0.9% NaCl and plate 100 μL of each dilution on SD-LW. Incubate the plate at 30 °C until the colonies appear (~3–4 days).
21. Spread the remaining resuspended cells on multiple 150 mm SD-LWHA/ 3-AT plates (300 μL resuspended cells per plate). Incubate the plate at 30 °C until the colonies are ~2 mm in diameter (~4–6 days).
22. Using a sterile toothpick, transfer (restreak of 2–3 mm) all the grown colonies from the SD-LWHA/ 3-AT plates to new 150 mm diameter SD-LWHA/3-AT plates to isolate the colonies and confirm the results. Incubate the plate at 30 °C until the colonies appear (~4–6 days). Repeat the re-streaking step (on SD-LWHA/3-AT plates) twice to eliminate false positive interactors.
23. Count the total number of colonies (co-transformants) in SD-LW plates (**step 20**). Calculate the "total number of co-transformants" and "transformation efficiency" by using the following equations:

$$\text{Total number of transformants} = \text{no.of co} - \text{transformants on SD} - \text{LW} \times \text{dilution factor} \times 10 \times 4.8$$

$$\text{Transformation efficiency (clones/gDNA)} = \text{Total no.of transformants}/30\ \mu\text{g}$$

For an efficient library transformation, the transformation efficiency should be higher than 1×10^5 clones/μg DNA.

3.3.2 Recovery of Prey Plasmid

Proceed with the prey plasmid recovery step for all the colonies that were obtained after the re-streaking process (twice) outlined in the library transformation protocol (**step 22**). Isolate the plasmid using standard mini-preparation protocol based on the alkaline lysis method described by Sambrook and Russell (2006) with minor modifications as mentioned in the protocol [19].

1. Inoculate 10 mL of SD-LW liquid media with a positive yeast colony and incubate at 30 °C with shaking at 200 rpm overnight.
2. Centrifuge each overnight culture at 4000× g for 5 min at room temperature to collect the cells in the pellet. Discard the supernatant.
3. After resuspension of the pellet in alkaline lysis solution I, add a scoop full (~100 μL) of acid-washed glass beads to each tube. Vortex the tubes for 5 minutes.
4. Proceed with the standard mini-preparation protocol to obtain isolated plasmid DNA.

5. Resuspend the plasmid in 25 μL of sterile, nuclease-free water.
6. Transform chemically competent *E. coli* DH5α cells with 10 μL of the extracted plasmid using standard protocols for preparation and transformation of competent *E. coli* using calcium chloride as described by Sambrook and Russel (2006) [20].
7. Plate the cells on LB plates supplemented with 100 μg/mL ampicillin. Incubate at 37 °C, overnight.
8. Select four transformed colonies from each plate. Isolate the plasmid and carry out a restriction digest analysis to determine the size of the prey insert. This will determine if all the recovered prey clones from the positive colony contain the same cDNA. If all the minipreps contain the same insert size, choose one miniprep and proceed with a "bait dependency test" (see below). If more than one size of the insert is observed, it indicates that the original positive yeast colony contained more than one species of prey plasmid. To determine the actual prey plasmid that encodes the interacting prey, all the clones should be subjected to a "bait dependency test." This should be followed for all the LB plates (each containing the bacterial cells harboring the cDNA-prey plasmid construct) obtained from **step 7**.

3.3.3 Bait-Dependency Test

In a bait-dependency test, all the isolated prey plasmids are transformed with the bait plasmid into NMY51 yeast cells and plated on the SD-LWHA selection plate to reconfirm the direct pairwise interaction between the bait and prey proteins. Only those preys which produce colonies on SD-LWHA when co-expressed with the original bait and do not yield colonies when co-transformed with a non-interacting bait (non-related bait) are considered true positives and considered for further analysis.

1. Co-transformation (pairwise-interaction method) of the bait vector (original bait or non-interacting bait) and the recovered prey plasmid containing the cDNA is to be carried out by following the protocol described in Subheading 3.1.1.
2. For each transformation reaction, plate 100 μL onto one 100 mm diameter SD-LW and one 100 mm diameter SD-LWHA selective plate. Incubate the plates at 30 °C until the colonies appear (~3–4 days for SD-LW and ~ 4–6 days for SD-LWHA).
3. Using a sterile toothpick, transfer (restreak of 2–3 mm) one colony from each positive SD-LWHA/ 3-AT plate to two new 100 mm diameter SD-LWHA/3-AT plates. Incubate the plate at 30 °C until colonies appear (~4–6 days). Store one plate at 4 °C and use the other for further analysis.

3.3.4 Qualitative ß-Galactosidase Activity (Colony-Lift Filter Assay)

For testing the ß-Galactosidase activity resulting from the expression of the *LacZ* reporter gene in the positive colonies, a colony-lift filter assay is performed by following the protocol described by Breeden and Nasmyth (1985) [21].

1. Using a sterile toothpick, transfer (restreak of 2–3 mm) all the grown colonies from the SD-LWHA/ 3-AT plates obtained from step 3 under Subheading 3.3.3 to a new 150 mm diameter SD-LW plate. Incubate the plate at 30 °C until colonies appear (~3–4 days).
2. Presoak a sterile Whatman No. 5 filter (size equivalent to the plate containing the transformants) in freshly prepared Z buffer/X-gal solution.
3. Using forceps, put a dry Whatman No.5 filter (size equivalent to the plate containing the transformants) on the surface of the SD-LW containing the colonies for 15 min. Gently rub the filter paper using the side of the forceps. This will allow cells from the yeast colonies to cling to the filter paper (*see* **Note 6**).
4. When the filter paper is wet, carefully lift it from the plate and transfer (colony side facing up) to liquid nitrogen for 10 s.
5. Transfer the filter paper into a Petri dish with the colony side facing up and allow it to thaw at room temperature.
6. Using forceps, carefully place the thawed filter paper, colony side up, on the presoaked Whatman No.5 filter (from **step 2**). Avoid trapping air bubbles in-between the filters.
7. Incubate at 30 °C and check periodically until a blue color develops on the colonies (~30 min to 8 h) (*see* **Note 7**).
8. Mark the corresponding positive colonies from the original selection plate (in **step 1**) and proceed with further identification.

Growth on SD-LWHA in the bait dependency test and a positive result in the X-Gal filter test confirm the results obtained in the library screening (*see* **Note 8**). Sequence all the positive prey plasmids using appropriate prey vector-specific primers. For the prey clones in pPR3-N vector, the FP-pPR3-N and RP-pPR3-N primers were used as forward primer and reverse primer, respectively, to determine the sequence of the inserts.

4 Notes

1. The transformation efficiency greatly depends on the concentration of PEG and the quality of single-stranded carrier DNA. To avoid evaporation of water, close the reagent bottle containing 50% PEG. Alternatively, always use a freshly prepared 50% PEG reagent. Store the carrier DNA at −20 °C and denature twice by boiling for 5 min just before use.

2. The primary inoculation of the yeast cell should be done by picking colonies from a fresh re-streaked plate (should be less than 2 weeks old). For obtaining high transformation efficiency of competent cells, the OD_{546} should not exceed 0.8 (the stage at which the majority of the yeast cells have undergone 2 cell divisions).
3. Sometimes due to inherent low expression (because of *CYC1p*) or low level of bait fusion protein extraction, there may be a failure in the detection of the bait protein expression in the Western blot expression test. In such cases, it is recommended to proceed with the bait functional assay as described under Subheading 3.1.1. If the bait functional assay yields a positive result, the yeast cells expressing the fusion bait protein can be used for the library screening procedure.
4. The percentage of growth may vary greatly depending on the expression and stability of the bait protein. It is recommended to incubate the plate for 4–6 days before analyzing the results.
5. Choosing the optimum concentration of 3-AT to use for library screening is often difficult. Although the current protocol suggests a pilot screening with a 3-AT concentration of up to 10 mM, the screening may be extended to higher concentrations (up to 50 or even 100 mM). Choosing a low stringent concentration of 3-AT (with a relatively low background) in order to detect low-affinity interactions (more often involving transiently interacting partners) may promote the appearance of more false negative colonies on the selective plates. On the other hand, using a high stringency 3-AT concentration during selection may mask true but weak interactions, thereby resulting in the detection of very few interactors in a screening process. Thus, the optimum threshold concentration of 3-AT should be determined on a case-by-case basis.
6. While the Whatman No.5 filter is sitting on the surface of the SD-LW containing the colonies, poke holes [2, 3] through the filter and into the SD-LW solid medium at asymmetric locations using a needle. This will enable the filter to be matched with the position of the colonies on the agar plate from which they were lifted.
7. Although it is recommended to wait up to 8 h for blue color development in the X-Gal assay, most interactions (strong and weak interactions) tend to result in color development by 1–2 h.
8. Apart from the bait dependency test, the interaction may also be further reconfirmed by swapping the bait and prey to eliminate the possibility of false positives (if any). The swapped bait and prey constructs are subjected to a pairwise interaction assay by following the protocol described under Subheading 3.3.3.

References

1. Berggard T, Linse S, James P (2007) Methods for the detection and analysis of protein-protein interactions. Proteomics 7(16): 2833–2842
2. von Mering C, Krause R, Snel B et al (2002) Comparative assessment of large-scale data sets of protein-protein interactions. Nature 417(6887):399–403
3. Yanagida M (2002) Functional proteomics; current achievements. J Chromatogr B Analyt Technol Biomed Life Sci 771(1–2):89–106
4. Phizicky EM, Fields S (1995) Protein-protein interactions: methods for detection and analysis. Microbiol Rev 59(1):94–123
5. Stoevesandt O, Taussig MJ, He M (2009) Protein microarrays: high-throughput tools for proteomics. Expert Rev Proteomics 6(2): 145–157
6. Fields S, Song O (1989) A novel genetic system to detect protein-protein interactions. Nature 340(6230):245–246
7. Alifano P, Fani R, Lio P et al (1996) Histidine biosynthetic pathway and genes: structure, regulation, and evolution. Microbiol Rev 60(1): 44–69
8. Gedvilaite A, Sasnauskas K (1994) Control of the expression of the ADE2 gene of the yeast Saccharomyces cerevisiae. Curr Genet 25(6): 475–479
9. Gong W, Shen YP, Ma LG et al (2004) Genome-wide ORFeome cloning and analysis of Arabidopsis transcription factor genes. Plant Physiol 135(2):773–782
10. Rolland T, Tasan M, Charloteaux B et al (2014) A proteome-scale map of the human interactome network. Cell 159(5):1212–1226
11. Yu H, Braun P, Yildirim MA, Lemmens I et al (2008) High-quality binary protein interaction map of the yeast interactome network. Science 322(5898):104–110
12. Dutta S, Teresinski HJ, Smith MD (2014) A split-ubiquitin yeast two-hybrid screen to examine the substrate specificity of atToc159 and atToc132, two Arabidopsis chloroplast preprotein import receptors. PLoS One 9(4): e95026
13. Pandey S, Assmann SM (2004) The Arabidopsis putative G protein-coupled receptor GCR1 interacts with the G protein alpha subunit GPA1 and regulates abscisic acid signaling. Plant Cell 16(6):1616–1632
14. Hershko A (2005) The ubiquitin system for protein degradation and some of its roles in the control of the cell division cycle. Cell Death Differ 12(9):1191–1197
15. Johnsson N, Varshavsky A (1994) Split ubiquitin as a sensor of protein interactions in vivo. Proc Natl Acad Sci U S A 91(22): 10340–10344
16. Stagljar I, Korostensky C, Johnsson N et al (1998) A genetic system based on split-ubiquitin for the analysis of interactions between membrane proteins in vivo. Proc Natl Acad Sci U S A 95(9):5187–5192
17. Thaminy S, Auerbach D, Arnoldo A et al (2003) Identification of novel ErbB3-interacting factors using the split-ubiquitin membrane yeast two-hybrid system. Genome Res 13(7):1744–1753
18. Sambrook J, Russell D (2001) Molecular cloning, a laboratory manual. Cold Spring Harbor Laboratory Press, Cold Spring Harbor, NY
19. Sambrook J, Russell DW (2006) Preparation of plasmid DNA by alkaline lysis with SDS: Minipreparation. CSH Protoc 2006(1)
20. Sambrook J, Russell DW (2006) Preparation and transformation of Competent E. coli using calcium chloride. CSH Protoc 2006(1)
21. Breeden L, Nasmyth K (1985) Regulation of the yeast HO gene. Cold Spring Harb Symp Quant Biol 50:643–650

Chapter 5

Preparation and Utilization of a Versatile GFP-Protein Trap-Like System for Protein Complex Immunoprecipitation in Plants

Danish Diwan and Karolina M. Pajerowska-Mukhtar

Abstract

Protein complex immunoprecipitation (co-IP) is an *in vitro* technique used to study protein–protein interaction between two or more proteins. This method relies on affinity purification of recombinant epitope-tagged proteins followed by western blotting detection using tag-specific antibodies for the confirmation of positive interaction. The traditional co-IP method relies on the use of porous beaded support with immobilized antibodies to precipitate protein complexes. However, this method is time-consuming, labor-intensive, and provides lower reproducibility and yield of protein complexes. Here, we describe the implementation of magnetic beads and high-affinity anti-green fluorescent protein (GFP) antibodies to develop an *in vitro* GFP-protein trap-like system. This highly reproducible system utilizes a combination of small sample size, versatile lysis buffer, and lower amounts of magnetic beads to obtain protein complexes and aggregates that are compatible with functional assays, Western blotting, and mass spectrometry. In addition to protein–protein interactions, this versatile method can be employed to study protein–nucleic acid interactions. This protocol also highlights troubleshooting and includes recommendations to optimize its application.

Key words Protein–protein interaction, Protein complex immunoprecipitation, Magnetic beads, GFP antibody, Western blotting, Mass spectrometry, Protein–nucleic acid interaction

1 Introduction

In biological systems, proteins are a class of highly regulated biomolecules that rarely work independently. To carry out essential roles in the cell in diverse biological systems, proteins often interact with each other in an organized manner [1–10]. Characterization of protein–protein interaction through high-throughput biological and computational methods such as *in silico* protein interaction analysis, yeast two-hybrid library screening, and phage display library screening can provide a global insight into the protein interactome of an organism[11–18]. However, all of these methods

Shahid Mukhtar (ed.), *Protein-Protein Interactions: Methods and Protocols*, Methods in Molecular Biology, vol. 2690, https://doi.org/10.1007/978-1-0716-3327-4_5,

require independent validation due to the limitations that these systems cannot mimic the native environment [11]. Protein–protein interaction also largely relies on the physical properties, modification, and state of proteins; therefore, it is important to preferentially substantiate protein–protein interactions in native or close to native organisms [7, 19].

In plants, achieving high throughput protein interaction experiments including pull-down assay coupled with mass spectrometry requires plants overexpressing recombinant tagged proteins. This protein overexpression is accomplished via transgenic plants generated through Agrobacterium-mediated transformation that requires 6–8 months of laborious screening of plants to identify a stable plant line overexpressing a gene of interest [20]. Similar strategies are utilized to test small-scale binary or multi-complex interactions [21]. Various transient *in-planta* techniques can be used to study protein–protein interaction including protoplast two hybrids, split YFP (yellow fluorescent protein) assay, split luciferase assay, and Forster resonance energy transfer [22, 23]. Most of these techniques are restricted to *in vivo* visualization to detect interaction and do not provide flexibility to isolate proteins or protein complexes *ex vivo* for further characterization and functional analysis [24].

In contrast, protein immunoprecipitation (IP) and protein-complex immunoprecipitation (co-IP) have been extensively adopted by the plant research community as reliable and powerful tools to study protein–protein interaction *ex vivo* [25]. This method relies on the transient expression of recombinant epitope-tagged proteins in *Nicotiana benthamiana*, followed by protein extraction and immunoprecipitation of protein complexes using beaded support and analysis of protein complexes [26].

The method described in this book chapter utilizes an advanced approach to perform protein co-IP utilizing magnetic beads and polyclonal α-green fluorescent protein (α-GFP) antibody to develop an *in vitro* GFP-protein trap-like system. To achieve this, proteins of interest are cloned in a plant expression vector to produce *in planta* epitope-tagged recombinant protein. Amongst the combination, either the bait or the prey must be GFP or GFP derivative tagged in any one of these formats: GFP-bait/ bait-GFP, GFP-prey/ prey-GFP. The resulting plant expression vector constructs are introduced into Agrobacterium cells using a suitable transformation method. An appropriate infiltration mixture of Agrobacterium cells is then prepared and infiltrated in Nicotiana leaf tissue [27, 28]. Once the recombinant proteins are expressed, the samples are harvested and homogenized. A suitable protein extraction buffer is added to obtain a crude lysate of plant proteins. Meanwhile, GFP antibody-coupled magnetic beads are prepared by coupling with a highly validated anti-GFP polyclonal antibody. The crude lysate of plant proteins is then incubated with anti-GFP

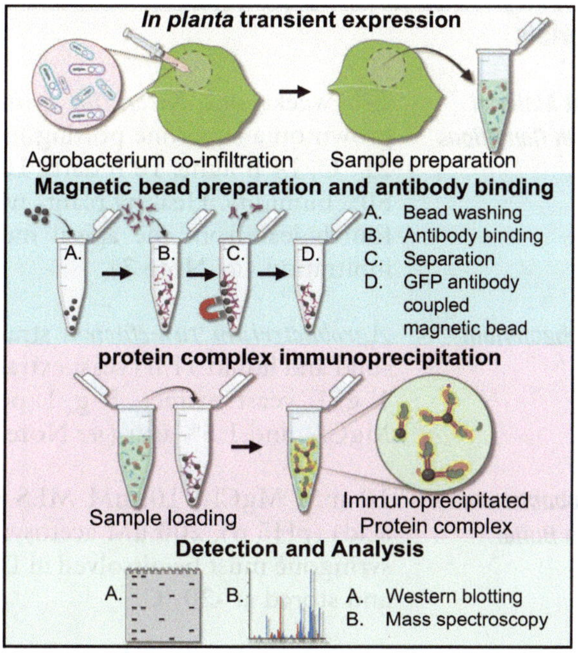

Fig. 1 A schematic diagram representing the major steps involved in preparation and utilization of GFP-protein trap-like system for protein complex immunoprecipitation in plants

coupled magnetic immunoprecipitation beads to precipitate GFP-tagged proteins and protein complexes. These protein complexes can then be analyzed for protein–protein interactions. A schematic diagram demonstrating an overview of this method is shown in Fig. 1.

The method described in this book chapter has advantages over the traditional methods for protein complex immunoprecipitation. The use of magnetic beads significantly reduces the washing and immunoprecipitation time to 30 min as compared to the use of porous beaded support, which requires at least 3 h. This reduces the chances of nonspecific protein elution, protein degradation, and disassembly of protein complexes and promotes the isolation of functionally active protein complexes that can be used for downstream applications.

This GFP-protein trap-like approach is highly flexible because it can be used to capture large protein complexes comprised of a bait protein tagged with any GFP derivative. This is achieved by developing the system using a polyclonal GFP antibody. Recently, an Arabidopsis subcellular non-membranous organelle containing complexes of proteins, metabolites, and RNA was isolated by capturing a GFP6-bait protein using a similar system [29]. This validates the strength and capacity of GFP-protein trap-like system in advanced applications.

2 Materials

2.1 Plant Material and Growth Conditions

4–5 weeks old *Nicotiana benthemiana* plants from seeds were grown on a superfine potting mix. The plants were maintained at 21 °C, 14 h light, 10 h dark, 100 μmol/m^2/s light intensity, and 50% humidity. Healthy plants with two large leaves, the third and fourth leaf from the apical meristem should be considered for infiltration (*see* **Note 3**).

2.2 Agrobacterium culture

Agrobacterium tumefaciens strain GV3101 must be cultured on solid and liquid YEB (yeast extract beef) media: 5 g/L beef extract, 1 g/L yeast extract, 5 g/L peptone, 5 g/L sucrose, 0.5 g/L $MgCl_2$, and 1.5% agar (*see* **Note 1**).

2.3 Agrobacterium Infiltration Buffer

10 mM $MgCl_2$, 10 mM MES (2-(N-morpholino)ethanesulfonic acid) (pH5.6), 200 μM acetosyringone (Millipore Sigma). Acetosyringone must be dissolved in DMSO, dispensed in small aliquots, and stored at -20 °C.

2.4 Protein Extraction Buffer

150 mM tris HCL (pH 7.5), 150 mM NaCl, 5 mM EDTA (ethylenediaminetetraacetic acid), 5% glycerol, 2% PVPP (polyvinylpolypyrrolidone), 10 mM DTT (dithiothreitol) (0.1–1%) NP40, 0.5 mM PMSF (phenylmethylsulfonyl fluoride) (Millipore Sigma), 1X protease inhibitor cocktail for plant cell lysate (Millipore Sigma), 10 μM MG132 proteosome inhibitor (Millipore Sigma). PMSF must be dissolved in isopropanol, dispensed in small aliquots, and stored at −20 °C (*see* **Note 6**).

2.5 Lysate Dilution Buffer

The buffer composition is the same as protein extraction buffer without DTT and PVPP.

2.6 Antibody

anti-GFP rabbit IgG antibody (Life Technologies) (*see* **Note 5**).

2.7 Antibody Dilution Buffer

PBS or TBS (pH 7.4) with 0.02% Tween 20. 1X buffer without Tween 20 can be prepared autoclaved and stored at 4 °C for up to 6 months. Tween 20 should be added to the buffer prior to use.

2.8 Magnetic Beads

Dynabeads protein A (ThermoFisher Scientific).

2.9 Bead Washing Buffer

150 mM Tris (pH 7.5), 250 mM NaCl, 5 mM EDTA, 5% glycerol, 0.5 mM PMSF.

2.10 Elution and Sample Loading Buffer

NU PAGE LDS sample buffer 4X (Invitrogen).

3 Methods

3.1 Agrobacterium Co-infiltration

1. Generate plant expression construct by cloning a gene of interest in plant expression vectors designed to generate tagged protein. To study protein–protein interaction using immunoblotting, use a combination of two or more vectors for recombinant expression of the tagged protein to generate bait and prey proteins. One vector in the combination must be an expression vector coding for the corresponding GFP/ GFP variant-tagged protein. There are various plant expression vectors available for *in-planta* expression of recombinant GFP/ GFP variant-tagged protein. These vectors utilize either classical cloning or a modern-day Gateway cloning system. The commonly utilized vectors include pGWB5 (35S promoter C-sGFP), pGWB6 (35S promoter N-sGFP), or pMDC43 (2X35S promoter N-GFP6) [30]. The other plant expression vector can feature any other tag than GFP; a collection of Gateway binary vectors for protein tagging and expression is described in the Nakagawa vector manual [31].
2. Agrobacterium transformation: Transform plant expression vector in Agrobacterium competent cells by using the standard freeze-thaw method. Briefly, add 3–5 μg of vector to 50 μL of Agrobacterium competent cells and maintain the cells on ice for 15 min. Snap-freeze the cells in liquid nitrogen followed by a heat shock at 45 °C for 45 sec. Recover the Agrobacterium cells using YEB liquid media at 28 °C for 3 hours (h) at 200 rpm. After incubation, the cells should be plated on YEB solid media containing rifampicin, gentamicin, and the selection antibiotic required for the plant expression vector. Transformed Agrobacterium colonies should appear within 2–3 days of incubation. Set 3 mL overnight liquid cultures of transformed Agrobacterium to proceed for further steps (*see* **Note 2**).
3. Infiltration mixture: Pellet down Agrobacterium cells by centrifugation at 3000 rpm for 10 min at room temperature (RT). Wash the Agrobacterium cells using 2 mL infiltration buffer and collect the cells by centrifugation at 3000 rpm for 10 min at RT. Resuspend the cells in 2 mL infiltration buffer and incubate in the dark for 2 h. Post incubation measures the OD_{600} of the cells and mixes equal OD of both prey and bait-expressing cells (and the corresponding control). The OD_{600} of the resulting mixture should not exceed the OD_{600} value of 1 (*see* **Note 4**).
4. Nicotiana leaf infiltration: Water plants 2 h before infiltration. Infiltrate the prepared Agrobacterium infiltration mixture using a needless infiltration syringe. Poke a hole on the adaxial

side of the leaf and infiltrate by gently applying pressure while supporting the syringe from the abaxial side using your fingers. After infiltration, mark the zone of infiltration using a marker.

3.2 Sample Preparation

1. Collect the sample 48 h post infiltration in a standard 2 mL homogenization tube with a 4 mm metal bead. Collect approximately 400 mg sample from the marked zone of infiltration. Flash freeze the sample in liquid nitrogen and homogenize the sample in a bead beater to obtain a fine powder. Keep the sample frozen till further use.
2. Freshly prepare ice-cold protein extraction buffer and add 400 μL of buffer to 400 mg of the powdered leaf tissue sample. Vortex the mixture and incubate on ice for 15 min.
3. Centrifuge the sample at 13,000 rpm for 30 mins. Separate the supernatant into two parts: 100 μL input fraction and 300 μL bead fraction. Dilute the bead fraction with 300 μL of ice-cold lysate dilution buffer (diluted protein lysate).

3.3 Magnetic Bead Preparation

1. Resuspend Dynabead protein A slurry completely by attaching the vial on a rotatory mixer for 5 min at RT. Once resuspended, aliquote 20 μL/reaction volume of the bead slurry in a 2 mL vial (*see* **Note** 7).
2. Separate the magnetic beads using a magnetic rack and remove the supernatant. Immediately add 100 μL of antibody dilution buffer. Maintain the magnetic beads on ice until the next step.

3.4 Antibody Dilution and Binding

1. Add 3 μg of anti-GFP rabbit IgG antibody to 200 μL of antibody dilution buffer, and mix by vortexing briefly.
2. Place the prepared magnetic bead vial on the magnetic stand, and remove the supernatant. Add 200 μL of the diluted antibody mixture to the magnetic beads. Incubate the mixture on a rotating mixer for 30 min at RT. Place the vial on a magnetic stand, and remove the supernatant.
3. Add 200 μL of antibody dilution buffer and wash by gently tapping. Place the magnetic beads-antibody complex containing vial on a magnetic stand and remove the supernatant.

3.5 Sample Loading and Protein Complex Immunoprecipitation

1. Add diluted protein lysate to the magnetic beads-antibody complex, and incubate the mixture on a rotatory mixer for 30 min at RT.
2. Post incubation, place the tube containing the magnetic bead-antibody-protein complex on a magnetic rack and remove the supernatant.

3.6 Bead Washing

1. Add 500 μL of bead washing buffer to the magnetic beads and incubate on a rotatory mixer for 5 min at RT. Separate the beads from the wash solution using the magnetic rack. Discard the wash solution. Repeat the wash step 2 times.
2. Resuspend the resulting magnetic bead-antibody-protein complex in 100–150 μL Antibody dilution buffer and proceed to the elution step.
3. Elution: Add 1X LDS sample buffer and heat the mixture in a dry bath for 10 min at 70 °C.
4. Detection and analysis: The resulting input and magnetic bead-antibody-protein complex samples can be analyzed using western blotting by probing with an anti-GFP antibody to identify whether the recombinant bait protein is expressed in plants. A prey-specific antibody or an antibody targeting the recombinant tag expressed with the prey can be used for the detection of positive interaction. To detect a multi-protein, the denatured protein complex can be used to perform Mass spectrometry.

4 Notes

1. All the reagents and buffers should be prepared fresh using sterile Milli-Q water and molecular biology-grade chemicals. Source and catalog numbers are mentioned for the chemicals that are distinctive to this protocol.
2. Freshly transformed Agrobacterium should be used for better expression of the protein. Agrobacterium cells obtained from glycerol stock may result in lower levels of expression, especially for difficult-to-express proteins. The concentration of acetosyringone in the infiltration media can be optimized for such difficult proteins.
3. The age and health of the plant are extremely important for a successful recombinant protein expression. It is recommended to use only two leaves per plant for infiltration and the zone of infiltration on the leaves should be around 3 cm in diameter. The use of flowering plants must be avoided. The *in-planta* expression pattern of both bait and prey protein should be optimized using Western blotting before setting up a co-IP experiment, which saves time and resources.
4. Including Agrobacterium expressing a silencing suppressor protein, such as the tomato bushy stunt virus p19 along with the prey and bait protein in the infiltration media can increase levels of protein expression [32].

5. This system can be used to generate versatile protein traps using antibodies other than anti-GFP. However, the incubation time of the antibody and magnetic beads must be optimized. If using a lower affinity antibody, reverse co-IP strategy can also be used. The lower concentration antibody can be pre-incubated in the diluted protein lysate followed by the addition of pre-washed magnetic beads followed by the protocol described above. However, the time of incubation must be optimized.
6. Protein extraction and bead washing buffer is the key to a successful co-IP experiment. A higher concentration of NP-40 in the buffer may abolish the interaction between proteins. Therefore, it is recommended to optimize the concentration. The suggested buffer in the protocol is Tris-based. Alternative buffering systems such as HEPES or phosphate can also be used. The specified concentration of NaCl in the wash buffer is compatible with most applications; however, the concentration can be increased to reduce nonspecific proteins or decreased to prevent washing off desired proteins.
7. Majority of the steps in this protocol are suggested to be performed at room temperature. If working with sensitive proteins the steps can be performed at 4 °C, the incubation times will vary and must be optimized.

Acknowledgments

This research was funded by the National Science Foundation (IOS-2038872).

References

1. Wessling R, Epple P, Altmann S et al (2014) Convergent targeting of a common host protein-network by pathogen effectors from three kingdoms of life. Cell Host Microbe 16(3):364–375. https://doi.org/10.1016/j.chom.2014.08.004
2. Smakowska-Luzan E, Mott GA, Parys K et al (2018) An extracellular network of Arabidopsis leucine-rich repeat receptor kinases. Nature 553(7688):342–346. https://doi.org/10.1038/nature25184
3. Mukhtar MS, Carvunis AR, Dreze M et al (2011) Independently evolved virulence effectors converge onto hubs in a plant immune system network. Science 333(6042):596–601. https://doi.org/10.1126/science.1203659
4. Mott GA, Smakowska-Luzan E, Pasha A et al (2019) Map of physical interactions between extracellular domains of Arabidopsis leucine-rich repeat receptor kinases. Sci Data 6: 190025. https://doi.org/10.1038/sdata.2019.25
5. Mishra B, Sun Y, Howton TC et al (2018) Dynamic modeling of transcriptional gene regulatory network uncovers distinct pathways during the onset of Arabidopsis leaf senescence. NPJ Syst Biol Appl 4:35. https://doi.org/10.1038/s41540-018-0071-2
6. Mishra B, Sun Y, Ahmed H et al (2017) Global temporal dynamic landscape of pathogen-mediated subversion of Arabidopsis innate immunity. Sci Rep 7(1):7849. https://doi.org/10.1038/s41598-017-08073-z
7. Mishra B, Kumar N, Shahid Mukhtar M (2022) A rice protein interaction network reveals high centrality nodes and candidate

pathogen effector targets. Comput Struct Biotechnol J 20:2001–2012. https://doi.org/10.1016/j.csbj.2022.04.027

8. Mishra B, Kumar N, Mukhtar MS (2021) Network biology to uncover functional and structural properties of the plant immune system. Curr Opin Plant Biol 62:102057. https://doi.org/10.1016/j.pbi.2021.102057
9. Mishra B, Kumar N, Mukhtar MS (2019) Systems biology and machine learning in plant-pathogen interactions. Mol Plant Microbe Interact 32(1):45–55. https://doi.org/10.1094/MPMI-08-18-0221-FI
10. McCormack ME, Lopez JA, Crocker TH et al (2016) Making the right connections: network biology and plant immune system dynamics. Current Plant Biology 5:2–12
11. Lopez J, Mukhtar MS (2017) Mapping protein-protein interaction using high-throughput yeast 2-hybrid. Methods Mol Biol 1610:217–230. https://doi.org/10.1007/978-1-4939-7003-2_14
12. Kumar N, Mishra B, Mukhtar MS (2022) A pipeline of integrating transcriptome and interactome to elucidate central nodes in host-pathogens interactions. STAR Protoc 3(3):101608. https://doi.org/10.1016/j.xpro.2022.101608
13. Kumar N, Mishra B, Mehmood A et al (2020) Integrative network biology framework elucidates molecular mechanisms of SARS-CoV-2 pathogenesis. iScience 23(9):101526. https://doi.org/10.1016/j.isci.2020.101526
14. Klopffleisch K, Phan N, Augustin K et al (2011) Arabidopsis G-protein interactome reveals connections to cell wall carbohydrates and morphogenesis. Mol Syst Biol 7:532. https://doi.org/10.1038/msb.2011.66
15. Gonzalez-Fuente M, Carrere S, Monachello D et al (2020) EffectorK, a comprehensive resource to mine for Ralstonia, Xanthomonas, and other published effector interactors in the Arabidopsis proteome. Mol Plant Pathol 21(10):1257–1270. https://doi.org/10.1111/mpp.12965
16. Garbutt CC, Bangalore PV, Kannar P et al (2014) Getting to the edge: protein dynamical networks as a new frontier in plant-microbe interactions. Front Plant Sci 5:312. https://doi.org/10.3389/fpls.2014.00312
17. Arabidopsis Interactome Mapping C (2011) Evidence for network evolution in an Arabidopsis interactome map. Science 333(6042):601–607. https://doi.org/10.1126/science.1203877
18. Ahmed H, Howton TC, Sun Y et al (2018) Network biology discovers pathogen contact points in host protein-protein interactomes. Nat Commun 9(1):2312. https://doi.org/10.1038/s41467-018-04632-8
19. Liu X, Merchant A, Rockett KS et al (2015) Characterization of Arabidopsis thaliana GCN2 kinase roles in seed germination and plant development. Plant Signal Behav 10(4):e992264. https://doi.org/10.4161/15592324.2014.992264
20. Moreno AA, Mukhtar MS, Blanco F et al (2012) IRE1/bZIP60-mediated unfolded protein response plays distinct roles in plant immunity and abiotic stress responses. PLoS One 7(2):e31944. https://doi.org/10.1371/journal.pone.0031944
21. Pajerowska-Mukhtar KM, Wang W, Tada Y et al (2012) The HSF-like transcription factor TBF1 is a major molecular switch for plant growth-to-defense transition. Curr Biol 22(2):103–112. https://doi.org/10.1016/j.cub.2011.12.015
22. Ehlert A, Weltmeier F, Wang X et al (2006) Two-hybrid protein-protein interaction analysis in Arabidopsis protoplasts: establishment of a heterodimerization map of group C and group S bZIP transcription factors. Plant J 46(5):890–900. https://doi.org/10.1111/j.1365-313X.2006.02731.x
23. Afrin T, Seok M, Terry BC et al (2020) Probing natural variation of IRE1 expression and endoplasmic reticulum stress responses in Arabidopsis accessions. Sci Rep 10(1):19154. https://doi.org/10.1038/s41598-020-76114-1
24. Verchot J, Pajerowska-Mukhtar KM (2021) UPR signaling at the nexus of plant viral, bacterial, and fungal defenses. Curr Opin Virol 47:9–17. https://doi.org/10.1016/j.coviro.2020.11.001
25. Munoz A, Castellano MM (2018) Coimmunoprecipitation of interacting proteins in plants. Methods Mol Biol 1794:279–287. https://doi.org/10.1007/978-1-4939-7871-7_19
26. Diwan D, Liu X, Andrews CF et al (2021) A quantitative Arabidopsis IRE1a Ribonuclease-dependent in vitro mRNA cleavage assay for functional studies of substrate splicing and decay activities. Front Plant Sci 12:707378. https://doi.org/10.3389/fpls.2021.707378
27. Afrin T, Costello CN, Monella AN et al (2022) The interplay of GTP-binding protein AGB1 with ER stress sensors IRE1a and IRE1b modulates Arabidopsis unfolded protein response and bacterial immunity. Plant Signal Behav 17(1):2018857. https://doi.org/10.1080/15592324.2021.2018857

28. Liu X, Sun Y, Korner CJ et al (2015) Bacterial leaf infiltration assay for fine characterization of plant defense responses using the Arabidopsis thaliana-Pseudomonas syringae Pathosystem. J Vis Exp 104. https://doi.org/10.3791/53364
29. Kosmacz M, Gorka M, Schmidt S et al (2019) Protein and metabolite composition of Arabidopsis stress granules. New Phytol 222(3): 1420–1433. https://doi.org/10.1111/nph.15690
30. Curtis MD, Grossniklaus U (2003) A gateway cloning vector set for high-throughput functional analysis of genes in planta. Plant Physiol 133(2):462–469. https://doi.org/10.1104/pp.103.027979
31. Nakagawa T, Kurose T, Hino T et al (2007) Development of series of gateway binary vectors, pGWBs, for realizing efficient construction of fusion genes for plant transformation. J Biosci Bioeng 104(1):34–41. https://doi.org/10.1263/jbb.104.34
32. Garabagi F, Gilbert E, Loos A et al (2012) Utility of the P19 suppressor of gene-silencing protein for production of therapeutic antibodies in Nicotiana expression hosts. Plant Biotechnol J 10(9):1118–1128. https://doi.org/10.1111/j.1467-7652.2012.00742.x

Chapter 6

Tandem Affinity Purification (TAP) of Interacting Prey Proteins with FLAG- and HA-Tagged Bait Proteins

Teck Yew Low and Pey Yee Lee

Abstract

Proteins often interact with each other to form complexes and play functional roles in almost all cellular processes. The study of protein–protein interactions is therefore critical to understand protein function and biological pathways. Affinity Purification coupled with Mass Spectrometry (AP-MS) is an invaluable technique for identifying the interaction partners in protein complexes. In this approach, the protein of interest is fused to an affinity tag, followed by the expression and purification of the fusion protein. The affinity-purified sample is then analyzed by mass spectrometry to identify the interaction partners of the bait proteins. In this chapter, we detail the protocol for tandem affinity purification (TAP) based on the use of the FLAG (a fusion tag with peptide sequence DYKDDDDK) and hemagglutinin (HA) peptide epitopes. The immunoprecipitation using dual-affinity tags offers the advantage of increasing the specificity of the purification with lower nonspecific-background interactions.

Key words Affinity purification, Mass spectrometry, Protein, protein interactions, FLAG tag, HA tag

1 Introduction

Though only ~20,000 human genes encode proteins, it has been postulated that the diversity and functions of the proteome can be further expanded by forming protein–protein interactions (PPIs), whereby an upper bound of ~650,000 physical PPIs among the human proteome have been mathematically simulated [1]. Disentangling these PPIs is an important step toward deciphering the genotype–phenotype relationship [2, 3]. Among the biochemical methods, Affinity Purification coupled to Mass Spectrometry (AP-MS) is the most widely used high-throughput method for PPI study [4–6].

In AP-MS, a target protein of interest (bait protein) is selectively purified with a specific antibody or other affinity reagents. As a result, many of its direct or indirect interacting protein partners can be co-purified from a cell or tissue lysate at the same time.

Shahid Mukhtar (ed.), *Protein-Protein Interactions: Methods and Protocols*, Methods in Molecular Biology, vol. 2690,
https://doi.org/10.1007/978-1-0716-3327-4_6,

These co-purified proteins comprise the bona-fide, specific interactors of the baits and are named "preys," as well as a host of nonspecific-background proteins that constitute the contaminants [7–11]. In the second step, which will not be covered in this chapter, co-purified proteins are identified and quantified using MS, because unlike the bait protein, these co-purified proteins are not known a priori. An AP-MS experiment usually comprises: (i) incubating pre-cleared lysate with beads conjugated with the bait or epitope tag-specific antibodies; (ii) washing to remove nonspecific proteins; (iii) elution of the purified proteins; and (iv) identification of the eluted proteins with MS. Hence, the AP-MS technique is sensitive, high-throughput, and hypothesis-free.

To facilitate the AP procedure, a bait protein is often genetically fused to an epitope tag and expressed in a carefully chosen system that carries the most optimum biological context. Such epitope tags may be short peptides or proteins that are recognizable by commercially available antibodies, including FLAG, c-myc, hemagglutinin (HA), polyHis, and streptavidin [12]. These epitope tags can be fused in single or multiple copies, as well as in tandem with different tags for multiple rounds of purification, namely tandem affinity purification (TAP). In this chapter, we will describe a tandem affinity purification (TAP) method using FLAG- and HA-tags.

In comparison to immunoprecipitation, epitope tagging provides generic affinity handles for performing AP, but with the downside that such tags may interfere with the functions and solubility of the bait protein, apart from the possibility that the ectopic expression from transient transfections could cause the baits to be mis-folded and mis-localized.

Another challenge in AP-MS is the co-purification of high-abundance, nonspecific-background proteins. Thus, it is a common practice to incorporate controls that can distinguish these contaminants [7, 13]. Such controls can consist of the expression of empty vectors, the use of antibody isotypes, and knockdown and knockout of the endogenous baits for co-immunoprecipitation (co-IP). Besides, TAP-tagging, as described in this chapter, permits multiple washing and elution steps, thus reducing nonspecific interactions.

2 Materials

2.1 Cell Transfection and Harvest

1. Thermocycler machine
2. Biological safety cabinet
3. CO_2 incubator
4. Sterile disposable serological pipettes, 5 mL and 10 mL
5. Micropipettes

6. High-speed refrigerated microcentrifuge, tabletop
7. Microcentrifuge tubes, 1.5 or 2.0 mL
8. Falcon tubes, 15 mL and 50 mL
9. 6-well culture plates
10. Expression vector
11. HEK 293 cells
12. Dulbecco's Modified Eagle Medium/Nutrient Mixture F-12 (DMEM/F-12)
13. 10% fetal bovine serum (FBS)
14. Penicillin–streptomycin, 100 U/mL
15. Polyethylenimine (PEI) "max"
16. Trypsin
17. Phosphate-buffered saline (PBS): 10 mM phosphate, 2.7 mM potassium chloride, and 137 mM sodium chloride, pH 7.4
18. Dry ice/ethanol bath

2.2 Cell Lysis

1. Lysis buffer: 8 M Urea, 50 mM Tris-HCl, pH 7.4, with 150 mM NaCl, 1 mM EDTA (ethylenediaminetetraacetic acid), 1% Triton X-100, 1X protease inhibitor
2. 20 mL or 30 mL syringe
3. 20G1 needle

2.3 Affinity Purification

2.3.1 FLAG IP

1. Poly-prep chromatography column: 9 cm high, 2 mL bed volume (0.8 × 4 cm), empty polypropylene column, 10 mL reservoir
2. Tris-buffered saline (TBS): 50 mM Tris-HCl, with 150 mM NaCl, pH 7.4
3. Anti-FLAG M2 affinity gel
4. 0.1 M glycine HCl, pH 3.5
5. 0.15 M NaCl
6. 3X FLAG peptide (MDYKDHDGDYKDHDIDYKDDDDK)

2.3.2 HA IP

1. Micro bio-spin chromatography columns: empty polypropylene spin columns, 0.8 mL bed volume
2. Rotating platform
3. 0.1 M glycine–HCl, pH 2.5
4. 1 M Tris-HCl, pH 8
5. Anti-HA beads
6. Tris-buffered saline containing 0.05% Tween-20 (TBS-T)
7. 1 mg/mL HA peptide in TBS

2.4 SDS-PAGE

1. SDS-PAGE gel casting and electrophoresis apparatus
2. Staining tray
3. Ultrapure water
4. Acrylamide/bis, 30%, 37.5:1
5. 1.5 M Tris-HCl, pH 8.8
6. 0.5 M Tris-HCl, pH 6.8
7. 10% SDS
8. N,N,N′,N′-tetramethylethylene-diamine (TEMED)
9. Ammonium persulfate (APS)
10. Laemmli sample buffer (2×)
11. Isopropanol
12. Molecular mass protein markers
13. SDS-PAGE running buffer: 0.025 M Tris-HCl, pH 8.3, 0.192 M glycine, 0.1% SDS
14. Coomassie blue or Silver stain

3 Methods

3.1 Cell Transfection and Harvest

1. Clone the gene of interest carrying FLAG and HA tags at the N-terminus (Fig. 1) into the expression vector by polymerase chain reaction (PCR). Verify the correct construct by sequencing.
2. Culture HEK 293 cells in DMEM:F12 + 10% FBS and penicillin–streptomycin (100 U/mL) to reach ~80–90% confluency (*see* **Note 1**).
3. Seed 3×10^5 cells in 6-well culture plates on the following day, with 1.5 mL medium in each well, and then incubated at 37 °C in a 5% CO_2 incubator overnight.
4. Mix plasmid DNA and 1 mg/mL polyethylenimine (PEI) "max" in a 1:3 weight ratio with 500 μL of DMEM (*see* **Note 2**). In a separate tube, prepare a transfection reaction using an empty vector with 2X FLAG-2X HA tags only as a control for the TAP experiment.
5. Incubate the mixture for 20 min at room temperature.
6. Add the mixture to each culture plate well and incubate at 37 °C in a 5% CO_2 incubator.
7. Verify successfully transfected cells by analyzing the expression of the protein of interest by SDS-PAGE.
8. Grow confluent flasks of the clones.

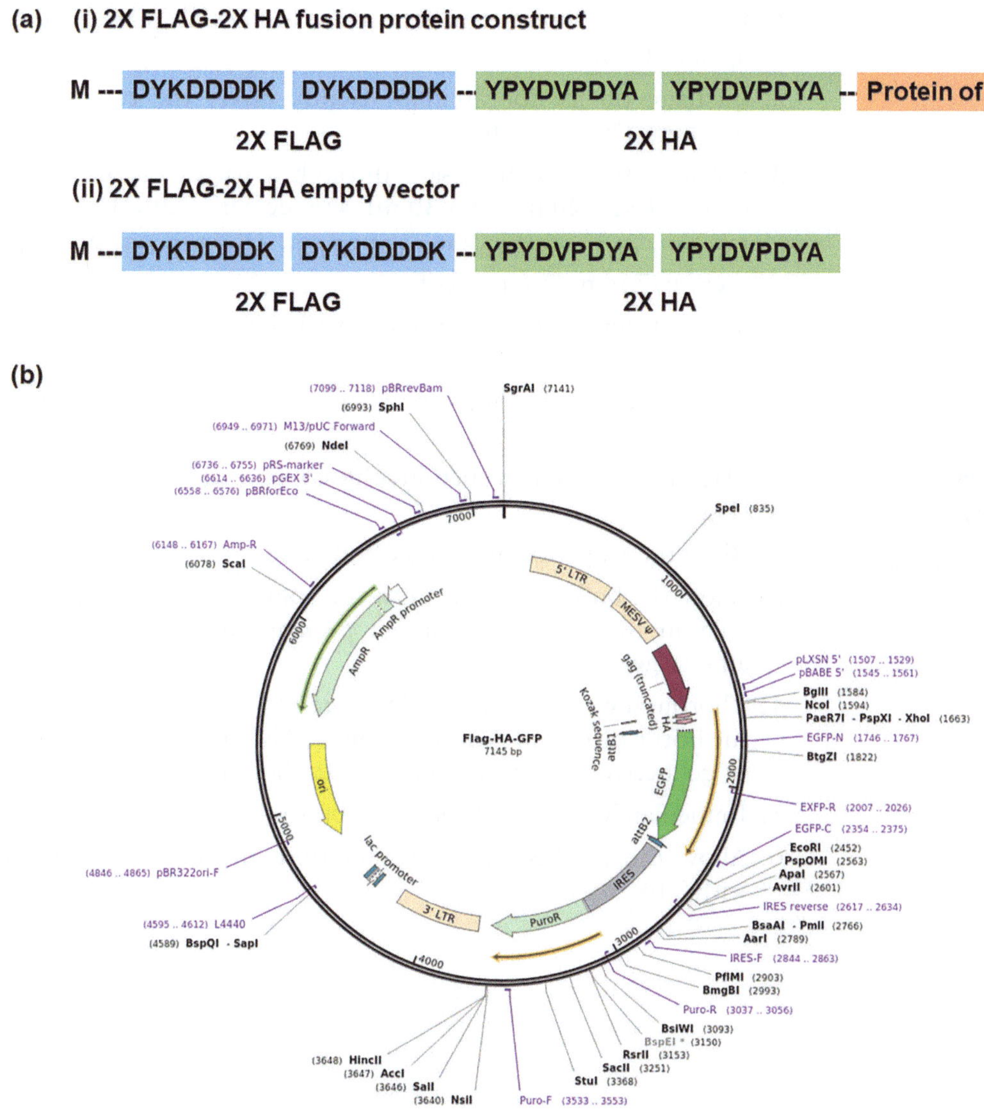

Fig. 1 (**a**) Schematic representation of (i) 2X FLAG-2X HA fusion protein construct and (ii) 2X FLAG-2X HA empty vector. (**b**) An example of a FLAG-HA-tagged green fluorescent protein (GFP) construct (Adapted from https:/www.addgene.org/22612/)

9. Trypsinize the cells (or scrape into ice-cold PBS) and centrifuge the cell suspension at 5000× g for 30 min at 4 °C.
10. Resuspend the cell pellets in a small volume of ice-cold PBS and combine the cells into a single 50 mL tube.
11. Centrifuge again at 5000× g for 30 min to pellet the cells. Repeat this PBS wash step three times to remove cell debris and trypsin.
12. Freeze using a dry ice/ethanol bath or at −20 °C in a freezer. Store at −80 °C.

3.2 Cell Lysis

1. Fully resuspend the cell pellets in lysis buffer by pipetting them up and down.
2. Incubate on ice for 30 min. Gently invert the tube every 6–7 min. Do not vortex.
3. Syringe 10 times the lysate through a narrow-gauge blunt needle. Use a 20 mL or a 30 mL syringe with a 20G1 needle.
4. Incubate on ice for an additional 30 min. Gently invert the tube every 6–7 min. Do not vortex.
5. Spin 30 min at 4 °C, 16,000× g to remove cell debris and insoluble materials.
6. Transfer supernatant into one 50 mL conical tube. Keep on ice.

3.3 Affinity Purification

3.3.1 Column Packing in FLAG IP

1. Place the empty chromatography column on a firm support (Fig. 2).
2. Rinse the empty column twice with TBS.
3. Allow the buffer to drain from the column and leave residual TBS in the column to aid in packing the anti-FLAG M2 affinity gel.
4. Thoroughly suspend the resin by gentle inversion. Make sure the bottle of anti-FLAG M2 affinity gel is a uniform suspension of gel beads. Remove an appropriate aliquot for use.
5. Immediately transfer the suspension to the column.
6. Allow the gel bed to drain and rinse the vial of the resin with TBS.

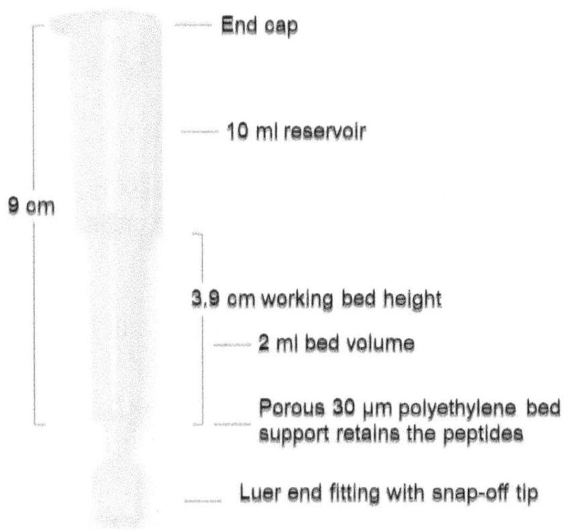

Fig. 2 Poly-prep chromatography column (Adapted from https:/www.bio-rad.com/en-my/product/poly-prep-chromatography-columns)

7. Add the rinse to the top of the column and allow it to drain again. The gel bed will not form channels when an excess solution is drained under normal circumstances, but do not let the gel bed run dry.
8. Wash the gel by loading three sequential column volumes of 0.1 M glycine HCl, pH 3.5. Avoid disturbing the gel bed while loading. Let each aliquot drain completely before adding the next. Do not leave the column in glycine HCl for longer than 20 min.
9. Wash the resin with five-column volumes of TBS to equilibrate the resin for use. Do not let the bed run dry (*see* **Note 3**). Allow a small amount of buffer to remain on the top of the column.

3.3.2 Sample Loading in FLAG IP

1. For proper binding of a FLAG fusion protein, the sample loaded should be in a buffer containing 0.15 M NaCl and pH 7.
2. Load the sample onto the column under gravity flow and let drain.
3. Fill the column completely several times for large volumes. Multiple passes over the column will improve the binding efficiency.

3.3.3 Washing in FLAG IP

1. Wash the column with 10–20 column volumes of TBS. This should remove any proteins that are not bound to the M2 antibody. Allow the column to drain completely.

3.3.4 Elution in FLAG IP

1. Elute the bound FLAG fusion protein with five one-column volumes of a solution containing 100 μg/mL of 3X FLAG peptide in TBS.
2. Collect the elution fragments in 1.5 mL Eppendorf tubes.

3.3.5 HA IP

1. Prepare the anti-HA antibody beads in a 1.5-mL Eppendorf tube and wash once using 0.1 M glycine–HCl (pH 2.5) followed by two washes with 1 M Tris-HCl (pH 8) (*see* **Note 4**).
2. Transfer the whole FLAG-tagged protein elution into the anti-HA antibody beads.
3. Gently mix for 6 h to overnight at 4 °C using a rotating platform.
4. Transfer the mixture to a micro bio-spin chromatography column (Fig. 3) and let the unbound protein flow through by gravity.
5. Wash the column twice using 0.5 mL of TBS containing 0.05% Tween-20 (TBS-T) and allow the buffer to pass through the column by gravity. Retain small aliquots for analysis.

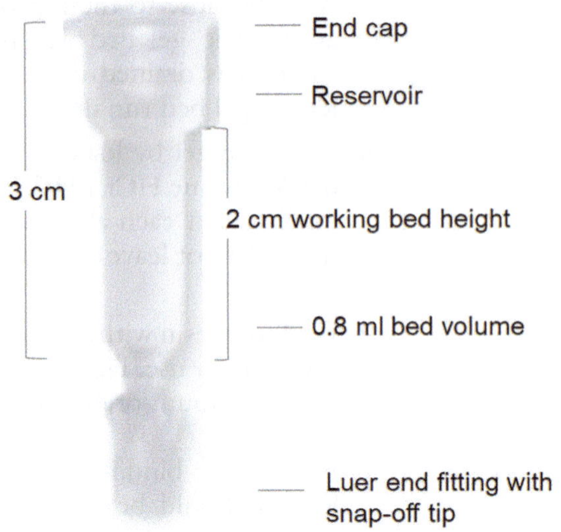

Fig. 3 Micro bio-spin chromatography column (Adapted from https:/www.bio-rad.com/en-my/product/bio-spin-6-micro-bio-spin-153-6-columns?ID=2b94 d889-0cc3-4839-8297-16591ae8f155)

6. Collect the anti-HA beads by centrifugation at 3000× g for 1 min at 4 °C and close the cap at the bottom of the column to the anti-HA beads to prevent additional flowthrough.
7. Add one volume of dry beads (~10 μL) at 1 mg/mL HA peptide in TBS and incubate the mixture in the column for 5 min at room temperature.
8. Open the cap at the bottom of the column and collect the eluate by centrifugation at 1000× g for 1 min at 4 °C using a refrigerated microcentrifuge.
9. Repeat two more times by incubating the anti-HA beads in 1 mg/mL HA peptide in TBS (*see* **Note 5**) for 5 min before collecting the eluate by centrifugation for 1 min at 4 °C using a microcentrifuge.
10. Combine elutions collected and retain a small 50- to 100-μL aliquot to evaluate the purification efficiency.

3.4 SDS-PAGE

1. Prepare the separation gel (10%). Mix in the following order as indicated in Table 1 (*see* **Note 6**).
2. Pour gel, leaving ~2 cm below the bottom of the comb for the stacking gel. Make sure to remove bubbles.
3. Layer the top of the gel with isopropanol. This will help to remove bubbles at the top of the gel and will also keep the polymerized gel from drying out.
4. Leave for ~30 min to allow the gel to polymerize.

Table 1
Preparation of separation gel

Reagents	Volume
Ultrapure water	4.1 mL
Acrylamide/bis (30% 37.5:1)	3.3 mL
Tris-HCl (1.5 M, pH 8.8)	2.5 mL
SDS, 10%	100 μL
N,N,N′,N′-tetramethylethylene-diamine (TEMED)	10 μL
Ammonium persulfate (APS), 10%	32 μL

Table 2
Preparation of stacking gel

Reagents	Volume
Ultrapure water	6.1 mL
Acrylamide/bis (30% 37.5:1)	1.3 mL
Tris-HCl (0.5 M, pH 6.8)	2.5 mL
SDS, 10%	100 μL
N,N,N′,N′-tetramethylethylene-diamine (TEMED)	10 μL
Ammonium persulfate (APS), 10%	100 μL

5. Remove the isopropanol and wash out the remaining traces of isopropanol with distilled water.
6. Prepare the stacking gel (4%). Mix in the following order as indicated in Table 2.
7. Pour stacking gel on top of the separating gel.
8. Add combs to make wells. In ~30 min, the stacking gel should become completely polymerized.
9. Clamp gel into apparatus and fill both buffer chambers with gel running buffer according to the instructions for the specific apparatus.
10. Prepare protein samples by boiling equal volumes in 2× Laemmli Sample Buffer for 5 min.
11. Load samples and molecular mass protein markers into wells for separation by electrophoresis.
12. Run the gel at 100 V until the dye front migrates into the running gel (~10 min) and increase to 200 V until the dye front reaches the bottom of the gel.

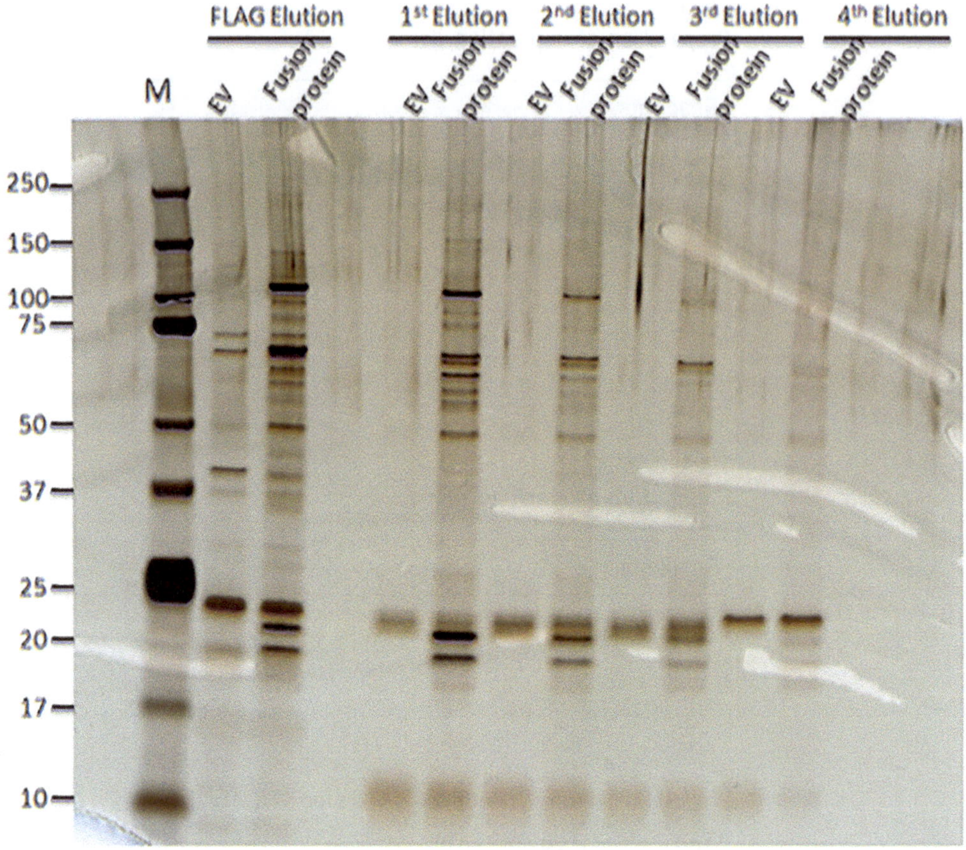

Fig. 4 Silver-stained SDS-PAGE gel of the elution samples of 2X FLAG-2X HA-tagged fusion protein from the TAP experiment. Nonspecific proteins bound to the anti-FLAG affinity resin were observed in the immunoprecipitation using the empty vector only. Sequential elution using anti-HA antibody that recognizes the second epitope significantly reduces the background proteins and increases the specificity. M: MW protein marker; EV: empty vector; fusion protein: 2X FLAG-2X HA-tagged fusion protein

13. Remove the gel from the apparatus and remove the spacers and glass plates. Place the gel into a small tray.
14. Perform staining using Coomassie blue or silver stain according to the manufacturer's protocol (Fig. 4) (*see* **Note 7**).

4 Notes

1. The cell type used depends on the specific needs of the protein of interest. HEK293 are immortalized human embryonic kidney cells. The addition of the Ad5 E1A gene to the HEK genome allows for high levels of recombinant protein production, specifically proteins within plasmid vectors containing the cytomegalovirus (CMV) promoter.

2. In general, use 1 μg of DNA per 1 mL of culture to be transfected. PEI and DNA should each be diluted into 1/20 of the total culture volume before being mixed. As a positive control for the transfection, you may separately transfect a vector expressing a fluorescent protein.
3. When the column runs dry, this can introduce air bubbles into the column, which prevents efficient binding and elution of the proteins from the column.
4. The amount of anti-HA antibody beads required is dependent on the quality of the beads. Use the least amount of anti-HA antibody beads necessary to immunoprecipitate the HA-tagged protein completely. In general, use 20 μL of anti-HA antibody beads (50% slurry) for 100 μL of FLAG-tagged protein eluate.
5. The HA-tagged proteins and their protein interactors can also be eluted with (i) 8 M urea lysis buffer or (ii) Laemmli buffer for direct analysis by SDS-PAGE.
6. After adding TEMED and APS to the SDS-PAGE separation gel solution, the gel will polymerize quickly, so add these two reagents last when ready to pour.
7. For proteins with low expression, silver staining may be used for more sensitive detection.

Acknowledgments

The authors would like to acknowledge the Higher Education Center of Excellence (HICoE) Grant (JJ-2021-004) awarded by the Ministry of Higher Education of Malaysia to TYL.

References

1. Stumpf MPH, Thorne T, de Silva E, Stewart R, An HJ, Lappe M et al (2008) Estimating the size of the human interactome. Proc Natl Acad Sci U S A 105:6959–6964
2. Bludau I, Aebersold R (2020) Proteomic and interactomic insights into the molecular basis of cell functional diversity. Nat Rev Mol Cell Biol 21:327–340
3. Low TY, Heck AJ (2016) Reconciling proteomics with next generation sequencing. Curr Opin Chem Biol 30:14–20
4. Low TY, Syafruddin SE, Mohtar MA, Vellaichamy AA, Rahman NS, Pung YF et al (2021) Recent progress in mass spectrometry-based strategies for elucidating protein–protein interactions. Cell Mol Life Sci 78:5325–5339
5. Dunham WH, Mullin M, Gingras AC (2012) Affinity-purification coupled to mass spectrometry: basic principles and strategies. Proteomics 12:1576–1590
6. Kovanich D, Low TY, Zaccolo M (2023) Using the proteomics toolbox to resolve topology and dynamics of compartmentalized cAMP signaling. Int J Mol Sci 24:4667
7. Low TY, Peng M, Magliozzi R, Mohammed S, Guardavaccaro D, Heck AJ (2014) A systems-wide screen identifies substrates of the SCFβTrCP ubiquitin ligase. Sci Signal 7:rs8
8. Antonova SV, Haffke M, Corradini E, Mikuciunas M, Low TY, Signor L et al (2018) Chaperonin CCT checkpoint function in basal transcription factor TFIID assembly. Nat Struct Mol Biol 25:1119–1127
9. D'Annibale S, Kim J, Magliozzi R, Low TY, Mohammed S, Heck AJ et al (2014) Proteasome-dependent degradation of

transcription factor activating enhancer-binding protein 4 (TFAP4) controls mitotic division. J Biol Chem 289:7730–7737

10. Magliozzi R, Low TY, Weijts BG, Cheng T, Spanjaard E, Mohammed S et al (2013) Control of epithelial cell migration and invasion by the IKKβ- and CK1α-mediated degradation of RAPGEF2. Dev Cell 27:574–585

11. Kim J, D'Annibale S, Magliozzi R, Low TY, Jansen P, Shaltiel IA et al (2014) USP17- and SCFβTrCP--regulated degradation of DEC1 controls the DNA damage response. Mol Cell Biol 34:4177–4185

12. Vandemoortele G, Eyckerman S, Gevaert K (2019) Pick a tag and explore the functions of your pet protein. Trends Biotechnol 37:1078–1090

13. Mellacheruvu D, Wright Z, Couzens AL, Lambert JP, St-Denis NA, Li T et al (2013) The CRAPome: a contaminant repository for affinity purification-mass spectrometry data. Nat Methods 10:730–736

Chapter 7

Affinity Purification-Mass Spectroscopy (AP-MS) and Co-Immunoprecipitation (Co-IP) Technique to Study Protein–Protein Interactions

Prabu Gnanasekaran and Hanu R. Pappu

Abstract

Affinity purification-Mass spectroscopy (AP-MS) is a biochemical technique to identify the novel protein–protein interaction that occurs in the most relevant physiological conditions, whereas co-immunoprecipitation (Co-IP) is used to study the interaction between two known protein partners that are expressed in the native physiological conditions. Both AP-MS and Co-IP techniques are based on the ability of the interacting partners to pull-down with protein of interest. In this chapter, we have explained the AP-MS and Co-IP methods to study protein–protein interactions in the plant cells.

Key words AP-MS, Co-IP, Immunoglobulin A (IgA) beads, GFP-trap resins

1 Introduction

Affinity purification-mass spectroscopy (AP-MS) is a technique to discover protein–protein interaction in a relevant physiological condition [1]. In this approach, the protein of interest is tagged with an affinity tag and the interacting proteins that bind to the target protein are indirectly captured by purifying the target protein by affinity chromatography [2, 3]. For example, protein A is fused to green fluorescent protein (GFP), and the GFP-A-bound proteins are purified using GFP-trap beads. Interacting protein complexes are purified and subjected to liquid chromatography-mass spectroscopy for protein identification. Co-immunoprecipitation (Co-IP) is a biochemical technique to study interaction between two known proteins in a relevant physiological condition [4]. In this technique, the proteins that are bound to the target protein are indirectly captured by immunoprecipitation using target protein-specific antibody in conjugation with Protein A/G beads [5, 6]. Interacting protein that binds to the target protein is confirmed by western blotting.

Shahid Mukhtar (ed.), *Protein-Protein Interactions: Methods and Protocols*, Methods in Molecular Biology, vol. 2690,
https://doi.org/10.1007/978-1-0716-3327-4_7,

2 Materials

2.1 AP-MS

1. Transient expression constructs:

   pCAMBIA1302 vector (contain coding sequence of GFP protein).

   pCAMBIA1302-A: Generate this construct by cloning the open reading frame (ORF) coding for the protein of interest, A, in frame with the GFP coding sequence.
2. *Agrobacterium tumefaciens* strain GV3101 competent cells.
3. Infiltration buffer containing 10 mM MES ((2-(N-morpholino)ethanesulfonic acid)) (pH 5.7), 10 mM $MgCl_2$, 100 μM acetosyringone.
4. Three- to four-week-old *Nicotiana benthamiana* plants.
5. Protein extraction buffer containing 25 mM Tris (pH 7.4), 100 mM NaCl, 5% glycerol, 5 mM 2-mercaptoethanol, 1 mM phenylmethylsulfonyl fluoride, and protease inhibitor cocktail.
6. GFP-trap beads.
7. Mass spectroscopy facility for protein identification.

2.2 Co-IP

1. Transient expression constructs:

   pXSN-HA vector (contain coding sequence of hemagglutinin [HA] tag).

   pXSN-Myc vector (contain coding sequence of Myc tag).

   pXSN-HA-X: Generate this construct by cloning the ORF coding for the protein of interest, X, in frame with the HA coding sequence.

   pXSN-Myc-Y: Generate this construct by cloning the ORF coding for the protein of interest, Y, in frame with the Myc coding sequence.
2. *A. tumefaciens* strain GV3101 competent cells.
3. Infiltration buffer containing 10 mM MES (pH 5.7), 10 mM $MgCl_2$, and 100 μM acetosyringone.
4. Three- to four-week-old *N. benthamiana* plants.
5. Negative control combination:

   pXSN-HA vector + pXSN-Myc vector.

   pXSN-HA-X + pXSN-Myc vector.

   pXSN-HA vector + pXSN-Myc-Y.

6. Test combination:

   pXSN-HA-X + pXSN-Myc-Y.

7. Protein extraction buffer containing 25 mM Tris (pH 7.4), 100 mM NaCl, 5% glycerol, 5 mM 2-mercaptoethanol, 1 mM phenylmethylsulfonyl fluoride, and protease inhibitor cocktail.
8. Anti-HA antibody.
9. Anti-Myc antibody.
10. Protein A beads.
11. Equilibration buffer containing 50 mM Tris (pH 7.4), 300 mM NaCl, 5% glycerol, 5 mM 2-mercaptoethanol, and 1 mM phenylmethylsulfonyl fluoride.
12. Wash buffer containing 50 mM Tris (pH 7.4), 300 mM NaCl, 5% glycerol, 5 mM 2-mercaptoethanol, and 1 mM phenylmethylsulfonyl fluoride.

3 Methods

3.1 AP-MS

1. Transform the expression constructs into *Agrobacterium tumefaciens* strain GV3101 cells.
2. Grow the *Agrobacterium* cells carrying transient expression construct at 28 °C and 200 rpm shaking for overnight in Luria Broth media containing kanamycin and rifampicin.
3. Harvest the overnight grown *Agrobacterium* cells and make the desired infiltration combination in infiltration buffer (using 0.5 optical density [OD] of cells corresponding to each construct).
4. Mix gently and incubate in dark for 2–3 h.
5. Infiltrate the cells containing GFP/GFP-A expression constructs onto the lower side of the leaves of *N. benthamiana* plants.
6. After 48 h post-infiltration, harvest the infiltrated leaves and extract the proteins using protein extraction buffer.
7. Preclear the GFP-trap beads with total protein extracted from mock-infiltrated *N. benthamiana* leaves.
8. Purify the GFP-A-bound protein complexes with precleared GFP-trap beads. Similarly, GFP-bound protein complexes are also purified.
9. Subject the protein complexes to mass spectroscopy analysis for protein identification. Identify the interacting partners of protein-A by filtering out the GFP-bound proteins (*see* **Note 1**).

3.2 Co-IP

1. Co-infiltrate the leaves of *N. benthamiana* plants with *Agrobacterium* cells harboring test and negative control plasmids.
2. After 48 h post-infiltration, harvest the infiltrated leaves and extract the proteins using protein extraction buffer.
3. Add 100 μL of Protein A bead to each of the four microcentrifuge tubes.
4. Add 1 mL of ice-cold equilibration buffer to each tube, incubate at 4 °C for 5 min, centrifuge at 4 °C and 1000 rpm for 5 min, pipette out the supernatant carefully without drying the beads, and discard it. Repeat this step three to five times.
5. Add equal concentrations of different combination proteins extracted containing HA + Myc, HA-X + Myc, HA + Myc-Y, and HA-X + Myc-Y into different tubes containing pre-equilibrated beads, add anti-HA antibody, and adjust the volume to 1 mL using equilibration buffer.
6. Incubate the tubes at 4 °C for 4 h with end-over-end mixing on the rotating shaker.
7. Centrifuge the tubes at 4 °C, 1000 rpm for 5 min and discard the supernatant without disturbing the bead.
8. Add 1 mL of ice-cold wash buffer to each tube, incubate at 4 ° C for 5 min, and then centrifuge at 4 °C and 1000 rpm for 5 min and discard the supernatant. Repeat this step five times.
9. Elute the bound protein by using a buffer with higher pH or higher salt concentration.
10. Analyze the eluted samples by sodium dodecyl sulfate (SDS)-polyacrylamide gel electrophoresis (PAGE) and western blotting (*see* **Note 2**).

4 Notes

1. Interacting partners identified by AP-MS should be confirmed by other interaction studies.
2. In Co-IP experiment, if Myc-Y protein gets immunoprecipitated with HA-X protein and not with HA alone, this confirms the interaction between protein X and Y.

References

1. Pardo M, Choudhary JS (2012) Assignment of protein interactions from affinity purification/mass spectrometry data. J Proteome Res 11(3): 1462–1474
2. Morris JH, Knudsen GM, Verschueren E, Johnson JR, Cimermancic P, Greninger AL et al (2014) Affinity purification-mass spectrometry and network analysis to understand protein-protein interactions. Nat Protoc 9(11): 2539–2554
3. Zhai Y, Gnanasekaran P, Pappu HR (2021) Identification and characterization of plant-

interacting targets of tomato spotted wilt virus silencing suppressor. Pathogens (Basel, Switzerland) 10(1):27

4. Yang JW, Fu JX, Li J, Cheng XL, Li F, Dong JF et al (2014) A novel co-immunoprecipitation protocol based on protoplast transient gene expression for studying protein–protein interactions in rice. Plant Mol Biol Report 32(1): 153–161
5. Lin JS, Lai EM (2017) Protein-protein interactions: co-immunoprecipitation. Methods Mol Biol 1615:211–219
6. Muñoz A, Castellano MM (2018) Coimmunoprecipitation of interacting proteins in plants. Methods Mol Biol 1794:279–287

Chapter 8

Co-immunoprecipitation-Based Identification of Effector–Host Protein Interactions from Pathogen-Infected Plant Tissue

Mamoona Khan and Armin Djamei

Abstract

Protein–protein interactions play an essential role in host–pathogen interactions. Phytopathogens secrete a cocktail of effector proteins to suppress plant immunity and reprogram host cell metabolism in their favor. Identification and characterization of effectors and their target protein complexes by co-immunoprecipitation can help to gain a deeper understanding of the functions of individual effectors during pathogenicity and can also provide new insights into the wiring of plant signaling pathways or metabolic complexes. Here we describe a detailed protocol to perform co-immunoprecipitation of effector–target protein complexes from plant extracts with an example of the *Ustilago maydis*/maize pathosystem for which we also provide a fungal protoplast transformation and maize seedling infection protocols.

Key words Effector, Co-IP, Protein complexes, Plant pathogen, *Ustilago maydis*

1 Introduction

Co-immunoprecipitation (co-IP) is a frequently used method and a powerful technique to identify protein–protein interactions from the native environment by using target protein-specific antibodies [1–3]. The interacting proteins that are bound together in the complex with the target protein are indirectly captured and co-immunoprecipitated from a crude lysate of tissue. These protein complexes can then be analyzed to identify new interacting partners, and the function of the target protein by mass spectrometry, or to verify *in planta* interactions by sodium dodecyl sulfate–polyacrylamide gel electrophoresis (SDS-PAGE) and immunoblotting [3, 4].

Protein–protein interactions play an essential role in host–pathogen interaction [5]. Plant pathogenic fungi have diverse lifestyles resulting in distinct strategies to interact with their host plants comprising necrotrophic, biotrophic, and hemibiotrophic

Shahid Mukhtar (ed.), *Protein-Protein Interactions: Methods and Protocols*, Methods in Molecular Biology, vol. 2690,
https://doi.org/10.1007/978-1-0716-3327-4_8,

strategies; they also differ in the range of plant species they can infect ranging from one to many hundreds [6, 7]. Phytopathogens that colonize living plant tissues secrete a cocktail of effector proteins to suppress plant immunity and reprogram host cell metabolism in their favor. Protein effectors are most often secreted via the conventional endoplasmic reticulum–Golgi apparatus secretory pathway. They act either in the plant cell apoplast or are translocated in the host cytoplasm where they localize to different subcellular organelles, interact with the target proteins, and may change their stability, activity or localization, etc. [8–11]. The expression profile of effectors is tightly tuned to the different infection stages and may be affected by the cell type and/or organ being infected [9, 12]. One way by which effector functions are usually inferred is via host-sided interaction partners, as many effectors show low conservation on the sequence level because of high selection pressure to evade host recognition [9]. Here we describe a co-IP protocol particularly suitable to identify protein complexes bound by a bait protein, for example, a fungal effector from infected host plant tissue. One prerequisite is that either the bait protein can be tagged by a suitable epitope of choice and is still functional or specific antibodies against the bait protein are available. We exemplify a detailed protocol for performing co-IP for effectors of the basidiomycetes fungus *Ustilago maydis*, which infects maize; however, it could also be used for other effectors as baits to isolate associated host protein complexes from the infected plant material.

2 Materials

1. *Ustilago maydis* mating-type compatible strains FB1 and FB2 (*see* **Note 1**).
2. Maize seeds, e.g., Early Golden Bantam (EGB) (*see* **Note 2**).
3. Double distilled water.
4. Yeast extract pepton saccharose light (YEPSL) medium: Take 4 g of yeast extract, 4 g of peptone, and 20 g of sucrose and fill up to 1 L with double distilled water. Autoclave at 121 °C for 15 min.
5. Potato dextrose (PD) agar medium: Take 24 g of Potato Dextrose Broth, 20 g of Bacto agar, and fill up to 1 L with double distilled water. Autoclave at 121 °C for 15 min; after autoclaving, pour approximately 30 mL of media into each round petri dish of 9.4 cm diameter.
6. Regeneration agar: Take 10 g of yeast extract, 20 g of peptone, 20 g of sucrose, 182.2 g of sorbitol, and 15 g of Bacto agar and fill up to 1 L with double distilled water. Autoclave at 121 °C for 15 min.

7. Sodiumcitrate sorbitol (SCS) buffer: To prepare SCS buffer, add solution II (20 mM citric acid, 1 M sorbitol) to solution I (20 mM sodium citrate, 1 M sorbitol) until pH 5.8 is reached (the estimated ratio is 1:5). Aliquot to 400 mL and autoclave at 121 °C for 15 min.
8. Sorbitol tris calcium cloride (STC) buffer: 1 M sorbitol, 10 mM Tris-HCl with pH 7.5, and 100 mM $Cacl_2$.
9. SCS-glucanex: Add 10 mg of glucanex powder (lysing enzymes from *Trichoderma harzianum*) to 1 mL of SCS, mix by vortexing, and filter sterile (*see* **Note 3**).
10. STC-PEG: Dissolve 40% polyethylene glycol (PEG) 3300 (v/v) in STC Buffer. Filter-sterilize it and store in 50 mL Falcons.
11. IP-buffer: 50 mM Tris-HCL with pH 8.0, 150 mM NaCl, 10% glycerol, and 0.1% Triton X-100; store at 4 °C.
12. Extraction buffer: IP-buffer supplemented with 2% polyvinylpolypyrolidone (PVPP), 1 mM ethylene diamine tetraacetic acid (EDTA), 1 × protease inhibitor tablet per 100 mL, and should be added just before use; store at 4 °C.
13. 1 × SDS loading buffer: 50 mM Tris-HCl with pH 6.8, 1% SDS, 1 mM EDTA, 10% glycerol, 0.05% bromophenol blue, and 50 mM dithiothreitol (DTT).
14. 15 mg/mL heparin solution (filter sterilized).
15. Plastic pots of 20 cm diameter.
16. Sand for grinding.
17. Soil for maize growth.
18. 1 mL syringes.
19. Needles (16 G, 40 mm).
20. 15 mL Falcon tubes.
21. 50 mL Falcon tubes.
22. Sterile bench.
23. Rotary Shaker.
24. Centrifuge with swing-out rotor.
25. Temperature-controlled climate chambers or greenhouse.
26. Round Petri dishes (9.4 cm diameter).
27. Glass test tubes.
28. Erlenmeyer flasks.
29. Pre-cooled Mortar and pestle.
30. 1.5 mL and 2 mL reaction tubes.
31. Liquid nitrogen for grinding plant material.
32. Anti-hemagglutinin (anti-HA) agarose.

33. HA-peptide.
34. μMACS™ HA isolation Kit.
35. μMACS separator.
36. μ-Columns.
37. MACS MultiStand.

3 Methods

3.1 Experimental Design

3.1.1 Selection of Promoter of Interest for the Co-IP

The expression of effector proteins varies depending on the different infection stages and may be affected by the cell type and/or the organ being infected. Generally, using the endogenous promoter and knowing about the expression pattern of the effector candidate gene is the optimal scenario considering biology. Nevertheless, in many cases, absolute promoter strength and effector expression level are too low to isolate sufficient amounts of protein from the infected tissue and stronger promoters of similar expression patterns over time might be a practical solution. In the end, the strength of interaction between the bait protein and its interaction partner(s) and the amounts of overall bait protein are critical to determining the success of the co-IP (*see* **Note 4**). In the case of *U. maydis*, transcriptional profiles of putative effectors during biotrophy are available [13, 14]. If the effector of interest is expressed during early biotrophy but overall weakly expressed, consider using already established strong biotrophy promoters like Cmu1 [15] or Pit2 [10] for the expression of effector in the case of *U. maydis*.

3.1.2 Selection of Epitope Tag for the Co-IP

The selection of epitope tag used to pull down the effector is another important criterion deciding the success of co-IP. A large tag can interfere with the folding, translocation, and hence the function of effectors of *U. maydis* [16]. Here, we present a protocol for co-IP where the effector is attached to the high-affinity short-peptide influenza hemagglutinin (HA) tag; however, the other small high-affinity tags such as an MYC-tag (EQKLISEEDL), a FLAG-tag (DYKDDDDK), or V5-tag (GKPIPNPLLGLDST) can be used if the respective antibody or antibody agarose is available. If a functional assay needs to be tested to determine whether the effector–epitope tag fusion protein is still biologically functional, it is worthwhile to perform this assay before co-IP. Such an assay could be a complementation assay of the virulence defect but can be also another functional readout in the plant. Usually, the epitope tag is added to the C-terminal part of the effector protein as the N-terminal signal peptide gets cleaved off, but there might be cases where the epitope is added N-terminally after the predicted signal peptide cleavage site.

3.1.3 Negative Control

A negative control is highly recommended during the co-IP. A secreted green fluorescent protein (GFP) or secreted mCherry expressed under the same promoter as the effector and attached to the same affinity tag can be used as a negative control for co-IP (*see* **Note 5**).

The expression vector used for cloning effector of interest should be compatible with the pathogen of choice, e.g., for *U. maydis*. In this case, a golden gate-compatible derivative of the *p123* expression vector can be used [17] that inserts the expression cassette at the *ip* locus by homologous recombination and hence causes resistance of transformed *U. maydis* strains against carboxin [18].

The coming steps will be exemplified for the usage of this protocol in the *U. maydis*/maize interaction and need adaptation for the respective host/pathogen system.

3.2 Preparation of DNA for Protoplast Transformations

Precisely, 5 μg of linearized DNA is required for the transformation of *U. maydis* protoplasts. Linearize the plasmid DNA with a restriction enzyme that digests the expression vector-only once in the carboxin resistance for at least 2–3 h at 37 °C. In the case of *p123* expression, vector restriction enzymes SspI or AgeI can be used but consider that the inserted promoter or gene might contain these restriction sites as well (*see* **Note 6**).

3.3 Preparation of Ustilago maydis FB1 and FB2 Competent Cells and Their Transformation with Plasmid DNA

1. The prerequisite of this step is to have freshly grown colonies of wild-type FB1 and FB2 strains of *U. maydis* for a pre-culture inoculation on PD agar plate.
2. Take 4 mL of YEPSL medium in two sterile glass test tubes and inoculate under a sterile bench a single fungal colony of FB1 and FB2; incubate overnight at 28 °C with constant agitation at 200 rpm.
3. The next day, use the pre-culture to inoculate 50 mL of the YEPSL in a 250 mL sterile Erlenmeyer flask to an optical density at 600 nm (OD_{600}) of 0.2 cells and let the cells grow at 28 °C with constant agitation at 200 rpm until OD_{600} reaches 0.6–0.8. To avoid a lag phase due to temperature shift, prewarm the medium before inoculation to 28 °C.
4. Pellet the cell cultures by centrifugation in 50 mL falcon tubes at 1500× g in a swing-out centrifuge for 10 min at room temperature (RT).
5. Wash the pellet with 20 mL of SCS buffer per 50 mL falcon by carefully pipetting up and down.
6. Pellet the cell cultures by centrifugation again at 1500× g in a swing-out centrifuge for 10 min at RT.

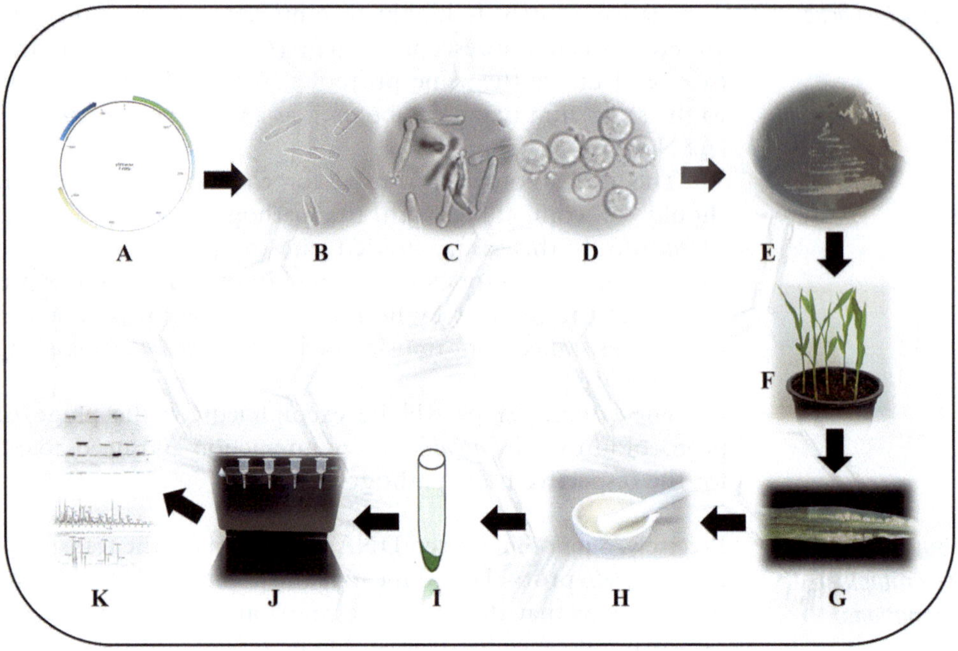

Fig. 1 Schematic representation of main steps of co-immunoprecipitation from expression of plasmid to analysis of protein complexes during studying effector–host protein interaction: (**a**) plasmid vector needed for fungal transformation, (**b**) wild-type *Ustilago maydis* cells, (**c**) lollipop-like protoplasting cells, (**d**) *U. maydis* protoplasts, (**e**) *U. maydis* individual colonies re-streaked out, (**f**) seven-day-old maize seedlings ready for infection, (**g**) *U. maydis*-infected maize leaf, (**h**) Mortar and pestle pre-chilled in liquid nitrogen, (**i**) tube containing clarified plant extract with debris at the bottom, (**j**) μ-Columns placed in μMACS separator fixed on MACS MultiStand, and (**k**) possible ways of analyzing protein complexes after co-IP

7. Discard the supernatant and resuspend the pellet by pipetting up and down in the 2 mL SCS-glucanex for cell wall digestion and let the cells incubate in SCS-glucanex for 7–10 min. Observe cells under the microscope. If two-thirds of the cells are looking like a lollipop (Fig. 1c), add 10 mL SCS (*see* **Note** 7).
8. Add 20 mL of SCS buffer to each falcon and centrifuge again at 900× g in a swing-out centrifuge for 10 min at 4 °C.
9. Wash the pellet with 20 mL of pre-cooled SCS buffer in each falcon by pipetting up and down. Resuspend the pellet very carefully as the cells are fragile now.
10. Pellet the cell cultures by centrifugation again at 900× g in a swing-out centrifuge for 10 min at 4 °C.
11. Repeat steps 9 and 10 two more times.
12. Wash the pellet with 10 mL of pre-cooled STC buffer in each falcon by pipetting up and down.
13. Pellet the cell cultures by centrifugation again at 900 × g in a swing-out centrifuge for 10 min at 4 °C.

14. Resuspend the pellet carefully in 0.5 mL of cold STC and observe under a microscope; the protoplasts should be looking round-shaped now (Fig. 1d).
15. Aliquot to 100 μL in 1.5 mL Eppendorf tubes. Store on ice until use or put in a −80 °C freezer for long-term storage.
16. Add 1 μL of heparin solution and a linearized plasmid (5 μg in maximum 20 μL) to the 100 μL of FB1 and FB2 protoplasts each and mix gently by pipetting (*see* **Note 8**).
17. Incubate on ice for 15 min.
18. Add 500 μL pre-cooled STC-PEG solution to the cells and pipette up and down two times carefully.
19. Incubate on ice for another 15 min exactly.
20. Plate the protoplasts on the regeneration agar plates containing appropriate antibiotics (*see* **Note 9**) by distributing the protoplasts suspension equally on the plate with the help of a 1 mL pipette.
21. Incubate the plates agar-side down to allow drying for one day, and then four more days at 28 °C with agar-side up to get colonies (*see* **Note 10**).
22. Re-streak single colonies that may appear on the regeneration agar plate on the PD agar plate containing appropriate antibiotics (Fig. 1e).

3.4 Growing Maize Seedlings for Infection

1. Prepare three round pots of 20 cm diameter (19 cm height) with soil for each effector construct that needs to be tested. Add three pots for each negative control.
2. Sow six seeds of appropriate maize variety EGB or B73 (Fig. 1f), 1 cm below the soil. For uniform seed germination, the soil should be kept moderately wet with water to ensure even humidity conditions after potting.
3. Place potted plants in a temperature-controlled greenhouse or climate chamber with the following conditions: 28 °C and 14 h of approximately 25,000–90,000 lux of illumination and 22 °C during the 10 h night period [19].

3.5 Infection of Maize Seedlings with FB1 and FB2 Strains

1. Prepare FB1, and FB2 inoculum separately, as described in Subheading 3.3, by performing **steps 1–4** for each effector strain and control strain.
2. Then discard the supernatant and resuspend the cell pellet by pipetting up and down in 50 mL of sterile double-distilled water and centrifuge again at 1500× g in a swing-out centrifuge for 10 min at RT.

3. Repeat **step 2** to remove all the traces of the YEPSL medium and finally resuspend the pellet in sterile water to an $OD_{600} = 2.0$.
4. Mix equal volumes of mating-type FB1 and FB2 strains expressing the same construct just before infecting maize seedlings so that the final OD_{600} of each strain in inoculum is equal to 1.0.
5. Use 300–500 μL of cell suspension of *FB1* × *FB2* strains into the leaf whorl of seven-day-old maize seedlings with a syringe, approximately 2 cm above the soil (*see* **Note 11**).
6. Let the plants grow further under the same growth conditions for two to five days (*see* **Note 12**).

3.6 Co-Immunoprecipitation of an Effector–Target Protein Complex from Infected Maize

1. Harvest plant samples by cutting only the areas with infection symptoms (*see* **Note 13**) (Fig. 1g), place infected plant material in 15 mL falcon tubes that were pre-chilled in liquid nitrogen, and shock freeze them in liquid nitrogen. The sample can be processed directly now or they can be stored at −80 °C.
2. Chill mortar and pestle with liquid nitrogen (Fig. 1h), add one scoop of washed sand into pre-cooled mortar and pestle along with frozen plant material, and grind the plant material to a fine powder (*see* **Note 14**).
3. Aliquot 1 g of plant powder in a pre-cooled 15 mL falcon tube. The samples can be directly used for protein extraction or can be stored at −80 °C (*see* **Note 15**).
4. Add 2 mL of ice-cold extraction buffer (or lysis buffer from μMACS™ HA isolation Kit; *see* **Note 16**) and vortex for 20 s immediately to make a homogeneous suspension of the powder in the buffer and put it back on ice (*see* **Note 17**).
5. Repeat **steps 2–5** for all construct combinations.
6. Centrifuge the falcon tubes in a pre-cooled centrifuge at 3000× g 4 °C for 15 min to remove the debris.
7. Transfer supernatants to a new 2 mL pre-cooled Eppendorf tube and centrifuge again at 8000× g 4 °C for 15 min to remove any debris left (*see* **Note 18**).
8. Repeat **step** 7.
9. Transfer the supernatant into a new 2 mL Eppendorf tube.
10. Take 75 μL of plant extract from each sample out into new 1.5 mL Eppendorf tubes; this will serve as *input control* for the co-IP (*see* **Note 19**). Add 25 μL of 4X SDS loading buffer and boil the sample at 95 °C for 5 min. These input control samples can be stored at −20 °C until further processed with SDS-PAGE and western blotting.

11. For immunoprecipitation of the protein complexes from the rest of the plant extracts from **step 9**, there are two possibilities depending on the choice of beads used (*see* **Note 20**).

3.6.1 Immunoprecipitation of Protein Complexes Using Agarose Beads

1. To equilibrate beads, add 50 μL of anti-HA agarose per sample in 2 mL pre-chilled IP-buffer in an Eppendorf tube (*see* **Note 21**) and then centrifuge the tubes at 800× g for 30 s and discard supernatant carefully without disturbing the beads.
2. Resuspend anti-HA agarose into 50 μL of IP-buffer for each sample (*see* **Note 22**).
3. Take 500 μL of plant extract from **step 9** in Subheading 3.6 into the Eppendorf tube and make the volume up to 1.95 mL by adding the IP-buffer.
4. Transfer 1.95 mL diluted plant extract to the Eppendorf tube containing 50 μL of equilibrated anti-HA agarose and incubate the Eppendorf tubes on a rotatory wheel for 2 h at 4 °C.
5. Pellet the beads by centrifugation at 800× g for 30 s and discard supernatant carefully without disturbing the beads.
6. Add 1 mL of pre-chilled IP-buffer for washing, invert two to three times gently, centrifuge at 800× g for 30 s at 4 °C, and discard the supernatant carefully without disturbing the bead (*see* **Note 23**).
7. Repeat **step 6** three more times, and proceed to **step 8** or **9** depending on the aim of the co-IP.
8. Elute the bound proteins by adding 100 μL 1 × SDS buffer to the beads, and heat for 10 min at 95 °C. Centrifuge at 800× g for 30 s, and transfer the supernatant containing the eluted proteins to a fresh tube gently without touching the agarose beads (*see* **Notes 24** and **25**).
9. There are at least three different possibilities to elute the immunoprecipitated proteins under non-denaturing conditions:

   A: Add 100 μL of 0.1 M glycine with pH 2.5 to the beads, and incubate for 10 min at 4 °C with mild shaking. Centrifuge at 800× g for 30 s and transfer the supernatant containing the eluted proteins to a fresh tube gently without touching the agarose beads. Repeat elution and pool the two eluates. Immediately neutralize the eluates by adding equal volume (200 μL) of Tris-HCl buffer with pH 8.0 (*see* **Note 26**).

   B: Add 100 μL of IP-buffer containing 150 ng per μL HA-peptide and incubate at 4 °C for 20 min with mild shaking. Centrifuge at 800× g for 30 s at 4 °C and transfer the supernatant containing the eluted proteins to a fresh tube gently without touching the agarose beads (*see* **Note 27**).

   C: Elution of immunoprecipitated proteins can be performed by bead digestion with trypsin (*see* **Note 28**).

3.6.2 Immunoprecipitation of Protein Complexes Using Magnetic Beads

1. Take 2 mL of plant extract from **step 9** in Subheading 3.6 (*see* **Note 17**), add 30 μL of μMACS anti-HA tag microbeads, mix by inverting the tube, then place the Eppendorf tubes on a rotatory wheel for 2 h at 4 °C.
2. Place μ-Columns in the μMACS separator (Fig. 1j), and equilibrate the column with a 200 μL lysis buffer supplied with the μMACS™ HA isolation Kit.
3. Apply 500 μL of each sample to a separate μ-Column by using a 1 mL pipette, and let the fluid run out by gravity flow.
4. Repeat this step until no sample is left in the Eppendorf tube.
5. Wash the column four times with a 200 μL wash buffer I supplied with the μMACS™ HA isolation Kit.
6. Wash the column with 100 μL of wash buffer II supplied with the μMACS™ HA isolation Kit to remove all the traces of residual salt and detergents and proceed to **step** 7 or **10** depending on the aim of the Co-IP.
7. Aliquot 140 μL of elution buffer provided in the Kit or 1 × SDS buffer for each sample, and incubate at 95 °C in a 1.5 mL Eppendorf tube for 5 min (*see* **Notes 24** and **25**).
8. Add 20 μL of pre-heated elution buffer to each column and incubate for 10 min at room temperature. Then add the rest of 60 μL of pre-heated elution buffer, and let it run by gravity flow and collect flow-through into a 1.5 mL reaction tube (*see* **Note 29**).
9. Repeat **step 8** and pool the two eluates and analyze by SDS-PAGE and western blotting.
10. There are at least two different possibilities to elute the immunoprecipitated proteins under non-denaturing conditions.

    A: Add 20 μL of wash buffer II containing 150 ng per μL HA-peptide to the column and incubate for 10 min at room temperature. Then add the 60 μL of wash buffer II containing 150 ng per μL HA-peptide and collect the flow-through into a 1.5 mL reaction tube. Repeat this step and collect the elution fraction and pool the two eluates.

    B: Elution of immunoprecipitated proteins can be performed by bead digestion with trypsin (*see* **Note 28**).

4 Notes

1. The progenitor strain SG200 [20] can also be used; however, the use of FB1 and FB2 strains has an advantage since they have two copies of the *Promoter:effector:tag* cassette in the genome after mating in comparison to SG200, which has only one copy.

2. Any other variety of available maize can be used.
3. About 2 mL of SCS-glucanex solution per strain is needed. The SCS-glucanex solution needs to be prepared just before use and cannot be stored.
4. Proper experimental conditions must be determined individually for each protein–protein interaction, e.g., the life cycle of the pathogen, the expression pattern of the effector, and the expression pattern of the promoter.
5. Please consider that GFP and mCherry are not secreted proteins; signal peptides at their N-terminals need to be added, to secrete them to the apoplast. Here it would be highly recommended to use the signal peptide either of the effector for which co-IP will be performed or of an already demonstrated effector, e.g., cmu1 as previously used [15].
6. Please check the expression cassette for the presence of any digestion sites of SspI, or AgeI. If there is one, mutagenizing may be considered, or find an alternative digestion site accordingly.
7. Wild-type *U. maydis* cells are elongated rod-like (Fig. 1b). From now onward, store the cells always on ice and cool down the SCS buffer and the STC-buffer to 4 °C.
8. For each effector plasmid that will be subjected to co-IP, there will be two tubes of protoplast: one each with FB1 and FB2 protoplast.
9. At this stage, freshly prepare regeneration agar plates. For this boil regeneration agar in a microwave, cool down to 50–60 °C, add the appropriate amount of antibiotic, e.g., carboxin, and pour 10 mL bottom layer. Store regeneration agar at 55 °C incubator. When the bottom layer is solidified, add 15 mL regeneration agar without antibiotics. This type of plate makes sure that the fragile protoplasts are not coming directly in contact with antibiotics on the first day after plating and have time to express the resistance gene.
10. It is important to incubate agar-side up for at least one day so that extra STC-PEG is absorbed in the medium leaving protoplast on the surface.
11. In case of seedling infections, hold the syringe in an oblique position and pierce sheaths of leaves in a way that the needle stays halfway in the center of the stem cylinder. Do not push the needle through the stem. Press the syringe to inject the cell suspension until the inoculum is coming out from the upper end of the whorl of leaves [19, 21].
12. If cmu1 promoter is used, the fifth day after infection is the right time to harvest plant material; on day 5 small galls are visible on the leaves (Fig. 1g). In the case of *pit2* promoter,

plant material can be harvested on day 3. Alternatively, the plant material can be harvested according to the expression pattern of the given promoter.

13. Care should be taken in harvesting infected areas only; the non-infected plant material can dilute the expressed effector.
14. Extreme care should be taken that the plant material does not thaw; otherwise, protease and other hydrolytic enzymes might deteriorate the quality of your sample.
15. Infected leaf areas of ten leaves will give a lot of ground material in the case of *U. maydis*, the rest of which could be stored at −80 °C as a backup for technical repetition. Nevertheless, it is important to grind several infected tissue pieces together as a pool to reduce variation.
16. Lysis buffer from μMACS™ HA isolation Kit should only be used if μMACS anti-HA microbeads for co-IP is the choice from **step 12** onward.
17. All the steps of co-IP should be performed at 4 °C until otherwise stated as heating of samples can lead to degradation of proteins.
18. Any debris present in the samples can result in clogging of the column.
19. Input control will serve as a measure to test that the effector protein was expressed in the given co-IP samples by performing SDS-PAGE and western blotting.
20. Agarose beads have a higher binding capacity due to their porous surface but magnetic beads are replacing them in IP/co-IP and other small-scale affinity procedures because they have lower non-specific binding. However, the costs of these beads may be considered when choosing the type of beads as magnetic beads are usually more costly and require special columns and tailored stands for washing and elution.
21. The agarose beads should be resuspended well before use by tapping the side of the vial several times and pipetting up and down using a 1 mL pipette with a cut tip. Care should be taken not to damage the beads.
22. If 200 μL of beads are originally used, e.g., for four samples, add 200 μL of IP-buffer.
23. Leave about 50 μL of the IP-buffer at the bottom of the tube to avoid disturbing the beads. Vortexing or shaking of beads should be avoided to prevent the disruption of the protein–protein interactions of the target complex.
24. This is elution under denaturing conditions, as SDS buffer reduces and denatures proteins; hence, it is often an effective way to separate the proteins from the beads, but it is only used

if SDS-PAGE is the detection method of choice. In case the detection is done via native PAGE protocol, or enzyme assays with your isolated proteins, then consider eluting the co-immunoprecipitated protein under non-denaturing conditions. High SDS concentrations also seem to interfere with mass spectrometry approaches, so clarify first the downstream analysis before deciding on how to elute from the protein complex.

25. For western blotting of your samples, consider the antibodies that are going to be used. One problem of this elution method is that the antibody or at least parts of it bound to the beads, are often eluted with the proteins of interest, and cause interference as bands during gel analysis.

26. The low pH of glycine weakens the interaction between the antibody and the beads and, hence, causes elution of proteins. However, some proteins will still denature or lose enzymatic activity under these conditions, and some proteins may not dissociate from beads with this method. The eluted sample should be immediately neutralized with Tris-HCl with pH 8.0 to avoid irreversible unfolding.

27. Type of peptide depends on the epitope tags that are used for co-IP.

28. Check with the mass spectrometry department or service provider.

29. Put the collection tubes in a rack under the column and collect the eluate only when the colorless IP-buffer has run out. This step needs to be performed one by one for each sample.

Acknowledgments

We thank Dr. Natalia De Sousa Teixeira E. Silva for providing the *U. maydis* cells for microscopy and Dr. Aladar Pettkó-Szandtner for helpful comments on the co-IP protocol. Our research is supported by funding from the German Research Foundation (DFG) under Germany's Excellence Strategy – EXC-2070 – 390732324 (PhenoRob) and DJ 64/5-1 and the Austrian Science Fund (FWF) (I 3033-B22).

References

1. Avila JR, Lee JS, Torii KU (2015) Co-immunoprecipitation of membrane-bound receptors. The Arabidopsis Book 2015(13):e0180
2. Sciuto MR, Warnken U, Schnolzer M et al (2018) Two-step coimmunoprecipitation (TIP) enables efficient and highly selective isolation of native protein complexes. Mol Cell Proteomics: MCP 17(5):993–1009. https://doi.org/10.1074/mcp.O116.065920
3. Lin J-S, Lai E-M (2017) Protein–protein interactions: co-immunoprecipitation. In:

Journet L, Cascales E (eds) Bacterial protein secretion systems: methods and protocols. Springer, New York, pp 211–219. https://doi.org/10.1007/978-1-4939-7033-9_17

4. Iqbal H, Akins DR, Kenedy MR (2018) Co-immunoprecipitation for identifying protein-protein interactions in Borrelia burgdorferi. In: Pal U, Buyuktanir O (eds) Borrelia burgdorferi: methods and protocols. Springer, New York, pp 47–55. https://doi.org/10.1007/978-1-4939-7383-5_4
5. Nicod C, Banaei-Esfahani A, Collins BC (2017) Elucidation of host-pathogen protein-protein interactions to uncover mechanisms of host cell rewiring. Curr Opin Microbiol 39:7–15. https://doi.org/10.1016/j.mib.2017.07.005
6. Lo Presti L, Lanver D, Schweizer G et al (2015) Fungal effectors and plant susceptibility. Annu Rev Plant Biol 66:513–545. https://doi.org/10.1146/annurev-arplant-043014-114623
7. Newman TE, Derbyshire MC (2020) The evolutionary and molecular features of broad host-range necrotrophy in plant pathogenic fungi. Front Plant Sci 11:591733. https://doi.org/10.3389/fpls.2020.591733
8. Navarrete F, Grujic N, Stirnberg A et al (2021) The Pleiades are a cluster of fungal effectors that inhibit host defenses. PLoS Pathog 17(6):e1009641. https://doi.org/10.1371/journal.ppat.1009641
9. Uhse S, Djamei A (2018) Effectors of plant-colonizing fungi and beyond. PLoS Pathog 14(6):e1006992. https://doi.org/10.1371/journal.ppat.1006992
10. Mueller AN, Ziemann S, Treitschke S et al (2013) Compatibility in the Ustilago maydis-maize interaction requires inhibition of host cysteine proteases by the fungal effector Pit2. PLoS Pathog 9(2):e1003177. https://doi.org/10.1371/journal.ppat.1003177
11. Saado I, Chia K-S, Betz R et al (2022) Effector-mediated relocalization of a maize lipoxygenase protein triggers susceptibility to Ustilago maydis. Plant Cell. https://doi.org/10.1093/plcell/koac105
12. Okmen B, Doehlemann G (2014) Inside plant: biotrophic strategies to modulate host immunity and metabolism. Curr Opin Plant Biol 20:19–25. https://doi.org/10.1016/j.pbi.2014.03.011
13. Lanver D, Muller AN, Happel P et al (2018) The biotrophic development of Ustilago maydis studied by RNA-Seq analysis. Plant Cell 30(2):300–323. https://doi.org/10.1105/tpc.17.00764
14. Zuo W, Depotter JRL, Gupta DK et al (2021) Cross-species analysis between the maize smut fungi *Ustilago maydis* and *Sporisorium reilianum* highlights the role of transcriptional change of effector orthologs for virulence and disease. New Phytol 232(2):719–733. https://doi.org/10.1111/nph.17625
15. Djamei A, Schipper K, Rabe F et al (2011) Metabolic priming by a secreted fungal effector. Nature 478(7369):395–398. https://doi.org/10.1038/nature10454
16. Lo Presti L, Zechmann B, Kumlehn J et al (2017) An assay for entry of secreted fungal effectors into plant cells. New Phytol 213(2):956–964. https://doi.org/10.1111/nph.14188
17. Navarrete F, Gallei M, Kornienko AE et al (2022) TOPLESS promotes plant immunity by repressing auxin signaling and is targeted by the fungal effector Naked1. Plant Commun 3(2):100269. https://doi.org/10.1016/j.xplc.2021.100269
18. Keon JPR, White GA, Hargreaves JA (1991) Isolation, characterization and sequence of a gene conferring resistance to the systemic fungicide carboxin from the maize smut pathogen, Ustilago maydis. Curr Genet 19(6):475–481. https://doi.org/10.1007/bf00312739
19. Khan M, Djamei A (2022) Performing infection assays of Sporisorium reilianum f. sp. Zeae in Maize. Methods Mol Biol 2494:291–298. https://doi.org/10.1007/978-1-0716-2297-1_20
20. Bölker M, Genin S, Lehmler C et al (1995) Genetic regulation of mating and dimorphism in Ustilago maydis. Can J Bot 73(S1):320–325. https://doi.org/10.1139/b95-262
21. Redkar A, Doehlemann G (2016) Ustilago maydis virulence assays in maize. Bio Protoc 6(6):e1760. https://doi.org/10.21769/BioProtoc.1760

Chapter 9

Co-immunoprecipitation for Assessing Protein–Protein Interactions in *Agrobacterium*-Mediated Transient Expression System in *Nicotiana benthamiana*

Ji Chul Nam, Padam S. Bhatt, Sung-Il Kim, and Hong-Gu Kang

Abstract

The characterization of protein–protein interactions (PPI) often provides functional information about a target protein. Yeast-two-hybrid (Y2H) and luminescence/fluorescence-based detections, therefore, have been widely utilized for assessing PPI. In addition, a co-immunoprecipitation (co-IP) method has also been adopted with transient protein expression in *Nicotiana benthamiana* (*N. benthamiana*) infiltrated with *Agrobacterium tumefaciens*. Herein, we describe a co-IP procedure in which structural maintenance of chromosome 1 (SMC1), identified from a Y2H screening, was verified as an interacting partner for microchidia 1 (MORC1), a protein well known for its function in plant immunity and epigenetics. SMC1 and MORC1 were transiently expressed in *N. benthamiana* when infiltrated by *Agrobacterium* with the respective genes. From this approach, we identified a region of SMC1 responsible for interacting with MORC1. The co-IP method, of which outputs are mainly from immunoblot analysis, provided information about target protein expression as well, which is often useful for troubleshooting. Using this feature, we showcased a PPI confirmation from our SMC1–MORC1 study in which a full-length SMC1 protein was not detectable, and, therefore, a subsequent truncated mutant analysis had to be employed for PPI verification.

Key words *Nicotiana benthamiana*, Co-immunoprecipitation, Transient expression, Protein, protein interactions

1 Introduction

Analysis of protein–protein interaction (PPI), an essential process in biological characterization [1], has provided critical insight into many research questions. A wide range of tools for assessing PPI has been developed accordingly [1, 2]. Yeast-two-hybrids (Y2Hs) have surveyed PPI in a targeted or high-throughput fashion. In particular, a large-scale Y2H analysis at the systems level has been instrumental in learning complicated protein complex networks, termed interactomes, in *Arabidopsis* and other species [3, 4]. For instance, Y2H analysis of immunity-related proteins from

Shahid Mukhtar (ed.), *Protein-Protein Interactions: Methods and Protocols*, Methods in Molecular Biology, vol. 2690, https://doi.org/10.1007/978-1-0716-3327-4_9,

Arabidopsis and two pathogens, *Pseudomonas syringae* and *Hyaloperonospora arabidopsidis*, suggested that pathogen-encoded effectors likely converge on a limited set of highly connected hub proteins in the immune system [5]. Luminescence- and fluorescence-based tools have also been widely adopted for analyzing PPI as they provide in vivo and *in planta* information [6], including bimolecular fluorescence complementation (BiFC), fluorescence resonance energy transfer (FRET), and bioluminescence resonance energy transfer (BRET). BiFC, for example, provides information about not only PPI but also its interaction site at the subcellular level [7, 8]. However, it was noted that background signals from BiFC could lead to low sensitivity, false positives, and reduced confidence in true positives [9], while FRET and BRET are technically demanding [6].

Co-immunoprecipitation (co-IP), when performed as a targeted approach, mostly involves two genes of interest. These targeted genes are expressed at a high level, driven by a highly/constitutively active promoter, and their analysis output is produced by immunoblot analysis. It is often easier to troubleshoot co-IP over other PPI tools because immunoblot analysis, an integral part of the analysis process, provides information about the expression and integrity of target genes. Thanks to this feature, we have often been successful in verifying PPI from proteins that were known to be difficult to detect, including NB-LRR (nucleotide binding-leucine rich repeat) types of R genes [10, 11], suggesting that co-IP could be a very powerful tool in detecting PPI if the target gene is sufficiently expressed.

Expression of heterologous genes in plants for co-IP is mostly performed on protoplasts [12, 13], cell cultures [14], or intact plants [15]. Protoplasts and cell cultures involve the sterilization of plant tissues and thus demand more stringent quality control. Therefore, gene delivery to intact plants has become the most popular choice, with *Agrobacterium tumefaciens* as the most common vehicle for this delivery, termed agroinfiltration [16]. Note that this common method requires the susceptibility of target plants to *A. tumefaciens*, as the genetic background of infiltrated plants often dictates the outcome of transient expression [17]. Among intact plants that have been used for infiltrated delivery by *A. tumefaciens*, *Nicotiana benthamiana* has demonstrated susceptibility to a broad range of *A. tumefaciens* strains with readily measurable expression levels [15], explaining its popular adoption as a host plant.

One of the significant advantages of intact plant delivery is the capacity to characterize a phenotype triggered by the expression of a target gene. For instance, a hypersensitive response (HR), in the form of immunity-related programmed cell death, was replicated in agroinfiltrated *N. benthaminana* transiently expressing an NB-LRR R gene and its cognate effector. These useful characteristics led to an important discovery about R gene-mediated resistance in plants, in which disruption of the interdomain interaction initiates defense responses [18, 19]. A comparable system also showed that activation of NB-LRR proteins leads to the physical dissociation of microchidia 1 (MORC1), a chromatin-remodeling protein in *Arabidopsis thaliana* [10]. When an HR is triggered from the transient expression, its development can be controlled by using an inducible promoter since estradiol [20] and glucocorticoid [21] inducible systems have been successfully used in the *N. benthamiana* transient expression system [10, 11].

MORC1 is a conserved GHKL (gyrase, HSP90, histidine kinase, MutL)-type ATPase and is required for effector-triggered immunity, pathogen-associated molecular pattern (PAMP)-triggered immunity, basal resistance, non-host resistance, and systemic-acquired resistance [10, 11, 22]. To characterize PPI associated with this important immunity protein, we identified 14 MORC1-interacting proteins via Y2H screening using MORC1 as bait (unpublished data). One of the MORC1-interacting proteins was found to be SMC1—structural maintenance of chromosome 1. Interestingly, an earlier report showed that defective in meristem silencing 3 (DMS3), an SMC protein, indirectly interacts with MORC6 [23], a MORC1 homolog. SMC proteins are required for holding sister chromatids together during mitosis and meiosis [24–26]. In addition, DMS3 and MORC1/6 are known to regulate chromatin topology. Thus, we tested if SMC1 and its homologs, SMC2 and DMS3, physically interact with MORC1 using targeted Y2H analysis. While all of the full-length SMCs displayed little interaction with MORC1, SMC1-partial (838-1123 aa) identified from the Y2H screening showed the interaction (Fig. 1). This observation suggests that these full-length SMCs have a domain inhibiting the physical interaction with MORC1.

To identify the SMC1 domain interacting with MORC1 and remove a putative inhibiting domain, we generated constructs for SMC1-truncated mutants (Fig. 2a) under an estradiol-inducible promoter (Fig. 2a): SMC1-1 (1-344 aa), SMC1-2 (345-958 aa), and SMC1-3 (959-1218 aa). These constructs produced SMC1-truncated proteins in agroinfiltrated *N. benthamiana* (Fig. 2b); note that full-length SMC1 failed to be expressed for an unknown reason (data not shown). Interestingly, SMC1-2, which contains a hinge domain [27, 28], displayed interactions with MORC1; green fluorescent protein (GFP) was used as a negative control (Fig. 2b).

In this chapter, we describe the detailed procedure for this PPI confirmation process via a co-IP approach, which identified the interacting domain (Fig. 2b), and suggest that SMC1 may have a putative domain interfering with the MORC1 interaction. This outcome highlights the importance of Co-IP as a tool in

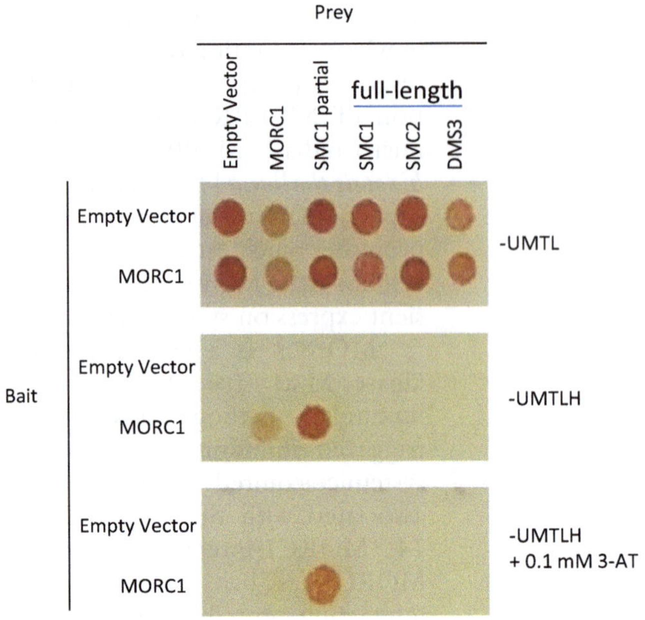

Fig. 1 Interaction analysis between Microchidia 1 (MORC1) and structural maintenance chromosomes (SMCs), and defective in meristem silencing (DMS) by using yeast-two-hybrid. MORC1 in pB27 was used as bait. SMC1, SMC2, and DMS3 in pP6 were used as prey. The plasmids were transformed into *Saccharomyces cerevisiae* carrying a *HIS3* reporter gene under the control of a LexA DNA-binding domain. Transformants were plated onto minimal media, -Ura/-Met/-Trp/-Leu (-UMTL), and -Ura/-Met/-Trp/-Leu/-His (-UMTLH) without and with 0.1 mM 3-AT (3-amino-1,2,4-triazole)

characterizing PPI identified from an independent PPI method, particularly of a target gene whose sufficient expression is challenging to attain.

2 Materials

2.1 Plant Material and *Agrobacterium tumefaciens*

1. *Agrobacterium tumefaciens* strain GV2260 carrying binary constructs.
2. Four- to six-week-old *Nicotiana benthamiana* grown at 22 °C and with a 16 h light period.

2.2 Equipment

1. A tabletop centrifuge (Sorvall, ST40R).
2. Microcentrifuge (Fisherbrand accuSpin Micro 17).
3. Vortex mixer.
4. Cold room at 4 °C.
5. Spectrometer (BIO-RAD SmartSpec Plus Spectrophotometer, Cat#: 170-2525).

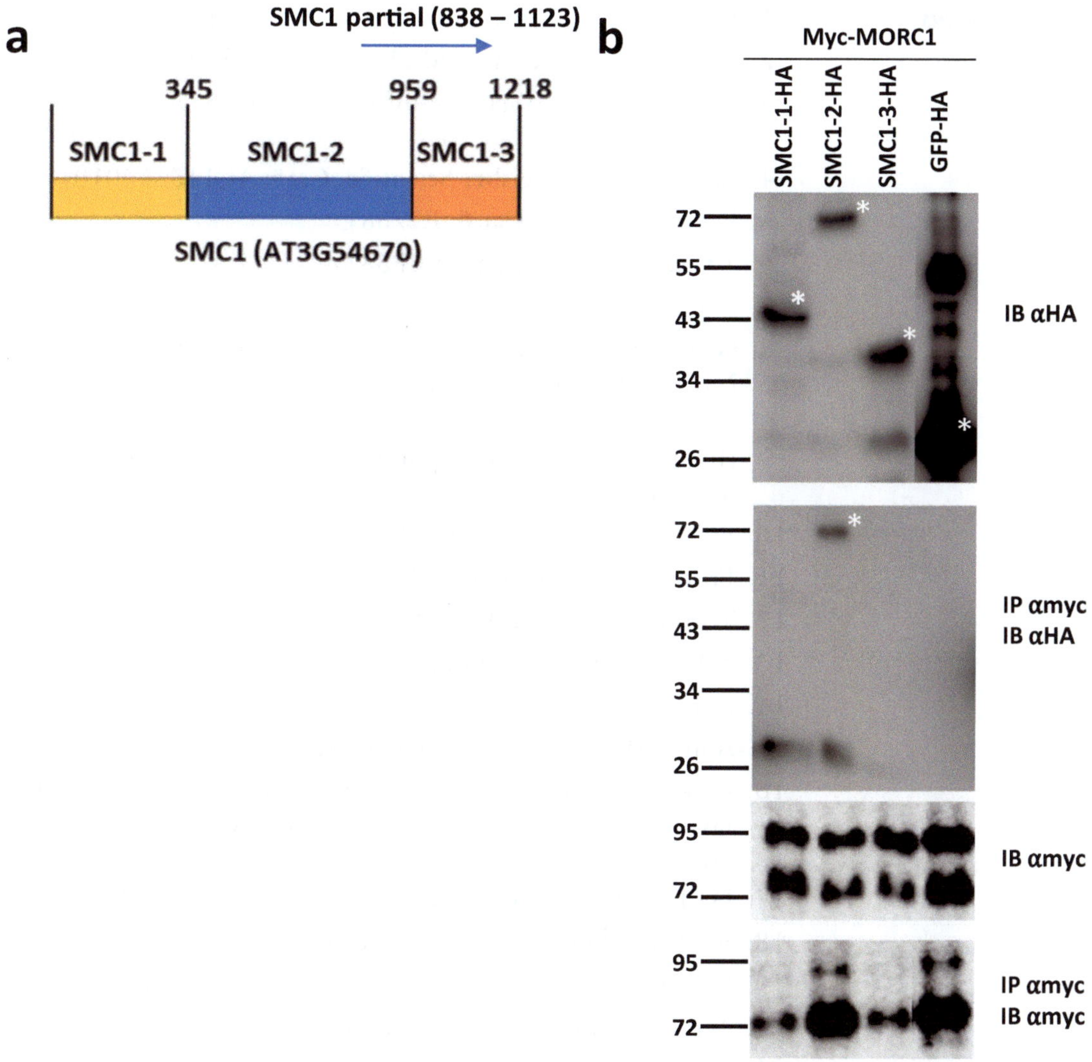

Fig. 2 SMC1 interacts with MORC1 *in planta*. (**a**) Schematic representation of the truncation mutants of SMC1: SMC1-1 (1-344aa), SMC1-2 (345-958aa), SMC1-3 (959-1218aa), and SMC1-partial (838-1123aa). (**b**) Transient expression of *Myc-MORC1*, *SMC1-HA*, and the *SMC1* mutants in agroinfiltrated *N. benthamiana*. Soluble extracts from the plants were subject to immunoblotting (IB) with αHA and αMyc or immunoprecipitation (IP) with αMyc, followed by IB with αHA and αMyc. Size markers are shown on the right of the panel in kDa. Asterisks indicate the expected sizes of the SMC/GFP proteins

6. Rotator (Fischer Scientific, Cat#: 88-861-051).
7. Up/down paint shaker for homogenizing (Harbil 5G).
8. −80 °C deep freezer.
9. An orbital shaker.
10. Millipore Milli-Q ultrapure water system.

2.3 Reagents

1. Deionized H_2O (measured at 18.2 MΩ-cm).
2. LB media (Lysogeny broth, Fisher Scientific, Cat #: BP9723).
3. 2-(*N*-morpholino) ethanesulfonic acid (MES; Fisher Scientific, Cat #: BP300)
4. Acetosyringone (Fisher Scientific, Cat#: AC115540010).
5. Estradiol (Sigma, Cat #: E2758).
6. Tween-20 (Fisher, Cat #: BP337).
7. Liquid nitrogen.
8. GTEN (Glycerol Tris Ethylenediaminetetraacetic acid (EDTA) NaCl) buffer (150 mM NaCl, 1 mM EDTA, 25 mM Tris 7.5, and 10% glycerol).
9. IP buffer (GTEN buffer plus 0.15% NP-40 and 5 mM dithiothreitol [DTT]).
10. Polyvinylpolypyrrolidone (PVPP; Sigma, Cat #: P6755).
11. Protease Inhibitor Cocktail (Sigma, Cat#: P9599).
12. Triton X-100 (EMD, Cat#: TX1568).
13. NAP-5 Sephadex columns (GE).
14. Anti-mouse IgG-agarose beads (Sigma, Cat#: A6531).
15. Anti-c-Myc agarose affinity gel (Sigma, Cat#: A7470).
16. 4× sodium dodecyl sulfate (SDS) sample buffer (200 mM Tris-HCl [pH 6.8], 8% SDS, 0.4% bromophenol blue, 40% glycerol)
17. DTT (dithiothreitol) (Fisher Scientific, Cat#: BP172-5).
18. EZ-Run^{TM} Protein Marker (Fisher Scientific, Cat#: BP36001).
19. Gel-loading tip (Fisher Scientific, Cat#: 02-707-138).
20. Acrylamide/bis (30% 37.5:1; Bio-Rad).
21. 1.5 M Tris-HCl (pH 8.8).
22. *N*,*N*,*N*′,*N*′-tetramethylethylene-diamine (TEMED) (Bio-Rad).
23. Ammonium persulfate (MP Biomedicals, Cat#: 193988).
24. PVDF membrane (Immobilon, Cat#: IPVH00010).
25. Methanol (Fisher Scientific, Cat#: A452).
26. Phosphate-buffered saline (PBS) (Sambrook et al., 1989).
27. PBS-T (PBS with 3% Tween-20).
28. 3B1M Buffer (3% BSA [Fisher Scientific], 1% nonfat dry milk [Carnation] in 1× PBS).
29. Horseradish peroxidase conjugated (HRP) anti-Myc antibody (Santa Cruz Biotechnology, Cat#: sc-40).

30. HRP-conjugated anti-hemagglutinin (anti-HA) antibody (Sigma, Cat#: 12013819001).
31. ECL plus western blotting solution (Fisher Scientific, Cat#: 32132X3).

3 Methods

3.1 Transient Expression in Nicotiana benthamiana

1. The coding sequence of target genes was cloned into plasmid pER8 and transformed into *A. tumefaciens* strain GV2260 via electroporation (*see* **Note 1**). pBA-HcPro was used to enhance transient expression by suppressing gene silencing as previously described [11].
2. *A. tumefaciens* cultured in LB liquid media for one day at 28 °C was subcultured in LB (1:100) with 10 mM MES at pH 5.7 and incubated at 28 °C for another day.
3. Cultured cells were spun down at 2500× *g* for 5 min at room temperature and resuspended with 1 mL of 10 mM MES at pH 5.7; this washing step was repeated one more time.
4. After the wash, cells were resuspended in 1 mL of infiltration medium (IM; 10 mM MES/pH 5.7 with 10 mM Acetosyringone).
5. Cell density was adjusted to 0.5 of optical density at 600 nm (OD_{600}) by adding IM and being further incubated at room temperature in a rotary incubator at 250 rpm for around 2 h.
6. Final mix of IM was prepared by mixing the IM containing all the target genes at an equal ratio (*see* **Note 2**). For example, if two genes are intended for expression together with a gene-silencing suppressor (such as pBA-HcPro), three cultures are mixed equally in volume.
7. Leaves of *N. benthamiana* were infiltrated with the final mix of IM via a needleless syringe, and the infiltrated plant was returned to a growth chamber.
8. If an inducible construct was used, an inducing chemical, such as 30 μM estradiol for pER8 constructs in 0.1% Tween-20, was sprayed the next day.
9. Agroinfiltrated plants were incubated for two days regardless of the presence of an inducing chemical.

3.2 Preparing Protein Extract and Co-IP

1. Agroinfiltrated leaves were homogenized with mortar and pestle in liquid nitrogen.
2. Homogenized samples were further ground for at least 3 min in 750 μL Extraction Buffer; while Extraction Buffer is frozen

when initially added, it should become liquid by the time this step is completed.

3. Next, 0.5 mL of supernatant from homogenized protein extracts after centrifugation at 5000× *g* at 4 °C for 5 min was then subject to size exclusion chromatography by using Illustra NAP-5 Sephadex G-25 columns (*see* **Note 3**).
4. Eluates were collected in 1.5 mL Eppendorf tubes. Then 20 μL of anti-mouse IgG agarose beads were added and incubated at 4 °C for 2 h with a mild rotation using a rotator to perform precleaning to reduce background (<30 rpm).
5. Target antibody beads, such as anti-c-Myc agarose affinity gel, were washed three times with 1 mL of IP buffer during the incubation described above (20 μL per sample; *see* **Note 4**).
6. Anti-mouse IgG-agarose beads for precleaning were removed by spinning at 5000× *g* at 4 °C for 5 min.
7. An aliquot of 50 μL was saved for input.
8. Then 20 μL of agarose beads conjugated with a target antibody in solid volume were added and incubated at 4 °C overnight with mild rotation using a rotator (<30 rpm).
9. IPed proteins were collected by spinning at 5000× *g* at 4 °C.
10. Washing was performed six times with 1 mL of IP buffer.
11. The final washing step was carried out by incubating for 10 min at room temperature, in which the washing solution was removed as much as possible from the antibody beads (*see* **Note 4**).
12. Finally, 30 μL of 4× SDS sample buffer with DTT (66 mg/mL) was added to release the bound proteins for immunoblotting analysis.

3.3 Immunoblotting Analysis

1. Samples were boiled for 5 min and centrifuged at 21,100× *g* for 1 min.
2. Then 5 μL or less of supernatant was loaded into sodium dodecyl sulfate–polyacrylamide gel electrophoresis (SDS-PAGE) along with a protein size marker.
3. SDS-PAGE was run as previously described [29].
4. Gel was transferred onto a polyvinylidene fluoride (PVDF) membrane (Immobilon).
5. After the completed transfer, the membrane was soaked in methanol, washed with PBS three times for 5 min, and subjected to a final wash with 3B1M Buffer for 30 min at room temperature.

6. The membrane was then incubated at 4 °C in 3B1M Buffer with a target antibody overnight with gentle shaking (<30 rpm) (*see* **Note 5**).
7. The membrane was washed with PBS, PBS-T, and PBS buffer twice, each for 5 min for a total washing of 30 min.
8. The membrane was developed with enhanced chemiluminescence (ECL) plus solution and imaged using an imaging station (Azure Biosystems, C600); alternatively, a conventional film can also be used.

4 Notes

1. The coding sequences of the *SMC* and *MORC1* genes were cloned into pER8 by using AscI sites. Triple HA or sextuple Myc epitopes were added at the 3′ and 5′ ends, respectively.
2. IM Solutions were equally mixed with a 1:1:1 ratio (SMC-HA [or HA-GFP]: Myc-MORC1: pBA-HcPro).
3. The column was pre-washed three times with 2.5 mL of IP buffer.
4. The solution was removed with a gel-loading tip as much as possible; the tip was in contact with the bottom of the tube.
5. Also, 1:5000 dilution was used for HRP-conjugated anti-Myc and anti-HA antibodies.

Acknowledgments

We thank Angela H. Kang for critical comments on the manuscript. This work is supported by National Science Foundation Grant (IOS-1553613) to HGK.

References

1. Struk S, Jacobs A, Sánchez Martín-Fontecha E et al (2019) Exploring the protein–protein interaction landscape in plants. Plant Cell Environ 42:387–409
2. Syafrizayanti BC, Hoheisel JD et al (2014) Methods for analyzing and quantifying protein–protein interaction. Expert Rev Proteomics 11:107–120
3. Carvunis AR, Charloteaux B, Dreze M et al (2011) Evidence for network evolution in an arabidopsis interactome map. Science 333
4. Braun P, Aubourg S, Van Leene J et al (2013) Plant protein interactomes. Annu Rev Plant Biol 64:161–187
5. Mukhtar MS, Carvunis AR, Dreze M et al (2011) Independently evolved virulence effectors converge onto hubs in a plant immune system network. Science 333:596–601
6. Morsy M, Gouthu S, Orchard S et al (2008) Charting plant interactomes: possibilities and challenges. Trends Plant Sci 13:183–191
7. Citovsky V, Lee LY, Vyas S et al (2006) Subcellular localization of interacting proteins by bimolecular fluorescence complementation in planta. J Mol Biol 362:1120–1131
8. Kerppola TK (2006) Visualization of molecular interactions by fluorescence complementation. Nat Rev Mol Cell Biol 7:449–456

9. Gookin TE, Assmann SM (2014) Significant reduction of Bi FC non-specific assembly facilitates in planta assessment of heterotrimeric G-protein interactors. The Plant J 80:553–567
10. Kang HG, Oh C-S, Sato M et al (2010) Endosome-associated CRT1 functions early in resistance gene–mediated defense signaling in Arabidopsis and tobacco. Plant Cell 22:918–936
11. Kang HG, Kuhl JC, Kachroo P et al (2008) CRT1, an Arabidopsis ATPase that interacts with diverse resistance proteins and modulates disease resistance to turnip crinkle virus. Cell Host Microbe 3:48–57
12. Jia N, Zhu Y, Xie F (2018) An efficient protocol for model legume root protoplast isolation and transformation. Front Plant 9:670
13. Lin HY, Chen JC, Fang SC (2018) A protoplast transient expression system to enable molecular, cellular, and functional studies in Phalaenopsis orchids. Front Plant 9:843
14. Kwon TH, Seo JE, Kim J et al (2003) Expression and secretion of the heterodimeric protein interleukin-12 in plant cell suspension culture. Biotechnol Bioeng 81:870–875
15. Norkunas K, Harding R, Dale J et al (2018) Improving agroinfiltration-based transient gene expression in Nicotiana benthamiana. Plant Methods 14:1–14
16. Tyurin AA, Suhorukova AV, Kabardaeva KV et al (2020) Transient gene expression is an effective experimental tool for the research into the fine mechanisms of plant gene function: advantages, limitations, and solutions. Plan Theory 9:1187
17. Wang YC, Yu M, Shih PY et al (2018) Stable pH suppresses defense signaling and is the key to enhance agrobacterium-mediated transient expression in Arabidopsis seedlings. Sci Rep 8:1–9
18. Moffett P, Farnham G, Peart J et al (2002) Interaction between domains of a plant NBS–LRR protein in disease resistance-related cell death. The EMBO J 21:4511–4519
19. Collier SM, Moffett P (2009) NB-LRRs work a "bait and switch" on pathogens. Trends Plant Sci 14:521–529
20. Zuo J, Niu QW, Chua NH (2000) An estrogen receptor-based transactivator XVE mediates highly inducible gene expression in transgenic plants. The Plant J 24:265–273
21. Aoyama T, Chua NH (1997) A glucocorticoid-mediated transcriptional induction system in transgenic plants. The Plant J 11:605–612
22. Kang HG, Woo Choi H, Von Einem S et al (2012) CRT1 is a nuclear-translocated MORC endonuclease that participates in multiple levels of plant immunity. Nat Commun 3:1–11
23. Lorković ZJ, Naumann U, Matzke AJM et al (2012) Involvement of a GHKL ATPase in RNA-directed DNA methylation in Arabidopsis thaliana. Curr Biol 22:933–938
24. Bolaños-Villegas P (2021) The role of structural maintenance of chromosomes complexes in meiosis and genome maintenance: translating biomedical and model plant research into crop breeding opportunities. Front Plant 563
25. Uhlmann F (2016) SMC complexes: from DNA to chromosomes. Nat Rev Mol Cell Biol 17:399–412
26. Jeppsson K, Kanno T, Shirahige K et al (2014) The maintenance of chromosome structure: positioning and functioning of SMC complexes. Nat Rev Mol Cell Biol 15:601–614
27. Schultz J, Milpetz F, Bork P et al (1998) SMART, a simple modular architecture research tool: identification of signaling domains. PNAS 95:5857–5864
28. Hirano M, Hirano T (2002) Hinge-mediated dimerization of SMC protein is essential for its dynamic interaction with DNA. The EMBO J 21:5733–5744
29. Sambrook J, Fritsch EF, Maniatis TE (1989) Molecular cloning: a laboratory manual. 2. Cold Spring Harbor Laboratory Press, NY

Chapter 10

Detection of Protein–Protein Interactions Using Glutathione-S-Transferase (GST) Pull-Down Assay Technique

Prabu Gnanasekaran and Hanu R. Pappu

Abstract

Pull-down assay is a technique to analyze direct protein–protein interaction under in vitro condition. Also, this technique is appropriate for investigating the direct interaction between two purified proteins. Glutathione-s-transferase (GST) protein is a widely used affinity tag for affinity purification. In this chapter, we explain the widely used GST pull-down assay to identify the protein–protein interaction between purified proteins.

Key words Glutathione-s-transferase, Affinity purification, Direct protein, protein interaction, Pull-down

1 Introduction

Glutathione-s-transferase (GST) pull-down assay is a widely used technique to detect the direct and specific interaction between two proteins or protein subunits in vitro [1]. It involves affinity purification of tagged protein or protein complexes from the biological sample. To study the interaction between proteins of interest, for instance, protein X is fused to a GST-tag and other putative interacting partners are fused to any affinity tag (histidine, maltose-binding protein [MBP] tag, etc.). Affinity tags bind specifically to the ligands immobilized on the beads. GST alone or GST-tagged protein binds specifically to the glutathione beads [2, 3]. For instance, while performing an interaction study between GST-tagged protein X and MBP-tagged protein Y, glutathione beads (binds specifically to GST or GST-tagged proteins) are used. If there is an interaction between GST-X and MBP-Y fusion proteins, during affinity purification using glutathione beads, MBP-Y protein gets pulled-down along with GST-X protein. If

Shahid Mukhtar (ed.), *Protein-Protein Interactions: Methods and Protocols*, Methods in Molecular Biology, vol. 2690, https://doi.org/10.1007/978-1-0716-3327-4_10,

there is no interaction between GST-X and MBP-Y fusion proteins, during affinity purification using glutathione beads, MBP-Y protein could not get pulled-down along with GST-X protein.

2 Materials

2.1 Protein Expression and Purification

1. Protein expression constructs:
   - pGEX-6p2 plasmid (contains coding sequence of GST protein).
   - pMAL-C2X plasmid (contains coding sequence of MBP protein).
   - pGEX-6p2-X: Generate this construct by cloning the open reading frame (ORF) coding for the protein of interest X in frame with the GST-tag in the pGEX vector.
   - pMAL-C2X-Y: Generate this construct by cloning the ORF coding for the protein of interest Y in frame with the MBP-tag in the pMAL vector.
2. *Escherichia coli* protein expression strain BL21 (DE3) competent cells.
3. Glutathione-sepharose resin.
4. Amylose resin.
5. Isopropyl β-D-1-thiogalactopyranoside (IPTG).
6. Luria Broth.
7. Lysis buffer containing 50 mM Tris (pH 7.4), 100 mM NaCl, 5% glycerol, 5 mM 2-mercaptoethanol, and 1 mM phenylmethylsulfonyl fluoride.
8. Wash buffer containing 50 mM Tris (pH 7.4), 300 mM NaCl, 5% glycerol, 5 mM 2-mercaptoethanol, and 1 mM phenylmethylsulfonyl fluoride.
9. GST protein elution buffer containing 50 mM Tris (pH 7.4), 100 mM NaCl, 5% glycerol, 5 mM 2-mercaptoethanol, 1 mM phenylmethylsulfonyl fluoride, and 20 mM reduced glutathione.
10. MBP protein elution buffer containing 50 mM Tris (pH 7.4), 100 mM NaCl, 5% glycerol, 5 mM 2-mercaptoethanol, 1 mM phenylmethylsulfonyl fluoride, and 20 mM reduced glutathione.

2.2 In Vitro GST Pull-Down

1. Glutathione-sepharose resin.
2. Equilibration buffer containing 50 mM Tris (pH 7.4), 100 mM NaCl, 5% glycerol, 5 mM 2-mercaptoethanol, and 1 mM phenylmethylsulfonyl fluoride.

3. Wash buffer containing 50 mM Tris (pH 7.4), 300 mM NaCl, 5% glycerol, 5 mM 2-mercaptoethanol, and 1 mM phenylmethylsulfonyl fluoride.
4. Elution buffer containing 50 mM Tris (pH 7.4), 100 mM NaCl, 5% glycerol, 5 mM 2-mercaptoethanol, 1 mM phenylmethylsulfonyl fluoride, and 20 mM reduced glutathione.

2.3 Detection of Interacting Proteins

1. Sodium dodecyl sulfate–polyacrylamide gel electrophoresis (SDS-PAGE) gel.
2. Vertical electrophoresis apparatus.
3. Semi-dry blotting apparatus.
4. Polyvinylidene fluoride (PVDF) protein transfer membrane.
5. Anti-GST antibody.
6. Anti-MBP antibody.

3 Methods

3.1 Protein Expression and Purification

1. Transform the protein expression vector (pGEX-6p2 and pMAL-C2X) and construct (pGEX-6p2-X and pMAL-C2X-Y) plasmids into *E. coli* strain BL21-DE3 [4].
2. Grow the bacterial cells carrying protein expression construct at 37 °C and 200 rpm shaking till the optical density of nm 600 (OD_{600}) reaches approximately 0.5.
3. Add 0.5 mM IPTG to induce the expression of protein and incubate at 37 °C and 200 rpm shaking for 5–8 h. Harvest the cells by centrifuging at 4 °C and 5000 rpm for 5 min and discard the supernatant.
4. Add ten volumes of lysis buffer to the cell pellet and dissolve it.
5. Purify the GST and GST-X proteins from corresponding bacterial cell lysate using glutathione-sepharose resin. Elute the protein with GST protein elution buffer containing 20 mM reduced glutathione. Remove the glutathione from the protein solution by dialysis using a typical molecular weight cut-off dialysis membrane tube [5, 6].
6. Purify the MBP and MBP-Y proteins from corresponding bacterial cell lysate using amylose resin. Elute the protein with MBP protein elution buffer containing 20 mM maltose.

3.2 In Vitro GST Pull-Down

1. Add 100 μL of glutathione-sepharose resin to each of the four microcentrifuge tubes.
2. Add 1 mL of ice-cold equilibration buffer to each tube, incubate at 4 °C for 5 min, centrifuge at 4 °C and 1000 rpm for 5 min, and pipette out the supernatant carefully without drying the beads and discard it. Repeat this step five times.

3. Add equimolar concentration of desired protein combination (GST and MBP; GST and MBP-Y; GST-X and MBP; and GST-X and MBP-Y proteins) and the required amount of lysis buffer into different tubes that have pre-equilibrated glutathione-sepharose beads. Incubate the tubes at 4 °C for 4 h with end-over-end mixing on the rotating shaker.
4. Centrifuge the tubes at 4 °C and 1000 rpm for 5 min and discard the supernatant without disturbing the bead.
5. Add 1 mL of ice-cold wash buffer to each tube, incubate at 4 °C for 5 min, and centrifuge at 4 °C and 1000 rpm for 5 min and discard the supernatant. Repeat this step five times.
6. Elute the GST/GST-fusion protein and any other bound protein with 50 μL of elution buffer (*see* **Note 1**).

3.3 Detection of Interacting Proteins

1. Load 10% of the input and pull-down proteins eluted from control and test combinations (GST and MBP; GST and MBP-Y; GST-X and MBP; and GST-X and MBP-Y proteins) on the sodium dodecyl sulfate (SDS)-polyacrylamide gel electrophoresis (PAGE) gel and run on vertical electrophoresis apparatus. Stain the gel with Coomassie Brilliant blue and analyze the results.
2. Detection of protein associated with GST-fusion protein could be further confirmed by western blotting. Transfer the protein from SDS-PAGE gels to PVDF protein transfer membrane using semi-dry blotting apparatus and perform immunoblotting with anti-GST antibody and anti-MBP antibody (*see* **Notes 2** and **3**).

4 Notes

3. Similar results are obtained while analyzing the GST pull-down assay elutes by SDS-PAGE and immunoblotting. In SDS-PAGE, presence of expected protein is confirmed by comparing its size with the protein ladder, while in immunoblotting presence of expected protein is confirmed using specific antibodies.
4. If MBP-Y protein gets pulled-down with GST-X protein and not with GST alone, it suggests that protein X and Y has direct and specific interaction (Fig. 1).
5. If MBP-Y protein does not get pulled-down with GST-X protein, it suggests that there is no direct interaction between protein X and Y (Fig. 1).

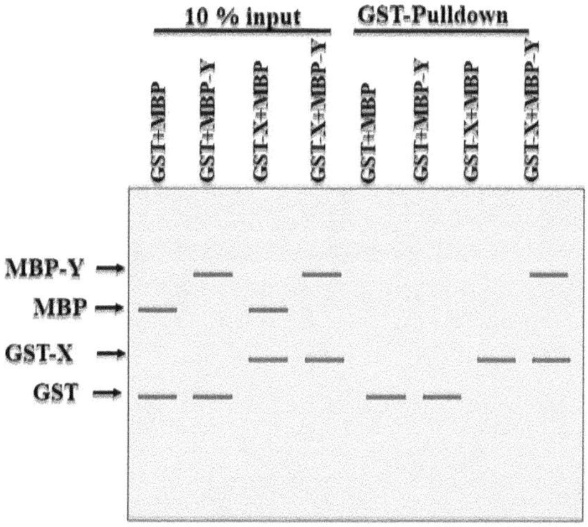

Fig. 1 Illustration of GST pull-down assay on SDS-PAGE demonstrating an interaction between proteins X and Y with proper negative controls

References

1. Kaelin WG Jr, Pallas DC, DeCaprio JA et al (1991) Identification of cellular proteins that can interact specifically with the T/E1A-binding region of the retinoblastoma gene product. Cell 64(3):521–532
2. Rodriguez MM, Ron D, Touhara K et al (1999) RACK1, a protein kinase C anchoring protein, coordinates the binding of activated protein kinase C and select pleckstrin homology domains in vitro. Biochemistry 38(42): 13787–13794
3. Dekker LV, Parker PJ (1997) Regulated binding of the protein kinase C substrate GAP-43 to the V0/C2 region of protein kinase C-delta. J Biol Chem 272(19):12747–12753
4. Gnanasekaran P, Ponnusamy K, Chakraborty S (2019) A geminivirus betasatellite encoded betaC1 protein interacts with PsbP and subverts PsbP-mediated antiviral defence in plants. Mol Plant Pathol 20(7):943–960
5. Gnanasekaran P, Gupta N, Ponnusamy K et al (2021) Geminivirus betasatellite-encoded betaC1 protein exhibits novel ATP hydrolysis activity that influences its DNA-binding activity and viral pathogenesis. J Virol 95(17):e0047521
6. Ruhel R, Mazumder M, Gnanasekaran P et al (2021) Functional implications of residues of the B' motif of geminivirus replication initiator protein in its helicase activity. FEBS J 288(22): 6492–6509

Chapter 11

Bimolecular Fluorescence Complementation (BiFC) Assay to Visualize Protein–Protein Interactions in Living Cells

Prabu Gnanasekaran and Hanu R. Pappu

Abstract

Bimolecular fluorescence complementation (BiFC) assay is a method to visualize the protein–protein interaction in living cells. This technique is based on ability of the non-fluorescent fragment of fluorescent protein to form fluorescent complex when they are fused to two interacting proteins. In this chapter, we describe the widely used split yellow fluorescent protein (YFP) system to visualize the protein–protein interaction in plant cells.

Key words BiFC, Reconstituted fluorescence, In vivo interactions, YFP

1 Introduction

Bimolecular fluorescence complementation (BiFC) is an important technique to verify protein–protein interaction in vivo in the most native system. BiFC is based on the reconstitution of non-fluorescent N- and C-terminal fragments of fluorescent protein to functional fluorescent protein. Green fluorescent protein (GFP) and its variants such as yellow fluorescent protein (YFP), cyan fluorescent protein (CFP), and Venus fluorescent protein could be used for BiFC assay. Several host systems such as animal cell lines, protoplast cultures, plant leaves, and onion epidermal cells can be used as in vivo systems to study the interactions [1–3]. To study the interaction, proteins of interest, protein X and protein Y, are fused to N-terminal and C-terminal fragments of the fluorescent protein, for instance, YFP. If there is an interaction between X and Y proteins, N- and C-terminal fragments of YFP reconstitute to form functional YFP, resulting in the exhibition of YFP fluorescence while observed under the confocal fluorescent microscope (Fig. 1). If there is no interaction between X and Y

Shahid Mukhtar (ed.), *Protein-Protein Interactions: Methods and Protocols*, Methods in Molecular Biology, vol. 2690,
https://doi.org/10.1007/978-1-0716-3327-4_11,

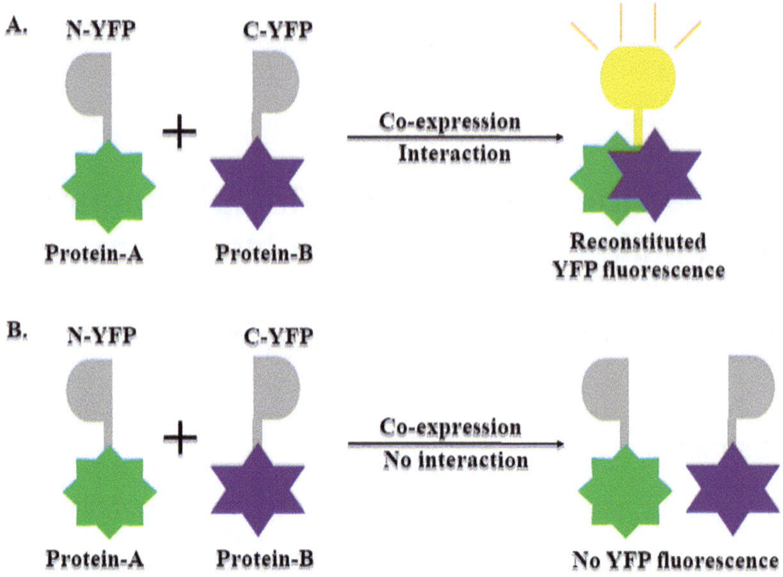

Fig. 1 Schematic representation of principle of BiFC assay

proteins, N- and C-terminal fragments of YFP do not reconstitute to form functional YFP, resulting in no YFP fluorescence while observed under the confocal fluorescent microscope (Fig. 1).

2 Materials

1. BiFC constructs:
   - pSITE-nEYFP-N1 plasmid (contains coding sequence of N-terminal fragment of YFP protein).
   - pSITE-nEYFP-C1 plasmid (contains coding sequence of C-terminal fragment of YFP protein).
   - pSITE-nEYFP-N1-X: Generate this construct by cloning the open reading frame (ORF) coding for the protein of interest X in frame with the N-terminal fragment of YFP protein in the pSITE-nEYFP-N1 vector.
   - pSITE-nEYFP-C1-Y: Generate this construct by cloning the ORF coding for the protein of interest Y in frame with the C-terminal fragment of YFP protein in the pSITE-nEYFP-C1 vector.
2. *Agrobacterium tumefaciens* strain GV3101 competent cells.
3. Infiltration buffer:
   - 10 mM MES (2-(N-morpholino)ethanesulfonic acid) (pH 5.7).
   - 10 mM $MgCl_2$.
   - 100 μM acetosyringone.

4. Three- to four-week-old *Nicotiana benthamiana* plants.
5. Negative control combination: The common challenge of BiFC assay is an occasional false positive due to self-assembly of the N- and C-terminal fluorescent proteins. This will be overcome by using appropriate negative controls and by comparing the fluorescence in test and control samples both quantitatively and qualitatively. Appropriate negative control [4] combinations are as follows:
   - pSITE-nEYFP-N1 vector + pSITE-nEYFP-C1 vector.
   - pSITE-nEYFP-N1-X + pSITE-nEYFP-C1 vector.
   - pSITE-nEYFP-N1 vector + pSITE-nEYFP-C1-Y.
6. Test combination:
   - pSITE-nEYFP-N1-X + pSITE-nEYFP-C1-Y.

3 Methods

1. Transform the BiFC constructs into *Agrobacterium tumefaciens* strain GV301 cells.
2. Grow the *Agrobacterium* cells carrying BiFC construct at 28 °C and 200 rpm shaking for overnight in Luria Broth media containing spectinomycin and rifampicin.
2. Harvest the overnight grown *Agrobacterium* cells and make the desired infiltration combination in infiltration buffer (using 0.5 optical density of 600 nm [OD_{600}] of cells of each construct).
3. Mix gently and incubate in dark for 2–3 h.
4. Infiltrate the cells containing BiFC constructs on the abaxial (lower) side of the leaves of *N. benthamiana* plants.
5. After 48 h post-infiltration, visualize the lower epidermal cells of infiltrated *N. benthamiana* leaves under the confocal fluorescent microscope. To visualize the reconstitution of YFP fluorescence, use 514 nm excitation and 530–570 nm emission filters [5] (see **Notes 1–4**).

4 Notes

1. If YFP fluorescence is observed in the leaves co-infiltrated with pSITE-nEYFP-N1-X + pSITE-nEYFP-C1-Y, this implies the reconstitution of YFP due to the interaction between X and Y proteins.
2. If no YFP fluorescence is observed in the leaves co-infiltrated with pSITE-nEYFP-N1-X + pSITE-nEYFP-C1-Y, this implies there is no interaction between X and Y proteins.

3. YFP fluorescence is not expected from the leaves that are co-infiltrated with negative control combination pSITE--nEYFP-N1 vector + pSITE-nEYFP-C1 vector, pSITE-nEYFP-N1-X + pSITE-nEYFP-C1 vector, or pSITE-nEYFP-N1 vector + pSITE-nEYFP-C1-Y.
4. To perform BiFC assay in animal cell lines, use BiFC vectors and transfection procedures that are specific to animal cells.

References

1. Kerppola TK (2008) Bimolecular fluorescence complementation (BiFC) analysis as a probe of protein interactions in living cells. Annu Rev Biophys 37:465–487
2. Schütze K, Harter K, Chaban C (2009) Bimolecular fluorescence complementation (BiFC) to study protein-protein interactions in living plant cells. Methods Mol Biol 479:189–202
3. Gnanasekaran P, Ponnusamy K, Chakraborty S (2019) A geminivirus betasatellite encoded betaC1 protein interacts with PsbP and subverts PsbP-mediated antiviral defence in plants. Mol Plant Pathol 20(7):943–960
4. Horstman A, Tonaco IA, Boutilier K et al (2014) A cautionary note on the use of split-YFP/BiFC in plant protein-protein interaction studies. Int J Mol Sci 15(6):9628–9643
5. Gupta N, Reddy K, Gnanasekaran P, Zhai Y et al (2022) Functional characterization of a new ORF βV1 encoded by radish leaf curl betasatellite. Frontiers. Plant Sci:13

Chapter 12

Detecting Protein–Protein Interactions Using Bimolecular Fluorescence Complementation (BiFC) and Luciferase Complementation Assays (LCA)

Pepijn Bais, Louai Alidrissi, and Ikram Blilou

Abstract

In multicellular organisms, establishing the full body plane involves cell-cell signaling where protein associations are important for the diverse cellular functions within the cells. For the study of protein–protein interactions (PPI), bimolecular fluorescence complementation (BiFC) and luciferase complementation assays (LCA) have proven to be reliable tools that can be used to confirm the physical association of two proteins in a semi-in vivo environment. This chapter provides a detailed description of these two techniques using *Nicotiana benthamiana* as a semi-in vivo transient expression system. As an example, we will use the interaction of the two well-described transcription factors SHORT-ROOT (SHR) and SCARECROW (SCR), which are known as regulators of asymmetric cell division and stem cell specification in the root meristem of the model plant *Arabidopsis thaliana*. While the BiFC assay provides subcellular information by displaying a fluorescence signal, nuclear in this case, resulting from the reconstituted fluorophore, the LCA generates a quantitative readout of the SCR–SHR interaction. The combination of both assays provides information on the localization and strength of the PPI.

Key words PPIs, BiFC, LCA, Tobacco, Luciferase, YFP, Luminescence, Fluorescence

1 Introduction

In plants, where cell migration does not occur, cell type specification relies predominantly on intricate signaling mechanisms where protein–protein interaction (PPI) studies have been instrumental to understanding a myriad of biological processes within cells, tissue, and organs [1, 2]. Multi-omics approaches have contributed greatly toward understanding signaling pathways. Genomic and transcriptomic studies have led to the characterization of new traits in various species and the identification of gene function in many

Authors Pepijn Bais and Louai Alidrissi have equally contributed to this chapter.

Shahid Mukhtar (ed.), *Protein-Protein Interactions: Methods and Protocols*, Methods in Molecular Biology, vol. 2690, https://doi.org/10.1007/978-1-0716-3327-4_12,

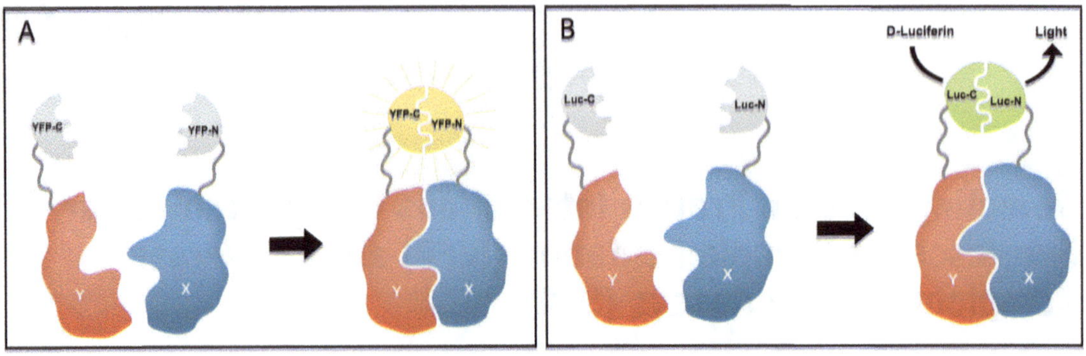

Fig. 1 Schematic representation of BiFC and LCA. (**a**) In BiFC, the C-terminal and N-terminal parts of the YFP fluorophore are reconstituted when proteins X and Y interact. The emitted fluorescence signal indicates protein interaction. (**b**) In LCA, when proteins X and Y interact, they reconstitute the functional luciferase enzyme. This luciferase is capable of converting D-Luciferin (= the substrate) to produce light

plant tissues [3, 4]. In this context, protein–protein interaction plays a key role in predicting the protein function in a given biological process. In recent years, spatial and temporal information of these associations at the cellular level has become of increasing interest to biologists [5]. Within a cell, PPIs are heavily influenced by changes caused by intrinsic and extrinsic cues. Therefore, many tools have been generated to study and understand PPI dynamics. Combining these approaches provides new insights that will help in understanding the biological relevance of PPIs [6]. PPIs can be studied either in vitro or in vivo. Subsequently, they can also be tested in a heterologous system to rule out the effect of endogenous co-factors on the signaling pathway.

In this chapter, we describe two methods widely used to study PPIs: bimolecular fluorescence complementation (BiFC) and luciferase complementation assays (LCA) (Fig. 1). Compared to other PPI techniques, BiFC provides information on the subcellular localization of PPIs, whereas LCA allows for both qualitative and quantitative assessment of the interactions [7, 8]. Throughout the years, BiFC and LCA have proven to be highly sensitive and robust methods to study PPIs [9, 10]. Both techniques are based on the reconstitution of functional protein reporters that are a fluorophore (in BiFC) or luciferase (in LCA), which have been split into two domains: an N-terminal domain and a C-terminal domain and fused to proteins of interest [11]. Reconstitution of reporter protein can only take place when both domains are in close proximity to each other, which occurs exclusively when two proteins of interest physically interact (Fig. 1). In BiFC, this yields a functional fluorescent protein whose light emission can be detected using a confocal microscope [12]. For LCA, luciferase complementation allows the conversion of D-Luciferin into light which is detected by a charged-coupled device (CCD) camera/luminometer [13]. Co-expression of potentially interacting proteins can be achieved by using protoplasts [13] or *Nicotiana benthamiana* epidermal cells using *Agrobacterium tumefaciens* to transfect the proteins of interest into the cell [14]. This allows the study of PPIs in plant tissue upon transient expression.

Below, we describe the experimental procedures for both BiFC and LCA. We illustrate these two techniques using the well-described interaction between the two developmental regulators SHORT-ROOT (SHR) and SCARECROW (SCR). SHR is a mobile transcription factor that plays essential roles in plant development [15]. In roots, SHR is synthesized in the vascular bundle and moves outward into the ground tissue [16]. SCR is expressed in the endodermis and delimits the radial movement of SHR through nuclear sequestration in nuclei of the endodermis tissue layer [17]. The spatial control of SHR activity is mediated by SCR and the C2H2 nuclear factors from the Indeterminate Domain proteins, also known as the BIRD proteins [18–20]. This restriction requires PPIs between SCR, the BIRD proteins, and SHR, and it is important for tissue boundaries delimitation and proper radial tissue patterning in *Arabidopsis* roots [18–20]. In the case of BiFC, the physical association between SCR and SHR is expected to result in YFP complementation [18, 19]. We also include a negative control comprising SCR and the breast cancer type 1 susceptibility protein (BRCA1) which have been shown not to interact [21]. For LCA, similar to the BiFC principle, protein interaction would bring the n-LUC and c-LUC domains in close proximity which results in luciferase reconstitution. This is subsequently detected as the luminescent light signal which serves to visualize and quantify PPIs. The workflow for both methods is described in Fig. 2.

2 Materials

2.1 Common Materials for BiFC and LCA

1. LB Agar plates (in 1 liter: 1.5 g agar, 10 g NaCl, 10 g tryptone, 5 g yeast extract) supplemented with corresponding antibiotic (s) (kanamycin: 50 μg/mL, rifampicin: 50 μg/mL).
2. LB liquid medium (in 1 liter: 10 g NaCl, 10 g tryptone, 5 g yeast extract) supplemented with corresponding antibiotic (s) (kanamycin: 50 μg/mL, rifampicin: 50 μg/mL).
3. 200 mM acetosyringone solution.
4. Sterile 10 mM $MgCl_2$ solution.
5. Sterile 0.5 M MES solution adjusted to pH 5.7.
6. Electrocompetent *Agrobacterium tumefaciens* strain GV3101.
7. Plasmids containing the proteins fused to the YFP N and C terminal.

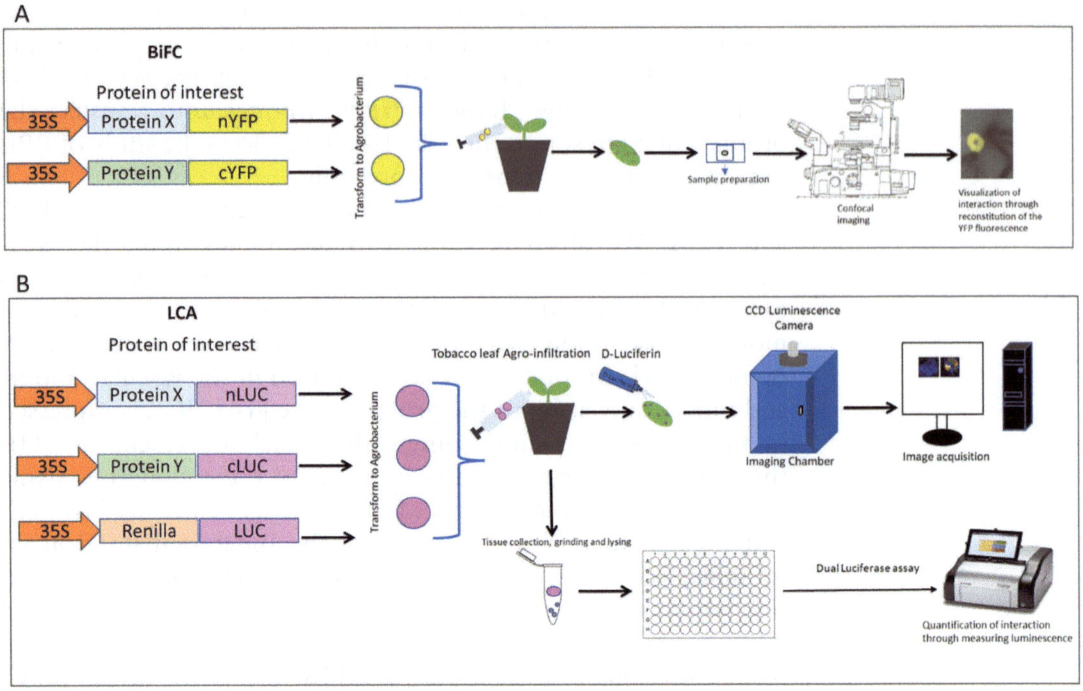

Fig. 2 Schematic representation showing the workflow for BiFC (**a**) and LCA (**b**). Adapted from [25]

8. p19 helper plasmid [22, 23] (*see* **Note 1**).
9. 4–5 weeks old healthy *Nicotiana benthamiana* plants
10. Sterile 90 × 15 mm disposable Petri dishes.
11. 15 and 50 mL conical falcon tubes.
12. Sterile 1 mL disposable syringes.
13. Nitrile gloves.
14. 0.1 cm gap Gene Pulser Electroporation cuvettes.
15. The growth chamber is set to a long day cycle (16 h light/8 h dark) at 22 °C.
16. MicroPulser electroporator.
17. The shaking incubator was set at 28 °C, 220 rpm.
18. Benchtop centrifuge.
19. Spectrophotometer with corresponding disposable cuvettes (optical density measured at OD_{600}).

2.2 Materials Specific to BiFC

1. Confocal LSM microscope (in this study, a Leica STELLARIS 8 FALCON microscopy system was used).
2. Slides and cover slides.

Table 1
Constructs used in this study

Construct name	Vector	Resistance in bacteria	References
SHR c-YFP	p35S_GW_VYCE	Kanamycin	[20]
SCR n-YFP	35S_GW_VYNE	Kanamycin	[20]
SHR n-Luc	pGWB-nLUC/(Plasmid #174050)	Kanamycin	[26]; this study
SCR c-Luc	pGWB-cLUC/(Plasmid #174051)	Kanamycin	[26]; this study
P19	pCB301-p19	Kanamycin	[22]
Renilla-LUC	*35S::Renilla:LUC*	Kanamycin	[24]

2.3 Materials Specific to LCA

1. Floating microtubes racks/floaters.
2. Thermal Protection Gloves for Low Temperature.
3. Vortex.
4. Orbital shaker.
5. Bench-top centrifuge for 2 mL microtubes.
6. Long forceps.
7. The plasmid containing proteins fused to LUC (N and C terminal) (Table 1).
8. 35S Renilla Luciferase constructs [24].
9. Metal beads.
10. TissueLyser II.
11. GloMax® Navigator Microplate Luminometer with fitted dual injector pump.
12. Dual-Luciferase® Reporter Assay System containing:
    - Passive lysis buffer (PLB) 5X.
    - Luciferase assay buffer II (LARII) containing dissolved Luciferase assay substrate (*see* **Note 2**).
    - 1x Stop & Glo® reagent.
13. D-Luciferin (D-Luciferin, Monosodium Salt; 100 mg).
14. Biopsy punch with a plunger (Integra Miltex, Disposable Biopsy punch with plunger, 4 mm, 33-34-P/25).
15. Bioluminescence in vivo imaging system consisting of imaging chamber, low-light CCD camera (cooled to –85 °C) (we used an Andor camera), a camera controller, refrigeration unit, acquisition computer, and a monitor in addition to the Andor Solis (i) software platform or any bioluminescence acquisition software.

3 Methods

3.1 Cloning

1. Clone the proteins of interest into a binary vector containing either the YFP or LUC parts.

   In this chapter we used SHR and SCR coding sequences (CDS).
2. Amplify the CDA from seedling's cDNA and subclone into a pDONR vector (we used pGEMTeasy 221) to generate an entry clone containing the proteins of interest.
3. Generate the expression clones by performing an LR recombination reaction between the entry clone and a Gateway destination vector (Table 1).
4. Use the 35S: Renilla:LUC for normalization and a p19 helper plasmid to inhibit silencing.
5. Transform each construct independently into Agrobacterium.

3.2 Infiltration of Nicotiana benthamiana Leaves

1. Grow *Nicotiana benthamiana* plants in a growth chamber or greenhouse for 4–5 weeks (*see* **Note 3**).
2. Transform each constructed vector into electrocompetent *A. tumefaciens* using standard electroporation methods.
3. Prepare LB medium supplemented with the corresponding antibiotics. Inoculate 5 mL of LB with a single positive colony and incubate in a shaking incubator at 28 °C for 2–3 days.
4. Use 500 μL of the cultures prepared in **step 3** to individually inoculate 15 mL of LB medium (in 50 mL falcon tubes) supplemented with 10 mM MES and 40 μM acetosyringone, and grow for 16 h at 28 °C (*see* **Note 4**).
5. Centrifuge the culture tubes at 3200× *g* for 10 min. Decant the supernatant.
6. Resuspend the remaining *Agrobacterium* pellets in 10 mM $MgCl_2$ solution and measure OD_{600} using a spectrophotometer.
7. Adjust the OD_{600} of each resuspended *Agrobacterium* culture by adding more $MgCl_2$ 10 mM solution to reach $OD_{600} = 1.5$ for n- and c- YFP and Luc vectors containing each protein, and reach $OD_{600} = 1.0$ for the P19 helper plasmid and 0.2 for rLuc.
8. Add acetosyringone to each resuspended culture to reach a final concentration of 200 μM. Keep the suspensions in the dark for 3 h without shaking.
9. For BiFC co-infiltration, mix the resuspended cultures containing the proteins of interest P19 helper plasmid in a 1:1:1 volume ratio, respectively.

10. For quantitative measurement of luciferase activity, mix the resuspended cultures containing n-Luc, c-Luc, P19 helper plasmid, and rLuc in a 1:1:1:0.2 volume ratio, respectively. Also infiltrate leaf with single plasmids containing r-Luc and c-Luc as control.
11. Infiltrate *Nicotiana benthamiana* epidermal cells by gently injecting the resuspended cultures into the abaxial side (= lower side) of the leaf using a needleless 3 mL syringe. Make sure that the whole surface of the leaf is infiltrated (*see* **Notes 5** and **6**; Fig. 2).
12. Place the infiltrated plants in growth chamber conditions. After 2–3 days, the leaves can be analyzed for fluorescence/luminescence using a confocal microscope or CCD camera, respectively (Fig. 3).

3.3 Imaging BiFC Samples

1. Cut a leaf disk using a biopsy punch with a plunger, or use a 15 mL falcon tube to cut the leaf disks by pressing a tube into the infiltrated leaf.
2. Mount the leaf disk in a glass slide containing 40–100 mL of water.
3. Image the samples using a confocal laser scanning microscope. In this study, we used a Leica SP8 the Stellaris Falcon fitted with an Objective HC PL APO 63x1,2 W CORR UVIS CS2.
4. YFP signal was excited at 514 nm, and the emission was between 525–550 nm.

3.4 Detection of LCA

3.4.1 Visualization of LCA in Leaf

1. Prepare a 200× Luciferin stock solution (30 mg/mL) in sterile water. Dilute in sterile water (1:200) prior to use.
2. Turn on the imaging system (*see* **Note 7**).
3. Once the imaging system is ready, carefully collect one infiltrated tobacco leaf from each combination (*see* **Note 8**).
4. Spray with D-Luciferin and incubate for 10 min at room temperature in the dark.
5. Put the plate inside a dark chamber containing the CCD imaging apparatus and capture luminescence by setting the light-field exposure time to 50 ms and the luminescence detection time to 1 min (*see* **Notes 9** and **10**).
6. Collect the signal using the bioluminescence in vivo imaging system following the manufacturer's instructions.
7. Save the raw data with the software format and export the images in TIFF or JPEG format.
8. Luminescence signal indication protein interactions can be observed in Fig. 3.

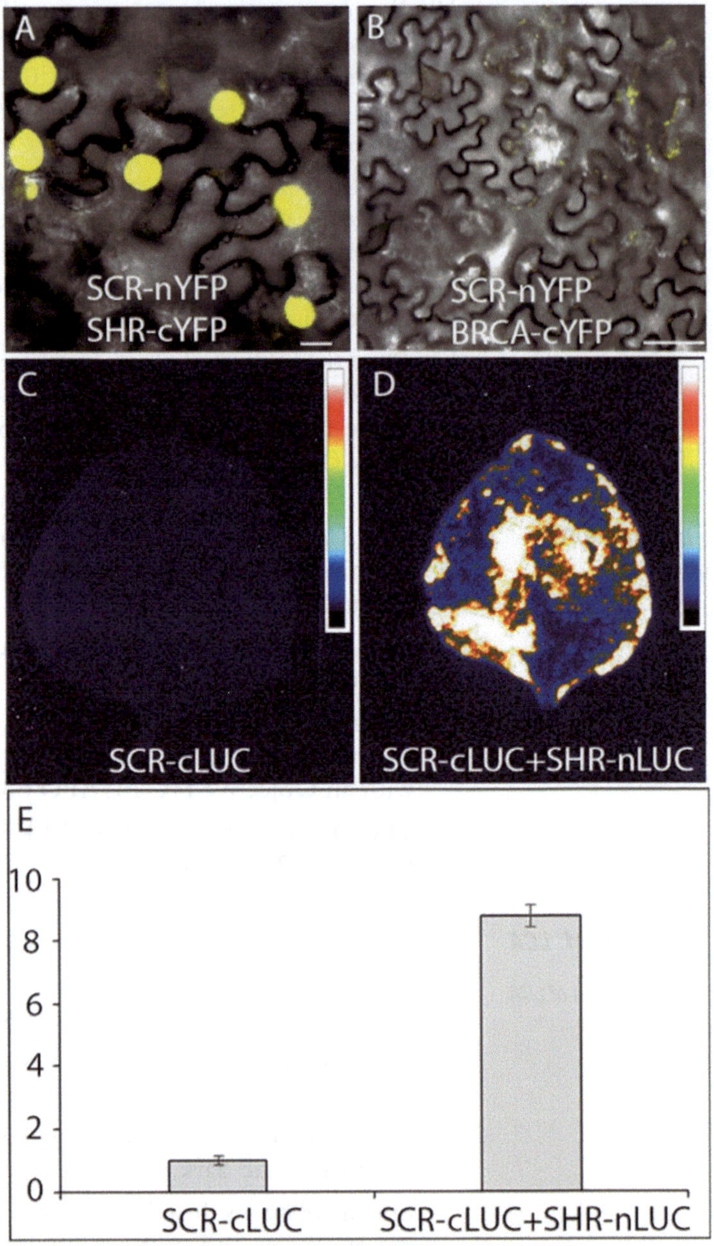

Fig. 3 PPI assays using BiFC and LCA. (**a**) Positive interaction was depicted by a nuclear signal upon YFP reconstitution. (**b**) No signal is observed using non-interacting proteins. (**c**) Luminescence image of a *Nicotiana benthamiana* leaf transformed with SCR-cLUC. (**d**) Luminescence image showing SHR-SCR interaction. The color bar indicates luminescence intensity from low (dark blue) to high (white). (**e**) Quantification of the interaction using Dual-Luciferase. The Y axis depicts the relative luciferase activity

3.4.2 Quantification of Protein Interaction Through Measuring Luciferase Activity

1. Pick an agro-infiltrated leaf and cut three leaf disks with a 15 mm diameter (*see* **Note 11**). Place the disks in a 2 mL Eppendorf tube containing 3 metal beads and freeze the tube in liquid nitrogen (*see* **Notes 12** and **13**).
2. Grind the frozen samples using a TissueLyser II machine set at 30 Hz for 45 s.
3. Quickly add 200 mL of PLB supplemented with 2 mM DTT and shake vigorously. Incubate samples for 10 min on ice.
4. During this time, prime the GloMax® Navigator Microplate Luminometer according to the manufacturer's instructions. Set the machine to inject 25 μL of LAR II followed by 25 μL of Stop & Glo® Reagent into each well for independent measurement of fLuc and rLuc activities. For each injection, we recommend setting the integration period at 10 s followed by 4 s of delay.
5. Vortex the samples and centrifuge at maximum speed for 3 min.
6. Transfer 75 μL of supernatant into a 96 luminometer plate and measure luciferase activity using the luminometer. The machine will generate a Word Excel file containing luminescence measurements for both fLuc and rLuc (*see* **Notes 14** and **15**).

3.4.3 Data Analysis

1. Calculate the relative ratio of fLUC/rLUC activity. This step is important as it normalizes variations introduced by different transformation efficiencies.
2. Calculate the average of the ratios from the three technical replicates for each combination.

3. Set the single control (protein of interest with fLUC or cLUC) $MgCl_2$ and without at 1, and normalize the rest of the samples against it. Calculate and normalize the standard deviations (Fig. 3).

4 Notes

1. The P19 helper plasmid encoding the P19 protein from Tomato Bushy Virus (TBSV) greatly facilitates the expression of foreign DNA in plant cells. The P19 protein is a suppressor of post-translational gene silencing (PTGS) in plants [23]. Under normal circumstances, plant cells neutralize foreign protein expression by activating PTGS. However, when the P19 helper plasmid is co-expressed in infiltrated cells, the siRNAs responsible for PTGS are inhibited [23]. In turn, the expression of your gene of interest is drastically increased. We,

therefore, recommend including P19 in your co-infiltration assays.

2. Make sure to always keep in the dark at −20 °C.
3. Watering of the soil should be done from the top down. Do not extensively soak the soil in water. Plants should have green, round leaves and be healthy. Plants that show signs of disease, stress, or pest infestation should be discarded.
4. The use of antibiotics becomes optional at this step.
5. Fluorescence/Luminescence signals are expressed at high levels up from 2 to 3 days after infiltration.
6. Nicotiana leaves that have fully expanded are used. We prefer infiltrating the third to fourth youngest leaves.
7. Some CCD cameras need at least 30 min to reach the temperature that will allow for imaging.
8. Always include a non-infiltrated leaf as a control.
9. Incubating in the dark helps to quench auto-fluorescence signals from chlorophyll.
10. It is recommended to use low temperature as it enhances the signal-to-noise ratio. Parameters vary according to the CCD camera model.
11. There is usually no expression in leaf veins, and it is recommended to avoid them.
12. It is recommended to use at least three technical replicas per sample.
13. Samples can be stored at −80 °C at this point.
14. Opening the luminometer lid can result in the loss of samples and data.
15. Repeat each experiment at least three times to generate independent biological replicates.

Acknowledgments

This study was supported by KAUST baseline research funding to I.B. The authors declare no competing financial interests.

References

1. Schwikowski B, Uetz P, Fields S (2000) A network of protein-protein interactions in yeast [Internet]. Nat Biotechnol 18. Available from: http://biotech.nature.com
2. Vidal M, Cusick ME, Barabási AL (2011) Interactome networks and human disease. Cell (Elsevier B.V.) 144:986–998
3. Morozova O, Marra MA (2008) Applications of next-generation sequencing technologies in functional genomics. Genomics 92:255–264
4. Wilhelm BT, Landry JR (2009) RNA-Seq-quantitative measurement of expression through massively parallel RNA-sequencing. Methods 48:249–257

5. Zhang Y, Gao P, Yuan JS (2010) Plant protein-protein interaction network and Interactome. Curr Genomics 11:40–46
6. Berggård T, Linse S, James P (2007) Methods for the detection and analysis of protein-protein interactions. Proteomics 7:2833–2842
7. Kerppola TK (2008) Bimolecular fluorescence complementation (BiFC) analysis as a probe of protein interactions in living cells. Ann Rev Biophys (Annual Reviews Inc) 37:465–487
8. Azad T, Tashakor A, Hosseinkhani S (2014) Split-luciferase complementary assay: applications, recent developments, and future perspectives. Anal Bioanal Chem 406:5541–5560
9. Chen H, Zou Y, Shang Y, Lin H, Wang Y, Cai R et al (2008) Firefly luciferase complementation imaging assay for protein-protein interactions in plants. Plant Physiol 146(2):368–376
10. Miller KE, Kim Y, Huh WK, Park HO (2015) Bimolecular fluorescence complementation (BiFC) analysis: advances and recent applications for genome-wide interaction studies. J Mol Biol (Academic) 427:2039–2055
11. Waadt R, Schlücking K, Schroeder JI, Kudla J (2014) Protein fragment bimolecular fluorescence complementation analyses for the in vivo study of protein-protein interactions and cellular protein complex localizations. Methods Mol Biol 1062:629–658
12. Hu CD, Chinenov Y, Kerppola TK (2002) Visualization of interactions among bZIP and Rel family proteins in living cells using bimolecular fluorescence complementation [Internet]. Mol Cell 9:789–798. Available from: https://doi.org/10.1016/S1097-2765(02)00496-3
13. Fujikawa Y, Kato N (2007) Split luciferase complementation assay to study protein-protein interactions in Arabidopsis protoplasts. Plant J 52(1):185–195
14. Tian G, Lu Q, Zhang L, Kohalmi SE, Cui Y (2011) Detection of protein interactions in plant using a gateway compatible bimolecular fluorescence complementation (BIFC) system. J Vis Exp 55:3473
15. Helariutta Y, Fukaki H, Wysocka-Diller J, Nakajima K, Jung J, Sena G et al (2000) The SHORT-ROOT gene controls radial patterning of the Arabidopsis root through radial Signaling. Cell 101:555
16. Nakajima K, Sena G, Nawy T, Benfey PN (2001) Intercellular movement of the putative transcription factor SHR in root patterning. Nature 413:307–311
17. Cui H, Levesque MP, Vernoux T, Jung JW, Paquette AJ, Gallagher KL et al (2007) An evolutionarily conserved mechanism delimiting SHR movement defines a single layer of endodermis in plants. Science (1979) 316(5823):421–425
18. Welch D, Hassan H, Blilou I, Immink R, Heidstra R, Scheres B (2007) Arabidopsis JACKDAW and MAGPIE zinc finger proteins delimit asymmetric cell division and stabilize tissue boundaries by restricting SHORT-ROOT action. Genes Dev 21(17):2196–2204
19. Long Y, Smet W, Cruz-Ramírez A, Castelijns B, de Jonge W, Mähönen AP et al (2015) Arabidopsis BIRD zinc finger proteins jointly stabilize tissue boundaries by confining the cell fate regulator SHORT-ROOT and contributing to fate specification. Plant Cell 27(4):1185–1199
20. Long Y, Stahl Y, Weidtkamp-Peters S, Postma M, Zhou W, Goedhart J et al (2017) In vivo FRET-FLIM reveals cell-type-specific protein interactions in Arabidopsis roots. Nature 548(7665):97–102
21. Horvath BM, Kourova H, Nagy S, Nemeth E, Magyar Z, Papdi C et al (2017) Arabidopsis RETINOBLASTOMA RELATED directly regulates DNA damage responses through functions beyond cell cycle control. EMBO J 36(9):1261–1278
22. Win J, Kamoun S (2004) pCB301-p19: a binary plasmid vector to enhance transient expression of transgenes by Agroinfiltration [Internet]. Available from: http://www.KamounLab.net
23. Rauschhuber C, Mueck-Haeusl M, Zhang W, Nettelbeck DM, Ehrhardt A (2013) RNAi suppressor P19 can be broadly exploited for enhanced adenovirus replication and microRNA knockdown experiments. Sci Rep 3:1363
24. Díaz-Triviño S, Long Y, Scheres B, Blilou I (2017) Analysis of a plant transcriptional regulatory network using transient expression systems. In: Methods in molecular biology. Humana Press Inc., pp 83–103
25. Aljedaani F, Rayapuram N, Blilou I (2021) A semi-in vivo transcriptional assay to dissect plant defense regulatory modules. In: Methods in molecular biology. Humana Press Inc., pp 203–214
26. Yu G, Xian L, Xue H, Yu W, Rufian JS, Sang Y et al (2020) A bacterial effector protein prevents mapk-mediated phosphorylation of sgt1 to suppress plant immunity. PLoS Pathog 16(9):e1008933

Chapter 13

Forster Resonance Energy Transfer (FRET) to Visualize Protein–Protein Interactions in the Plant Cell

Prabu Gnanasekaran and Hanu R. Pappu

Abstract

Forster resonance energy transfer (FRET) is an efficient method to visualize the protein–protein interaction in living cells. This technique is based on transfer of energy between two different fluorophores that are fused to two interacting proteins. In this chapter, we described the FRET assay to visualize the protein–protein interaction in plant cells.

Key words FRET, Energy transfer, Fluorescent acceptor, Fluorescent donor

1 Introduction

Forster resonance energy transfer (FRET) is a fluorescence-based approach for studying protein–protein interaction in the live system. FRET is based on energy transfer between the fluorescent donor (i.e., Cyan fluorescent protein (CFP)) and the fluorescent acceptor (i.e., Yellow fluorescent protein (YFP)). The absorption spectrum of the selected fluorescent acceptor must overlap with the emission spectrum of the selected fluorescent donor [1, 2]. Energy transfer occurs only if the FRET pair is in close proximity (1–10 nm) [3]. To study the interaction, proteins of interest, protein A and protein B, are fused to CFP and YFP respectively and expressed in plants. If there is an interaction between A and B proteins, then due to close proximity energy transfers from CFP to YFP. That is, if there is an interaction between A and B proteins, YFP fluorescence will be observed under the confocal fluorescent microscope while inducing with the CFP excitation filter (Fig. 1). If there is no interaction between A and B proteins, then due to lack of close proximity could not transfer energy from CFP to YFP. Thus, no YFP fluorescence will be observed under the confocal fluorescent microscope while inducing with the CFP excitation filter (Fig. 1).

Shahid Mukhtar (ed.), *Protein-Protein Interactions: Methods and Protocols*, Methods in Molecular Biology, vol. 2690, https://doi.org/10.1007/978-1-0716-3327-4_13,

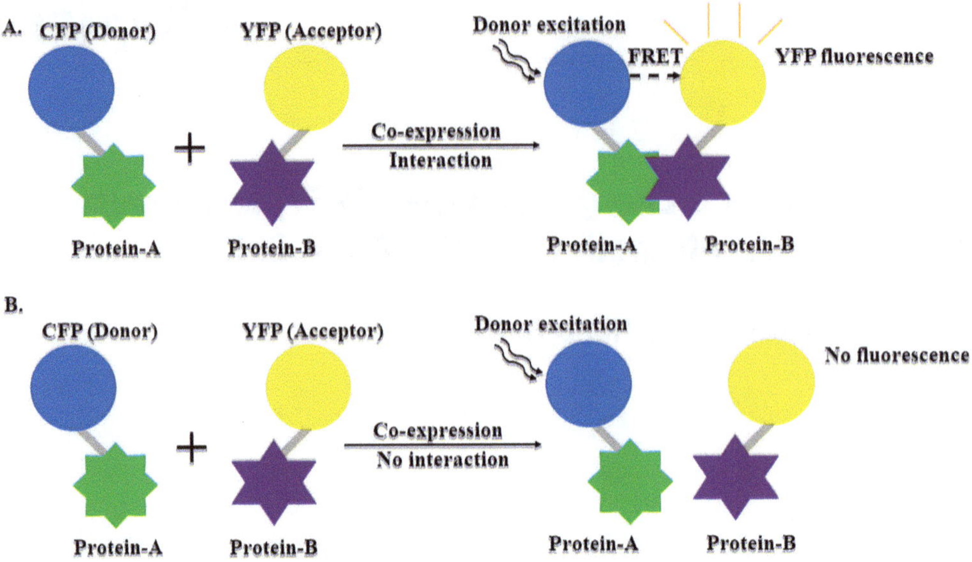

Fig. 1 Schematic representation of the principle of FRET assay

2 Materials

1. FRET constructs:

   pAN579 plasmid (contain coding sequence of CFP protein).

   pAN581 plasmid (contain coding sequence of YFP protein).

   pAN579-A: Generate this construct by cloning the open reading frame (ORF) coding for the protein of interest A in frame with the CFP coding sequence.

   pAN581-B: Generate this construct by cloning the ORF coding for the protein of interest B in frame with the YFP coding sequence.

2. *Agrobacterium tumefaciens* strain GV3101 competent cells.
3. Infiltration buffer:

   10 mM MES (pH 5.7).

   10 mM MgCl2.

   100 μM acetosyringone.

4. Three to four weeks old *Nicotiana benthamiana* plants.
5. Negative control combination:

   pAN579 vector + pAN581 vector.

   pAN579-A plasmid + pAN581 vector.

   pAN579 vector + pAN581-B plasmid.

6. Test combination:

   pAN579-A + pAN581-B.

3 Methods

1. Transform the FRET constructs into *Agrobacterium tumefaciens* strain GV3101 cells.
2. Grow the Agrobacterium cells carrying FRET construct at 28 °C and 200 rpm shaking for overnight in Luria Broth media with appropriate antibiotic selection.
3. Harvest the overnight grown Agrobacterium cells and make the desired infiltration combination in infiltration buffer (using 0.5 optical density (OD) of cells corresponding to each construct).
4. Mix gently and incubate in dark for 2–3 h.
5. Infiltrate the cells containing FRET constructs on the abaxial (lower) side of the leaves of *N. benthamiana* plants.
6. After 48 h post-infiltration, FRET analysis was performed using the epidermal cells of infiltrated *N. benthamiana* leaves under the confocal fluorescent microscope. CFP excitation filter (405 nm) and YFP emission filter (525 nm) are used to study these interactions (*see* **Notes 1** and **2**).

4 Notes

1. If YFP fluorescence is detected in the leaves co-infiltrated with pAN579-A + pAN581-B, this indicates there is an interaction between protein A and B.
2. If no YFP fluorescence is detected in the leaves co-infiltrated with pAN579-A + pAN581-B, this indicates there is no interaction between protein A and B.

References

1. Rios AF, Radoeva T, De Rybel B et al (2017) FRET-FLIM for visualizing and quantifying protein interactions in live plant cells. Methods Mol Biol 1497:135–146
2. Terryn C, Paës G, Spriet C (2018) FRET-SLiM on native autofluorescence: a fast and reliable method to study interactions between fluorescent probes and lignin in plant cell wall. Plant Methods 14(1):74
3. Bücherl CA, Bader A, Westphal AH et al (2014) FRET-FLIM applications in plant systems. Protoplasma 251(2):383–394

Chapter 14

Dynamic Proximity Tagging in Living Plant Cells with Pupylation-Based Interaction Tagging

Ruiqiang Ye, Zhuoran Lin, Kun-Hsaing Liu, Jen Sheen, and Sixue Chen

Abstract

Identification of protein–protein interactions (PPIs) and protein kinase substrates is fundamental for understanding how proteins exert biological functions with their partners and targets. However, it is still technically challenging, especially for transient and weak interactions involved in most cellular processes. The proximity-tagging systems enable capturing snapshots of both stable and transient PPIs. In this chapter, we describe in detail the methodology of a novel proximity-based labeling approach, PUP-IT (pupylation-based interaction tagging), to identify PPIs using a protoplast transient expression system. We have successfully identified potential kinase substrates by targeted screening and tandem mass spectrometry analysis.

Key words Protein interactions, Proximity-tagging system, PUP-IT, Mesophyll protoplast, Signal transduction

1 Introduction

Protein–protein interactions (PPIs) contribute to the regulation and execution of most cellular pathways and processes. The majority of PPIs are highly dynamic and temporary in nature, or conditional that rely on post-translational modifications or conformational changes [1, 2]. Although multiple approaches have been developed to map PPIs, identifying the transient and unstable interactions remains a challenge [3–5].

Proximity-tagging systems have emerged as reliable techniques that enable the discovery of neighboring proteins in cells, capturing both transient or weak and stable interactions [4, 5]. Generally, these methods use proximity labeling enzymes, such as biotin ligase or peroxidase [6–9], to introduce biotin tags to proximal proteins close to the protein of interest (bait) in living cells. The biotinylated proteins can then be isolated with streptavidin beads [7], Tamavidin 2-REV [6], or antibiotin antibody [8], followed by

Shahid Mukhtar (ed.), *Protein-Protein Interactions: Methods and Protocols*, Methods in Molecular Biology, vol. 2690,
https://doi.org/10.1007/978-1-0716-3327-4_14,

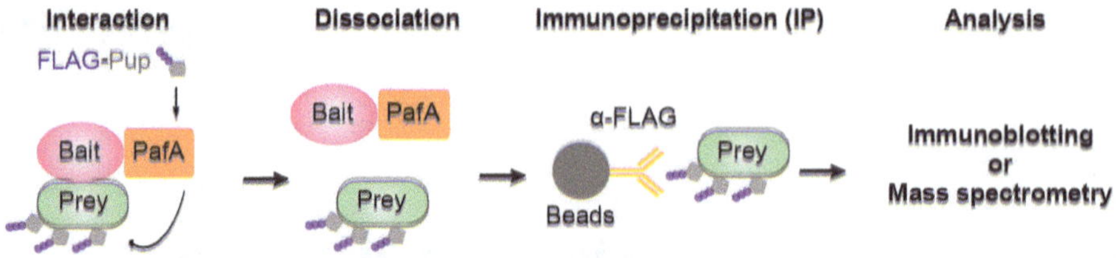

Fig. 1 Schematic presentation of the proximity-tagging system PUP-IT. The bait protein is fused with a bacterial Pup ligase (PafA), which conjugates the tagged Pup to a lysine residue of interacting proteins (prey). The Pup is triple-FLAG tagged at the N terminus. The 3XFLAG-Pup-labelled proteins could be immunoprecipitated with anti-FLAG antibody conjugated beads for immunoblotting or mass spectrometry analysis

immunoblotting and mass spectrometry analyses. However, living plant cells contain naturally biotinylated proteins [10, 11] and high endogenous peroxidases [12], which contribute to high background in proximity labeling by biotin ligase or peroxidase commonly used in mammalian cells.

Here, we introduce a newly developed proximity tagging method termed PUP-IT (pupylation-based interaction tagging) [13] to the plant research community. In this method, bait protein is fused with a bacterial Pup ligase (PafA), which conjugates the tagged Pup (64 amino acids) to a lysin residue in the prey protein in the presence of ATP (Fig. 1). This process resembles the ubiquitination process in plant and animal cells. The N terminus of Pup can be tagged to facilitate the purification of Pup-modified proteins (*see* **Note 1**). Compared to other proximity-tagging systems [14], the PUP-IT possesses additional advantages for plant research. First, unlike the biotinylated proteins, there are no preexisting Pup-modified proteins in the plant system, which provides a lower background of protein contaminants. Second, there is no need for extra biotin or H_2O_2 treatment, making it more suitable for in vivo and physiology studies. Third, as a relatively large molecule modification, Pup tagged proteins exhibit a significant size shift and retain a unique GGE motif on lysine, making it more specific for immunoblot and mass spectrometry analysis. However, the larger size of the enzyme and labeling tag may also affect the functions of the bait and prey protein. This needs to be considered when using PUP-IT or any other proximity-tagging/affinity-tagging systems. Additionally, the distribution of Pup should also be considered for studies of PPIs in organelles.

In this method, we perform the PUP-IT analysis in a protoplast transient expression system. The cell-based assay provides a versatile and quick approach to explore the function or substrate of a candidate protein [15]. The isolated plant cells retain almost all the cellular activities and processes, including transcription, translation, DNA repair, chromatin remodeling, cell wall biosynthesis, and other metabolic processes, as well as the ability to respond specifically to various signals, such as nutrient, hormone, stress, and elicitors [16, 17]. Thus, the protoplast assay can be employed to analyze the PPIs in these signaling and metabolic processes. The PUP-IT assay works very well in *Arabidopsis* protoplast system. We have successfully identified a substrate of Target of Rapamycin (TOR) by targeted screening [18] and more potential interactors by tandem mass spectrometry using the protocols described in this chapter. The drawback of cell-based assays is that they may lose some cell-type-specific responses and cause an ectopic effect of part of prey genes. As there is no need to deliver substrates as in the other proximity-tagging systems [6–9], the PUP-IT system can be easily transformed into plants with universal, inducible, or cell-type-specific promoters to investigate the spatiotemporal resolution of PPIs (*see* **Note 2**).

2 Materials

2.1 Plant Material and Growth Conditions

Arabidopsis plants (Col-0) grown in soil (Metro-Mix 360, SunGro Horticulture, Burton, OH, USA) or Jiffy-7 (Jiffy Products Ltd., Canada) in a growth chamber at 23/20 °C, 12 h/12 h light/dark, 60% relative humidity, and 75 μmol m^{-2} s^{-1} of light for 4~5 weeks. Leaves 5–7 are used for optimal mesophyll protoplast isolation [15, 19].

2.2 Construction of the Plant Expression Plasmids

1. Bait protein in the PUP-IT vector: Clone the coding DNA sequence of desired protein into the SpeI site of *pCambia-PUP-IT* vector (Addgene plasmid #186478) by Gibson assembly (NEB #E2621).
2. Candidate prey protein for targeted screening: Clone the coding DNA sequence of desired protein into the *pHBT-MYC* plant expression vector (available from the authors upon request).

2.3 Protoplast Isolation and Transfection

1. Enzyme solution: 1% cellulase R10, 0.25% macerozyme R10, 0.4 M mannitol, 20 mM KCl, 20 mM MES, pH 5.7. Heat the enzyme solution at 55 °C for 10 min to inactivate proteases and enhance enzyme solubility. Cool the solution to room temperature before adding 10 mM $CaCl_2$, 1 mM ß-mercaptoethanol, and 0.1% BSA. Filter the solution through a 0.45 μm syringe filter.
2. Razor blades.
3. Petri dish (100 × 20 mm or 150 × 25 mm).
4. Desiccator.
5. Nylon mesh (35–70 μm, Lab-Line Instruments, #190-158-00).
6. 30 mL Round-bottom polypropylene tube
7. Hemocytometer.

8. W5 solution: 154 mM NaCl, 125 mM $CaCl_2$, 5 mM KCl, 2 mM MES, pH 5.7.
9. WI solution: 0.5 M mannitol, 4 mM MES, pH 5.7, 20 mM KCl.
10. MMg solution: 0.4 M mannitol, 15 mM $MgCl_2$, 4 mM MES, pH 5.7.
11. 40% (w/v) polyethylene glycol (PEG) solution: To make 10 mL of PEG solution, dissolve 4 g of PEG 4000 (Sigma-Aldrich) into 3 mL of H_2O, 2.5 ml of 0.8 M mannitol, and 1 ml of 1 M $CaCl_2$.
12. 15 mL or 50 mL round-bottomed microcentrifuge tubes.

2.4 Immuno-precipitation

1. Extraction buffer: 25 mM Tris–HCl, pH 7.6, 150 mM NaCl, 5 mM $MgCl_2$, 1% Triton X-100, 0.1% SDS, 0.5 mM DTT, and fresh protease inhibitor cocktail.
2. Anti-FLAG M2 Agarose Beads (Sigma, A2220).
3. 3xFLAG peptide (Sigma, F4799).
4. Trichloroacetic acid (TCA).
5. Acetone.

2.5 Immunoblotting or Mass Spectrometry Analysis

1. Tris–glycine SDS–polyacrylamide gel.
2. Prestained molecular weight markers: Kaleidoscope markers (Bio-Rad, CA).
3. Tris-glycine running buffer: 125 mM Tris, 960 mM glycine, 0.5% (w/v) SDS.
4. Transfer buffer: 25 mM Tris, 190 mM glycine, 20% (v/v) methanol, 0.05% (w/v) SDS.
5. TBST buffer: 20 mM Tris–HCl, pH 7.6, 150 mM NaCl, 0.05% Tween®20.
6. Blocking and antibody incubation buffer: 1× TBST, 5% w/v nonfat dry milk.
7. HA-HRP (Sigma, 12,013,819,001, 1:5000), FLAG-HRP (Sigma, A8592, 1:5000), and MYC-HRP (Roche, 1-814-150, 1:1000).
8. Coomassie staining buffer: 0.2% Brilliant Blue G250 in 20% methanol, and 0.5% acetic acid.
9. Destaining buffer: 20% methanol, 0.5% acetic acid.
10. An ultraperformance liquid chromatography (UPLC) tandem mass spectrometry system (e.g., nanoEasy UPLC-Orbitrap Tribrid) and software for data analysis (e.g., Proteome Discoverer or MaxQuant, a freeware available at https://www.maxquant.org/).

3 Methods

3.1 Protoplast Isolation

1. Prepare fresh enzyme solution into a 100 × 20 mm or 150 × 25 mm petri dish (*see* **Note 3**).
2. Cut the middle part of well-expanded leaves (true leaf numbers five to seven) into 0.5–1 mm strips with a fresh sharp razor blade (*see* **Note 4**).
3. Immediately and gently transfer leaf strips into the petri dish and completely submerge them with enzyme solution with an inoculating loop (BD).
4. Cover the Petri dish with aluminum foil and apply vacuum infiltration for 30 min by using a desiccator connected to a building vacuum system.
5. Continue the digestion without vacuum and shaking for another 2.5–3 h at room temperature.
6. Gently shake the Petri dish by hand or use a shaker at 50 rpm for 1 min to release the protoplasts. Ideally, most leaf strips turn transparent, and the enzyme solution becomes green after this step.
7. Add an equal volume of W5 solution and filter the slurry with a wet 35–70 μm nylon mesh to remove undigested leaf tissues.
8. Centrifuge the flow-through at 100× *g* in a 30 mL round-bottomed tube for 1–2 min to pellet the protoplasts.
9. Remove as much supernatant as possible and gently resuspend the protoplast pellet with 10 mL of cold W5 solution.
10. Keep the protoplasts on ice for at least 30 min for recovery from isolation stress. Intact and healthy protoplasts should settle at the bottom of the tube by gravity.
11. Remove the W5 solution as much as possible and resuspend the protoplasts with MMg solution.
12. Count protoplasts using a hemocytometer under the light microscope and adjust the protoplasts in the MMg solution to a density of 2×10^5 cells/mL.

3.2 Protoplast Transfection

1. Prepare fresh 40% (w/v) PEG solution and high-quality plasmid DNA at 2 μg/μL (*see* **Note 5**).
2. For targeted screening of the prey protein, mix 100 μg of PUP-IT vectors with bait protein or control plasmids and 100 μg of candidate gene–MYC tag constructs in a 15 mL round-bottom microcentrifuge tube. For tandem mass spectrometry analysis, add 1000 μg of PUP-IT vector with bait protein into a 50ml round-bottom microcentrifuge tube (*see* **Note 6**).

3. Add 1 mL or 5 mL of protoplasts (100 μL protoplasts per 10–20 μg plasmid DNA) in MMg solution into the tube.
4. Add an equal volume (1.1 mL or 5.5 mL) of PEG solution and then mix thoroughly by gently tapping the tube.
5. Incubate the mixtures at room temperature for 5 min.
6. Dilute the transfection mixture with at least 3 volumes of W5 solution and mix well by gently rocking or inverting to quench the transfection.
7. Centrifuge at 100× *g* for 2 min at room temperature using a swinging-bucket centrifuge and remove the supernatant.
8. Resuspend the protoplasts gently with 5 mL or 20 mL of WI solution and transfer the cells into 100 × 20 mm or two 150 × 25 mm Petri dishes (*see* **Note 7**).
9. Incubate the protoplasts for 12 h at room temperature (22–25 °C) under low light (30–35 μmol m^{-2} s^{-1}) (*see* **Note 8**).
10. Harvest the protoplasts by centrifugation at 100× *g* for 2 min and remove the supernatant.
11. Freeze the samples immediately with liquid nitrogen and store them at −80 °C until further analysis.

3.3 Immuno-precipitation

1. Resuspend the frozen protoplasts in the extraction buffer (500 μL per 2 × 10^5 cells) on ice.
2. Vortex vigorously for 10 s and incubate on ice for 5–10 min to lysate the cells.
3. Distribute the extracts into 1.5 mL reaction tubes and sonicate them in an ice bath four times, 30 s on and 30 s off each time, at high setting using a Bioruptor UCD-200 (Diagenode).
4. Centrifuge at the maximum speed for 10 min at 4 °C and transfer all supernatant to the new tube.
5. Aliquot 30 μL of supernatant into a new tube as input sample, add 10 μL of 4 × sample buffer and boil for 5 min at 95 °C.
6. Add prewashed anti-FLAG M2 Agarose Beads to the rest supernatant (10 μL per 500 μL extracts), and incubate at 4 °C for 2 h.
7. Wash the beads four times with 1 mL extraction buffer on the rotor wheel for 5 min (transfer the beads into a new tube the first time).
8. Elute the proteins three times with an equal volume of 500 mg/mL of 3 × FLAG peptide at room temperature for 10 min with shaking at 1000 rpm.
9. For immunoblot analysis, add 10 μL of 4 × sample buffer to the eluted samples (30 μL) and boil for 5 min at 95 °C.

10. For mass spectrometry analysis, precipitate the elution with an equal volume of 20% TCA/acetone with 5 mM DTT on ice for 2 h and wash with acetone twice. Add 20 μL of 1 × sample buffer to the pellet and boil for 5 min at 95 °C.

3.4 Immunoblotting Analysis

1. Prepare 6–15% Tris-glycine SDS-polyacrylamide gel according to the size of the protein of interest.
2. Resolve 10–20 μL of input and eluted samples on the gel until the dye runs out.
3. Electrotransfer the proteins from the gel to a PVDF membrane.
4. Incubate the membrane in the blocking buffer for 1 h at room temperature under gentle shaking.
5. Incubate the membrane in the blocking buffer containing HRP-conjugated HA, FLAG, or MYC antibodies for 1–2 h at room temperature under gentle shaking.
6. Wash the membrane three times with TBST buffer for 5 min each at room temperature.
7. Proceed detection with ECL reagent. An example result is shown in Fig. 2a (*see* **Note 9**).

3.5 Prepare Samples for Mass Spectrometry Analysis

1. Prepare 4–12% gradient Tris–glycine SDS–polyacrylamide gel.
2. Load all the immunopurified samples (20 μL) to one well. Stop the electrophoresis when the bromophenol blue dye migrates approximately 1–2 cm into the gel.

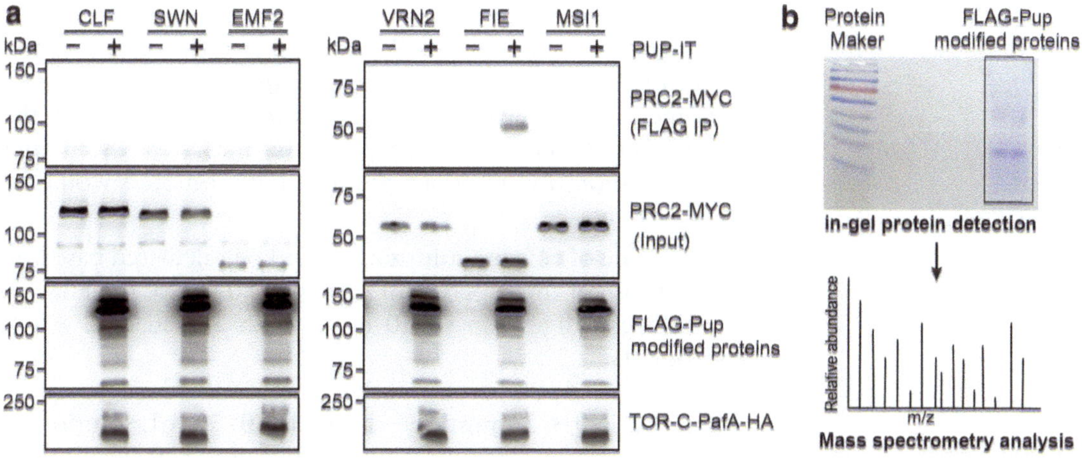

Fig. 2 Immunoblotting and mass spectrometry analyses of the PUP-IT labelling assay. (**a**) Immunoblot analysis of PUP-IT screening of MYC-tagged PRC2 components with TOR-C-PafA-HA. Only FIE interacts with TOR. (**b**) An example of FLAG-Pup modified proteins for mass spectrometry analysis. Proteins obtained from immunoprecipitation by anti-FLAG beads were separated by SDS-PAGE and stained with Coomassie Blue. Proteins excised from the gel were subjected to in-gel digestion with trypsin and analyzed by UPLC-tandem mass spectrometry

3. Rinse the gel 3 times with Milli-Q water to remove SDS and buffer salts.
4. Incubate the gel with Coomassie Brilliant Blue G250 staining buffer for 1 h with a gentle shake.
5. Gently shake the gel in the destaining buffer for at least 2 h with changes of this solvent every half hour until the desired background is achieved. An example result is shown in Fig. 2b.
6. Rinse the gel three times for 5 min with Milli-Q water.
7. Excise the gel containing protein into a clean tube. Freeze and store at −80 °C for further in-gel trypsin digestion and mass spectrometry analysis (*see* **Note 10**).
8. For UPLC-tandem mass spectrometry, please refer to the published method [20].

4 Notes

1. In the original version, the *E. coli* Pup protein was tagged with a bacteria-derived carboxylase domain, which can be biotin-labeled in vivo and further pulled down by streptavidin. The isolation of biotinylated proteins can be done under denaturing conditions to reduce the background. However, efficient labeling requires exogenous biotin treatment. The free biotin has the potential to interfere with streptavidin pull-down of pup-labeled proteins and needs extra steps to be removed. In this protocol, we fused the Pup protein with a commonly used triple FLAG tag, facilitating easy and efficient capture of Pup-modified proteins.
2. The transient expression PUP-IT vector is in the binary *pCambia1300* backbone, which can be directly used for *Agrobacterium*-mediated transformation. However, we found that the FLAG-Pup driven by the 35S promoter may be silenced after T2 generation. Endogenous or inducible promoters would be ideal options.
3. The volume of enzyme solutions should be calculated based on the purpose of the experiment. Usually, 10 mL of the enzyme solution can digest up to 40 leaves with a yield of 2×10^6 protoplasts.

   The 100 × 20 mm petri dish can hold 10–15 mL solution, while the 150 × 25 mm one can hold 30–50 mL solution. A workshop movie for protoplast isolation and transfection can be downloaded on the Sheen Lab Web site (http://genetics.mgh.harvard.edu/sheenweb/protocols_reg.html). Be gentle with the protoplasts during all the steps!

4. The selection of healthy and fully expended leaves is critical for the success and efficient transfection of protoplasts. Leaves five to eight from 4-week-old Arabidopsis plants are normally used. However, well-expended leaves from younger plants, such as leaves three to four from 3-week-old plants and leaves one to two from 2-week-old plants, may also be used.
5. The quality of the plasmid DNA is very important for high transfection efficiency. We routinely use the CsCl gradient for maxi-plasmid DNA preparation. The protocol is available on the Sheen Lab Web site (http://genetics.mgh.harvard.edu/sheenweb/protocols_reg.html). However, DNA preparation by commercial DNA maxiprep kits is acceptable but may lower protoplast transfection efficiency due to chemical carryover.
6. Protoplast transfection can be scaled up or down following the recommended DNA/protoplast ratio. Usually, 1 mL of cells (2×10^5) are sufficient for targeted screening, and at least 10 mL of cells (2×10^6) are required for mass spectrometry analysis.
7. The petri dish can be precoated with 5% (vol/vol) calf serum for 1–2 s to prevent the sticking of protoplasts to the plastic surface. The depth of the WI solution is approximately 0.1 mm to avoid hypoxia stress during protoplast incubation. The 100×20 mm petri dish is recommended for incubation of 5 mL of protoplasts/WI solution, and the 150×25 mm dish is recommended for incubation of 10–15 mL of protoplasts/WI solution.
8. Normally, 10–12 h is enough for PUP-IT labeling. However, the incubation time can still be optimized for a specific bait protein. The protoplasts could be treated with different stimuli during the incubation. For example, the flg22 elicitor can be added 1 h before harvest to trigger the innate immune response.
9. The Pup labeled prey protein would exhibit at least 11 kDa (one modification) size shift. The band shift could be detected in both input and immunoprecipitated samples for strong interaction but may only be detected after immunoprecipitation for weak interaction.
10. The mass spectrometry analysis can be performed as per the protocol described by Perron et al. [20]. There is an extra 243.10-Da mass (GGE modifications) on the lysine side-chain of Pup-modified peptides.

Acknowledgments

We thank former and current members of the Sheen Laboratory for their efforts in improving the *Arabidopsis* mesophyll protoplasts transient expression system. We thank Jenifer Bush for plant management. This work was supported by the NIH grants GM060493 and GM129093 to J.S. and R.Q.Y., National Natural Science Foundation of China grants 32170270 and 31870227 to K.-H. L., and USDA 2020-67013-31615/accession no. 1022409 grant to S.C. (co-PI) and Ping He (PI).

References

1. Wang S, Osgood AO, Chatterjee A (2022) Uncovering post-translational modification-associated protein-protein interactions. Curr Opin Struct Biol 74:102352. https://doi.org/10.1016/j.sbi.2022.102352
2. De Las RJ, Fontanillo C (2010) Protein-protein interactions essentials: key concepts to building and analyzing interactome networks. PLoS Comput Biol 6(6):e1000807. https://doi.org/10.1371/journal.pcbi.1000807
3. Elhabashy H, Merino F, Alva V et al (2022) Exploring protein-protein interactions at the proteome level. Structure 30(4):462–475. https://doi.org/10.1016/j.str.2022.02.004
4. Lin CPJ, Vela S et al (2021) Spatial and single-cell proteomics workshop report. J Proteomics Bioinformatics 14:553
5. Zhang T, Chen S, Harmon AC (2016) Protein-protein interactions in plant mitogen-activated protein kinase cascades. J Exp Bot 67(3):607–618. https://doi.org/10.1093/jxb/erv508
6. Motani K, Kosako H (2020) BioID screening of biotinylation sites using the avidin-like protein Tamavidin 2-REV identifies global interactors of stimulator of interferon genes (STING). J Biol Chem 295(32):11174–11183. https://doi.org/10.1074/jbc.RA120.014323
7. Cho KF, Branon TC, Udeshi ND et al (2020) Proximity labeling in mammalian cells with TurboID and split-TurboID. Nat Protoc 15(12):3971–3999. https://doi.org/10.1038/s41596-020-0399-0
8. Udeshi ND, Pedram K, Svinkina T et al (2017) Antibodies to biotin enable large-scale detection of biotinylation sites on proteins. Nat Methods 14(12):1167–1170. https://doi.org/10.1038/nmeth.4465
9. Hung V, Lam SS, Udeshi ND et al (2017) Proteomic mapping of cytosol-facing outer mitochondrial and ER membranes in living human cells by proximity biotinylation. elife 6. https://doi.org/10.7554/eLife.24463
10. Arora D, Abel NB, Liu C et al (2020) Establishment of proximity-dependent Biotinylation approaches in different plant model systems. Plant Cell 32(11):3388–3407. https://doi.org/10.1105/tpc.20.00235
11. Mair A, Xu SL, Branon TC et al (2019) Proximity labeling of protein complexes and cell-type-specific organellar proteomes in Arabidopsis enabled by TurboID. elife 8. https://doi.org/10.7554/eLife.47864
12. Hiraga S, Sasaki K, Ito H et al (2001) A large family of class III plant peroxidases. Plant Cell Physiol 42(5):462–468. https://doi.org/10.1093/pcp/pce061
13. Liu Q, Zheng J, Sun W et al (2018) A proximity-tagging system to identify membrane protein-protein interactions. Nat Methods 15(9):715–722. https://doi.org/10.1038/s41592-018-0100-5
14. Yang X, Wen Z, Zhang D et al (2021) Proximity labeling: an emerging tool for probing in planta molecular interactions. Plant Commun 2(2):100137. https://doi.org/10.1016/j.xplc.2020.100137
15. Yoo SD, Cho YH, Sheen J (2007) Arabidopsis mesophyll protoplasts: a versatile cell system for transient gene expression analysis. Nat Protoc 2(7):1565–1572. https://doi.org/10.1038/nprot.2007.199
16. Niu Y, Sheen J (2012) Transient expression assays for quantifying signaling output. Methods Mol Biol 876:195–206. https://doi.org/10.1007/978-1-61779-809-2_16

17. Sheen J (2001) Signal transduction in maize and Arabidopsis mesophyll protoplasts. Plant Physiol 127(4):1466–1475
18. Ye R, Wang M, Du H et al (2022) Glucose-driven TOR-FIE-PRC2 signalling controls plant development. Nature 609(7929): 986-993. https://doi.org/10.1038/s41586-022-05171-5
19. Jiang W, Bush J, Sheen J (2021) A versatile and efficient plant protoplast platform for genome editing by Cas9 RNPs. Front Genome Ed 3: 719190. https://doi.org/10.3389/fgeed.2021.719190
20. Perron N, Tan B, Dufresne CP et al (2023) Proteomics and phosphoproteomics of C_3 to CAM transition in the common ice plant. Methods Enzymol. https://doi.org/10.1016/bs.mie.2022.06.004

Chapter 15

Characterization of Small Molecule–Protein Interactions Using SPR Method

Binmei Sun, Jianmei Xu, Shaoqun Liu, and Qing X. Li

Abstract

Surface plasmon resonance (SPR) is an optical phenomenon being used to monitor molecular binding events. With the advantages of being label-free, real-time, and sensitive, SPR assays have become one of the most commonly used techniques to measure binding kinetics, affinity, specificity, and concentration of molecular interactions. In an SPR experiment to measure small molecule–protein interactions, the protein is immobilized on the biosensor surface, while the small molecule flows through the surface of the sensor chip. The interactions between the small molecules and proteins are monitored by subsequent changes in the refractive index and quantified with resonance units. In this chapter, we have utilized an SPR assay to study the interaction of flavonoids and the glucose-regulated protein 78. Assay steps are detailed for immobilization optimization, SPR instrument setup, operation, sample injection, and affinity data analysis.

Key words Biacore, Binding affinity, Molecular interaction, Molecular recognition, Sensor chip, Surface plasmon resonance

1 Introduction

Surface plasmon resonance (SPR) is a powerful optical technique for detecting the interactions of protein–protein, small molecule–protein, DNA–protein, DNA–DNA, and so on. With a series of advantages such as being label-free, real-time, and sensitive, SPR has been widely used in a variety of applications, such as life science, electrochemistry, chemical vapor detection, food, and environmental safety, and beyond [1–6]. SPR can occur when plane-polarized light hits a metal film under total internal reflection conditions, causing free electrons in the metal to produce surface plasmon (Fig. 1) [7].

In an SPR experiment described here, the ligand (i.e., protein) is immobilized on a sensor chip through an amine-coupling reaction. A solution containing the analyte (i.e., a small molecule) is injected and passed through the surface of the sensor chip. The

Shahid Mukhtar (ed.), *Protein-Protein Interactions: Methods and Protocols*, Methods in Molecular Biology, vol. 2690, https://doi.org/10.1007/978-1-0716-3327-4_15,

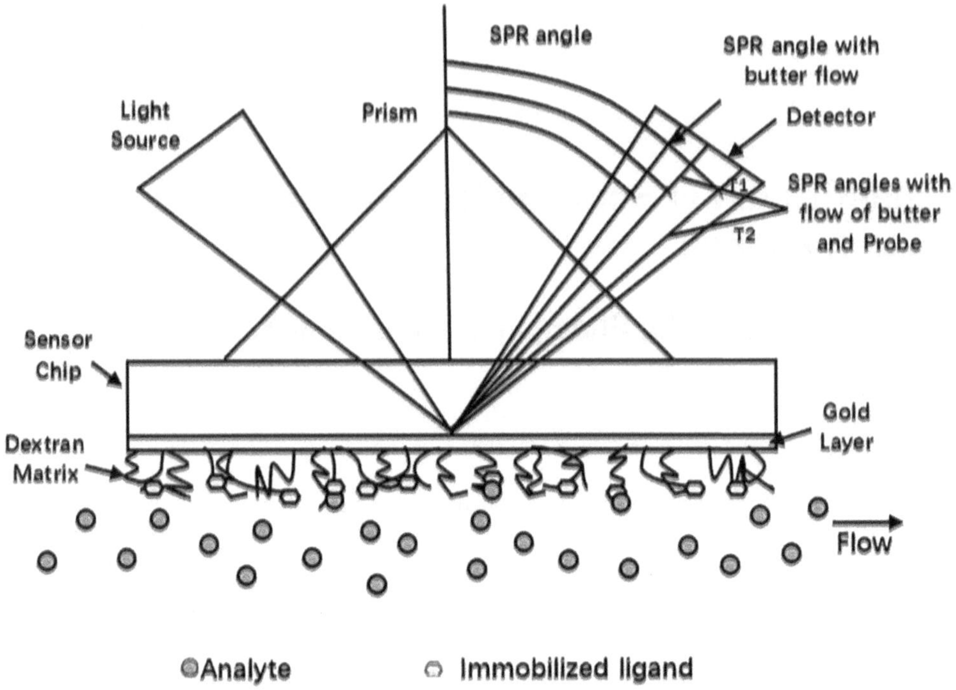

Fig. 1 Principles of SPR

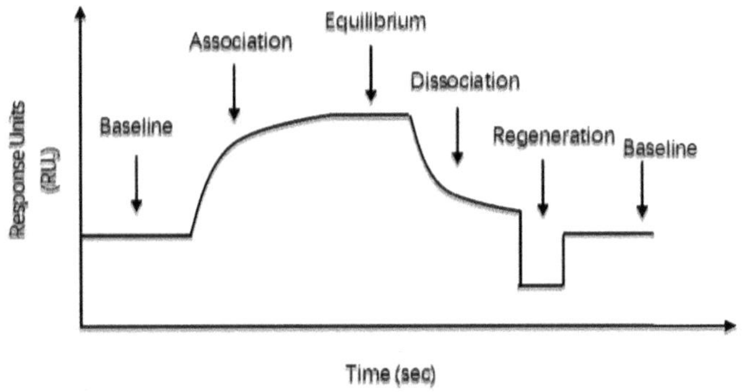

Fig. 2 Different phases of an SPR sensorgram. An initial baseline is obtained after the immobilization of the ligand. Following injection of analyte over the ligand, the increase in the RUs corresponds to the association phase. When the number of association and dissociation events is equal, the equilibrium level is maintained. The dissociation phase starts when stopping the analyte injection: the decrease in the RUs corresponds to the flowing out of the analyte from the ligand surface. The regeneration step aims at the full removal of the analyte and at reaching the baseline level

interactions between the ligand and analyte produce the change of refractive index on the gold surface, which is observed and quantified with resonance units (RUs) (Fig. 2). SPR experiments can measure the equilibrium binding constant or dissociation constant (a measure of binding affinity, yielding a straightforward characterization of the small molecule–protein interaction.

2 Materials

1. Series S CM5 sensor chip (*see* **Note 1**).
2. Amine coupling kit: contains 750 mg 1-ethyl-3-(3- dimethylaminopropyl) carbodiimide hydrochloride (EDC), 115 mg *N*-hydroxy succinimide (NHS), and 10.5 mL 1.0 M ethanolamine-HCl, pH 8.5. Dissolve EDC and NHS by adding 10.0 mL of filtered, deionized water to each vial and the concentration of EDC and NHS is 0.4 M and 0.1 M, respectively (*see* **Notes 2** and **3**).
3. Immobilization buffers: 10 mM sodium acetate buffers at pH 4.0, 4.5, 5.0, and 5.5.
4. Running buffers: phosphate-buffered saline (PBS) (pH 7.4) supplemented with 1% dimethyl sulfoxide (DMSO).
5. Ligand: recombinant mouse GRP78/heat shock 70 kDa protein 5 (HSPA5) (*see* **Note 4**).
6. Analyte: (-)-epigallocatechin gallate (EGCG) (*see* **Note 5**).
7. Plastic vials and rubber caps (type 2).
8. Biacore T100 using biomolecular interaction analysis (BIA) evaluation 3.1 software.

3 Methods

SPR experiments were performed on a Biacore T100 to determine binding curves. The immobilization involves activation of carboxymethyl groups on a dextran-coated chip by reaction with NHS, followed by covalent bonding of the ligand (i.e., protein) to the chip surface via amide linkages and blockage of excess activated carboxyl groups with ethanolamine [8]. Reference surfaces are prepared in the same manner, except that all carboxyl groups are blocked and no ligand is added. The small molecule analytes flow over the immobilized-ligand surface and the binding response of the analyte to the ligand is recorded. The final concentration of the bound ligand, expressed in RUs, is calculated by subtracting the reference RU from the ligand RU. PBS (pH 7.4) with 1% DMSO buffer is used as both running and analyte-binding buffer. At least five concentrations of the small molecule are needed. Generally, the interaction between small molecules and protein is too weak to capture the association rate constant (k_a) and dissociation rate constant (k_d), which is different from the strong interaction between protein and protein or DNA. Therefore, the affinity of small molecule–protein interaction can be determined from the dependence of steady-state binding levels on analyte concentrations.

3.1 Preparation for Immobilization (See Note 6)

1. Take a Series S CM5 sensor chip from 4 °C and keep it at room temperature for at least 1 h before the experiment.
2. Prepare 500 μL running buffer by diluting stock solution 10 times and supplementing with 1% DMSO. Filter the running buffer through a 0.22 μM membrane, followed by ultrasonic degassing (*see* **Notes** 7 and **8**).
3. Prepare a vial, containing 400 μL of 50 mM NaOH solution for regeneration.
4. Prepare a sample (typically 100 μL) by mixing the required amount of ligand (i.e., protein) with sodium acetate solution at a pH determined by the pH scouting. The pH range of the immobilization buffer is chosen to be below the estimated pI of the fusion protein (*see* **Note 9**). The final protein concentration after dilution on immobilization buffers should be in the range of 5–200 μg/mL; 20 μg/mL is recommended.
5. Insert a Series S CM5 sensor chip on the sensor chip port and dock the sensor chip using the Biacore T100 control software (*see* **Note 10**).
6. Eject a rack tray and place vials in the rack, assign RACK POSITIONS to samples and solutions.
7. Prime the system using running buffer at least three times to ensure that all system tubing is properly flushed with running buffer.
8. Click on APPLICATION WIZARD/SURFACE PREPARATION/pH SCOUTING/NEXT (*see* **Note 11**).
9. The result of a pH scouting run is shown in Fig. 3.

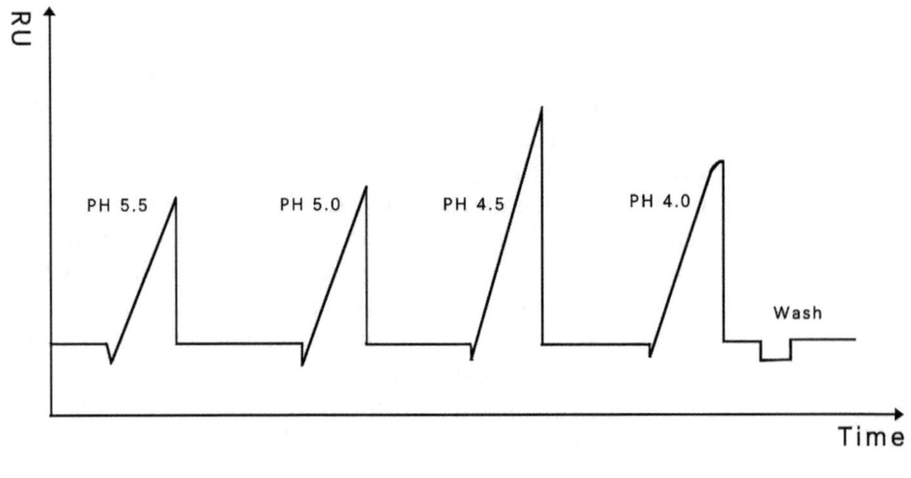

Fig. 3 Immobilization pH scouting

3.2 Immobilization of Ligand Using Amine Coupling

1. Prepare three vials, one containing 115 μL of EDC, the other one containing 115 μL of NHS, and another empty vial for mixing the EDC and NHS.
2. Prepare three vials, one containing 400 μL of running buffer, the second one containing 80 μL of ethanolamine solution for blocking the unreacted active sites on the chip, and the last one with 400 μL of 50 mM NaOH regeneration solution.
3. In the Biacore T100 control software dialogue window, use the application wizard control coupling level, and go to FILE/OPEN/APPLICATION WIZARD/SURFACE PREPARATION/IMMOBILIZATION.
4. Specify the sensor surface (CM5) and the immobilization method (Amine Coupling).
5. In the flow path (flow cell 2), enter the name of the ligand to coupling and the flow rate of 10 μL/min. Use flow cell 1 as a reference surface (*see* **Note 12**).
6. Inject a short pulse of ligand solution without activating the surface to determine the rate of binding of ligand to the surface.
7. Inject a pulse of wash solution to clean the surface.
8. If the binding rate as determined in **step 6** is suitable, i.e., the target level can be reached within a reasonable time frame, continue with **step 9**. Otherwise, abort the procedure.
9. Activate the surface with NHS/EDC (*see* **Note 13**).
10. Inject the ligand in pulses using a contact time based on the binding rate (*see* **Note 14**) so that the target level is expected to be reached after a few pulses (*see* **Note 15**). The injection procedure is stopped when the target level is reached or if:
    (i) The sample loop is empty.
    (ii) The target level is not reached within 20 injection pulses.
    (iii) The response level decreases during an injection pulse.
11. Inject ethanolamine to complete the immobilization procedure.
12. Reference surfaces (flow cell 1) are prepared in the same manner, except that all carboxyl groups are blocked, and no ligand is added. A typical amine coupling sensorgram is shown in Fig. 4.

3.3 Kinetics/Affinity Measurements (See Note 16)

1. Make analyte dilutions in running buffer with 200 μL of 10, 100, and 1000 μM of the analyte for a surface test.
2. Specify the flow path (2–1) if the ligand is immobilized.
3. Select three startup cycles with a running buffer to stabilize the baseline signal before starting the experiment.

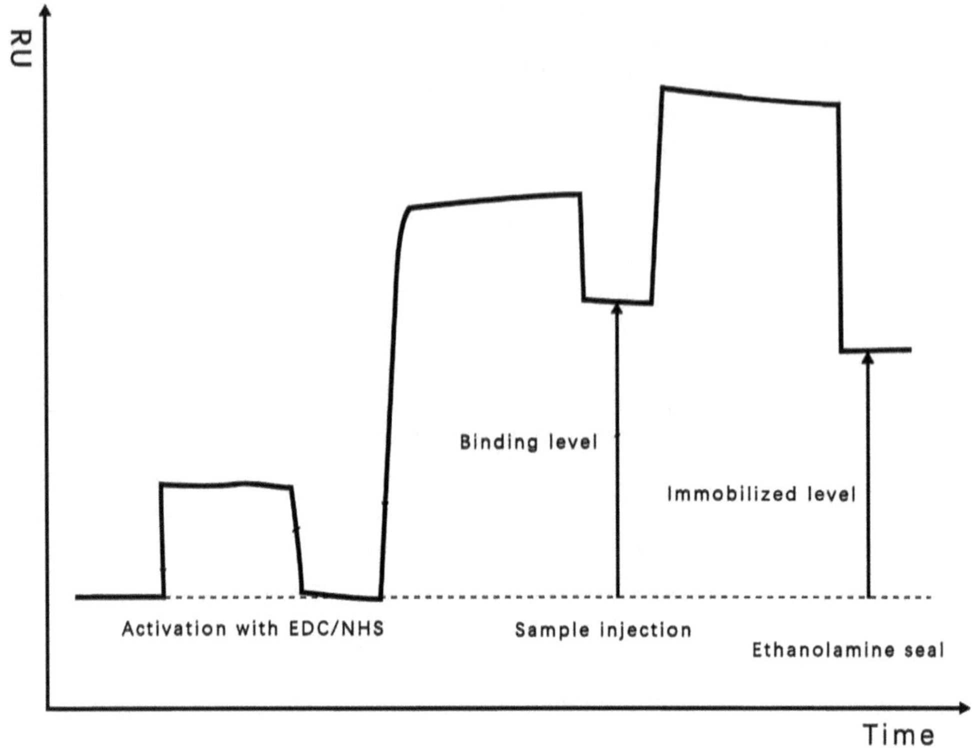

Fig. 4 A typical amine coupling sensorgram. The sensorgram shows the difference between the amount of ligand binding and the amount of ligand fixation

4. Place vials in corresponding positions, and start a manual run by setting the flow rate at 10 μL/min (*see* **Note 17**), injection time of 2 min, and dissociation time of 3 min. Select flow path 2–1 (the analyte will be injected overflow cells 1 and 2) (*see* **Note 18**). This step is called the surface test.
5. Make a series of analyte dilutions according to the surface test results (*see* **Note 19**), at least five concentrations, and one or two zero levels of analyte for double referencing. Select flow path 2–1 (the analyte will be injected into overflow cells 1 and 2).
6. Place vials in corresponding positions and start the experiment.

3.4 Kinetics/Affinity Data Analysis

1. Use BIA evaluation software. Select the data to be analyzed and create an overlay file by selecting the relevant curves. Align all the curves to the start of the injection.
2. Select AFFINITY (*see* **Note 19**) to start the steady-state affinity analysis evaluation wizard and perform the curve-fitting process. Set the baseline to zero at a point at the start of the injection in the most stable region. Perform a Y TRANSFORM command in the wizard panel to align all the desired curves.

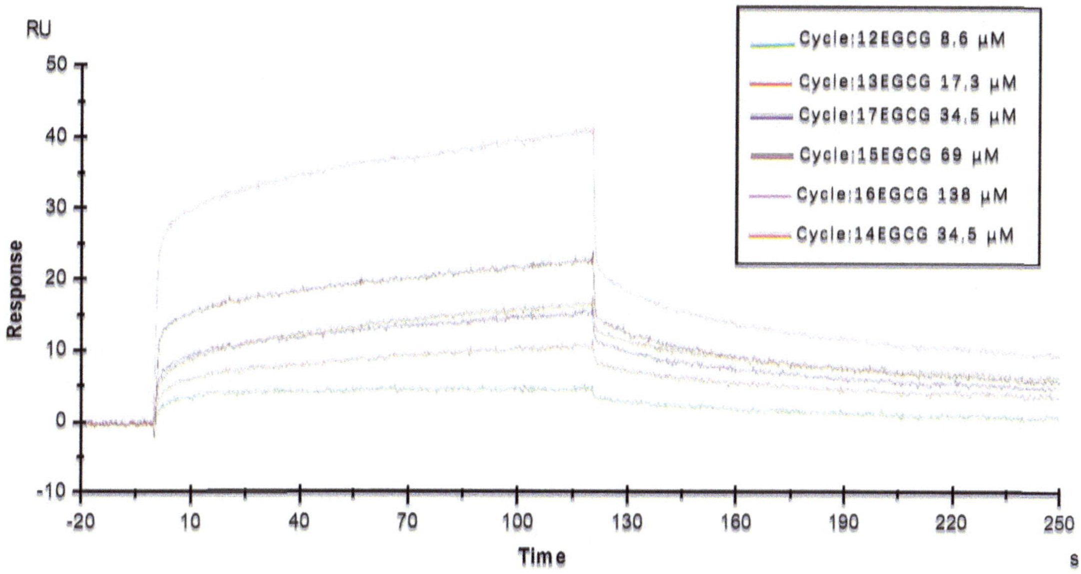

Fig. 5 Sensorgrams of EGCG interacting with the protein GRP78

3. Select the binding model (1:1 binding), which best fits the ligand-analyte interaction (*see* **Note 20**).
4. Select FIT and observe the progress.
5. After the fitting, the curves will appear as an overlay of the experimental and fitted data (Figs. 4 and 5).
6. Take out the CM5 sensor chip and shut down the Biacore system.

4 Notes

1. Series S sensor chips are suitable for use in Biacore T100, Biacore T200, Biacore A100, Biacore S200, Biacore S51, and Biacore 3000. The most versatile Series S sensor chip is the CM5 chip, which is the first choice for immunization via –NH, –SH, –CHO, –OH, or –COOH groups. Several sensor chips with different performance characteristics such as CM7, CM4, CM3, and C1 are available from GE Healthcare. These chips can be coupled or captured with different characteristics of biomolecules, please refer to the sensor surface handbook, or visit the official website (www.gelifesciences.com/biacore).

   In Biacore systems, all sensor chips are composed of glass and a gold-coated film, which is covered with a modified dextran connection matrix layer. The surface of the covering substrate determines the application of the sensing chip. In addition, to produce an SPR signal, a docking system is needed for implanting a sensor chip in the optical system.

2. Amine coupling chemistry is the most widely applicable approach for attaching biomolecules covalently to the surface [8]. For protein coupling, two common approaches can be used:
   - Covalent immobilization. Covalent coupling ligands through irreversible chemical bonds connect to the chip surface, typically by CM5, CM7, CM4, and CM3 of carboxymethyl groups on the sensor chip. Ligands will remain on the surface, and the noncovalent binding of analyte molecules will be removed by regeneration after each analysis cycle. It can use amino, sulfhydryl, or aldehyde groups on the ligands (such as through oxidation of glycol). Amine coupling is one of the most widely used methods.
   - Noncovalent immobilization. Biacore developed various types of modified gold thin film surfaces that could detect affinity-tagged proteins such as glutathione S-transferase (GST)-tagged, hexahistidine (His)-tagged, and maltose binding protein (MBP)-tagged proteins [9–11].
3. Dispense the EDC and NHS solutions separately in aliquots for storage to avoid repeated freezing and thawing and store the aliquots of EDC and NHS frozen at −18 °C or below. Ethanolamine may be stored at 2–8 °C.
4. By means of isoelectric focusing or using the software to determine or predict the isoelectric point (pI) of the protein. In general, it is not recommended to use a Biacore system for the analysis of proteins when the pI is higher than 7 or less than 3. The pH for amine coupling should not be less than 4 or so [7]. To obtain good data, pure and stable protein samples are critical. The purity of the ligand is very important to ensure binding specificity and capacity. The purity of the protein to be analyzed should reach 90% or greater. The protein should be active in PBS buffer. It is recommended to check the purity and the stability of the proteins at least 1 day before doing the experiment. This can be done by sodium dodecyl sulfate (SDS)-polyacrylamide gel electrophoresis (PAGE) and Coomassie blue staining. Low ionic strength is advantageous to the electrostatic interaction, and therefore, it is generally recommended to use a buffer with a 10–20 mM total cation concentration.
5. Compounds can be dissolved in PBS buffer containing 1% DMSO and then diluted into 10^{-3}, 10^{-4}, and 10^{-5} M for preliminary analysis (surface test); the concentration of the positive compounds in the preliminary analysis may be applicable at 10^{-4} M, for example, and then diluted to 10^{-4}, 5×10^{-6}, 2.5×10^{-6}, 1.25×10^{-6}, and 6.25×10^{-7} M with PBS buffer containing 1% DMSO for kinetic constant analysis.

6. Sample preparation is a key factor affecting the quality of dynamic measurement. Ideally, the ligand and analyte both should be of high purity, with properties of uniform, to simplify the interpretation of the results.
7. The SPR experiments require the use of filtered and degassed buffers.
8. HEPES (4-(2-hydroxyethyl)-1-piperazineethanesulfonic acid, pK 7.55) is applicable to many protein–protein interaction studies. Phosphate buffer (PBS) with DMSO is usually more suitable for small molecule analysis because the organic elements such as HEPES buffer can combine with ligands and interfere with the accurate detection of low molecular weight compounds. Many low molecular weight organic compounds, particularly compounds in pharmaceutical research and development work, have low water solubility, so an organic solvent such as DMSO is often added to increase the solubility. Biacore systems can be compatible with up to 10% DMSO and 10% dimethylformamide (DMF) in buffer. However, organic solvents can significantly change the refractive index of ontology of samples and buffer. For example, 1% DMSO can produce about 1200 RU response value level.
9. Before making activation and coupling, you must find the optimum coupling pH. When pH $>$ 3.5, the chip's surface glucan is negatively charged; while the protein is positively charged when pH is $<$ pI. The condition of pH $>$ 3.5 but $<$ pI is suitable for amine coupling immobilization. For many proteins, 10 mM sodium acetate buffer (pH 4.5) is suitable.
10. A dummy sensor chip must also be docked in the instrument in order for the procedure to be run when performing immobilization in the Surface Prep unit.
11. Immobilization pH scouting applies only to CM-series chips.
12. If you use the immobilization wizard template from the Biacore control software, do not mix the EDC and NHS solution. The software dialog box will ask you to prepare an empty vial to mix the two solutions. Blank surfaces prepared in this way are activated with NHS/EDC and deactivated with ethanolamine, without immobilizing ligand. For some applications, particularly those using low immobilized levels and low response levels, blank surfaces should preferably be prepared by immobilizing an equivalent amount of an irrelevant protein.
13. Blank surfaces prepared in this way are activated with NHS/EDC and deactivated with ethanolamine, without immobilizing ligand. For some applications, particularly those using low immobilized levels and low response levels, blank surfaces are preferably prepared by immobilizing an equivalent amount of an irrelevant protein.

14. Ligand coupling level determines the ligands and the combination of the analyte molecules; R_{max} describes the maximum protein binding Max of analyte binding ligand. If the relative sizes of ligand and analyte are known, the theoretical maximum binding capacity of the analyte on the surface of the chip can be obtained from the fixed amount of ligand on the chip (the relative response value increases). Assume a 1:1 stoichiometric relationship of binding reaction:

$$R\max = \frac{\text{analyte } MW}{\text{ligand } MW} \times R_L \times S_m$$

where analyte *MW*: molecular weight of the analyte. Ligand *MW*: molecular weight of the ligand. R_L: the amount of immobilized ligand in RUs. S_m: the stoichiometry of the reaction (number of binding sites per ligand).

15. Observation of the pattern of the sensor gram resulting from ligand immobilization and the analyte binding gives you information on the strength of the interaction. If the association and the dissociation are fast (1–2 min), it is difficult to estimate the association rate constant (k_a) and dissociation rate constant (k_d). This suggests that the experiment will allow only the estimation of the binding affinity of the complex. On the other hand, if the sensorgram shows slow association/dissociation phases (5 min for association and 10–60 min or more for dissociation), it is likely able to estimate the kinetics of the interaction.
16. Velocity varies among various analysis categories; a typical range is 5–40 μL/min.
17. This step is performed after ligand immobilization and baseline stability. The choice of the analyte concentration depends on the strength of the interaction. If you are testing a new interaction, it is recommended to start with micromolar concentration (e.g., 10–50 μM). If this leads to a high signal with a very slow dissociation, decrease your analyte concentration to the nanomolar range. In the case of antibody-antigen interactions, it is known that they are tight, and you can start with a nanomolar concentration (e.g., 100 nM).
18. In determining whether a ligand immobilization is satisfactory, remember that the level of immobilized ligand does not necessarily reflect the functional capacity of the surface. The important criterion is the analyte binding capacity.
19. An important application of Biacore systems is the determination of interaction kinetic constants and affinity. The determination of kinetics and affinity requires the detection of gradient concentrations of analytes. The kinetic (rate) constants can be obtained by fitting the sensorgrams with mathematical models describing the interaction mechanism. The

affinity constant can be obtained either from the steady-state binding level or from the ratio of the rate constant.

20. The dynamic analysis software uses 1:1 interaction model to fit the binding experimental data between analyte and ligand. A response versus concentration plot for the five samples of different concentrations is acquired and the affinity (KD) is read from the plot.

Acknowledgments

This work was financially supported in part by the USDA (HAW05020H and HAW05044R). B.S. received a fellowship from the graduate student overseas study program, South China Agricultural University, China (2018LHPY005) for studies at the University of Hawaii at Manoa.

References

1. Singh P (2016) SPR biosensors: historical perspectives and current challenges. Sensors Actuators B Chem 229:110–130
2. Piliarik M, Párová L, Homola J (2009) High-throughput SPR sensor for food safety. Biosens Bioelectron 24:1399–1404
3. Al-Qazwini Y, Arasu PT, Noor ASM (2011) Numerical investigation of the performance of an SPR-based optical fiber sensor in an aqueous environment using finite-difference time domain. In: 2011 2nd international conference on photonics, pp 1–4. https://doi.org/10.1109/ICP.2011.6106886
4. Miyazaki CM, Shimizu FM, Ferreira M (2017) Surface plasmon resonance (SPR) for sensors and biosensors. In: Nanocharacterization techniques. William Andrew, Norwich, pp 183–200. https://doi.org/10.1016/B978-0-323-49778-7.00006-0
5. Zhou L, Wang YH, Tong SS et al (2017) Single chain Fc-dimer-human growth hormone fusion protein for improved drug delivery. Biomaterials 117:24–31
6. Sun B, Tan D, Pan D et al (2021) Dihydromyricetin imbues antiadipogenic effects on 3T3-L1 cells via direct interactions with 78-kDa glucose-regulated protein. J Nutr 151:1717–1725
7. Concepcion J, Witte K, Wartchow C et al (2009) Label-free detection of biomolecular interactions using biolayer interferometry for kinetic characterization. Comb Chem High Throughput Screen 12:791–800
8. Fischer MJE (2010) Amine coupling through EDC/NHS: a practical approach. Methods Mol Biol 627:55–73
9. Jung SO, Ro H-S, Kho BH et al (2005) Surface plasmon resonance imaging-based protein arrays for high-throughput screening of protein-protein interaction inhibitors. Proteomics 5:4427–4431
10. Hutsell SQ, Kimple RJ, Siderovski DP et al (2010) High-affinity immobilization of proteins using biotin- and GST-based coupling strategies. Methods Mol Biol 627:75–90
11. Qu JH, Horta S, Delport F et al (2020) Expanding a portfolio of (FO-) SPR surface chemistries with the Co(III)-NTA oriented immobilization of His6-tagged bioreceptors for applications in complex matrices. ACS Sensors 5:960–969

Chapter 16

High-Throughput Protein–Protein Interactions Screening Using Pool-Based Liquid Yeast Two-Hybrid Pipeline

Benoît Castandet, Claire Lurin, Étienne Delannoy, and Dario Monachello

Abstract

Because of its adaptability to high-throughput approaches and a low operating cost, the yeast two-hybrid (Y2H) assay remains the most widely used one for high-throughput protein–protein interactions (PPI) mapping experiments. Here we provide a detailed protocol for a liquid culture-based high-throughput binary protein–protein Y2H screen pipeline of a pool of 50 proteins used as baits against a collection of ~12,000 Arabidopsis proteins encoded by sequence-verified open reading frames (ORFs).

Key words Protein–protein interaction, Yeast two-hybrid, Screening, High-throughput, Interactome network

1 Introduction

More than 20 years after the first release of the Arabidopsis genome [1], the functions of a significant portion of its encoded proteins are still not known [2]. Because physical interactions mediated by proteins are a critical element of any biological system, one method of choice to get insights into the putative function(s) of a protein is to identify its partners. This can be done with high-throughput methods that can produce highly reliable protein–protein interactions or proteome-scale PPI networks datasets, "proteins interactomes" [3–6].

Two major classes of experimental approaches are used for protein interaction mapping: analysis of direct interactions using binary methods such as yeast two-hybrid (Y2H) or split ubiquitin, and analysis of protein complexes through affinity purification followed by mass spectrometry. The Y2H system is based on the reconstitution of a transcription factor through the coexpression of two-hybrid (fusion) proteins in the same yeast cell. The bait protein X is fused to the Gal4p DNA binding domain (DB) that binds to the promoter sequence upstream of a reporter gene, while

Shahid Mukhtar (ed.), *Protein-Protein Interactions: Methods and Protocols*, Methods in Molecular Biology, vol. 2690,
https://doi.org/10.1007/978-1-0716-3327-4_16,

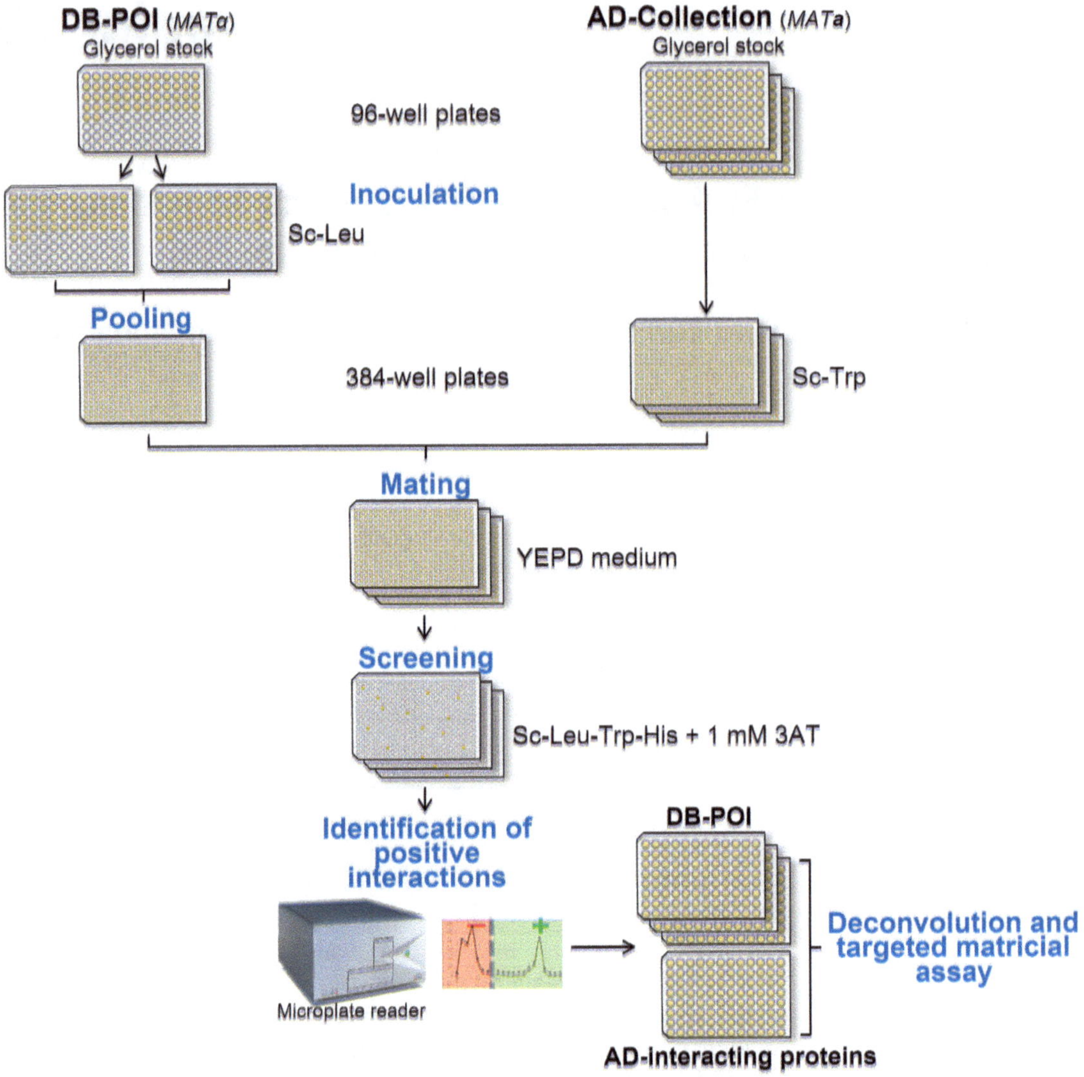

Fig. 1 Schematic overview of the liquid culture-based high-throughput binary protein–protein Y2H screen pipeline

the prey protein Y is fused to the Gal4p activation domain (AD) that recruits the RNA polymerase and activates transcription of the reporter gene. Protein–protein interaction can therefore be assessed by readout of phenotypic screens due to the activation of the reporter gene that usually allows positive growth selection or colorimetric detection. The liquid protocol described here (Fig. 1) is an adaptation of the one presented in Dreze et al. (2010) [7], where the same low copy number yeast expression vectors and the two Y2H strain backgrounds were used. A pool of 50 proteins of interest (POI) used as baits is screened against 12,000 Arabidopsis proteins encoded by sequence-verified ORFs (12k_space) [8].

A subset of 8,000 ORFs (8k_space) was previously used to generate the Arabidopsis Interactome 1 (AI-1) [9] and the Plant-Pathogen Immune Network 1 (PPIN-1) [10]. After identification of candidate interacting proteins, a matricial Y2H assay is performed by depooling the 50-baits POI and testing against individual candidate preys. A final sequencing step allows the unambiguous identification of both interacting proteins. The combination of a liquid pipeline with a pooling strategy allows the identification of interaction networks in an optimal automated procedure.

2 Materials

Prepare all media using deionized water. Prepare and store all reagents at room temperature (unless indicated otherwise). When indicated sterilize by autoclaving media for 20 min at 120 °C.

2.1 Reagents

1. Gateway® LR Clonase Enzyme Mix (Invitrogen).
2. DH5α competent cells.
3. Platinum® *Taq* DNA Polymerase (Invitrogen).
4. 25 mM dNTP mix.
5. 96-well plasmid miniprep plate kit.
6. Zymolase 20 T (21,100 U/g).
7. Salmon sperm DNA.

2.2 Yeast Strains, Plasmids, and Primers

1. *Saccharomyces cerevisiae* Y8800 *MAT*a and Y8930 *MAT*α harbor the following genotype: *leu2-3 112 trp1-901 his3-200 ura3-52 gal4D gal80D GAL2-ADE2 LYS2::GAL1-HIS3 MET2::GAL7-lacZ cyh2R*. The reporter genes *GAL2-ADE2* and *LYS2::GAL1-HIS3* are integrated into the yeast genome. Expression of the *GAL1-HIS3* reporter gene has to be tested with 1 mM 3AT (3-amino-1,2,4- triazole, a competitive inhibitor of the HIS3 gene product). By convention, the Y8800 *MAT*a and Y8930 *MAT*α strains are transformed with AD-Y and DB-X constructs, respectively.
2. Genes of interest are cloned into the pDest-AD (yeast selection marker: TRP1) and pDest-DB (yeast selection marker: LEU2) vectors [7].
3. Primer sequences:

   Gal4AD: 5′ TACCACTACAATGGATG 3′.

   Gal4DB: 5′ TCATCGGAAGAGAGTAG 3′.

   Gal4STOP: 5′ CACTTTGTACAAGAAAGC 3′.

2.3 Media and Solutions

1. LB (Lysogeny broth): mix 10 g of NaCl, 10 g of tryptone, and 10 g of yeast extract adjust pH to 7.0 with NaOH, add deionized H_2O to a final volume of 1 L, and autoclave. To prepare LB plates, add 20 g of agar per liter of LB broth and autoclave. Allow the medium to cool down below 45 °C, then add the appropriate antibiotics and pour the plates.
2. YEPD (yeast extract–peptone–dextrose) medium: Dissolve 10 g of yeast extract, 20 g of peptone, and 20 g of glucose in deionized H_2O. Adjust to a final volume of 1 L and autoclave. To prepare YEPD plates, add 20 g of agar per liter and autoclave.
3. 10× TE (Tris EDTA) Buffer: 100 mM Tris–HCl pH 8.0, 10 mM EDTA. Dissolve 15.759 g of Tris–Cl pH 8.0 and 2.92 g of EDTA pH 8.0 and add deionized H_2O to a final volume of 1 L and autoclave. Store at 4 °C.
4. 1 M LiAc (lithium acetate): Dissolve 5.1 g lithium acetate dihydrate in 50 mL deionized water and autoclave. Store at 4 °C.
5. 44% (w/v) PEG3350 (polyethylene glycol): Add 44 g polyethylene glycol (average molecular weight 3350) to 100 mL of deionized water. Stir until everything is dissolved and autoclave. Store at 4 °C.
6. TE/LiAc (Tris–EDTA–lithium acetate solution): TE/LiAc is prepared by ten-fold dilution of 10× TE and 1 M LiAc stocks to give 10 mM Tris–HCl (pH 8.0), 1 mM EDTA (pH 8.0), and 100 mM LiAc final concentration.
7. TE/LiAc polyethyleneglycol (PEG) solution supplemented with boiled carrier DNA. This solution is prepared by combining 8 volumes of 44% (w/v) PEG 3350 with 1 volume of 10× TE and 1 volume of 1 M LiAc. Add 200 μL of boiled carrier DNA to 10 mL of TE/LiAc/PEG solution.
8. Dropout amino acids mix (−Ade, −Ura − His, −Leu, −Trp): Add 10 g of each chemical: alanine, arginine, asparagine, aspartic acid, cysteine, glutamic acid, glutamine, glycine, isoleucine, lysine, methionine, phenylalanine, proline, serine, threonine, tyrosine, and valine in a container, mix all components thoroughly and store under dry conditions at 4 °C.
9. Sc (synthetic complete) medium: Dissolve 2.6 g of dropout amino acids mix, 3.4 g of Yeast Nitrogen Base without amino acids, 10 g of ammonium sulfate, and 20 g of glucose, adjust pH to 5.9 with NaOH, add deionized H_2O to a final volume of 1 L and autoclave. To prepare Sc plates, add 20 g of agar per liter and autoclave.

10. 65 mM Ade (adenine) stock solution: Dissolve 2.4 g of adenine hemisulfate salt, add deionized H_2O to a final volume of 200 mL, and filter-sterilize.
11. 20 mM Ura (uracil) stock solution: Dissolve 0.45 g of uracil, add deionized H_2O to a final volume of 200 mL, and filter-sterilize.
12. 100 mM His (histidine) stock solution: Dissolve 3.1 g of L-histidine, add deionized H_2O to a final volume of 200 mL, and filter-sterilize (store light protected).
13. 100 mM Leu (leucine) stock solution: Dissolve 2.64 g of L-leucine, add deionized H_2O to a final volume of 200 mL, and filter-sterilize.
14. 40 mM Trp (tryptophan) stock solution: Dissolve 1.63 g of L-tryptophane, add deionized H_2O to a final volume of 200 mL, and filter-sterilize. Store in the dark at 4 °C.
15. 1 M 3-AT (3-amino-1,2,4-triazole) stock solution: Dissolve 0.84 g of 3-amino-1,2,4-triazole in 10 mL of deionized water. Filter-sterilize. Divide into 1 mL aliquots and store at −20 °C.
16. Sc-selective media: Depending on the selective medium you want add 8 mL of Leu-Trp-Ura–His, 15 mL of Ade, and 1 mL of 3-AT from the stock solutions for a final volume of 1 L.
17. 0.1 M Sodium phosphate buffer (pH 7.4): Prepare 0.2 M dibasic sodium phosphate (dissolve 35.61 g of $Na_2HPO_4*2H_2O$, add deionized H_2O to a final volume of 1 L) and 0.2 M monobasic sodium phosphate (dissolve 27.6 g of $NaH_2PO_4*H_20$ add deionized H_2O to a final volume of 1 L). Mix 81% of 0.2 M dibasic sodium phosphate with 19% of 0.2 M monobasic sodium phosphate and dilute 1:1 with deionized H_2O.
18. Lysis buffer: Dissolve 2.5 mg/mL zymolase 20 T in 0.1 M sodium phosphate buffer (pH 7.4). Keep on ice.
19. 40% (v/v) Glycerol: Dilute pure glycerol in deionized H_2O to create a 40% glycerol solution and autoclave.

2.4 Disposables

1. 96-well and 384-well microplates with cover.
2. Pipette tips.
3. 50 mL polypropylene conical tubes.
4. 96 well PCR plate.
5. Standard microcentrifuge tubes, 1.5 mL and 2.2 mL.
6. Petri dishes.
7. 120 mm square Petri dishes.

8. 2 mL deep-well 96-well plates.
9. Sterile 50 mL plastic reservoirs.
10. Erlenmeyer flasks.
11. PCR microplate sealing films.
12. Microporous microplate sealing films.
13. Aluminum microplate sealing foil.
14. Combitips for multipette dispenser.

2.5 Equipment

1. Automated picking system or 96-Pin and 384-Pin hand replicators.
2. Microplate reagent dispenser or multichannel pipette.
3. Microplate reader.
4. Water bath.
5. Pipetman.
6. Tabletop centrifuge.
7. Plate centrifuge.
8. Incubator.
9. Thermocycler with programmable temperature stepping functionality, 96 well.
10. 96-well plate shaker.
11. Multipette dispenser.

3 Methods

Pools used in this Y2H liquid pipeline combine 50 different DB-X expressing yeast clones corresponding to the proteins of interest (POI) used as baits screened against the 12k_space AD-collection. This experimentally defined pool size provides an optimal compromise between screening efficiency (number of plates to be processed) and screen sensitivity (number of interactors identified). For high-throughput experiments, it is preferable to use sequence-independent recombinational cloning technologies [11].

3.1 Generation of Expression Plasmids by Recombinational Cloning

The 50 ORFs used as baits were transferred into pDest-DB and pDest-AD yeast expression vectors in a simple single-step recombination reaction (*see* **Note 1**). Completed recombination reactions are then transformed into *Escherichia coli*, grown for 18 h, and plasmids are isolated.

1. For each bait combine in a 96-well PCR plate:

   1 μL of entry clone (10 ng/μL).

   1 μL of destination vector (100 ng/μL).

1 μL of 5× clonase buffer.

1.6 μL 1× TE.

0.4 μL of cloning enzyme mix (keep this mix on ice) (*see* **Note 2**).

2. Homogenize by gently pipetting up and down.
3. Seal the plate with a PCR microplate film.
4. Incubate in a thermocycler at 25 °C for 18 h.
5. Add 0.2 μL of Proteinase K solution (provided together with the cloning kit).
6. Seal the plate with a PCR microplate film.
7. Incubate in a thermocycler at 37 °C for 15 min.
8. Do bacterial transformation or store at −20 °C.

3.2 Bacterial Transformation

1. Thaw competent DH5*α* cells on ice (*see* **Note 3**).
2. Add 100 μL of competent cells into the 96-well PCR plate containing the recombination reaction mix on ice.
3. Seal the plate with aluminum adhesive foil.
4. Incubate on ice-cold water for 20 min (*see* **Note 4**).
5. Heat shock cells in a water bath at 42 °C for 1 min.
6. Incubate on ice-cold water for 5 min.
7. Using a multichannel pipette transfers the cells into a 96-deep well plate precedently filled with 400 μL LB media.
8. Seal the plate using a microporous sheet to allow gas exchange.
9. Incubate on a 96-well plate shaker at 37 °C for 1 h.
10. Add 500 μL of LB media with 200 mg/L Ampicillin (*see* **Note 5**).
11. Seal the plate using a microporous sheet to allow gas exchange.
12. Incubate overnight on a 96-well plate shaker at 37 °C.
13. Remove 5 μL for subsequent analysis by PCR.
14. Keep bacterial cultures at 4 °C until PCR results are determined.
15. Remove 100 μL of the overnight culture, mix with 100 μL of 40% (v/v) autoclaved glycerol into a 96-well microplate, seal the plate with aluminum microplate foil, and store at −80 °C.
16. Use the remainder of the overnight culture for plasmid isolation with a 96-well plasmid miniprep plate kit following the instructions of the manufacturer.

3.3 Bacterial Culture PCR

1. Using a multichannel pipette dilute 5 μL of bacterial culture into 95 μL of sterile water and mix by pipetting up and down. Keep on ice.
2. Following the manufacturer's protocol scale down the reaction volume to 10 μL per well and prepare a PCR master mix (enough for 96-wells) containing Gal4AD or Gal4DB forward primer and Gal4STOP reverse primer. Keep on ice.
3. Using a multichannel pipette dispense 9 μL of the PCR master mix into each well of a 96-well PCR plate.
4. Using a multichannel pipette add 1 μL of the diluted bacterial culture as a DNA template.
5. Seal the plate with a PCR microplate film.
6. Place the PCR plate on a thermocycler and run the following program:

   Step 1: Denaturation at 94 °C for 10 min.

   Step 2: Denaturation at 94 °C for 30 s.

   Step 3: Annealing at 50 °C for 30 s.

   Step 4: Elongation at 68 °C for × min (depending on the length of the longer ORF).

   Repeat Steps 2–3-4, 34 times.

   Step 5: Final elongation at 68 °C for 10 min.

   Step 6: Hold at 10 °C.
7. Once PCR reactions are completed, run 5 μL of PCR product on a 1% agarose. Successful recombination reactions will give a band corresponding to the ORFs sizes to which ~300 bp of vector sequences are added (because of the primer location on the plasmid).

3.4 Yeast Transformation

This 96-well scale yeast transformation protocol allows the transformation of up to 96 plasmids into the same yeast strain in parallel. pDest-DB and pDest-AD plasmids are transformed into yeast *S. cerevisiae* strain Y8930 (α-type) and Y8800 (**a**-type) respectively. In addition, transform the Y8800 strain with an empty pDest-AD vector for the subsequent identification of autoactivating DB-X hybrid proteins. In order to obtain maximum transformation efficiencies, solutions need to be freshly prepared from stock solutions.

1. Streak Y8800 and Y8930 strains from glycerol stock on separate YEPD Petri dishes and incubate at 30 °C for 72 h (*see* **Note 6**).
2. For each strain, inoculate 20 mL preculture of YEPD into a 2 L Erlenmeyer (*see* **Note** 7) and incubate at 30 °C on a shaker for 14–18 h.

3. Add 80 mL of YEPD media.
4. Incubate at 30 °C on a shaker for 1 h (*see* **Note 8**).
5. Boil carrier DNA (salmon sperm DNA) for 5 min then place on ice until needed.
6. Transfer the cultures into 50 mL polypropylene conical tubes and harvest cells by centrifugation at 800 × *g* for 5 min. Discard the supernatant and resuspend cells gently in 10 mL of sterile water.
7. Centrifuge at 800 × *g* for 5 min and discard the supernatant.
8. Resuspend cells in 10 mL of TE/LiAc solution, centrifuge at 800 × *g* for 5 min, and discard the supernatant.
9. Resuspend cells in 2 mL of TE/LiAc solution, then add 10 mL of TE/LiAc/PEG solution (*see* **Note 9**) supplement with 200 μL of boiled carrier DNA.
10. Transfer the cell into a 50 mL sterile reservoir.
11. Dispense 120 μL into each well of a round-bottom 96-well PCR plate.
12. Add 10 μL of plasmid DNA (*see* **Note 10**) to the competent yeast and mix by pipetting up and down several times (*see* **Note 11**).
13. Seal the 96-well PCR plate with a PCR microplate foil and incubate at 30 °C for 30 min in a thermocycler.
14. Subject to heat shock in a 42 °C water bath for 15 min.
15. Centrifuge 5 min at 800 × *g* in a plate centrifuge. Carefully remove the supernatant (*see* **Note 12**).
16. Add 100 μL of sterile water and resuspend cell pellets by pipetting up and down.
17. Seal the plate with aluminum adhesive foil.
18. Centrifuge 5 min at 800 × *g*, then carefully remove 90 μL of water.
19. Resuspend cell pellets by pipetting up and down.
20. Using a multichannel pipette spot 5 μL of cell suspension onto appropriate selective 120 mm square Petri dishes media (Sc-Trp for AD-Y, Sc-Leu for DB-X agar media).
21. Incubate at 30 °C for 72 h.
22. Using a multichannel pipette pick transformed yeast colonies into 96-well microplates containing 100 μL of selective media (Sc-Trp for AD-Y, Sc-Leu for DB-X). Inoculate the Y8800 strain carrying the empty pDest-AD vector into a 50 mL polypropylene conical tube filled with 10 mL of Sc-Trp.
23. Incubate on a shaker at 30 °C for 72 h.

24. Remove 5 μL for subsequent lysis and analysis by PCR.
25. Transfer the 10 mL culture of the Y8800-empty pDest-AD into a 50 mL sterile reservoir.
26. Using a multichannel pipette transfer 100 μL of the culture into each well of a 96-well microplate.
27. Prepare archival stocks by combining 100 μL of the yeast cultures with 100 μL of 40% (v/v) autoclaved glycerol into a 96-well microplate. Seal the 96-well plates with an aluminum microplate foil and store them at −80 °C.

3.5 Yeast Cell Lysis

1. Prepare lysis buffer considering 15 μL volume per well. Keep on ice.
2. Using a multipette dispenser aliquot 15 μL of lysis buffer into the wells of a 96-well PCR plate. Keep on ice.
3. Add 5 μL of each yeast cell culture into the wells of the 96-well PCR plate.
4. Seal the 96-well plate with a PCR film.
5. Place the PCR plate on a thermocycler and run the following program:

   Step 1: 37 °C for 15 min.

   Step 2: 95 °C for 5 min.

   Step 3: Hold at 10 °C.
6. Add 80 μL of filter-sterilized water to each well.
7. Do yeast lysate PCR or store at −20 °C.

3.6 Yeast Lysate PCR

1. Following the manufacturer's protocol, scale down the reaction volume to 10 μL per well and prepare a PCR master mix (enough for 96-wells) containing Gal4AD or Gal4DB forward primer and Gal4STOP reverse primer. Keep on ice.
2. Using a multichannel pipette dispense, 9 μL of the PCR master mix into each well of a 96-well PCR plate.
3. Using a multichannel pipette, add 1 μL of the diluted yeast cell lysate as a DNA template.
4. Seal the plate with a PCR microplate film.
5. Place the PCR plate on a thermocycler and run the following program:

   Step 1: Denaturation at 94 °C for 5 min.

   Step 2: Denaturation at 94 °C for 30 s.

   Step 3: Annealing at 50 °C for 30 s.

   Step 4: Elongation at 68 °C for × min (depending on the length of the longer ORF).

   Repeat Steps 2–3-4, 34 times.

Step 5: Final elongation at 68 °C for 10 min.

Step 6: Hold at 10 °C.

6. Once PCR reactions are completed, run 5 μL of PCR product on a 1% agarose. Successful reactions will give a band corresponding to the ORFs sizes to which ~300 bp of vector sequences are added (because of the primer location on the plasmid).

3.7 Identification of Autoactivating DB-X Hybrid Proteins

There is a possibility that DB-X hybrid proteins activate transcription of the Y2H reporter genes regardless of the presence of any AD-Y. In the Y2H system, DB-X autoactivators can be identified and removed by performing a prephenotypic screen for reporter gene activation. Autoactivator baits can be alternatively screened in an AD-fusion configuration. Identification of autoactivators is achieved in diploid yeast strains obtained by mating DB-X yeast strains with the Y8800 yeast strain transformed with the AD encoding plasmid containing no insert (empty pDest-AD).

1. Thaw glycerol stocks plate of the 50 individual DB-bait yeast strains and of the Y8800 strain carrying the empty pDest-AD vector.
2. Carefully vortex the glycerol stocks plates (*see* **Note 13**).
3. Using the colony picker or a hand replicator, inoculate the DB-bait yeast and the Y8800 strain transformed with an empty pDest-AD vector plate into 96-well microplates filled with 100 μL of fresh Sc-Leu and Sc-Trp selective media, respectively.
4. Seal all plates with adhesive tape and return glycerol stocks to −80 °C.
5. Incubate at 30 °C for 48 h on a shaker (*see* **Note 14**).
6. Using the colony picker or a hand replicator successively transfers the DB-baits and the Y8800 with the empty pDest-AD cultures by replica-plate onto a YEPD media plate. Wash the hand replicator between two plates using consecutive baths of H_2O and absolute ethanol. In the end, sterilize the replicator with a flame.
7. Incubate at 30 °C for 24 h.
8. Replica-plate the YEPD plate into a 100 μL Sc-Leu-Trp diploids media plate and into a 100 μL Sc-Leu-Trp-His +1 mM 3AT media plate. Wash the hand replicator between two plates using consecutive baths of H_2O and absolute ethanol and flame sterilize it.
9. Incubate at 30 °C for 5 days (*see* **Note 15**).

10. Score growth phenotypes. Using a microplate reader, measure the OD600 of the plates in order to identify yeast growth (usually an OD600 higher than 0.4 for a 5-day yeast culture and lower than 0.1 for no growth). All yeast diploid cells should grow in the Sc-Leu-Trp diploid media. Nonautoactivating yeast diploid cells are not able to activate the GAL1::HIS3 reporter gene hence should not grow on the Sc-Leu-Trp-His +1 mM 3AT media. Yeast strains showing a growth phenotype on this media are considered autoactivators. They are physically removed from the collection of DB-X baits and screened against the space DB-collection using their AD construction.

3.8 Construction of DB Baits Pool

This protocol describes the construction of one pool of 50 different DB-hybrid constructs individually transformed into Y8930 yeast strain.

1. Thaw glycerol stocks plate of the 50 individual DB-bait yeast strains.
2. Carefully vortex the glycerol stocks plate.
3. Using the colony picker or a hand replicator, inoculate two 96-well microplates filled with 150 μL of fresh Sc-L selective media.
4. Seal the plate with aluminum tape and return glycerol stocks to −80 °C.
5. Incubate at 30 °C for 72 h on a shaker.
6. Resuspend yeast cell cultures by careful vortexing the culture plates.
7. Using a microplate reader, measure the OD600 to ensure that growth is homogenous throughout each plate, thus that each DB-X yeast strain will be represented in the same proportion.
8. Transfer the contents of the two culture plates into a sterile reservoir. Mix thoroughly to ensure equal representation of all DB-bait yeast strains in the pool.
9. With a multichannel pipette, dispense 50 μL into the wells of a 384-well microplate.

3.9 AD-Collection Culture

1. Thaw glycerol stocks of the 12k_space AD-collection (*see* **Note 16**).
2. Carefully vortex the glycerol stocks plates.
3. Using the colony picker or a hand replicator inoculate into 32 384-well plates (*see* **Note 17**) filled with 50 μL of fresh Sc-Trp selective media. In order to avoid contamination, wash the hand replicator between two plates using consecutive baths of H_2O and absolute ethanol and flame sterilize it.

4. Seal all plates with aluminum tape and return glycerol stocks to −80 °C.
5. Incubate at 30 °C for 72 h.

3.10 Mating

1. Using the colony picker or a hand replicator replica plate, the DB-baits plate pools into 32 384-well mating plates filled with YEPD media.
2. Using the colony picker or a hand replicator replica plate, the 72-h culture plates corresponding to the AD-collection into the mating plates precedently inoculated with the DB pool. Wash the hand replicator between two plates using consecutive baths of H_2O and absolute ethanol and flame sterilize it.
3. Incubate the YEPD plates at 30 °C for 24 h.
4. Store the AD-collection culture plates at 4 °C (*see* **Note 18**).

3.11 Screening

1. Using the colony picker or a hand replicator replica plate onto 32 384-well diploid plates filled with 50 μL of fresh Sc-Leu-Trp diploids media and into 32 384-well screening plates filled with 50 μL of fresh Sc-Leu-Trp-His +1 mM 3AT screening media. Wash the hand replicator between two plates using consecutive baths of H_2O and absolute ethanol and flame sterilize it.
2. Incubate at 30 °C for 5 days.
3. Score growth phenotypes. Using a microplate reader measure the OD600 of the plates in order to identify primary positives. All yeast diploid cells should grow in the Sc-Leu-Trp diploid media. Only diploid yeast with interacting couples can grow in screening media.

3.12 AD-Interacting Proteins Rearray and Archival Stock

1. Thaw glycerol stocks of the AD-collection plates corresponding to the identified primary positive AD-interacting proteins or use the AD-collection culture plates precedently stored at 4 °C.
2. Carefully vortex the glycerol stocks plates.
3. Using the colony picker or sterile toothpicks rearray the primary positives by inoculating AD-interacting proteins yeast into 96-well plates filled with 100 μL of fresh Sc-Trp media.
4. Seal all plates with adhesive tape and return glycerol stocks to −80 °C.
5. Incubate at 30 °C for 72 h.
6. Prepare an archival glycerol stock by adding 100 μL of 40% (v/v) autoclaved glycerol.

3.13 Deconvolution and Targeted Matricial Assay Cultures

1. Thaw the glycerol stocks of the 50 individual DB-bait yeast strains and the identified primary positive AD-interacting proteins plates.
2. Carefully vortex the glycerol stocks plates.
3. Individually inoculate 50 distinct 50 mL polypropylene conical tubes containing 10 mL of fresh Sc-Leu selective media with 5 μL of the 50 individual DB-baits.
4. Using the colony picker or a hand replicator, inoculate the glycerol stocks of the identified primary positive AD-interacting proteins yeast plates into 96-well plates filled with 100 μL of fresh Sc-Trp selective media.
5. Seal all plates with aluminum tape and return glycerol stocks to −80 °C.
6. Incubate at 30 °C for 72 h.
7. Individually transfer the content of the 50 mL polypropylene conical tubes into 50 mL reservoirs.
8. With a multichannel pipette, dispense 100 μL into the wells of 50 distinct 96-well microplates.

3.14 Mating

1. Using the colony picker or a hand replicator, replica-plate the identified primary positive AD-interacting proteins cultures into 50 96-well mating plates filled with 100 μL of YEPD media.
2. Using the colony picker or a hand replicator, replica-plate the individual DB-baits cultures into the 96-well mating plates precedently inoculated with the AD-interacting proteins. Wash the hand replicator between two plates using consecutive baths of H_2O and absolute ethanol and flame sterilize it.
3. Incubate at 30 °C for 24 h.

3.15 Screening

1. Using the colony picker or a hand replicator replica plate onto 50 96-well diploid plates filled with 100 μL of fresh Sc-Leu-Trp and into 50 96-well screening plates filled with 100 μL of fresh Sc-Leu-Trp-His +1 mM 3AT media. Wash the hand replicator between two plates using consecutive baths of H_2O and absolute ethanol and flame sterilize it.
2. Incubate at 30 °C for 5 days.
3. Score growth phenotypes. Using a microplate reader measure the OD600 of the plates in order to identify primary positives. All yeast diploid cells should grow in the Sc-Leu-Trp diploid media. Only diploid yeast with interacting couples can grow in the screening media.

3.16 Diploids Rearray and Interacting Proteins Identification

1. Using the colony picker or sterile toothpicks rearray the identified positive diploid cells inoculating 96-well plates filled with 100 μL of fresh Sc-Leu-Trp-His +1 mM 3AT.
2. Incubate at 30 °C for 72 h.
3. Lyse cells according to 3.5.
4. Amplify the DB-X and AD-Y inserts of positive diploid cultures by yeast colony PCR according to 3.6 for subsequent ORF identification by end-read sequencing.
5. Prepare an archival glycerol stock by adding 100 μL of 40% (v/v) autoclaved glycerol.

4 Notes

1. For large (>5 kb) clones linearizing the entry clone vector may increase the efficiency by up to twofold.
2. We set up a more cost-effective protocol in which we use less than half of the LR Clonase™ enzyme mix compared to the standard procedure.
3. Transform recombinant vectors into a strain that is susceptible to the CcdB toxin to maximize cloning efficiency.
4. Use ice-cold water to optimize thermal exchanges within all the plate wells.
5. Add 500 μL of LB media with 200 mg/L Ampicillin in order to have 1 mL at the final concentration of 100 mg/L Ampicillin.
6. Always use freshly streaked yeast cells. This will result in the highest transformation efficiency.
7. Inoculate into 2-L Erlenmeyer flask in order to allow better gas exchange.
8. Many yeast transformation protocols emphasize the importance of reaching a final OD_{600} range of 0.4–0.8 with 4–6 h culture incubation before harvesting. We find out that transformation efficiency is always optimal just after 1-h incubation independently of the OD_{600} when starting from freshly streaked yeast cells and a saturated preculture.
9. PEG3350 should be taken out of its storage condition in advance, so it can warm up to room temperature before adding to the transformation mixture.
10. For an optimal transformation rate, a minimal plasmid DNA concentration of 100 ng/μL is recommended.
11. TE-LiAc-PEG solution is very viscous. It is very important to properly mix the plasmid and yeast in this solution.

12. Since it could be quite tricky to remove the supernatants without disturbing the pellets from a 96-well plate, tips could be previously marked by a reference point. This should be done by inserting the tips first at the bottom of an empty 96-well plate and simply drawing a mark on them. Marks on the tips will guide you in removing the supernatant without reaching the plate bottom and in keeping the multichannel parallel to the plate axis.
13. To shake 96-well and 384-well microtiter plates, we routinely use the Eppendorf MixMate Mixer, 1 min at 950 and 1750 rpm, respectively.
14. Plate agitation is not mandatory, liquid cultures will grow anyway.
15. Place a water tray in the incubator to preserve high humidity and to prevent media evaporation during the 5 days of incubation.
16. The entire Arabidopsis pDest-AD library will be contained in 127 96-well plates.
17. Inoculate four 96-well plates into one 384-well plates. Do the first plate inoculation by placing the hand replicator in the 384-well A1 position first then sequentially in A2, B1, and B2.
18. In order to avoid multiple thawing of the AD-collection glycerol stocks plates for the successively AD-interacting proteins rearray, store the AD-collection culture plates at 4 °C.

Acknowledgments

The IPS2 Interactomic platform benefited from the support of Saclay Plant Sciences-SPS (ANR-17-EUR-0007).

References

1. The Arabidopsis Genome Initiative (2000) Analysis of the genome sequence of the flowering plant Arabidopsis thaliana. Nature 408: 796–815
2. Zaag R, Tamby JP, Guichard C et al (2015) GEM2Net: from gene expression modeling to -omics networks, a new CATdb module to investigate Arabidopsis thaliana genes involved in stress response. Nucleic Acids Res 43: D1010–D1017
3. Braun P, Tasan M, Dreze M et al (2009) An experimentally derived confidence score for binary protein–protein interactions. Nat Methods 6(1):91–97
4. Cusick ME, Yu H et al (2009) Literature-curated protein interaction datasets. Nat Methods 6(1):39–46
5. Simonis N, Rual JF, Carvunis AR et al (2009) Empirically controlled mapping of the Caenorhabditis elegans protein–protein interactome network. Nat Methods 6(1):47–54
6. Venkatesan K, Rual JF, Vazquez A, Stelzl U, Lemmens I et al (2009) An empirical framework for binary interactome mapping. Nat Methods 6(1):83–90
7. Dreze M et al (2010) High-quality binary interactome mapping. Methods Enzymol 470:281–315

8. Weßling R, Epple P, Altmann S, He Y, Yang L et al (2014) Convergent targeting of a common host protein-network by pathogen effectors from three kingdoms of life. Cell Host Microbe 16(3):364–375
9. Arabidopsis Interactome Mapping Consortium (2011) Evidence for network evolution in an Arabidopsis Interactome map. Science 333(6042):601–607
10. Mukhtar MS, Carvunis AR, Dreze M, Epple P et al (2011) Independently evolved virulence effectors converge onto hubs in a plant immune system network. Science 333(6042): 596–601
11. Walhout AJ et al (2000) GATEWAY recombinational cloning: application to the cloning of large numbers of open reading frames or ORFeomes. Methods Enzymol 328:575–592

Chapter 17

Dynamic Enrichment for Evaluation of Protein Networks (DEEPN): A High Throughput Yeast Two-Hybrid (Y2H) Protocol to Evaluate Networks

Ali Zeeshan Fakhar, Jinbao Liu, and Karolina M. Pajerowska-Mukhtar

Abstract

Proteins are the building blocks of life, and a vast array of cellular processes is handled by protein–protein interactions (PPIs). The protein complexes formed via PPIs lead to tangled networks that, with their continuous remodeling, build up systematic functional units. Over the years, PPIs have become an area of interest for many researchers, leading to the development of multiple *in vitro* and *in vivo* methods to reveal these interactions. The yeast-two-hybrid (Y2H) system is a potent genetic way to map PPIs in both a micro- and high-throughput manner. Y2H is a technique that involves using modified yeast cells to identify protein-protein interactions. For Y2H, the yeast cells are engineered only to grow when there is a significant interaction between a specific protein with its interacting partner. PPIs are identified in the Y2H system by stimulating reporter genes in response to a restored transcription factor. However, Y2H results may be constrained by stringency requirements, as the limited number of colony screenings through this technique could result in the possible elimination of numerous genuine interactions. Therefore, DEEPN (dynamic enrichment for evaluation of protein networks) can be used, offering the potential to study the multiple static and transient protein interactions in a single Y2H experiment. DEEPN utilizes next-generation DNA sequencing (NGS) data in a high-throughput manner and subsequently applies computational analysis and statistical modeling to identify interacting partners. This protocol describes customized reagents and protocols through which DEEPN analysis can be utilized efficiently and cost-effectively.

Key words DEEPN, Y2H, PPI, Illumina Sequencing, System Biology, High-throughput

1 Introduction

Proteins, sometimes called "molecular machines," function by intricate dynamic interactors with other complex molecules such as DNA, RNA, and other proteins in living systems [1–9]. Multiple processes in living systems, including signaling, molecular trafficking, immunomodulation, metabolic activities, and further growth and developmental regulations, are accurately governed via proteins or protein–protein interactions (PPIs) [10]. Many *in vitro*

Shahid Mukhtar (ed.), *Protein-Protein Interactions: Methods and Protocols*, Methods in Molecular Biology, vol. 2690, https://doi.org/10.1007/978-1-0716-3327-4_17,

and *in vivo* techniques have been developed over time for studying PPIs. Biotechnological advancements make these newly devised methods readily available to biologists/researchers. Information into the system for potential new binding partners typically precedes a study of protein–protein interactions. Affinity-tagged proteins/affinity-based protein profiling, Y2H, and other quantitative proteomic-based approaches that can be combined, such as co-immunoprecipitation and affinity chromatography, can be utilized to study the potential PPIs [11–19]. Several techniques can validate the PPIs, such as co-immunoprecipitation, surface plasmon resonance (SPR), confocal microscopy for intracellular protein colocalization, and spectroscopic studies [20]. Despite being widely used, several technical limitations exist with the commonly practiced techniques that hinder the identification of essential interactors, which may only be present in a few copies per cell. As a result, data generated using these techniques can sometimes be biased, limiting the ability to compare these protein interactomes [20] precisely. For instance, the Y2H assay is a potent genetic way to map PPIs at both a micro-level and high-throughput manner [21, 22]. PPIs are identified in the Y2H system by stimulating reporter genes in response to a restored transcription factor [23]. Since only a few distinct yeast colonies are isolated and examined in a single experiment, the results from Y2H tests are insufficient. While fewer colonies must be evaluated, hundreds of fundamental interactions may be eliminated by a stringent requirement for interaction with binding partners for greater degrees of transcriptional activation [24]. A matrix-formatted approach can be employed to address this problem and achieve comprehensive proteome coverage. In this approach, an array consisting of an individual prey is digitally analyzed. Nevertheless, this strategy is neither readily available nor cost-effective for an individual researcher [20, 25, 26]. Alternatively, we can perform other assays using arrayed prey libraries in batches, and interacting clones can be assessed through parallel high-throughput sequencing [27]. The Y2H-based DEEPN approach has been previously developed for large-scale PPI studies in a single plate/batch format [24]. Moreover, DEEPN can eliminate most of the restrictions, including the need for infrastructure, cost-effectivity, bioinformatics expertise, instrumentation, and optimization. The details of DEEPN can be found in [24]; here layout of this previously published system in a step-by-step protocol is presented. Y2H-cDNA prey plasmids are chosen to associate with a particular bait plasmid, and DEEPN tracks their abundance in a library population. In addition, Y2H interactions can predict the relative affinities of a specific prey recombinant for one bait protein compared to another. To achieve this, it is possible to utilize various bait plasmids to target distinct prey populations within a Y2H-cDNA library with similar starting populations [24, 28]. The DEEPN approach has the advantage of identifying interactions that are strong for one bait but weak for

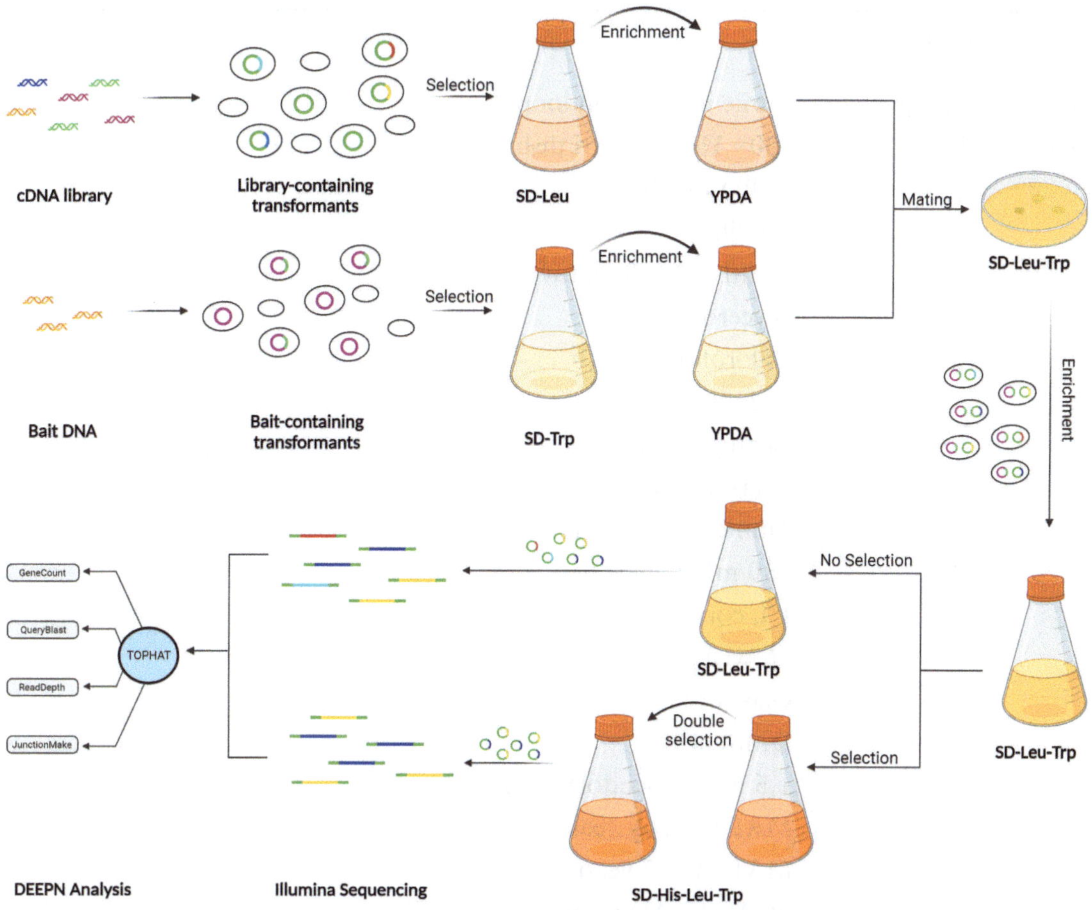

Fig. 1 Experimental schematic of DEEPN. cDNA prey library and Bait DNA are integrated into antibiotic-resistance-containing vectors, followed by transformation into respective strains. Both strains are screened with the corresponding selection medium and further enriched by a non-selection medium. After mating between prey-containing transformants and bait-containing transformants on a solid double selection medium, the diploids possessing both vectors were amplified with a similar but liquid double selection medium. Furthermore, the enriched diploids are split into two groups. In the non-selection group, the diploids are grown on the same selection medium for the detection of prey cDNA abundance. In the selection group, two more rounds of extra selection are applied to sift out strong prey-bait interactions. Both groups are sequenced and subsequently analyzed by assigned modules within DEEPN software

another (Fig. 1). Moreover, by selecting a limited read area or competition space, the population growth can be limitless since DEEPN analyzes a fixed number of reads. In addition, the *Gene-Count* module determines the aligned reads for every candidate gene, indicating the degree of enrichment for that particular gene [28]. In earlier research, it was discovered that DEEPN could differentiate between interactions specific to the active, GTP-bound conformation of Rab5 and those associated with its inactive, GDP-bound conformation [24]. This protocol describes customized reagents and protocols through which DEEPN analysis can be utilized efficiently and cost-effectively.

2 Materials

2.1 Y2H Library Construction

Sterile inoculation loop.

100× 15 mm Petri dish.

YEPD Broth.

Parafilm.

15 mL centrifuge tubes.

Glycerol.

Chloroform.

Phenol.

Isoamyl alcohol.

Beta-mercaptoethanol.

Ammonium acetate.

250 mL conical flask.

1 L conical flask.

96-well plate.

Glass beads.

DMSO.

1.5 mL Eppendorf tubes.

MATA PJY69 yeast.

MATα Y187 yeast.

pGADT7 AD vector.

pGBKT7 DNA-BD vector.

EcoRl enzyme.

BamHl enzyme.

Tryptophan antibiotic.

Yeast transformation kit.

Synthetic-defined (SD) media.

Yeast-rich media (YPD).

Yeast-buffered-rich media (bYPDA).

TWIRL sample buffer.

Zymolyase 100 T (ThermoFisher) 10 mg/mL in Buffer (50 mM K_2PO_4 pH 7.5/50% glycerol).

RNAse A, DNAse protease-free stock.

Synthetic gene fragment.

Gibson Assembly Master Mix (Biolabs).

High-Fidelity DNA Polymerase 2× PCR Master Mix (Ampliqon).

Primers: to amplify inserts of mouse cDNA library in Trp1 vector.

PCR purification Kit (Monarch).

E. coli competent cells.

Normalized Mate and Plate Universal Mouse cDNA Library (Takarabio).

DNA Single Index Kit (Takarabio, Cat. R400697).

2.2 Antibodies

Monoclonal anti-HA antibodies (Thermo Fisher Scientific, Cat. 26183).

Polyclonal anti-myc antibodies (Thermo Fisher Scientific, Cat. R951–25).

Monoclonal 6x His-Tag antibodies (Thermo Fisher Scientific, Cat. R960–25).

Monoclonal V5-Tag antibodies (Thermo Fisher Scientific, Cat. R960–25).

Monoclonal Ubiquitin antibodies (Thermo Fisher Scientific, Cat. MA5–28514).

2.3 Special Equipment

Thermal Incubator.

Refrigerator.

Spectrophotometer.

−80 °C Freezer.

PCR machine (thermocycler).

Centrifuge.

Vortex machine.

Shaking thermal incubator.

Water bath.

Gel Documentation System.

Illumina HiSeq2500.

2.4 Composition of All Buffers, Media, and Solutions

Prepare all buffers, media, and solutions using autoclaved deionized H_2O and keep the reagents at room temperature.

2.4.1 Synthetic Defined (SD) Media for Yeast

Yeast Nitrogen Base (Ammonium sulfate) without amino acids.

Dextrose (final conc. 2%).

Adenine (200 mg/L).

Arginine (20 mg/L).

Aspartic acid (100 mg/L).

Glutamate monosodium (100 mg/L).

Histidine (200 mg/L).

Leucine (60 mg/L).

Phenylalanine (50 mg/L).

Serine (375 mg/L).

Threonine (200 mg/L).

Tryptophan (200 mg/L).

Tyrosine (30 mg/L).

Valine (150 mg/L).

Uracil (200 mg/L).

Omit Leucine.

Tryptophan/Histidine.

1.5% Bacto Agar for solid plates.

Add all the components in a conical flask and bring up the volume to 1 L by adding deionized H_2O and sterilize by autoclaving (*see* **note 1**). Add bacto agar and pour into sterile Petri plates (≈25–30 mL) of autoclaved media for solid plates.

2.4.2 Yeast-Rich Media (YPD)

Peptone (20 g/L).

Yeast extract (10 g/L).

Sugar (dextrose/glucose) (20 g/L).

Agar (18 g/L).

Add and mix all the components in a conical flask and bring up the volume to 1 L by adding deionized H_2O and sterilize by autoclaving. Add agar and pour into sterile Petri plates (≈25–30 mL) of autoclaved media for solid plates.

2.4.3 Buffered Yeast-Rich Media (bYPDA)

Peptone (20 g/L).

Yeast extract (10 g/L).

Glucose (20 g/L).

Adenine (200 mg/L).

HCl.

Add and mix all the components (except HCl) to make the required media volume. Then add HCl to adjust the pH of the media to 3.7 and filter-sterilize it.

2.4.4 TWIRL Sample Buffer

Urea (8 M).

SDS (4%).

Glycerol (10%).

Tris–HCl (50 mM).

Bromophenol blue (0.02%).

Add and mix all the components in autoclaved deionized H_2O. Then add the HCL to adjust the pH to 6.8.

3 Methods

3.1 Development of Plasmids Containing Gal4-DNA-Binding Domain

The synthetic DNA fragment was cloned into the pGBKT7 plasmid harboring kanamycin and tryptophan-resistant genes. The vector carries a TRP1 nutritional marker for selection in yeast.

1. The synthetic DNA fragment was PCR amplified using the Gibson Assembly Master Mix following manufacturer's protocol [29].
2. Visualize the desired PCR amplified product on 1% agarose gel.
3. Purify the PCR product using the Monarch PCR purification kit.
4. Digest the Kanamycin-resistant plasmid pGBKT7 using BamH1 and EcoR1 restriction enzymes.
5. Mix the purified PCR synthetic DNA fragment and plasmid in Gibson Assembly Reaction and incubate in a thermocycler at 50 °C for 1 h.
6. The heat shock method was used for DNA Transformation into *E. coli* competent cells, discussed below:
7. Take out the vial of *E. coli* competent cells stored at −80 °C and thaw them on ice for 10–15 min.
8. Mix 10 ng of DNA into competent cells by pipetting while incubating it on ice for 20–30 min.
9. Put the competent cells voile on a pre-heated water bath/heat block at 42 °C for 45–60 s (*see* **Notes 1–6**).
10. Immediately put it back on ice for 2–5 min.
11. Pour 1 mL of LB growth media to the vial and incubate on a shaker at 37 °C for 1 h.
12. After incubation, spread the transformed cells onto an LB agar media plate having kanamycin or an appropriate antibiotic and incubate at 37 °C in a shaker overnight.
13. The following day, pick the single bacterial colony and inoculate it into selection (kanamycin-LB) media and again incubate it in a shaker at 37 °C overnight.
14. Measure the OD_{600} of the culture using a spectrophotometer to ensure it falls within the range of 0.8–1.
15. Isolate the plasmid using the alkaline lysis method [30].
16. Set up the colony PCR to verify the transformed colonies. Prepare 3–5 reactions per sample by adding 50 pmoles of forward and reverse primers, 5 μg of template DNA, and 12.5 μL of High-Fidelity DNA Polymerase 2x Master Mix, and make up the final volume to 25 μL using nuclease-free water.

3.2 Expression of Fusion Proteins Containing Gal4 DNA-Binding Domain

The TRP1-carrying bait plasmids were transformed into the yeast MATA strain PJ69-4A, which is ideal for observing the yeast two-hybrid (Y2H) interactions. After transformation, the resultant strain is mated with the Y187 strain that contains the Y2H library.

1. Transform yeast cells PJ69-4A with the bait plasmid to confirm the presence of the Gal4-DNA DBD-fusion protein following the previously published protocol [21].
2. Grow transformed PJ69-4A cells overnight on 1 mL SD media lacking tryptophan (SD-Trp).
3. Next day, dilute the grown cells with two volumes of YPD media and grow for 1 h at 30 °C.
4. Centrifuge the cells and resuspend the pellet into 1 mL of 0.2 N NaOH.
5. Incubate at 25 °C for 5 min.
6. Centrifuge and re-pellet the cells, and discard the NaOH.
7. Add 100 μL of TWIRL buffer for resuspension.
8. Incubate the sample for 5 min at 70 °C for lysis.
9. Analyze/quantify the lysate using SDS-PAGE and immunoblotting [31] using anti-myc antibodies.

3.3 Self-Activation Test

3.3.1 Test Selection Conditions for Yeast Two-Hybrid (Y2H) Interaction

The Gal4-DBD-fusion proteins must be tested for circumstances that favor potential Y2H interactions. For this purpose, the bait and library prey plasmids are placed inside the same diploid background after the mating.

1. Y187 strain was transformed with empty pGBKT7 vector-only prey plasmid expressing the Gal4-transcriptional activation domain using the above protocol. For instance, the pGADT7 vector with LEU2 gene for selection in yeast and ampicillin resistance for sample in bacteria.
2. Mate the Trp + PJ69-4A carrying bait plasmid with Leu + Y187 carrying prey through patching them on the YPD plate.
3. Incubate the plates overnight at 30 °C.
4. Next day, streak the yeast cells using a sterile loop on the SD media plate lacking both Leu and Trp.
5. Isolate the single PJ69-4A/Y187 Mata/diploid colonies from the SD-Leu plate.
6. Make testing plates with different combinations: (i) SD-Leu-Trp, SD-Trp-Leu-His and (ii) SD-Leu-Trp-His with 0.1 to 10 mM 3AT (3-amino-triazole).
7. Patch the diploids on the plates and let them grow overnight.
8. Pellet and resuspend the grown cells in sterile water.

9. Measure the OD and maintain it at $OD_{600} = 0.5$ by further resuspension (if required).
10. Perform the serial dilution at 1:10.
11. Spot 4 μL from each dilution on pre-prepared test plates (*see* **note 2**).
12. Let the cells grow by incubating the plates at 30 °C for 2–3 days.

3.3.2 Mating and Selection

It can be difficult to mate with the Y187 strain. Therefore, the following optimized protocol must be followed to preserve the library's complexity.

- **Day 1**

  1. Inoculate the *(Trp)* pGBKT7 bait plasmid carrying PJ69-4A yeast strain into 23 mL of freshly prepared SD-Trp media.
  2. Inoculate Y187 yeast cells with LEU2-carrying "prey" library into 125 mL of freshly prepared SD-Leu media.
  3. For overnight growth, put the cultures in a 30 °C shaker incubator at 200 rpm.

- **Day 2**

  1. Measure the OD_{600}, which should be between 1.0 and 1.5.
  2. Centrifuge and pellet the cells of the Y187 strain containing the library plasmids and PJ69-4A cells in separate 50 mL sterile conical tubes.
  3. Resuspend the cell pellet in 10 mL of sterile water.
  4. Transfer into new 50 mL sterile culture tubes and re-pellet by centrifugation.
  5. Resuspend the PJ69-4A and Y187 cells into the 4 mL of bYPDA media separately.
  6. Adjust the pH to 3.7 by adding 1 N HCl.
  7. For mating, take a sterile 50 mL conical tube and add 1 mL of each, i.e., PJ69-4A, Y187, and bYPDA, into it.
  8. For incubation, put the reaction mixture in a shaker incubator at 30 °C at 90–130 rpm for 90 min (*see* **Note 3**).
  9. Centrifuge the cells and resuspend the pellet into 2 mL of bYPDA.
  10. Spread the entire 2 mL of bYPDA resuspension on a 150 mm YPD-agar plate and overnight incubate at 30 °C.

- **Day 3**

  1. Harvest the cells using a cell scrapper and rinse with 10–12 mL of SD-Leu-Trp media.
  2. Inoculate the harvested cells from the YPD plates into 12 mL SD-Leu-Trp media in a 50 mL sterile conical tube.

3. Pellet the cells by centrifugation and resuspend the pellet into 40 mL of SD-Leu-Trp media.
4. To measure the diploid cells in the reaction mixture, dilute 4 μL of sample into 200 μL of SD-Trp-Leu media.
5. Gently mix by pipetting and plate it onto an SD-Leu-Trp selection plate and incubate at 30 °C for two days.
6. Inoculate the remaining 40 mL of resuspension culture into 500 mL of SD-Leu-Trp media.
7. Incubate the flasks for 36 h at 30 °C (or more) in a shaker/incubator at 200 rpm until the OD_{600} reaches 2.0.

- **Day 4**

1. Observe the growth in each flask by measuring the OD_{600}.

- **Day 5**

At this stage, we split the yeast culture into two for selection and non-selection. The titer plates ought to be ready for analysis by this time, enabling verification of enough mating efficiency to sustain library complexity (*see* **Note 4**).

1. Take the 20 mL of grown culture and inoculate it into 750 mL of SD-Leu-Trp (non-selection) media.
2. Take another 20 mL of grown culture and inoculate it into 750 mL of SD-Trp-Leu-His (selection) media with minimal 3AT concentration (*see* **Note 5**).
3. Gently mix the new 770 mL cultures and measure the initial OD_{600}.
4. Let the cultures grow separately in the shaker at 30 °C at 200 rpm until it reaches the OD_{600} of 2.0.

- **Day 6**

1. Confirm that the OD_{600} of non-selection media is at 2.0.
2. Take 10 mL of culture and pellet it by centrifugation. Cells can be stored at −20 °C or proceed for sequencing at this stage.

- **Day 7**

1. Measure the OD_{600} of the SD-Leu-Trp-His (selection) media at 2.0.
2. Take 2 mL out and inoculate into 75 mL of freshly made SD-Leu-Trp-His media with minimum 3AT to avoid background growth.
3. Let the cells grow until they reach OD_{600} of 2.0 by placing them in a shaker incubator at 30 °C at 200 rpm for 30–60 h.

- **Days 8 and 9**

  1. Check the cell growth by measuring the saturation of selection (SD-Leu-Trp-His) cultures at OD_{600} of 2.0.
  2. Take 10 mL of culture and pellet it by centrifugation. Cells can be stored at −20 °C or proceed for sequencing at this stage.

3.4 Preparation of DEEPN Samples for Illumina Sequencing

3.4.1 DNA Extraction/Purification

1. Gently resuspend pellets into 500 μL of 50 mM Tris and 20 mM EDTA.
2. Transfer it to a 1.5 mL sterile Eppendorf tube.
3. Add and mix 10 μL of Zymolase stock and 3 μL of Beta-Mercaptoethanol.
4. Incubate the sample at 37 °C for 24–36 h.
5. Add a mixture of phenol/chloroform/isoamyl alcohol (25: 24:1) and carry out ethanol precipitation.
6. Add 100 μL of 50 mM Tris/20 mM EDTA in the precipitated pellet and resuspend it.
7. Add 1 μL of RNaseA from stock and incubate for 1 h at 37 °C.
8. Precipitate it again using ethanol and resuspend it into 100 μL of 5 mM Tris/2 mM EDTA.
9. Quantify the DNA using a spectrophotometer at 280 nm absorbance (*see* **Note 6**).

3.4.2 PCR Purification of cDNA Fragments

1. Set up two individual PCR reactions for selection and non-selection samples.
2. Set up the PCR for each reaction mixture. Add 1 μL of forward and reverse primers, 5 μg of template DNA, and 25 μL of PCR master mix, and make up the final volume to 50 μL using nuclease-free water.
3. Run the agarose gel electrophoresis on the PCR product (as discussed above).
4. For PCR product purification, the Monarch PCR purification kit was used.

3.4.3 Illumina Sequencing

1. For Illumina sequencing, shear the 400–550 ng PCR-purified product into 300 bp average length fragments using Covaris E220 Ultrasonicator.
2. Indexed Sequencing libraries were generated using the DNA Single Index Kit for the Illumina sequencing.
3. Pool up the Indexed libraries and run the samples on Illumina HiSeq2500.

3.5 DEEPN

DEEPN is a novel approach to analyzing the protein–protein interaction in Y2H (*see* **Note 7**). DEEPN is created to handle and interpret sequencing data from the Illumina platform, which generates reads between 110 and 140 bp. The distinct sides of a paired-end sequence are treated as two unique sequences by DEEPN, so both single and parried-end sequences are acceptable. Process the . fastq sequence files using Tophat2, which produces the mapped and unmapped sequence files in *SAM* format. DEEPN requires the mapped and unmapped sequence files to analyze and process the data. DEEPN analyzes the dataset using the following modules:

.fastq! Tophat2/Bowtie! Gene Count! Junction Make! Blast Query! Read Depth.

Gene Count is used to evaluate each gene's number of reads, wherein Stat *Maker* ranks these evaluated reads. While screening of unmapped reads from the junction between Gal4 AD and cDNA fragment is carried through *Junction Make.* To identify the insert fragment, it also runs the *Blastn*. Query Blast is used to determine the frequency and position of each junction within a gene. At last, *Read Depth* generates a plot for the mapped data of a cDNA fragment.

4 Notes

1. Filter-sterilized amino acids can be added after autoclaving when the media cools to 50 °C.
2. For the best interaction, observe the growth in the presence of His regardless of 3AT.
3. This is necessary to prevent the cells from sedimentation.
4. A minimum of one million total diploids are advised for this workflow.
5. Normally, 3AT prevents yeast growth and its concentration should be between 0.1 and 1 mM.
6. 5 μg of DNA is required for PCR amplification.
7. Several other screening experiments have been devised concurrently with the technical development of the Y2H system. Y2H libraries can be created using genomic DNA, cDNA, normalized cDNA, full-length cDNA, or open reading frame (ORF) libraries, depending on the research's goals and available materials [32]. The development of mating type a and type two-hybrid strains with the bait and prey plasmids is an alternative strategy to the original Y2H [33]. In modern Y2H testing, the mating type a is still extremely popular. Moreover, complementary assays such as YFP fluorescent-based protein complementation assay can be used as a validation technique.

The fluorescence-activated cell sorting can restore functional YFP if the two proteins interact. The nucleic acid programmable protein array (NAPPA) is another validation assay. For instance, two proteins were expressed in a linked transcription/translation reticulocyte lysate and fused to a glutathione S-transferase (GST) tag and a HA epitope tag, respectively [34]. An anti-GST antibody fixed at the bottom of a 96-well microtiter plate captures the GST-tagged protein [35, 36].

Acknowledgments

This research was funded by National Science Foundation (IOS-2038872).

References

1. Wessling R, Epple P, Altmann S et al (2014) Convergent targeting of a common host protein-network by pathogen effectors from three kingdoms of life. Cell Host Microbe 16(3):364–375. https://doi.org/10.1016/j.chom.2014.08.004
2. Smakowska-Luzan E, Mott GA, Parys K et al (2018) An extracellular network of Arabidopsis leucine-rich repeat receptor kinases. Nature 553(7688):342–346. https://doi.org/10.1038/nature25184
3. Mukhtar MS, Carvunis AR, Dreze M et al (2011) Independently evolved virulence effectors converge onto hubs in a plant immune system network. Science 333(6042):596–601. https://doi.org/10.1126/science.1203659
4. Mott GA, Smakowska-Luzan E, Pasha A et al (2019) Map of physical interactions between extracellular domains of Arabidopsis leucine-rich repeat receptor kinases. Sci Data 6:190025. https://doi.org/10.1038/sdata.2019.25
5. Mishra B, Sun Y, Howton TC et al (2018) Dynamic modeling of transcriptional gene regulatory network uncovers distinct pathways during the onset of Arabidopsis leaf senescence. NPJ Syst Biol Appl 4:35. https://doi.org/10.1038/s41540-018-0071-2
6. Mishra B, Sun Y, Ahmed H et al (2017) Global temporal dynamic landscape of pathogen-mediated subversion of Arabidopsis innate immunity. Sci Rep 7(1):7849. https://doi.org/10.1038/s41598-017-08073-z
7. Mishra B, Kumar N, Shahid Mukhtar M (2022) A rice protein interaction network reveals high centrality nodes and candidate pathogen effector targets. Comput Struct Biotechnol J 20:2001–2012. https://doi.org/10.1016/j.csbj.2022.04.027
8. Mishra B, Kumar N, Mukhtar MS (2021) Network biology to uncover functional and structural properties of the plant immune system. Curr Opin Plant Biol 62:102057. https://doi.org/10.1016/j.pbi.2021.102057
9. Mishra B, Kumar N, Mukhtar MS (2019) Systems biology and machine learning in plant-pathogen interactions. Mol Plant-Microbe Interact 32(1):45–55. https://doi.org/10.1094/MPMI-08-18-0221-FI
10. Marchand A, Van Hall-Beauvais AK, Correia BE (2022) Computational design of novel protein–protein interactions–An overview on methodological approaches and applications. Curr Opin Struct Biol 74:102370
11. McCormack ME, Lopez JA, Crocker TH et al (2016) Making the right connections: network biology and plant immune system dynamics. Current Plant Biology 5:2–12
12. Lopez J, Mukhtar MS (2017) Mapping protein-protein interaction using high-throughput yeast 2-hybrid. Methods Mol Biol 1610:217–230. https://doi.org/10.1007/978-1-4939-7003-2_14
13. Kumar N, Mishra B, Mukhtar MS (2022) A pipeline of integrating transcriptome and interactome to elucidate central nodes in host-pathogens interactions. STAR Protoc 3(3):101608. https://doi.org/10.1016/j.xpro.2022.101608
14. Kumar N, Mishra B, Mehmood A et al (2020) Integrative network biology framework elucidates molecular mechanisms of SARS-CoV-2 pathogenesis. iScience 23(9):101526. https://doi.org/10.1016/j.isci.2020.101526

15. Klopffleisch K, Phan N, Augustin K et al (2011) Arabidopsis G-protein interactome reveals connections to cell wall carbohydrates and morphogenesis. Mol Syst Biol 7:532. https://doi.org/10.1038/msb.2011.66
16. Gonzalez-Fuente M, Carrere S, Monachello D et al (2020) EffectorK, a comprehensive resource to mine for Ralstonia, Xanthomonas, and other published effector interactors in the Arabidopsis proteome. Mol Plant Pathol 21(10):1257–1270. https://doi.org/10.1111/mpp.12965
17. Garbutt CC, Bangalore PV, Kannar P et al (2014) Getting to the edge: protein dynamical networks as a new frontier in plant-microbe interactions. Front Plant Sci 5:312. https://doi.org/10.3389/fpls.2014.00312
18. Arabidopsis Interactome Mapping C (2011) Evidence for network evolution in an Arabidopsis interactome map. Science 333(6042):601–607. https://doi.org/10.1126/science.1203877
19. Ahmed H, Howton TC, Sun Y et al (2018) Network biology discovers pathogen contact points in host protein-protein interactomes. Nat Commun 9(1):2312. https://doi.org/10.1038/s41467-018-04632-8
20. Berggård T, Linse S, James P (2007) Methods for the detection and analysis of protein–protein interactions. Proteomics 7(16):2833–2842
21. Lopez J, Mukhtar MS (2017) Mapping protein-protein interaction using high-throughput yeast 2-hybrid. In: Plant genomics. Springer, pp 217–230
22. Brückner A, Polge C, Lentze N et al (2009) Yeast two-hybrid, a powerful tool for systems biology. Int J Mol Sci 10(6):2763–2788
23. Mehla J, Caufield JH, Uetz P (2015) The yeast two-hybrid system: a tool for mapping protein–protein interactions. Cold Spring Harbor Protocols 2015 (5):pdb. top083345
24. Pashkova N, Peterson TA, Krishnamani V et al (2016) DEEPN as an approach for batch processing of yeast 2-hybrid interactions. Cell Rep 17(1):303–315
25. Rao VS, Srinivas K, Sujini G et al. (2014) Protein-protein interaction detection: methods and analysis. International journal of proteomics 2014
26. Fields S, Song O-k (1989) A novel genetic system to detect protein–protein interactions. Nature 340(6230):245–246
27. Vidalain P-O, Boxem M, Ge H et al (2004) Increasing specificity in high-throughput yeast two-hybrid experiments. Methods 32(4):363–370
28. Peterson TA, Stamnes MA, Piper RC (2018) A yeast 2-hybrid screen in batch to compare protein interactions. JoVE (Journal of Visualized Experiments) 136:e57801
29. Gibson DG, Young L, Chuang R-Y et al (2009) Enzymatic assembly of DNA molecules up to several hundred kilobases. Nat Methods 6(5):343–345
30. Ehrt S, Schnappinger D (2003) Isolation of plasmids from E. coli by alkaline lysis. In: E. coli Plasmid Vectors. Springer, pp 75–78
31. Walker JM (2002) SDS polyacrylamide gel electrophoresis of proteins. In: The protein protocols handbook. Springer, pp 61–67
32. Gentleman R, Huber W (2007) Making the most of high-throughput protein-interaction data. Genome Biol 8(10):1–10
33. Huang H, Jedynak BM, Bader JS (2007) Where have all the interactions gone? Estimating the coverage of two-hybrid protein interaction maps. PLoS Comput Biol 3(11):e214
34. Sutandy FR, Qian J, Chen CS et al (2013) Overview of protein microarrays. Curr Protoc Protein Sci 72 (1):27.21:21-27.21. 16
35. Miersch S, LaBaer J (2011) Nucleic acid programmable protein arrays: versatile tools for array-based functional protein studies. Curr Protoc Protein Sci 64 (1):27.22:21-27.22. 26
36. Fung E (2004) Protein arrays: Methods and protocols, vol 264. Springer

Chapter 18

An Interactome Assay for Detecting Interactions between Extracellular Domains of Receptor Kinases

Jente Stouthamer, Sergio Martin-Ramirez, and Elwira Smakowska-Luzan

Abstract

Interactions between extracellular domains (ECDs) are crucial for many physiological processes in the cell, most importantly perception of its environment. However, studying these often-transient interactions can be challenging. Here we describe a method that allows for in vitro detection of extracellular domain interactions through an oligomerization-based cell surface interaction (CSI) assay. In a CSI, bait- and prey-tagged proteins are produced and secreted by *Drosophila* S2 cells to ensure proper folding and post-translational modifications. Subsequently, Bait (FC fragment) and Prey (pentamer domain and alkaline phosphatase) tags allow the detection of interactions in protein A-coated 96 wells plates through a colorimetric readout. Due to the easy detection of interactions this approach can be used for high-throughput screening and mapping of extracellular interaction networks.

Key words Receptor kinases, ECDs, Protein–protein interactions, Signaling, S2 cell protein expression, Interactome assay, High-throughput screen

1 Introduction

Receptor kinases (RKs) located at the plasma membrane are a necessary component for cells to sense their environment. Sensing occurs through the extracellular domains (ECDs) of these receptors and often involves interactions between RKs [1]. Furthermore, interactions can be modulated in the presence of different ligands, post-translational modifications (PTMs), or other signaling molecules [2, 3]. RKs together with other proteins form complex signal transduction networks in cells which are needed to mount an appropriate response to outside stimuli.

Although a large amount of protein–protein interaction (PPI) data were generated in the past decade, extracellular proteins are greatly underrepresented in these data sets due to technical challenges. ECDs of RKs require specific conditions for their expression and secretion, such as an oxidizing environment (for disulfide

Shahid Mukhtar (ed.), *Protein-Protein Interactions: Methods and Protocols*, Methods in Molecular Biology, vol. 2690, https://doi.org/10.1007/978-1-0716-3327-4_18,

bonds) and specific post-translational modifications (predominantly glycosylation) for their proper folding and function [4]. Understanding how ECD interactions drive receptor complex formation is challenging because they are transient and occur with moderate- to low-affinity (KD in the nM–μM range) [4]. To overcome these bottlenecks, an oligomerization-based method (CSI) was developed for in vitro detection of ECD interactions in a high throughput manner [4–6].

First, proteins are produced in insect cells (*Drosophila melanogaster* derived cell line, S2) a eukaryotic system, thus ensuring a proper folding and the presence of crucial PTMs needed to obtain functional protein. Furthermore, by tagging proteins with an N-terminal signaling peptide (from the *Drosophila* BiP protein), S2 cells can secrete the protein into the medium, allowing for easy isolation of relatively pure protein (compared to cell lysates). Protein can be used directly in the assay without the need for further purification steps.

The RK ECDs are expressed and secreted from the insect cells, as bait- and prey-tagged proteins (Fig. 1a). Bait ECDs are fused to an Fc antibody fragment, which dimerizes and can bind to 96-well protein A-coated plate. Prey ECDs have a pentamerization-COMP domain to enhance interactions and an alkaline phosphatase (AP) for facile enzymatic detection using colorimetric substrates. During the assay bait ECDs are captured in 96-well protein A-coated plate and arrayed against prey ECDs fused to an alkaline phosphatase enzyme (Fig. 1b). By applying a substrate for alkaline phosphatase, after complex formation occurred, positive interactions versus negative interactions can be scored by measuring absorbance at 650 nm (Fig. 1c). The amplitude of the signal can be related to the strength of the interaction.

With the CSI assay even low-affinity interactions between ECDs of different receptor kinases can be measured in a high-throughput manner. A big advantage of this assay is the fact that it can be applied to any protein family of interest with minimal adjustments. Furthermore, the modulation of interaction by ligands or other bioactive molecules can be studied [5]. If many interactions are studied, it is possible to build interaction networks and determine interaction hubs of the sub-networks. However, while working with large interactome data sets it is essential to determine the rate of false positive and negative interactions. Interaction between proteins of interest has to be validated with different in vitro and in vivo methods.

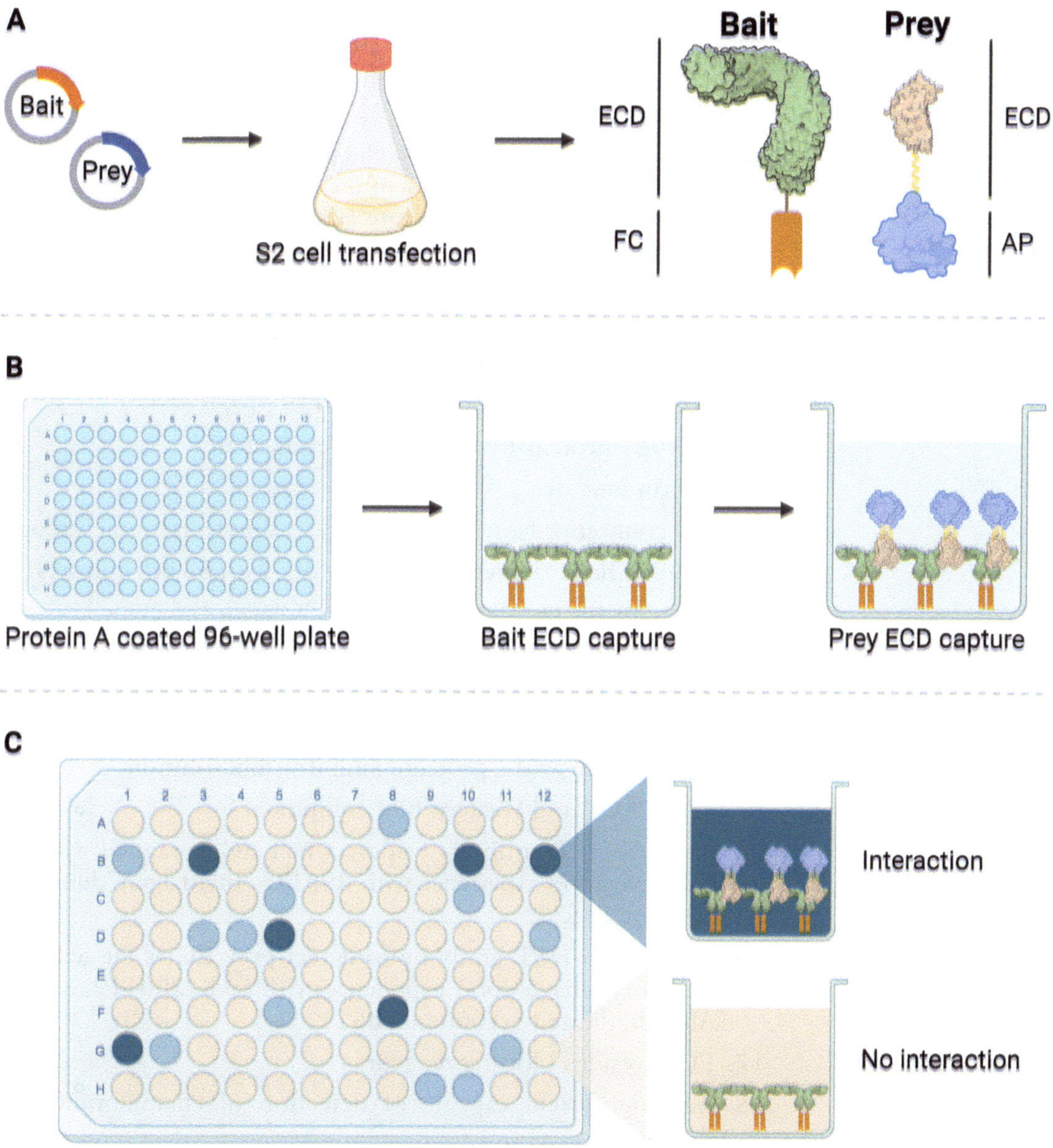

Fig. 1 Schematic overview of the CSI workflow. (**a**) ECDs with a bait or prey tag are expressed in *Drosophila* S2 cells. (**b**) Baits are captured on a protein A-coated plate and tested against preys. (**c**) Conversion of a colorimetric substrate by alkaline phosphatase (prey) present in the well indicates the presence of an interaction

2 Materials

2.1 Secreted Expression of Extracellular Domains

1. pECIA2 and pECIA14 plasmids with cDNA of interest cloned in between the *Drosophila* BiP signaling peptide sequence and the HRV-3C site, no stop or start codon (addgene #47032 and #47051 respectively) (*see* **Note 1**). The pECIA2 plasmid

encodes a C-terminal FC fragment followed by a V5 and His tag while pECIA14 encodes a C-terminal pentameric COMP protein and an alkaline phosphatase, followed by a FLAG and His tag.

2. *Drosophila melanogaster* S2 cell culture for protein expression. Cell culture should be used in the log phase of growth. Either grown as a suspension culture (for larger volumes) or grown adherently. Cell culture should be grown on a protein-free medium, preferably without antibiotics.
3. 125 mL culture flasks with baffles, for example, Pyrex.
4. Shaker incubator, capable of 27–28 °C.
5. Regular incubator, capable of 21 °C.
6. ESF 921 protein-free medium or equivalent.
7. Trypan blue dye.
8. Cell counter or hemocytometer and a simple light microscope.
9. Laminar flow cabinet to sterilely work with S2 cell cultures.
10. Reagents for transient transfection: Qiagen Effectene kit, consisting of EC buffer, enhancer, and Effectene reagent.
11. 6-well plates for cell culture or T-75 flasks.
12. 500 mM $CuSO_4$ stock solution.
13. 10% Sodium azide, protease inhibitor mix solution.
14. SDS-PAGE and Western blot equipment.
15. Anti-V5-HRP antibody to detect bait ECDs tagged with V5 (we have good experience with R96125, Invitrogen, diluted 1: 5000 in TBST).
16. Anti-FLAG-HRP antibody to detect prey ECDs tagged with FLAG (we have good experience with SAB4200119, Sigma-Aldrich, diluted 1:1000 in TBST).
17. ECL western blotting reagent.
18. Imaging system capable of detection of chemiluminescent signals.
19. KPL Blue Phos Microwell phosphatase substrate.
20. Regular flat-bottomed 96-well plates.
21. Plate reader capable of measuring absorbance at 650 nm in 96-well plates.

2.2 Cell Surface Interaction Assay

1. Bait and prey ECDs produced in 3.1.
2. Tris-buffered saline with Tween-20 (TBST) (25 mM Tris pH 7.5, 150 mM NaCl, 0.1% Tween-20).
3. Interaction buffer (25 mM Tris pH 7.5, 150 mM NaCl, 1 mM $MgCl_2$, 1 mM $CaCl_2$).

4. Protein A-coated 96-well plates (15,154, Pierce, Thermo Fischer).
5. Multichannel pipettes (20 and 200 μL).
6. Multichannel reservoirs.
7. Adhesive 96-well plate seals.
8. KPL Blue Phos Microwell phosphatase substrate.
9. Orbital shaker at room temperature (RT) and at 4C.
10. Multi-mode plate reader.

3 Methods

3.1 Secreted Expression of Extracellular Domains

1. For all work with *Drosophila* S2 cell cultures, work sterilely in a Laminar flow cabinet unless specified otherwise (*see* **Note 2**). S2 cells are grown in ESF 921 protein-free medium in 125 mL Erlenmeyer flasks with baffles at 27 °C while shaking at ~110 RPM (*see* **Note 3**). Twice a week the cells are passaged into fresh medium, to maintain log phase growth (*see* **Note 4**).
2. Cells are counted, either using a cell counter or manually using a hemocytometer and simple brightfield microscope. Trypan Blue dye should be added to assess the viability of the cells (*see* **Note 5**). It is important to check on the morphology and health of the cells before use. If insect cells are grown in suspension culture, they are small and round (Fig. 2). Dividing

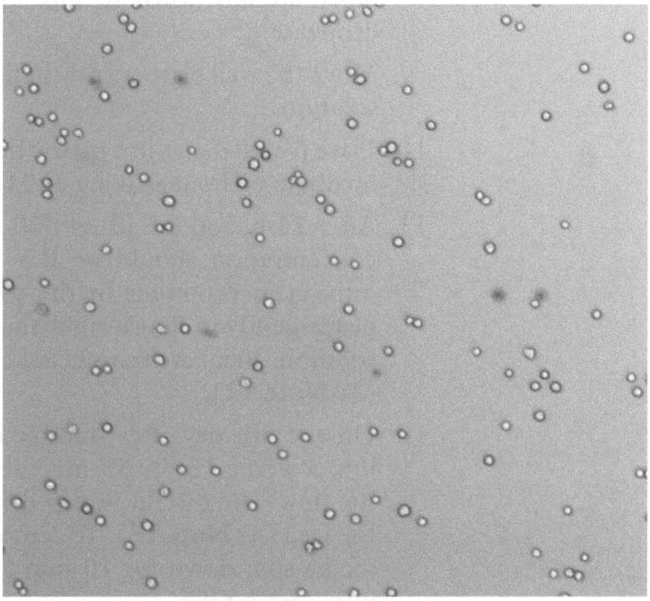

Fig. 2 S2 cells grown in suspension culture should have a round morphology. Cells are stained with trypan blue to assess the viability of the culture

cells appear in small clusters, indicating cells are in the log phase. Trypan blue dye penetrates the membranes of dead cells and stains them completely, for live cells only the membrane is stained, creating a dark outline. For good expression, it is important that the viability of the cells is above 90%.

3. Transfections are done adherently in 6-well plates. Each well should contain 3 mL S2 cell culture with a density of ~$1*10^6$ cells/mL (*see* **Note 6**). If doing multiple wells make the total volume needed for all samples in 50 mL tubes or another sterile container. Resuspend by swirling or pipetting carefully before distributing as cells sediment quickly.
4. To express genes of interest in pECIA2 and pECIA14 (bait and prey plasmids) using Qiagen Effectene reagent. First, dilute 0.5 μg of plasmid in 10 μL MQ in a microcentrifuge tube (50 ng/μL), this can be done beforehand outside of the flow hood (*see* **Notes 7–10**).
5. In the laminar flow cabinet: Add 90 μL EC buffer (Qiagen kit) to the DNA (10 μL).
6. Add 4 μL enhancer to each sample, vortex briefly, and wait for 2–3 min.
7. Add 12.5 μL Effectene, vortex 10 seconds. Incubate for at least 15 min, to allow Effectene-DNA complexes to form. (For time efficiency, cells could be distributed over wells plates during this waiting step).
8. Add 600 μL of protein-free culture medium to the sample, pipet up and down to mix. Add the entire sample to cells dropwise.
9. Move the well plates gently in figure eight shape to disperse the solution.
10. Place 6-well plates in a plastic box (to prevent evaporation) and incubate without shaking at 21 °C (*see* **Note 11**).
11. After 24 h, add 5.5 μL of 500 mM $CuSO_4$ to each well (final concentration should be 0.5–1 mM) to induce the metallothionein promoter in the pECIA2 and 14 vectors. Move plates gently in figure eight shape to immediately disperse the solution. Incubate transfected cells at 21 °C for another 3 days (*see* **Note 12**).
12. On the 5th day after transfection collect media by pipetting into a sterile 15- or 50-mL tube (this can be done outside of the flow). Be careful not to disturb the cells at the bottom of the well (*see* **Note 13**). To remove the remaining cells from the media spin down for 10 min at 1000×g. Decant supernatant into a fresh tube.

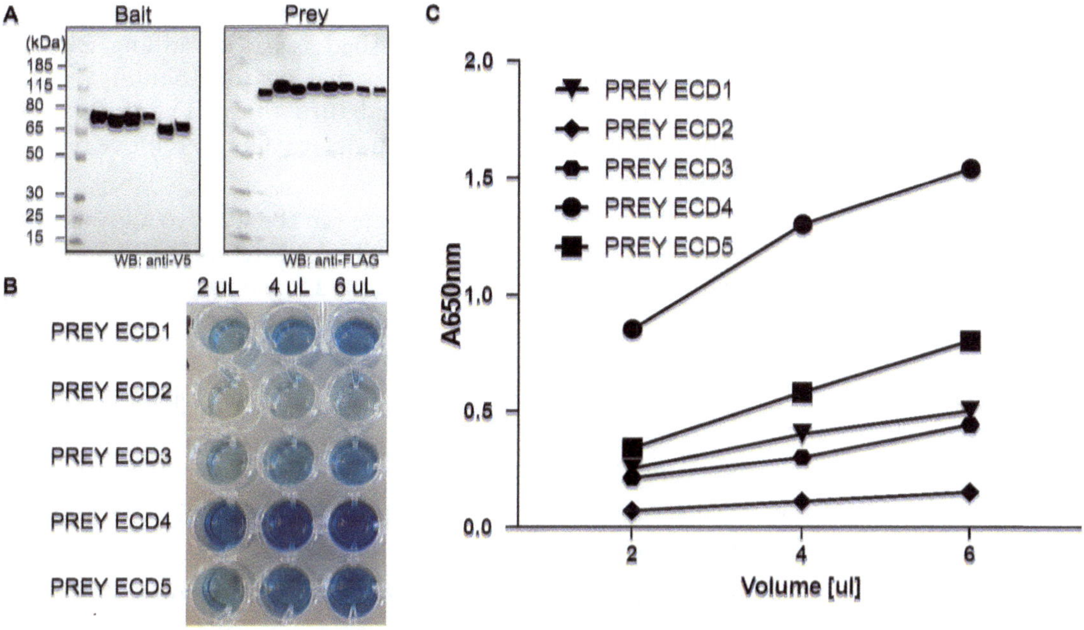

Fig. 3 Quality assessment of protein produced with S2 cell expression system. (**a**) Western blots of bait and prey samples detected with anti-V5-HRP or anti-FLAG-HRP antibodies, respectively. (**b**) Substrate conversion by alkaline phosphatase for different protein concentrations as a measure for the expression level of the prey. (**c**) Absorbance measured at 650 nm for AP activity tested in B

13. Add 5.5 μL 10% (w/v) sodium azide, protease inhibitor cocktail solution per 1 mL protein solution, and mix well. Store proteins at 4 °C, proteins can be stably kept this way for up to a year.
14. Check protein expression by western blot using the anti-V5 and anti-FLAG antibodies against the tags present in bait and prey proteins, respectively (Fig. 3a).
15. For prey ECDs substrate conversion by alkaline phosphatase can also be used as a measure of protein concentration (Fig. 3b, c). To set up an AP-activity test, pipet 2, 4, and 6 μL prey into wells of a 96-well plate. Add 100 μL KPL Blue Phos Microwell phosphatase substrate to each well and incubate for 3 h at RT. Measure substrate at 650 nm. Well-expressed preys generally have an absorbance between 0.5 and 1.5 for 6 μL after 3 h incubation (*see* **Note 14**).

3.2 Cell Surface Interaction Assay

Bait and prey proteins produced in previous steps can be used for the protein–protein interaction assay. Before starting a CSI, it is important to design the whole experiment thoroughly, this includes plate layouts as well as proper negative (only prey proteins) and positive controls (known interacting partners). It is recommended to include controls on each plate, to account for differences between batches of plates and handling. The general method, for a single plate, is described as follows:

1. Before proceeding with the interactome assay the protein A-coated plate should be washed twice by adding 200 μL TBST buffer to each well and shaking for 15 min at room temperature. To remove liquid turn the plate upside down above the sink. Tap the plate on tissue paper to remove residual liquid.
2. First, the bait ECDs are incubated in the wells to allow attachment of the FC tag to the protein A-coated plate. Pipet 50 uL of each bait into a well (*see* **Notes 15** and **16**) and add 50 μL interaction buffer to each well (*see* **Note 17**).
3. Seal the plate and leave overnight at 4 °C while shaking gently.
4. Next day remove all the bait from each well by pipetting, ensuring each well is treated the same (the same number of pipetting steps).
5. Add 50 μL of prey to each well and 50 μL interaction buffer (*see* **Notes 18** and **19**).
6. Seal the plate and incubate for 3 h at room temperature while shaking.
7. Remove all proteins by pipetting (*see* **Note 20**). Then to ensure the complete removal of the proteins, invert the plate on a tissue and tap gently (*see* **Note 21**).
8. Add 100 uL KPL substrate (prepared according to the manufacturer's instructions) to each well, if bubbles are observed they should be removed (*see* **Note 22**).
9. Reseal the plate and incubate for 3 h at room temperature while shaking.
10. To determine substrate conversion by alkaline phosphatase, measure the absorbance at 650 nm using a plate reader. Take a picture of the plate for future reference (Fig. 4a).
11. To analyze small data sets, first, subtract the signal coming from the negative control as a background (Fig. 4b). Next, a simple statistical analysis can be performed to assess significance. It is important to account for technicalities such as plate edge effects caused by handling of the plates, pipetting errors, and the tendency of certain proteins to stick to the well.
12. For high-throughput screens generating large data sets, refer to the Method Sections of Ozkan et al. (2014), Smakowska-Luzan et al. (2018), and Mott et al. (2019). In short, all interactions should be measured reciprocally making it possible to normalize over bait and prey. A two-way median polish is performed to generate Z scores, followed by the removal of interactions that do not confidently interact in both orientations.

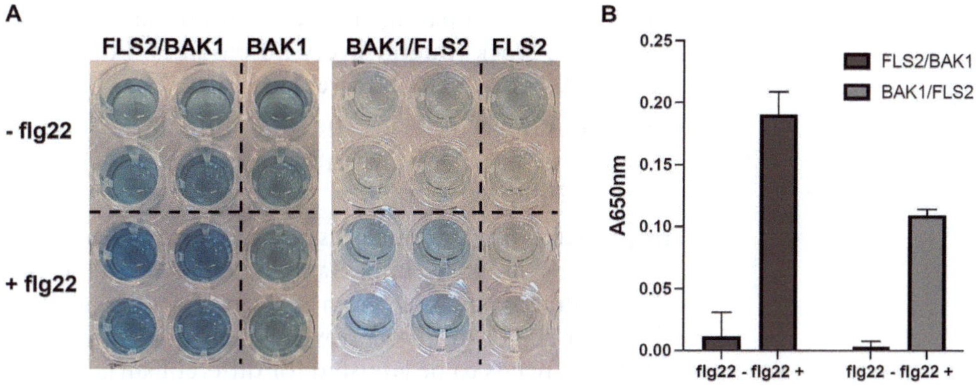

Fig. 4 Cell surface interaction assay demonstrating that flg22 enhances interaction between FLS2 and BAK1. (**a**) Interaction between the ECDs of FLS2 and BAK1 in the presence (+) and absence (–) of flg22. Interactions were tested in both orientations of bait/prey (FLS2-bait/Bak1-prey and the reverse). Only prey was included as a negative control. Substrate conversion by prey-bound AP after 3 h. (**b**) Analyzed absorbance measurements of samples in A. The bar graph displays negative control subtracted absorbances and standard deviations

4 Notes

1. The insert can be introduced into the pECIA plasmid through gateway cloning. We also have good experience with linearizing the vector by PCR, using a high-fidelity polymerase (FW CTGGAGGTGCTGTTCCAGGGAC , RV CCCGAGCGAGAGGCCAACAAAG) and then introducing the insert through HiFi DNA assembly (NEB).
2. The cells are grown without antibiotics, thus extra care should be taken to prevent contamination (sterile technique in the laminar flow cabinet and proper cleaning/sterilization of the glass flasks after each use). If needed antibiotics (Gentamycin or Penicillin-Streptomycin, 50 μg/mL) can be added, although this may negatively affect expression levels.
3. When using reusable (glass) culture flasks, it is important to clean and sterilize them thoroughly between each use. Setting up a cleaning protocol can be helpful when used by multiple people. Avoid growing cultures in the same flask for >1 week to prevent salt deposits in the flask which compromise the sterility upon reuse.
4. Passage cells when the culture reaches a density of 25–35 $*10^6$ cells/mL. Prepare 50 mL of (preferably preheated) medium at RT in a sterile Erlenmeyer flask with baffles. Seed cells into a new medium to a cell density of $2*10^6$ cells/mL. Avoid reaching the maximum cell density of $50*10^6$ cells/mL to prevent stress which can seriously impair growth and protein expres-

sion. Keep track of passage numbers and restart culture freshly from a cryo stock roughly every 30 passages. For more information follow the manufacturer's instructions.

5. When counting, make a 10× dilution of the cells in the medium, then add 1:1 trypan blue (e.g., 20 μL cells, 180 μL medium, 200 μL trypan blue).
6. For proper growth and expression confluency (surface coverage) of the cells is very important. When grown adherently the cells should coat the bottom in an even layer. A total of ~$3*10^6$ cells cover the bottom of a 6-well plate sufficiently. This number of cells needs to be adjusted for different surface sizes, e.g., if the expression is carried out in a T-75 flask instead.
7. If you are expressing the same protein multiple times it may be easier to make one larger 50 ng/μL dilution in advance and take a 10 μL aliquot when needed.
8. The quality of the plasmid DNA is very important for successful transfection and protein expression. Specialized high-purity kits are recommended; however, we have good experience with regular miniprep kits too.
9. The sterility of the plasmid is usually not a concern, and plasmid isolation can be done outside the flow hood.
10. The ratio between DNA and reagents can be optimized for different plasmids, for more information see the Qiagen Effectene manual.
11. Temperature is lowered to 21° to increase the stability of the plant-derived protein. However, as this is protein dependent, tests may be done to optimize the expression.
12. Harvest time should be optimized for different proteins of interest, depending on the ease of expression and stability of the protein, a longer- or shorter-expression time may be beneficial. We recommend testing expression over the time course of 3 consecutive days to experimentally determine at which time point the protein has the highest expression without apparent degradation products (with WB).
13. The insect cells are semi-adherent and will have formed a layer on the bottom of the well, and the layer should look even and thin. A thick layer with cells in solution or a very patchy layer with few cells might point to a confluency issue.
14. If protein concentration is insufficient, spin filters with an appropriate molecular weight cut-off can be used to concentrate. Mix intermittently while concentrating to avoid saturation of the filter.

15. Depending on the interaction strength and the protein expression levels different volumes of proteins may be needed to obtain a sufficient signal. In the case of lowly expressed proteins or transient interactions that are tested low protein concentrations may influence the signal intensity. However, if the interaction is strong a high signal can be achieved with little protein. It may be beneficial to globally adjust the amount of protein added to each well, up to 100 μL, depending on what is being studied.
16. For larger screens, proteins can be distributed over 2 mL deep well plates in advance.
17. To ensure proper liquid removal at each step there must be no bubbles in the wells at any step.
18. The final volume of prey ECD also can be adjusted depending on the expression level and expected strength of interaction.
19. In this step, it is also possible to include a ligand or bioactive molecule that may modulate the interaction.
20. For protein removal from wells washing steps are possible; however, for transient interactions this already reduces the signal significantly.
21. It is important not to let the plate dry out, as this may negatively affect the proteins. Furthermore, plates usually dry unevenly resulting in more edge effects observed in the analysis.
22. To avoid the formation of air bubbles, it may be useful to use the reverse pipetting setting on the multichannel pipette, if available.

Acknowledgments

This work was supported by The NWO Talent Programme Vidi grant VI.Vidi.193.074 to E.S.L.

References

1. Belkhadir Y, Yang L, Hetzel J, Dangl JL, Chory J (2014) The growth defense pivot: crisis management in plants mediated by LRR-RK surface receptors. Trends Biochem Sci 39(10):447–456. https://doi.org/10.1016/j.tibs.2014.06.006
2. Song W, Han Z, Wang J, Lin G, Chai J (2017) Structural insights into ligand recognition and activation of plant receptor kinases. Curr Opin Struct Biol 43:18–27. https://doi.org/10.1016/j.sbi.2016.09.012
3. Hohmann U, Lau K, Hothorn M (2017) The structural basis of ligand perception and signal activation by receptor kinases. Annu Rev Plant Biol 68(1):109–137. https://doi.org/10.1146/annurev-arplant-042916-040957
4. Özkan E, Carrillo Robert A, Eastman Catharine L, Weiszmann R, Waghray D, Johnson Karl G, Zinn K, Celniker Susan E, Garcia KC (2013) An extracellular Interactome of immunoglobulin and LRR proteins reveals receptor-

ligand networks. Cell 154(1):228–239. https://doi.org/10.1016/j.cell.2013.06.006

5. Smakowska-Luzan E, Mott GA, Parys K, Stegmann M, Howton TC, Layeghifard M, Neuhold J, Lehner A, Kong J, Grünwald K, Weinberger N, Satbhai SB, Mayer D, Busch W, Madalinski M, Stolt-Bergner P, Provart NJ, Mukhtar MS, Zipfel C, Desveaux D, Guttman DS, Belkhadir Y (2018) An extracellular network of Arabidopsis leucine-rich repeat receptor kinases. Nature 553(7688):342–346. https://doi.org/10.1038/nature25184

6. Mott GA, Smakowska-Luzan E, Pasha A, Parys K, Howton TC, Neuhold J, Lehner A, Grünwald K, Stolt-Bergner P, Provart NJ, Mukhtar MS, Desveaux D, Guttman DS, Belkhadir Y (2019) Map of physical interactions between extracellular domains of Arabidopsis leucine-rich repeat receptor kinases. Scientific Data 6(1):190025. https://doi.org/10.1038/sdata.2019.25

Chapter 19

Next-Generation Yeast Two-Hybrid Screening to Discover Protein–Protein Interactions

J. Mitch Elmore, Valeria Velásquez-Zapata, and Roger P. Wise

Abstract

Yeast two-hybrid is a powerful approach to discover new protein–protein interactions. Traditional methods involve screening a target protein against a cDNA expression library and assaying individual positive colonies to identify interacting partners. Here we describe a simple approach to perform yeast two-hybrid screens of a cDNA expression library in batch liquid culture. Positive yeast cell populations are enriched under selection and then harvested en masse. Prey cDNAs are amplified and used as input for next-generation sequencing libraries for identification, quantification, and ranking.

Key words Protein–protein interactions, Yeast two-hybrid, Interaction screening, Sequencing, cDNA library

1 Introduction

Protein–protein interactions (PPI) are an essential part of most cellular activities. Understanding the *what*, *when*, *where*, and *how* of these interactions is a major goal of systems biology approaches to model cell function. Identification of a protein's interaction partners can help to elucidate its function, regulation, and localization within the cell, as well as understand how gene variants contribute to phenotypes. A plethora of low-, medium- to high-throughput assays have been developed to identify PPI [1, 2]. Of these, yeast two-hybrid (Y2H) screening using pre-constructed, full-length open reading frame (ORF) libraries has been a method of choice to reconstruct genome-scale interactomes in model organisms [3]. Generally, Y2H involves fusing a transcription factor DNA-binding (DB) domain to protein X of interest (DB-X, "the bait") and the activation domain (AD) to a putative interacting protein Y (AD-Y, "the prey") and expressing the two hybrid proteins in yeast. If X and Y physically interact, transcription factor

Shahid Mukhtar (ed.), *Protein-Protein Interactions: Methods and Protocols*, Methods in Molecular Biology, vol. 2690, https://doi.org/10.1007/978-1-0716-3327-4_19,

activity is reconstituted, which activates the expression of prototrophic marker(s) that are used to select for the growth of yeast cells.

When ORF libraries are unavailable, which is often the case for non-model organisms, custom Y2H cDNA libraries can be constructed from the specific organism(s) and condition(s) under study. Proteins of interest are screened against the library to identify novel interacting partners. Traditional Y2H screens of cDNA libraries involve picking individual positive clones from selection plates, isolating plasmids, and using Sanger sequencing to identify the ORFs that encode interacting proteins [4–7]. These steps are time- and labor-intensive and can limit the number of baits that can be screened. Furthermore, the identification of putative interactors is qualitative; there is no quantitative measure that can be used to rank interactors for follow-up experiments.

In the last decade, the decreasing cost of next-generation sequencing (NGS) has spurred the development of methods and systems that use NGS to score the results from one-by-many [8–13] and many-by-many [14–18] Y2H screens. Based on the experimental design, mapped read counts from the NGS data can be used as a proxy for the abundance of preys in samples enriched for positive interactions. Read counts can be compared to the background library or non-selected samples to derive an enrichment score and compared across baits to determine the specificity of the putative interactions [13, 19]. Additionally, sequence reads that span the translational fusion of prey cDNA with the AD domain can be used to ascertain if the native polypeptide was expressed *in frame* from the Y2H vector [13, 19]. These characteristics can be used to rank putative interactions for follow-up confirmation tests in yeast and orthogonal PPI assays [13].

1.1 Considerations for the cDNA Library

A high-quality and comprehensive cDNA library is necessary to ensure an adequate depth of screening. Commercial kits for the construction of cDNA libraries compatible with common Y2H systems are available, for example, from Thermo Fisher (#A11180, #PQ1000101) and Takara Bio (#634933, #630490, #630489). Poly(A) 3′-anchored, full-length enriched, and random-primed cDNA libraries have different strengths and weaknesses and should be considered based on project goals. For example, full-length enriched libraries will contain complete cDNA sequences but 5′ untranslated regions may include stop codons that are in-frame with the AD domain and will interfere with the expression of the AD-prey fusion protein.

It is important to quantify the cDNA library titer after transformation to ensure sufficient transcriptome representation, as well as qualify the library to determine the average insert size and percentage of clones with inserts [20, 21]. Additionally, we have found it useful to normalize the cDNA before library construction to reduce the levels of high-abundance transcripts and increase the representation of low-abundance cDNAs [22]. It may also be beneficial to express the cDNAs in all three reading frames to ensure that the native polypeptide is expressed in-frame with the activation domain, although this comes with the cost of diluting the "correct" clone in the final, combined library.

1.2 Choose Your Y2H System

This protocol assumes that the DB-X bait and AD-Y prey library are cloned into compatible Y2H vectors and transformed into compatible yeast strains of opposite mating types. We have employed the Y2H system that uses the yeast strains Y8800 (*MATa*)/Y8930 (*MAT*α) with the genotype *leu2-3112 trp1-901 his3-200 ura3-52 gal4*Δ*gal80*Δ *GAL2-ADE2 LYS2::GAL1-HIS3 MET2::GAL7-lacZ cyh2*^{R} and Y2H expression plasmids pDEST-AD (pPC86-based, *TRP1*)/pDEST-DB (pPC97-based, *LEU2*) described in [23]. This system has been used in interactome projects for *Arabidopsis thaliana* [24, 25], human [8, 26], worm [27], and yeast [28]. In this system, Y8930 pDEST-DB bait strains are selected on Synthetic Complete (SC) media lacking Leucine (SC-L), Y8800 pDEST-AD prey strains are selected on SC media lacking tryptophan (SC-W), diploid yeast containing both pDEST-DB and pDEST-AD are selected on SC media lacking leucine and tryptophan (SC-LW), and diploids expressing the *HIS3* reporter indicating a positive PPI are selected on SC media lacking leucine, tryptophan, and histidine (SC-LWH). The system has another prototrophic reporter, ADE2, allowing growth on media lacking adenine, but in this protocol, we use only the HIS3 reporter. Please refer to Dreze et al. [23] for a detailed description of the system and associated protocols. Although the protocol presented here uses a specific Y2H system, it can be adapted to other systems and selection schemes.

1.3 Watch for Autoactivators

Autoactivation of the reporter gene in the absence of a *bona fide* bait-prey interaction can be problematic for all Y2H screens as it can impede the identification of true biological interactions [23, 29–31]. Autoactivators can arise via several mechanisms and should be monitored closely [23]. Prior to screening, bait strains should be tested for autoactivation by mating to an empty prey vector strain [23]. If bait autoactivation is detected, a minimal concentration of 3-amino-1,2,4-triazole (3-AT), an inhibitor of the His3 reporter protein, can be added to the interactor media (SC-LWH) to prevent autoactive growth. Autoactivators in the prey library, although generally rarer than bait autoactivators, can be difficult to detect and manage prior to transforming yeast with the cDNA expression library. However, with adequate controls (i.e., use of an empty bait plasmid or a bait not expected to interact with any proteins expressed from the cDNA library) and/or by screening a diversity of baits, autoactivators can be identified during downstream analyses of the Y2H sequencing data. Once the problem sequences are known, methods exist to deplete these species during cDNA library or NGS library construction [32–34]. In addition, approaches that use negative selection to minimize or remove autoactivators in Y2H libraries have been described [31].

1.4 Protocol Overview

Here we describe a simple approach to screen baits of interest against a cDNA library and generate amplicons that can be used as input for NGS libraries. The protocol, adapted from [11], involves large-scale mating of a bait strain with a prey strain cDNA library, growing diploid yeast in batch liquid culture, and two rounds of enrichment of diploid yeast expressing putative positive interactions (Fig. 1). We also provide protocols for isolating plasmid from yeast populations *en masse*, and amplification of prey cDNA fragments for NGS. This approach is straightforward and cost-effective for screening a small-to-large number of baits against a cDNA library. The process takes about a week to complete and can be scaled up to screen multiple baits in parallel. We have routinely screened 12 baits at a time, with replicate screens in successive weeks. Along with appropriate data analyses (see accompanying Chapter 20 by Velásquez-Zapata et al.) [13, 19], this approach can distinguish true interactors from among the candidates with an 85–95% success rate using independent confirmation [13].

2 Materials

2.1 Library-Scale Mating of Bait Strain with Prey Library

1. Y8930 yeast strain(s) expressing pDEST-DB bait(s) of interest.
2. Frozen aliquot(s) of Y8800 yeast strain expressing pDEST-AD cDNA prey library.
3. 250 mL Erlenmeyer flasks.
4. Synthetic Complete liquid media lacking Leucine (SC-L, for selection of Y8930 pDEST-DB bait strain): 1.7 g/L Yeast Nitrogen Base without Amino Acids and Ammonium Sulfate, 4 g/L Ammonium sulfate, 1.3 g/L Amino Acid powder mix, pH 5.9, autoclaved. Add immediately before use: 50 mL/L 40% Dextrose, 50 mL/L 10 mM Adenine sulfate dihydrate, 8 mL/L 100 mM Histidine, 8 mL/L 40 mM Tryptophan.
5. Amino Acid powder mix: combine 6 g each of alanine, arginine, aspartic acid, asparagine, cysteine, glutamic acid, glutamine, glycine, isoleucine, lysine, methionine, phenylalanine, proline, serine, threonine, tyrosine, valine, uracil into a bottle with 5 clean marbles and shake vigorously to mix.
6. 40% Dextrose in H_2O, autoclaved.
7. 10 mM Adenine sulfate dihydrate in H_2O, autoclaved.

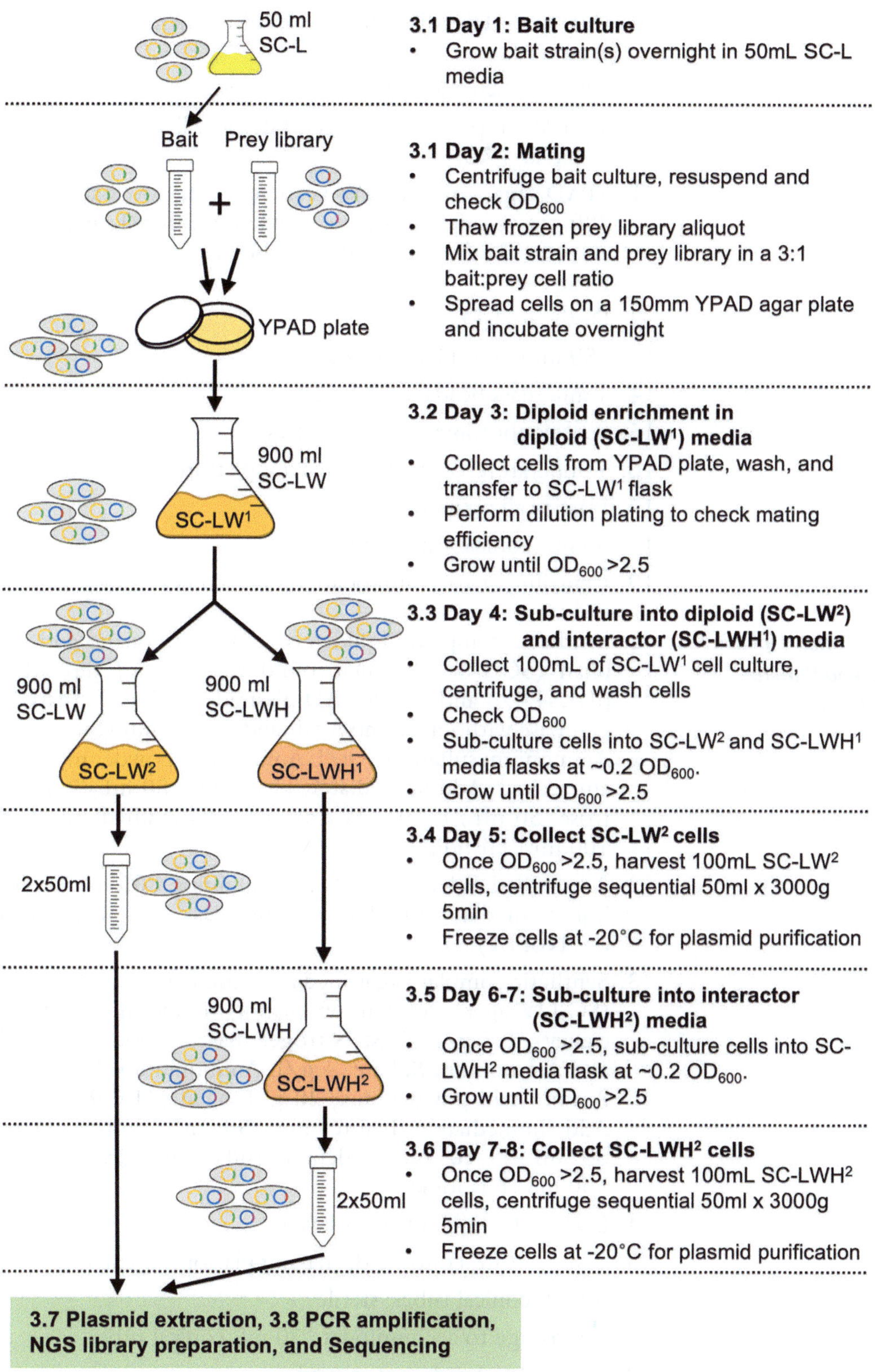

Fig. 1 General flow chart of Y2H-NGIS with a cDNA expression library in batch liquid culture

8. 100 mM Histidine in H_2O, filter sterilized. Store light protected.
9. 100 mM Leucine in H_2O, autoclaved.
10. 40 mM Tryptophan in H_2O, filter sterilized. Store light protected at 4 °C.
11. YPAD agar media: 1% yeast extract, 2% bacto-peptone, 200 mg/L adenine sulfate dihydrate, 20 g/L agar, pH 6.5, autoclaved. Add immediately before use: 50 mL/L 40% Dextrose.
12. Distilled H_2O, autoclaved.
13. 150 mm petri plates, sterile.
14. 4 mm glass beads, autoclaved.
15. Spectrophotometer and cuvettes.
16. Temperature-controlled, shaking platform incubator.
17. Plastic cell scraper.
18. 50 mL conical tubes, sterile.
19. Centrifuge to accommodate 50 mL conical tubes.

2.2 Enrich Diploids in Liquid Culture (SC-LW[1])

1. Synthetic complete liquid media lacking leucine and tryptophan (SC-LW[1], for first-round selection of diploids with pDEST-DB and pDEST-AD plasmids): 1.7 g/L Yeast Nitrogen Base without Amino Acids and Ammonium Sulfate, 4 g/L Ammonium sulfate, 1.3 g/L Amino Acid powder mix, pH 5.9, autoclaved. Add immediately before use: 50 mL/L 40% Dextrose, 50 mL/L 10 mM Adenine sulfate dihydrate, 8 mL/L 100 mM Histidine.
2. 2 L baffled Erlenmeyer flasks containing 800 mL of SC-LW liquid media, autoclaved. One flask per bait screened (*see* **Note 1**).
3. Synthetic complete agar media lacking leucine and tryptophan (SC-LW agar, for dilution plating to monitor mating efficiency): 1.7 g/L Yeast Nitrogen Base without Amino Acids and Ammonium Sulfate, 4 g/L Ammonium sulfate, 1.3 g/L Amino Acid powder mix, 20 g/L agar, pH 5.9, autoclaved. Add immediately before use: 50 mL/L 40% Dextrose, 50 mL/L 10 mM Adenine sulfate dihydrate, 8 mL/L 100 mM Histidine.
4. 0.9% NaCl in H_2O, autoclaved.
5. Temperature-controlled, shaking platform incubator.
6. 50 mL conical tubes, sterile.
7. Centrifuge to accommodate 50 mL conical tubes.
8. 4 mm glass beads, autoclaved.

9. 1.5 mL microcentrifuge tubes, sterile, labeled for dilution plating: 10^{-2} (990 μL 0.9% NaCl), 10^{-3} (900 μL 0.9% NaCl), 10^{-4} (900 μL 0.9% NaCl), 10^{-5} (900 μL 0.9% NaCl).
10. Large cell scraper, sterile.

2.3 Sub-Culture Yeast Populations in Diploid (SC-LW²) and Interactor Media (SC-LWH¹)

1. Synthetic complete liquid media lacking leucine and tryptophan (SC-LW², for second-round selection of diploids with pDEST-DB and pDEST-AD plasmids): 1.7 g/L Yeast Nitrogen Base without Amino Acids and Ammonium Sulfate, 4 g/L Ammonium sulfate, 1.3 g/L Amino Acid powder mix, pH 5.9, autoclaved. Add immediately before use: 50 mL/L 40% Dextrose, 50 mL/L 10 mM Adenine sulfate dihydrate, 8 mL/L 100 mM Histidine.
2. Synthetic complete liquid media lacking leucine, tryptophan, and histidine (SC-LWH¹, for first-round selection of diploids expressing the *HIS3* reporter): 1.7 g/L Yeast Nitrogen Base without Amino Acids and Ammonium Sulfate, 4 g/L Ammonium sulfate, 1.3 g/L Amino Acid powder mix, pH 5.9, autoclaved. Add immediately before use: 50 mL/L 40% Dextrose, 50 mL/L 10 mM Adenine sulfate dihydrate.
3. 2 L baffled Erlenmeyer flasks containing 800 mL of SC-LW liquid media, autoclaved. One flask per bait screened.
4. 2 L baffled Erlenmeyer flasks containing 800 mL of SC-LWH liquid media, autoclaved. One flask per bait screened.
5. 50 mL conical tubes, sterile.
6. Centrifuge to accommodate 50 mL conical tubes.
7. Spectrophotometer and cuvettes.

2.4 Harvest SC-LW² Cell Population

1. 50 mL conical tubes, sterile.
2. Centrifuge to accommodate 50 mL conical tubes.
3. Spectrophotometer and cuvettes.

2.5 Monitor SC-LWH¹ Cultures and Sub-Culture in SC-LWH² Media

1. Synthetic complete liquid media lacking leucine, tryptophan, and histidine (SC-LWH², for second-round selection of diploids expressing the *HIS3* reporter): 1.7 g/L Yeast Nitrogen Base without Amino Acids and Ammonium Sulfate, 4 g/L Ammonium sulfate, 1.3 g/L Amino Acid powder mix, pH 5.9, autoclaved. Add immediately before use: 50 mL/L 40% Dextrose, 50 mL/L 10 mM Adenine sulfate dihydrate.
2. 2 L baffled Erlenmeyer flasks containing 800 mL of SC-LWH liquid media, autoclaved. One flask per bait screened.
3. Spectrophotometer and cuvettes.

2.6 Monitor SC-LWH[2] Cultures and Harvest Cells

1. 50 mL conical tubes, sterile.
2. Centrifuge to accommodate 50 mL conical tubes.
3. Spectrophotometer and cuvettes.

2.7 Plasmid Isolation from Yeast

1. Lyticase (100× stock = 10,000 U/mL in 10 mM NaH_2PO_4, 50% glycerol, pH 7.5).
2. RNAse A (100× stock = 10 mg/mL in Molecular Biology Grade H_2O).
3. β-mercaptoethanol.
4. Solution RI: 100 mM NaH_2PO_4, 1.2 M sorbitol, 25 mM EDTA, pH 7.5. Add immediately before use: 30 mM β-mercaptoethanol, 100 U/mL lyticase, 0.1 mg/mL RNAse A.
5. Solution YI: 25 mM Tris-HCl pH 8, 100 mM EDTA, 50 mM Glucose, 0.1 mg/mL RNAse A.
6. Solution YII: 0.2 N NaOH, 1% SDS.
7. Solution YIII: 3 M NaOAc, pH 4.8.
8. Isopropanol (100%).
9. Ethanol (70%).
10. Miracloth.
11. 50 mL conical centrifuge tubes.

2.8 PCR Amplification of Prey Sequences

1. Advantage 2 Polymerase Mix and 10× buffer (*see* **Note 2**).
2. dNTP mix (10 mM each dNTP).
3. Oligonucleotide primers that flank cDNA inserts in the prey plasmid (*see* **Note 3**).
4. Molecular Biology Grade H_2O.

3 Methods

3.1 Library-Scale Mating of Bait Strain with Prey Library

1. Inoculate SC-L agar plates with Y8930 pDEST-DB bait strain (s) (*see* **Note 4**) from an −80 °C glycerol stock. Streak for isolation of single colonies and incubate ~2–3 days at 30 °C. Plates can be wrapped in parafilm and stored at 4 °C for at least one month.

[Day 1]

2. Pick 1–3 single colonies into 50 mL of SC-L liquid media [final OD_{600} = 0.05–0.15]. Grow overnight shaking at 225 rpm at 30 °C (expect 14–18 h to reach OD_{600} = ~3).

[Day 2]

3. Determine overnight culture OD_{600} using a spectrophotometer.
4. When bait cultures are ready (>2.5 OD_{600}), remove aliquots (one aliquot for each bait) of the Y8800 prey library from –80 °C and thaw on ice for 15 min (*see* **Note 5**).
5. Centrifuge 40 mL of the bait culture(s) in sterile 50 mL tubes at 1000 *g* for 5 min.
6. Discard the supernatant and resuspend pellets in 5 mL of sterile H_2O.
7. Determine the OD_{600} using 10^{-2} dilution (10 μL cells in 990 μL SC-L). Use a standard curve developed for the yeast strains being used (*see* **Note 6**).
8. Add the appropriate volume of bait cells to each labeled prey tube to achieve a 3:1 bait:prey cell ratio (*see* **Note** 7).
9. Mix gently and centrifuge at 1000 *g* for 5 min. Remove supernatant.
10. Resuspend the pellet in 1 mL sterile H_2O. Incubate the mating tubes on the sides for approximately 3 h shaking at 100 rpm at 30 °C.
11. For each bait screen, pour one 150 mm YPAD agar plate and leave covered in the hood to solidify until ready to use.
12. After the 3-h incubation of the mating tubes, centrifuge at 1000 *g* for 5 min. Remove supernatant.
13. Resuspend the pellet in 500 μL sterile H_2O.
14. For each screen, pipette the cells onto the labeled 150 mm YPAD "mating" agar plate with 6 to 8 sterile 4 mm glass beads.
15. Wash each tube with 250 μL sterile H_2O and add it to the associated YPAD plate. Shake the plate(s), using the beads to spread the cells. Remove the beads.
16. Let the plates air dry in a sterile hood for ~10–15 min until the surface is dry. Incubate the plates inverted for 16 h at 30 °C.

3.2 Enrich Diploids in SC-LW[1] Media

[Day 3]

1. Collect cells from YPAD mating plates. Specifically, pipette 5 mL of 0.9% NaCl onto the plate and use a cell scraper to gently dislodge cells. Tilt plates to collect cell solution with a pipette and add it to a 50 mL labeled tube.
2. Repeat washing of the plate 4 times with 5 mL 0.9% NaCl (total volume = 25 mL) to remove as many cells as possible (*see* **Note 8**).

3. Vortex gently and centrifuge the 50 mL tubes at 2000 *g* for 20 min. Remove supernatant.
4. Wash 1: Resuspend the pellet with 25 mL 0.9% NaCl. Vortex low and centrifuge the 50 mL tubes at 2000 *g* for 15 min. Remove supernatant.
5. Wash 2: Resuspend the pellet with 25 mL 0.9% NaCl. Vortex low and centrifuge the 50 mL tubes at 2000 *g* for 15 min. Remove supernatant.
6. Wash 3: Resuspend the pellet with 25 mL 0.9% NaCl. Vortex low and centrifuge the 50 mL tubes at 2000 *g* for 15 min. Remove supernatant.
7. Resuspend the pellet with 23.5 mL 0.9% NaCl (~25 mL final volume).
8. Remove 10 μL of cells from each sample and resuspend in the 10^{-2} dilution tubes containing 990 μL 0.9% NaCl.
9. Add the remaining ~25 mL of cells to 900 mL SC-LW^1 media in glass baffled flasks.
10. Wash tubes with 25 mL SC-LW and add the wash solution to the SC-LW^1 flasks.
11. Incubate the SC-LW^1 cultures shaking at 225 rpm at 30 °C for 18–24 h (or until OD_{600} > 2.5).
12. Perform dilutions: transfer 100 μL from the 10^{-2} dilution into the 10^{-3} dilution tube, then transfer 100 μL from the 10^{-3} dilution into the 10^{-4} dilution tube, and then transfer 100 μL from the 10^{-4} dilution into the 10^{-5} dilution tube. Be sure to vortex tubes to mix well prior to dilutions and plating.
13. Pipette 100 μL of the 10^{-3}, 10^{-4}, and 10^{-5} dilutions onto the corresponding plates. Use beads to spread the cells.
14. Incubate the dilution plates @ 30 °C for 2–3 days and then count the colonies formed on each. From these counts, estimate the total # of diploids formed for each bait (*see* **Note 9**).

3.3 Sub-Culture into Diploid (SC-LW^2) and Interactor Media (SC-LWH^1)

[Day 4]

1. Check OD_{600} of the overnight SC-LW culture with a 10^{-1} dilution (100 μL cells in 900 μL SC-LW) (*see* **Note 10**).
2. Harvest the cells once cultures have reached OD_{600} > 2.5.
3. Mix the culture well and pour/pipet 50 mL into a labeled 50 mL tube.
4. Centrifuge the cells at 2000 *g* for 10 min. Remove supernatant.
5. Mix the culture well and pour/pipet 50 mL into the same tube for a total of 100 mL of cell culture harvested.
6. Centrifuge the cells at 2000 *g* for 10 min. Remove supernatant.

7. Wash 1: Resuspend the pellet with 25 mL of SC-LWH media. Vortex gently and centrifuge the 50 mL tubes at 2000 *g* for 5 min. Remove supernatant.
8. Wash 2: Resuspend the pellet with 25 mL of SC-LWH media. Vortex gently and centrifuge the 50 mL tubes at 2000 *g* for 5 min. Remove supernatant.
9. Resuspend the cell pellet in 25 mL SC-LWH media.
10. Check OD_{600} with a 10^{-2} dilution (10 μL cells in 990 μL SC-LWH).
11. Subculture enough cells to start new 900 mL $SC\text{-}LW^2$ and $SC\text{-}LWH^1$ cultures at $OD_{600} = 0.15–0.20$ (*see* **Note 11**).
12. Incubate the $SC\text{-}LW^2$ flask shaking at 225 rpm at 30 °C for ~16–18 h (or until $OD_{600} > 2.5$). This culture should be ready to harvest on Day 5.
13. Incubate the $SC\text{-}LWH^1$ flask(s) shaking at 225 rpm at 30 °C for 60–72 h (or until $OD_{600} > 2.5$). This culture should be ready to harvest on Day 6 or 7 (*see* **Note 12**).

3.4 Monitor $SC\text{-}LW^2$ Cultures and Harvest Cells Once $OD_{600} > 2.5$

[Day 5]

1. Determine $SC\text{-}LW^2$ culture OD_{600} using a 10^{-1} dilution (100 μL cells in 900 μL SC-LW).
2. Once the $SC\text{-}LW^2$ cultures reach $OD_{600} > 2.5$, harvest two 100 mL samples.
3. Label two 50 mL tubes per sample (A and B).
4. Mix the culture well and pour/pipette 50 mL into the labeled 50 mL tube.
5. Centrifuge the cells at 3000 *g* for 5 min. Remove supernatant.
6. Mix the culture well and pour additional 50 mL culture into the 50 mL tube.
7. Centrifuge the cells 3000 *g* for 5 min. Remove supernatant.
8. Invert tubes on a paper towel to remove residual media and then freeze tubes at −20 °C. These non-selected controls will be used later to isolate plasmids for amplicon prep and NGS (*see* also Subheading 3.6). Note that only the "A" sample will be initially used for plasmid purification and NGS library prep. The "B" sample is a useful, and sometimes necessary, backup.

3.5 Monitor $SC\text{-}LWH^1$ Cultures and Subculture Cells into $SC\text{-}LWH^2$ Media Once $OD_{600} > 2.5$

[Day 6–7 based on the growth of cultures]

1. Determine the $SC\text{-}LWH^1$ culture OD_{600} using a 10^{-1} dilution (100 μL cells in 900 μL SC-LWH).
2. Once cultures reach $OD_{600} > 2.5$, subculture cells into a fresh 900 mL $SC\text{-}LWH^2$ culture.

3. Prepare SC-LWH^2 media: to 800 mL of SC media in a 2 L baffled flask, add 45 mL 40% Dextrose and 45 mL 10 mM Adenine.
4. Mix the SC-LWH^1 culture well and pipet enough cells into the SC-LWH^2 flasks to reach $OD_{600} = 0.15–0.2$.
5. Incubate the SC-LWH^2 flasks shaking at 225 rpm at 30 °C for 18–24 h (or until $OD_{600} > 2.5$).

3.6 Monitor SC-LWH^2 Cultures and Harvest Cells Once $OD_{600} > 2.5$

[Day 7–8]

1. Determine the SC-LWH^2 culture OD_{600} using a 10^{-1} dilution (100 μL cells in 900 μL SC-LWH).
2. Once the SC-LWH^2 cultures reach $OD_{600} > 2.5$, harvest two 100 mL samples.
3. Label two 50 mL tubes per sample (A and B).
4. Mix the culture well and pour/pipette 50 mL into each labeled 50 mL tube.
5. Centrifuge the cells at 3000 *g* for 5 min. Remove supernatant.
6. Mix the culture well and pour additional 50 mL culture into the 50 mL tube.
7. Centrifuge the cells 3000 *g* for 5 min. Remove supernatant.
9. Invert tubes on a paper towel to remove residual media and then freeze tubes at −20 °C. These selected samples will be used later to isolate plasmids for amplicon prep and NGS. Note that only the "A" sample will be initially used for plasmid purification and NGS library prep. The "B" sample is a useful, and sometimes necessary, backup.

3.7 Yeast Plasmid Isolation

1. Thaw frozen yeast cell pellets on ice.
2. Resuspend the pellet in 5 mL RI buffer.
3. Incubate for 2–3 h at 37 °C with gentle shaking at 65 rpm for cell wall digestion.
4. Centrifuge the tubes at 4000 *g* for 10 min. If the supernatant is still cloudy, centrifuge longer. Remove the supernatant with a pipette.
5. Resuspend the pellet in 4 mL Solution YI.
6. Add 10 mL Solution YII and mix by inversion several times.
7. Incubate for 15 min at 25 °C.
8. Add 8 mL cold Solution YIII to neutralize the lysate and invert the tubes several times to mix well.
9. Incubate for 30 min on ice.

10. Centrifuge at 9000 *g* for 15 min at 4 °C (*see* **Note 13**).
11. Transfer the supernatant to a new tube using Miracloth as a filter to catch any cell debris.
12. Add 0.6 volumes (~13.2 mL) of isopropanol. Invert the tube quickly several times and place it on ice.
13. Incubate for 30 min on ice.
14. Centrifuge 9000 *g* for 20 min at 4 °C. Carefully pour off the supernatant.
15. Gently add 5 mL of 70% Ethanol to the tube (add to the side).
16. Centrifuge 9000 *g* for 10 min at 4 °C. Carefully pour off the supernatant.
17. Repeat wash in **steps #15–16**.
18. Centrifuge tubes briefly and remove any excess ethanol with a pipette.
19. Dry the DNA pellet at 55 °C for about 15 min or until all ethanol is evaporated.
20. Resuspend the DNA pellet in 500–800 μL molecular biology grade H_2O, or for extended storage, 1× TE (pH -8.0), heat @ 65 °C for 5 min, aliquot and flash freeze.

3.8 PCR Amplification of Prey Sequences

1. Quantify the DNA yield from the yeast plasmid preps and dilute to 25 ng/μL.
2. For each sample, set up 4–8 20 μL PCR reactions with the following components: 11 μL PCR-grade H_2O, 2 μL 10× Advantage 2 PCR buffer, 4 μL 25 ng/μL DNA template, 0.5 μL 10 mM dNTP mix, 1 μL 10 μM 5′ primer, 1 μL 10 μM 3′ primer, 0.5 μL 50× Advantage 2 polymerase mix.
3. Run the PCR on a thermocycler with the following conditions: **STEP 1**: 95 °C 1:00; **STEP 2**: 95 °C 0:30; **STEP 3**: 70 °C 0: 30; **STEP 4**: 68 °C 4:00; **STEP 5**: GOTO **STEP 2**, 21 cycles (depending on empirical determination for the selected system); **STEP 6**: 10:00; **STEP** 7: 4 °C HOLD (*see* **Notes 14** and **15** for comments on number of cycles).
4. Pool the PCR reactions from the same sample.
5. Purify the PCR amplicons using DNA precipitation or column purification.
6. Perform DNA gel electrophoresis to quality-check the PCR reactions. Figure 2 shows the results from replicated screens using two different bait proteins with the PCR conditions described in **steps 2** and **3** above.
7. PCR amplicons can be used as input for NGS library construction (*see* **Note 15**).

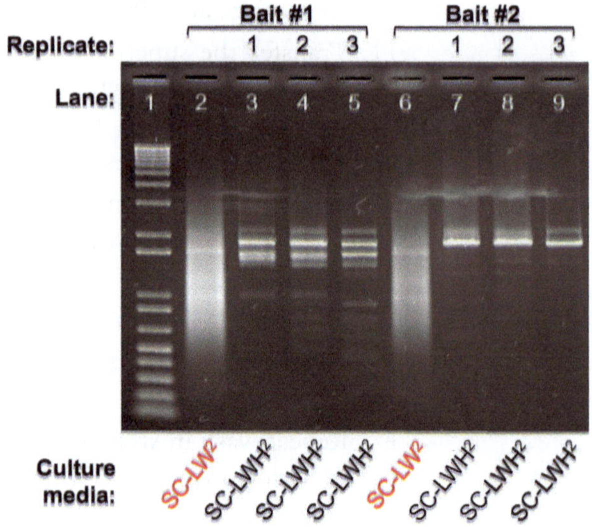

Fig. 2 Example of a DNA electrophoresis gel from replicated screens of two different bait proteins mated with a cDNA prey library constructed from a time course of barley infected with the powdery mildew pathogen. Lane 1, 1 kb DNA ladder; Lane 2, Bait #1 SC-LW^2 diploid culture; Lane 3, Bait #1 SC-LWH^2 interactor culture replicate 1; Lane 4, Bait #1 SC-LWH^2 interactor culture replicate 2; Lane 5, Bait #1 SC-LWH^2 interactor culture replicate 3; Lane 6, Bait #2 SC-LW^2 diploid culture; Lane 7, Bait #2 SC-LWH^2 interactor culture replicate 1; Lane 8, Bait #2 SC-LWH^2 interactor culture replicate 2; Lane 9, Bait #2 SC-LWH^2 interactor culture replicate 3. Non-selected samples are highlighted in red

4 Notes

1. The use of 2 L baffled flasks promotes culture aeration and cell growth. Larger culture volumes are also possible using baffled flasks.
2. We have had success using this DNA polymerase to generate PCR amplicons suitable for NGS library construction. Other high-yield and high-fidelity polymerases will also work.
3. It is important to design the oligo primers used to amplify the cDNA inserts in such a way that the 5′ fusion reads will capture both the AD prey vector sequence and the 5′ end of the cDNA. This will allow reconstruction of the AD-cDNA sequence and determine if the translational fusion of the cDNA-encoded polypeptide was in-frame with the AD protein. Researchers should consider the type of sequencing (single or paired-end, and length of reads) in order to determine the optimal distance the forward primer should anneal from the 5′ end of the cDNA. The 5′ fusion read should contain enough sequence from the AD and the cDNA that they can be reliably mapped to

each. For optimal use of the NGPINT/Y2H-SCORES informatics pipeline [13, 19], we recommend designing PCR primers as close to the junction as possible while leaving at least 25 nucleotides of vector sequence on either end of the cDNA insert. Paired-end sequencing provides more sequence information about the junctions and 5' cDNA ends and can facilitate unambiguous mapping of reads.

4. Prior to screening, baits should be tested for autoactivity by mating to an empty prey vector strain growing the resulting diploids on SC-LWH. If there is autoactive growth, it is possible that the bait can still be screened by adding a minimal amount of 3-AT to the media to prevent autoactive growth.

5. As a general rule, the yeast prey library aliquots should contain 5–10× the cells as the primary number of yeast transformants. For example, if the yeast library contained 1×10^7 primary transformants, then each aliquot should comprise greater than 5×10^7 cells. It is also important to know the number of viable cells in each aliquot for optimal mating efficiency (*see* **Note** 7).

6. It is helpful to generate a standard curve of OD_{600} vs. cell number for the yeast (haploid and diploid) strains used for Y2H. This will allow quick estimation of cell counts using OD measurements.

7. We and others have observed optimal mating efficiency using a 3:1 bait:prey ratio [35].

8. The tubes with the yeast cultures should be loosely capped as multiple plates are processed. The yeast cells are actively respiring, and pressure can build up in tightly capped tubes. It is advised to gently swirl the yeast in the tube and then cap it tightly before the centrifuge steps.

9. Generally, it is recommended that the number of diploids generated should be 5 times the library complexity to ensure sufficient interrogation of the library.

10. We have repeatedly observed a very high correlation ($R > 0.98$) in the read counts in diploid-selected SC-LWH^2 cultures from different bait screens performed at the same time. Thus, under most situations, it is probably not necessary to culture all samples in SC-LW^2, provided that the mating efficiency and growth rate are largely similar across the baits. If many baits are being screened, we recommend picking one sample at random to culture in SC-LW^2, which can serve as the background library for all baits in that replicate.

11. To minimize a bottleneck effect during subculturing, at least 5–10% of the cells in the source culture should be used in the new culture.

12. The growth rate of diploids containing different baits can vary in SC-LWH media. This will depend on the number of interacting proteins and their initial levels in the prey library.
13. Depending on the available centrifuge and rotors, if one needs to use slower centrifuge speeds, longer run periods are advised. For example, we have also used 6800 *g* (5270 RPMs—Beckman JS-5.3 rotor) for 20 min at 4 °C for **steps 10–16**. This rotor can process 28 samples at one time but is limited to 6800 *g*, thus the longer run time.
14. The annealing temperature should be optimized for each primer set. To avoid bias introduced through PCR overcycling, one should adjust the number of cycles to the minimum number necessary to generate enough product for NGS library preparation. This should be determined empirically for the system of choice, but 4–8 replicate PCR reactions with a range of 15–21 cycles is a good place to start. It is preferable to use more PCR reactions and pool the products, rather than increasing the number of cycles.
15. PCR-free whole-genome sequencing (WGS) library preparations are recommended to limit any further PCR artifacts. Again, it is better to gain the necessary quantity for WGS library preparation by the number of PCR reactions, instead of increasing cycles. See also reference [21].

Acknowledgments

The authors thank Gregory Fuerst, USDA-ARS Ames, IA, for the critical reading of the manuscript. Research supported in part by USDA-ARS Postdoctoral Research Associateship and USDA-National Institute of Food and Agriculture (NIFA)-Education and Literacy Initiative (ELI) Postdoctoral Fellowship **2017-67012-26086** to JME, Fulbright–Minciencias 2015 & Schlumberger Faculty for the Future fellowships to VVZ, and National Science Foundation–Plant Genome Research Program grant 13-39348, USDA-NIFA grant 2020-67013-31184 and USDA-Agricultural Research Service project 3625-21000-067-00D to RPW. The funders had no role in study design, data collection, and analysis, decision to publish, or preparation of the manuscript. Mention of trade names or commercial products in this publication is solely for the purpose of providing specific information and does not imply recommendation or endorsement by the USDA, NIFA, ARS, or the National Science Foundation. USDA is an equal opportunity provider and employer.

References

1. Benz C, Kassa E, Tjärnhage E, et al (2020) Chapter 1 Identification of cellular protein–protein interactions. In: Inhibitors of protein–protein interactions: small molecules, peptides and macrocycles. pp 1–39
2. Titeca K, Lemmens I, Tavernier J, Eyckerman S (2019) Discovering cellular protein-protein interactions: Technological strategies and opportunities. Mass Spectrom Rev 38(1): 79–111. https://doi.org/10.1002/mas.21574
3. Vidal M, Fields S (2014) The yeast two-hybrid assay: still finding connections after 25 years. Nat Methods 11(12):1203–1206. https://doi.org/10.1038/nmeth.3182
4. Gietz RD (2006) Yeast two-hybrid system screening. In: Xiao W (ed) Yeast protocol. Humana Press, Totowa, pp 345–371
5. Mohr K, Koegl M (2012) High-throughput yeast two-hybrid screening of complex cDNA libraries. In: Suter B, Wanker EE (eds) Two hybrid technologies: methods and protocols. Humana Press, Totowa, pp 89–102
6. Roberts GG, Parrish JR, Mangiola BA, Finley RL (2012) High-throughput yeast two-hybrid screening. In: Suter B, Wanker EE (eds) Two hybrid technologies: methods and protocols. Humana Press, Totowa, pp 39–61
7. Paiano A, Margiotta A, De Luca M, Bucci C (2019) Yeast two-hybrid assay to identify interacting proteins. Curr Protoc Protein Sci 95(1): e70. https://doi.org/10.1002/cpps.70
8. Yu H, Tardivo L, Tam S et al (2011) Next-generation sequencing to generate interactome datasets. Nat Methods 8(6):478–480. https://doi.org/10.1038/nmeth.1597
9. Lewis JD, Wan J, Ford R et al (2012) Quantitative Interactor Screening with next-generation Sequencing (QIS-Seq) identifies *Arabidopsis thaliana* MLO2 as a target of the *Pseudomonas syringae* type III effector HopZ2. BMC Genomics 13(1):8. https://doi.org/10.1186/1471-2164-13-8
10. Weimann M, Grossmann A, Woodsmith J et al (2013) A Y2H-seq approach defines the human protein methyltransferase interactome. Nat Methods 10(4):339–342. https://doi.org/10.1038/nmeth.2397
11. Pashkova N, Peterson TA, Krishnamani V et al (2016) DEEPN as an approach for batch processing of yeast 2-hybrid interactions. Cell Rep 17(1):303–315. https://doi.org/10.1016/j.celrep.2016.08.095
12. Erffelinck M-L, Ribeiro B, Perassolo M et al (2018) A user-friendly platform for yeast two-hybrid library screening using next generation sequencing. PLoS One 13(12): e0201270. https://doi.org/10.1371/journal.pone.0201270
13. Velásquez-Zapata V, Elmore JM, Banerjee S et al (2021) Next-generation yeast-two-hybrid analysis with Y2H-SCORES identifies novel interactors of the MLA immune receptor. PLoS Comput Biol 17(4):e1008890. https://doi.org/10.1371/journal.pcbi.1008890
14. Yachie N, Petsalaki E, Mellor JC et al (2016) Pooled-matrix protein interaction screens using Barcode Fusion Genetics. Mol Syst Biol 12(4):863. https://doi.org/10.15252/msb.20156660
15. Trigg SA, Garza RM, MacWilliams A et al (2017) CrY2H-seq: a massively multiplexed assay for deep-coverage interactome mapping. Nat Methods 14(8):819–825. https://doi.org/10.1038/nmeth.4343
16. Yang F, Lei Y, Zhou M et al (2018) Development and application of a recombination-based library versus library high- throughput yeast two-hybrid (RLL-Y2H) screening system. Nucleic Acids Res 46(3):e17. https://doi.org/10.1093/nar/gkx1173
17. Yang J-S, Garriga-Canut M, Link N et al (2018) rec-YnH enables simultaneous many-by-many detection of direct protein–protein and protein–RNA interactions. Nat Commun 9(1):3747. https://doi.org/10.1038/s41467-018-06128-x
18. Andrews SS, Schaefer-Ramadan S, Al-Thani NM et al (2019) High-resolution protein–protein interaction mapping using all-versus-all sequencing (AVA-Seq). J Biol Chem 294(30): 11549–11558. https://doi.org/10.1074/jbc.RA119.008792
19. Banerjee S, Velásquez-Zapata V, Fuerst G et al (2021) NGPINT: a next-generation protein–protein interaction software. Brief Bioinform 22 (4):bbaa351. https://doi.org/10.1093/bib/bbaa351
20. Fülle H-J (2003) Quality assessment of cDNA libraries. In: Ying S-Y (ed) Generation of cDNA libraries: methods and protocols. Humana Press, Totowa, pp 145–153
21. Yu Q, Hu Y, Su J et al (2020) Evaluation of a yeast two-hybrid library by high-throughput sequencing. J Proteome Res 19(8): 3567–3572. https://doi.org/10.1021/acs.jproteome.0c00189
22. Bogdanova EA, Barsova EV, Shagina IA et al (2011) Normalization of full-length-enriched cDNA. In: Lu C, Browse J, Wallis JG (eds)

cDNA libraries: methods and applications. Humana Press, Totowa, pp 85–98
23. Dreze M, Monachello D, Lurin C et al (2010) Chapter 12 – High-quality binary interactome mapping. In: Methods in enzymology. Academic Press, pp 281–315
24. Arabidopsis Interactome Mapping Consortium, Dreze M, Carvunis A-R, et al (2011) Evidence for network evolution in an Arabidopsis interactome map. Science 333 (6042): 601–607. https://doi.org/10.1126/science.1203877
25. Mukhtar MS, Carvunis A-R, Dreze M et al (2011) Independently evolved virulence effectors converge onto hubs in a plant immune system network. Science 333(6042):596–601. https://doi.org/10.1126/science.1203659
26. Rual J-F, Venkatesan K, Hao T et al (2005) Towards a proteome-scale map of the human protein–protein interaction network. Nature 437(7062):1173–1178. https://doi.org/10.1038/nature04209
27. Simonis N, Rual J-F, Carvunis A-R et al (2009) Empirically controlled mapping of the Caenorhabditis elegans protein-protein interactome network. Nat Methods 6(1):47–54. https://doi.org/10.1038/nmeth.1279
28. Yu H, Braun P, Yıldırım MA et al (2008) High-quality binary protein interaction map of the yeast interactome network. Science 322(5898):104–110. https://doi.org/10.1126/science.1158684
29. Vidalain P-O, Boxem M, Ge H et al (2004) Increasing specificity in high-throughput yeast two-hybrid experiments. Methods 32(4): 363–370. https://doi.org/10.1016/j.ymeth.2003.10.001
30. Mehla J, Caufield JH, Uetz P (2015) The yeast two-hybrid system: a tool for mapping protein–protein interactions. Cold Spring Harb Protoc 2015 (5):pdb.top083345. https://doi.org/10.1101/pdb.top083345
31. Shivhare D, Musialak-Lange M, Julca I et al (2021) Removing auto-activators from yeast-two-hybrid assays by conditional negative selection. Sci Rep 11(1):5477. https://doi.org/10.1038/s41598-021-84608-9
32. Gu W, Crawford ED, O'Donovan BD et al (2016) Depletion of Abundant Sequences by Hybridization (DASH): using Cas9 to remove unwanted high-abundance species in sequencing libraries and molecular counting applications. Genome Biol 17(1):41. https://doi.org/10.1186/s13059-016-0904-5
33. Archer SK, Shirokikh NE, Preiss T (2014) Selective and flexible depletion of problematic sequences from RNA-seq libraries at the cDNA stage. BMC Genomics 15(1):401. https://doi.org/10.1186/1471-2164-15-401
34. Culviner PH, Guegler CK, Laub MT (2020) A simple, cost-effective, and robust method for rRNA depletion in RNA-sequencing studies. mBio 11(2):e00010–e00020. https://doi.org/10.1128/mBio.00010-20
35. Soellick T-R, Uhrig JF (2001) Development of an optimized interaction-mating protocol for large-scale yeast two-hybrid analyses. Genome Biol 2(12):research0052.1. https://doi.org/10.1186/gb-2001-2-12-research0052

Chapter 20

Bioinformatic Analysis of Yeast Two-Hybrid Next-Generation Interaction Screen Data

Valeria Velásquez-Zapata, J. Mitch Elmore, and Roger P. Wise

Abstract

Yeast two-hybrid next-generation interaction screening (Y2H-NGIS) uses the output of next-generation sequencing to mine for novel protein–protein interactions. Here, we outline the analytics underlying Y2H-NGIS datasets. Different systems, libraries, and experimental designs comprise Y2H-NGIS methodologies. We summarize the analysis in several layers that comprise the characterization of baits and preys, quantification, and identification of true interactions for subsequent secondary validation. We present two software designed for this purpose, NGPINT and Y2H-SCORES, which are used as front-end and back-end tools in the analysis. Y2H-SCORES software can be used and adapted to analyze different datasets not only from Y2H-NGIS but from other techniques ruled by similar biological principles.

Key words Next Generation Interaction Screening (NGIS), Yeast two-hybrid (Y2H), Systems biology, ranking interactions, interaction prediction

1 Introduction

In the field of proteomics, the determination of protein–protein interactions (PPIs) is a fundamental step for the characterization of protein and cell function. PPI also sets the basis for systems biology investigations through the construction of interactome networks [1]. Yeast two-hybrid (Y2H) and its derivatives exploit yeast genetics to determine PPIs on a small or large scale [2]. The molecular basis of the technique consists of the utilization of the modularity of a transcription factor (TF) to test a protein interaction. The activation and DNA binding domains of a TF are expressed in a mutant yeast cell, each fused with a bait and prey protein, such that when the two interact, transcription and activation of a reporter gene ensue [3]. The activation of the reporter gene permits the survival of the cell under media selection, and then yeast growth becomes the measurable variable to identify PPIs. To conduct PPI mining on a genome-wide scale, a set of techniques has emerged, termed

Shahid Mukhtar (ed.), *Protein-Protein Interactions: Methods and Protocols*, Methods in Molecular Biology, vol. 2690,
https://doi.org/10.1007/978-1-0716-3327-4_20,

yeast-two-hybrid next-generation-interaction screening (Y2H-NGIS), which couples Y2H with high throughput sequencing [4]. Using massively parallel, next-generation sequencing (NGS) to quantitatively mine PPIs improves the scalability of the Y2H technique.

The experimental settings of Y2H-NGIS include different Y2H systems, number of baits and preys per screening, bait and prey library type, such as open-reading-frame (ORF) or cDNA, quantification technology, and controls [4]. Here we present a set of recommendations for the analysis of these datasets, placing an emphasis on batch designs coupled with NGS to quantify baits and preys. In our group, we developed two software to process Y2H-NGIS data. NGPINT processes NGS reads from designs with one or more individual bait(s) and a prey library per experiment, reconstruction of the prey fragments, visualization, and prey quantification. As a back-end software, Y2H-SCORES is used to rank the candidate interactions and can be used with any type of Y2H-NGIS dataset. Y2H-SCORES optimizes different aspects of the analysis, for example, count normalization, and use of different controls, and uses the biological information to calculate three scores that are used to rank candidate interactions. We explain how Y2H-SCORES can be customized to different datasets and what to consider to optimize analysis.

2 Materials

2.1 Computational Hardware

In the following example(s), the NGPINT pipeline was assessed on a computer cluster with Intel(R) Xeon E7-4860v2, with an allocation of six cores and 150 GB of RAM. The first mapping step of the pipeline is memory intensive, requiring more than 50 GB of RAM to run the toy dataset. It is also recommended to allocate 10 GB to test the pipeline with the toy dataset.

In the case of Y2H-SCORES, it was run on a computer Intel (R) core vPRO i7 with four cores and 32 GB of RAM. This pipeline can be run on a personal computer. For allocation, we suggest to have at least 8 GB of RAM and a storage of 2 GB for the output files.

2.2 Computational Software

NGPINT requires a Unix-like operating system, with a bash shell or similar; therefore, all commands have to be run in a terminal. It also requires conda or miniconda installed and it comes with an environment which should be used for installation of all the required tools. Y2H-SCORES is a series of R scripts, which should be run in the terminal in shell. All software requirements include R [5], and the packages: DESeq2 [6], reshape2 [7], tidyverse [8], psych [9], mass [10], optparse [11], and Biostrings [12].

3 Methods

3.1 Design the Y2H-NGIS Experiment

3.1.1 Identify Available Resources for Your Experiment

Several aspects should be considered when designing a Y2H-NGIS experiment, including the experimental setting, protocol, and analysis pipeline [13]. First, the resources available for the organism under study determine what type of bait and prey libraries can be used. Full-length ORF libraries offer many advantages, but they are only available for some model organisms like humans, mouse, and Arabidopsis [14–17]. These libraries are always expressed *in-frame* and offer the possibility of selecting the proteins that will be tested. They also offer the chance of building screening arrays for testing the interactions, giving complete control of the prey and bait abundance and composition [4]. The other option is given by cDNA libraries which can be used to study PPIs in any system and whose composition depends on the RNA source and transcript abundance [13, 18, 19]. Therefore, these libraries may require normalization to reduce the presence of high-abundance transcripts that can bias the test [20]. An advantage of cDNA libraries is the random generation of preys that may include different protein domains, increasing the possibility of identifying interactions that would not be found with full-length proteins [13].

3.1.2 Select Number of Baits and Preys

Another aspect to consider consists of the number of baits and preys that will be tested. Different variations in the technique allow one to test in single-screen scenarios such as one bait by a prey library (one-by-many), or multiple baits by a prey library (many-by-many), with various experimental and analysis requirements [14–16, 18, 19]. The multi-bait screening increases the throughput; however, it requires a more complex analysis pipeline in the imputation of the bait-prey interactions. The single-bait analysis has more front-end analytical options and requires lower sequencing depth to ensure coverage.

3.1.3 Select Number of Replicates, Sequencing Depth, and Controls

Once the type of library and throughput have been determined, other aspects of the experimental design should be selected, i.e., the number of replicates, sequencing depth, NGS platform, controls, and sampling time points [13]. Results of simulation experiments suggest that at least three replicates are necessary for the reproducibility of Y2H-NGIS analysis [13]. The stochastic nature of Y2H mating, imparted by the initial concentration of preys in the library, may induce high variability in the identification of true interactors. Therefore, in order to reach a complete dataset of interactors in the library, multiple replicates and re-tests should be evaluated.

As for sequencing depth, which is associated with the NGS platform, consider the rules for an RNA-Seq project, with an emphasis on the prey complexity and average prey size. Increasing the sequencing depth for non-selected samples (which are only selected for bait and preys and not the interaction) is also recommended as it is expected that they are more complex than selected samples [13]. Multiple NGS platforms are available for this type of project; for the experiments outlined here, we suggest using the short-read platform with a low base-error rate as prey reconstruction for binary validation of candidates is an essential step in the pipeline.

Different controls have been proposed for Y2H-NGIS experiments [14–19, 21–28]. Use a non-selected control and a specificity control that can be an empty bait vector or a control bait expected to not interact with any preys in the library [13]. The non-selected control, which is a sample with the selection of bait and prey but not their interactions, provides a baseline of prey concentration in the batch culture. The specificity control, which is a sample with selection for interaction using an empty or control bait, contains information about the specificity of the interactions across other baits. This control can facilitate the identification of auto-activators. As for the timepoint of sampling, there are methods that track the yeast growth [16] over time or measure the end of the exponential phase [14, 15, 17–19, 21–28]. The multiple timepoint method increases the precision and recall of the method; however, it also increases the labor and cost, depending on how many time points are taken [13].

3.2 Generate the Count Data Using the NGPINT V2 Pipeline

A general workflow for Y2H-NGIS analysis is shown in Fig. 1. The analysis pipeline starts by inputting Y2H-NGIS raw reads, performing quality control, and determining bait-prey assignments. Following these processes, mapping and quantification of total reads should be performed and, in the case of cDNA prey libraries, fusion reads should be assessed. Fusion reads span the junction between the prey plasmid and the prey sequence and thus provide essential information about the frame in which the prey is being translated. As the following step, alignments and FASTA files with the reconstructed prey fragments can then be used for secondary validation, which is particularly important when cDNA prey libraries are used during the screening. Count tables for total and fusion reads should be normalized and then modeled to infer the interactions. This can be achieved using a scoring method followed by a ranking process. Once candidates have been ranked then secondary validation can be performed. The general analysis pipeline is split in two, a front-end (colored in blue), which should generate count tables and reconstruct preys, and a back-end (in green), which should score the candidates. NGPINT V2 and Y2H-SCORES software are the front-end and back-end analysis tools for Y2H-NGIS datasets [13, 29].

NGPINT was originally designed for processing single-bait batch Y2H-NGIS data from cDNA libraries, but it is compatible with other single-bait designs. The main steps in this pipeline

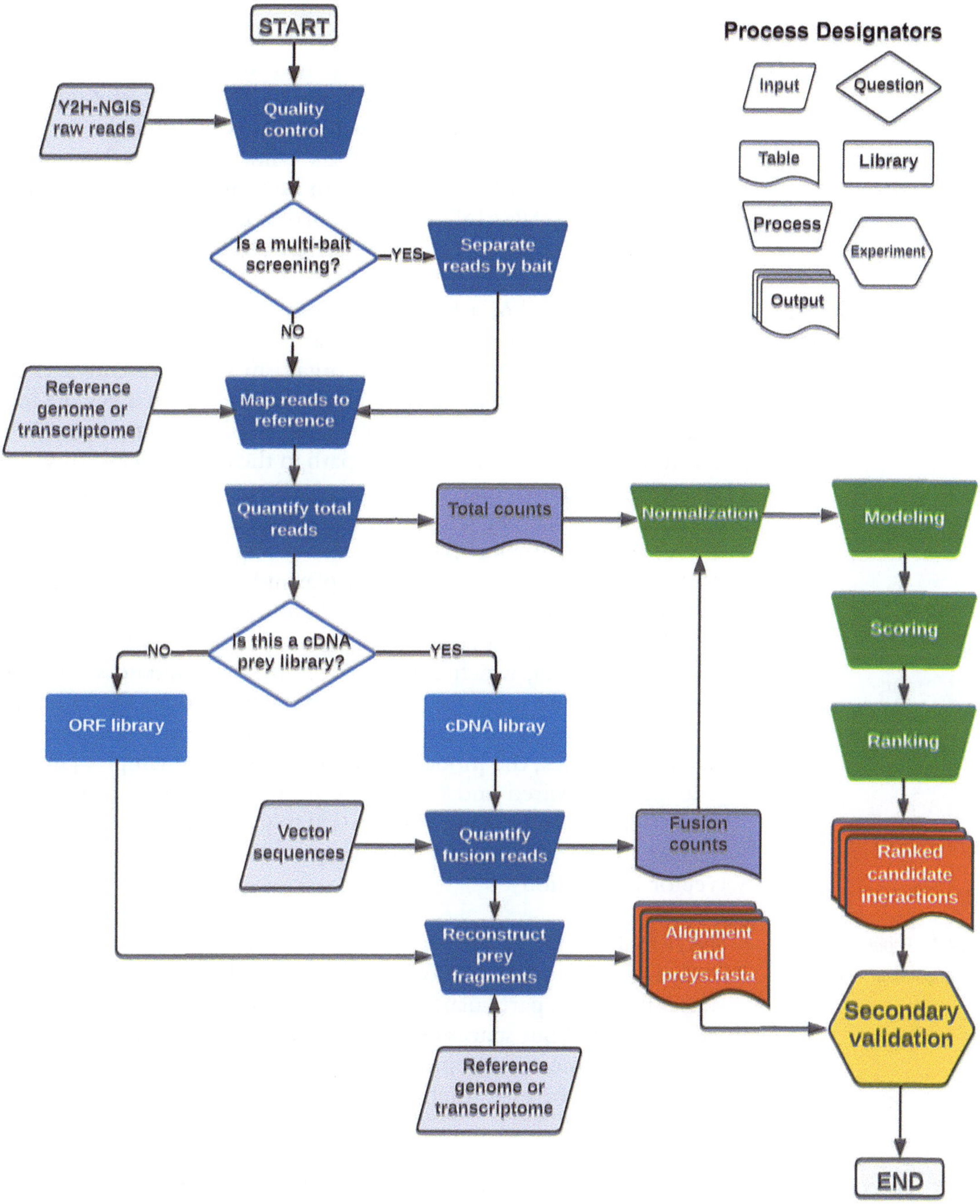

Fig. 1 General analysis workflow for Y2H-NGIS data. A typical analysis pipeline includes processing raw reads and using accessory data such as a reference genome or transcriptome to obtain fusion and total read counts. Back-end analysis consists of ranking interactions and performing secondary validation. Labels for each shape are shown, including inputs, processes, tables, library, questions, output, and experiment. Front-end (blue), back-end (green), and required experimental procedures (yellow, orange) are color-coded

include read trimming, read alignment to a reference genome or transcriptome, total read quantification as prey abundance, and detection of fusion reads that help in the determination of the translational reading frame of each prey [29]. Lastly, the pipeline reconstructs prey fragments which serve as a guide for primer design or sequence synthesis for binary validation. A full description of the software and how to use it can be found at https://github.com/Wiselab2/NGPINT with the current update described in this report at https://github.com/Wiselab2/NGPINT_V2.

3.2.1 Prepare Inputs to the NGPINT V2 Pipeline

Inputs to the pipeline are listed in a comma separated values (CSV) configuration file as shown in Fig. 2a. First, it requires a reference genome or transcriptome of the organism of interest in FASTA format, and an annotation file in gene transfer format (GTF) format. If several bait screenings are being used, generate a genome STAR index [30] and provide the path in the config file, so this step is not repeated with each run of the pipeline. To generate the index, activate the NGPINT V2 conda environment and run STAR from there, to avoid version conflicts. The config file also requires one FASTA file with the prey and bait plasmid sequences without any inserts up to the start of the forward and reverse primers.

The NGPINT V2 config file also requires entering the 5′ and 3′ vector sequences, which consist of the primer sequences used for amplification and up to the fusion with the prey insert. Other inputs to the pipeline include the FASTQ files of the non-selected and selected samples, the mode of running each set of samples [single-end (SE) or paired-end (PE)], the number of central processing units (CPUs), and the nucleotide used to generate different reading frames in the prey library (if none put the last nucleotide in the vector-prey fusion).

3.2.2 Populate the Configuration File

NGPINT V2 reads the information from a configuration file (also called metadata file) as shown in Fig. 2a. When building the configuration file is particularly important to keep the semicolon as the separator, without spaces between file names, and check the path to the files. Name the samples without special characters and with a code that can be easily identifiable of the type and replicate (e.g., S, NS, R1, R2, . . .).

3.2.3 Select Primer Sequences for Fusion Read Identification

Fusion reads are a characteristic product of cDNA libraries, containing information on the prey fragment and its translation frame. Figure 2b shows an example of fusion reads in a Y2H-NGIS dataset and the information they carry. NGPINT V2 identifies fusion reads and extracts frame information from the 5′ plasmid-prey junctions. For the pipeline to be able to use this information to rank interactors, design PCR primers as close to the junction as possible and leaving at least 25 bp at each side of the junction to avoid ambiguous mapping. Paired-end sequencing also provides more information from the junctions, as they can be captured at any end.

A **NGPINT configuration file**

Argument name	Argument description	Required/ Optional	Default Argument value	Argument value
input_ended	Single-end (SE) or Paired-end (PE)	1	SE	
input_fullpath	Path to S fastq files	1	None	
background_ended	Single-end (SE) or Paired-end (PE)	1	SE	
background_fullpath	Path to NS fastq files	1	None	
genome	Path to ref genome	1	None	
star_genome_index	Path to star gen index	0	None	
output_directory	Path to output	1	None	
plasmid_sequences	Path to plasmid fasta	1	None	
five_prime_vector	Seq of the 5' vector	1	None	
three_prime_vector	Seq of the 3' vector	1	None	
CPU	# of CPUs	0	1	
frame_of_TF_fusion	# bases to make prey fusion *in-frame*	0	0	
nucleotide_for_frame_shift	Base used for frame-shift	0	None	
min_trimmed_length	Min seq left after trimming	1	15	
force	Overwrites the output directory	1	0	
clean_up	Delete temp files	0	0	
gtf_annotation	Path to GTF file	1	None	
functional_annotation	Path to annotation	0	None	
transcriptome_index	Path to star transcriptome index	1	None	

B

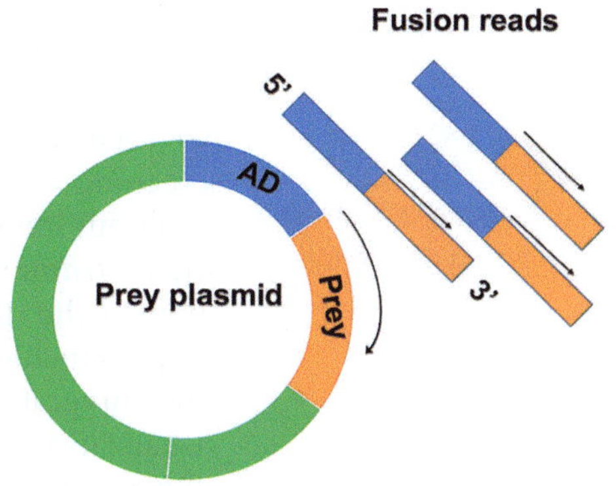

C **Total count table**

Prey	Total counts non-selected...	Total counts selected...
Prey 1		
Prey 2		
...		

Fusion count table

Gene	Transcript	Length...	Fusion counts in-frame selected...	Fusion counts selected...	Fusion counts in-frame non-selected...	Fusion counts non-selected...
Gene 1	*Prey 1.1*					
Gene 1	*Prey 1.2*					
...	...					

Prey reconstructed IGV alignments

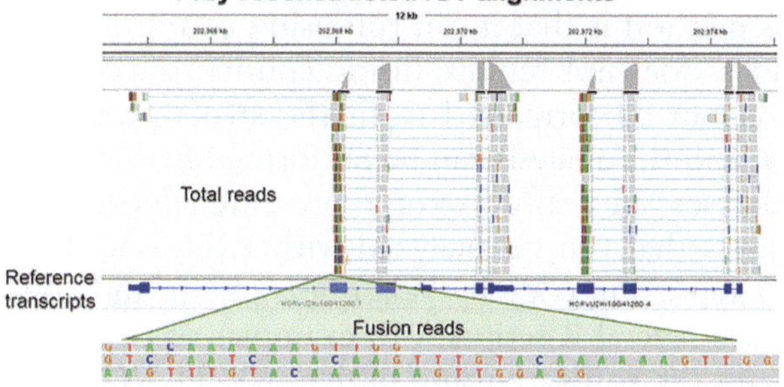

Prey reconstructed fasta

```
>Prey1.1
agaagagagagagagagagcaATGGCGGCGAGTGAACACAGATGCG
TGGGATGTGGTTTTAGGGTAAAGTCATTGTTCATTCAATACT
CTCCGGGTAACATTCGTCTCATGAAATGCGGAAATTGCAAG
GAAGTAGCAGATGAGTACATCGAGTGTGAACGCATGATTAT
AGCATTCCATAAatcatgcactaaagttattgaattagaccttactaa*

>Prey1.2
agaagagagagagagagagcaATGGCGGCGAGTGAACACAGATGCG
TGGGATGTGGTTTTAGGGTAAAGTCATTGTTCATTCAATACT
CTCCGGGTCAAGGAAGTAGCAGATGAGTACATCGAGTGTG
AACGCATGATTATAGCATTCCATAAatcatgcacttagaccttactaa*
...
```

Fig. 2 NGPINT V2 inputs and outputs. (**a**) Inputs to NGPINT V2 are listed in a configuration file including references, reads, vectors, and primers, among others. (**b**) Fusion reads are an important piece of information in Y2H-NGIS datasets generated from cDNA libraries and used to determine which reading frame preys are being expressed. The prey vector plasmid contains the activation domain (AD) and prey fragments, which when sequenced will generate the fusion reads. (**c**) Outputs of NGPINT V2 including counts files of total and fusion reads, visualization files for IGV, and a FASTA file with the reconstructed preys

3.2.4 Run NGPINT V2 and Identify the Outputs

To run the pipeline, follow the instructions in the GitHub repository. Use the following code in command line: ngpint -a configuration_file.csv.

Once the pipeline finishes running, different outputs are generated. Outputs of the NGPINT V2 pipeline include alignment files in BAM format, count tables of the preys identified in each bait screening, fusion read counts, and reconstruction of the prey fragments in FASTA format [29]. Figure 2c shows examples of each of these files which should be used as inputs for subsequent analyses and binary validation. The total count table is generated per bait screening using the Salmon algorithm [31]. It contains a column with the genes from where each prey was mapped to and a column per each of the non-selected and selected replicates per that bait screening. The fusion count table reports a row per transcript, with a column indicating the gene, sample, the number of fusion reads, and *in-frame* reads in each condition (selected or non-selected). To support the design of preys for binary interaction tests, additional outputs include alignment and sequence files with some metadata. The Integrative Genomics Viewer (IGV) software [32] allows the visualization of the alignments, including the fusion reads. It also allows comparison of the mapping across transcripts in order to determine the best matching isoform. The sequence file in FASTA format will provide the candidate prey fragments for primer design or synthesis, which can be used for secondary validation of the interaction [29].

3.3 Rank Interactions with the Y2H-SCORES Software

Once count tables have been generated, the next steps in the Y2H-NGIS analysis consist of scoring the candidate interactions. A second software for this purpose, designated Y2H-SCORES, takes the outputs from NGPINT and generates a ranked list of interactions per bait based on three scores (*see* **Note 1**) [13]. An advantage of this method is that it can run using count data from any software, and it does not require fusion counts to work. This means that inputs can be formatted to fit the structure that the software requires, and the analysis can be performed from any type of Y2H-NGIS dataset (in-house example presented **Note 2** and other count datasets that can be analyzed with Y2H-SCORES in **Note 3**). In the supporting code by Velásquez-Zapata and colleagues [13], there are several R scripts with examples of count table reformatting to run the software using different Y2H-NGIS datasets analyzed with other software https://github.com/Wiselab2/Y2H-SCORES/tree/master/Publication/Benchmarking.

The Y2H-SCORES software detects different inputs and runs three scores (enrichment, specificity, and *in-frame*) depending on the available information. The enrichment score depends on the availability of non-selected controls while the *in-frame* score requires fusion counts of at least the selected samples. The specificity score can be calculated if more than one bait was used in the study, or if a control with an empty or random bait was used. For more details about running the software please refer to the manual in GitHub https://github.com/Wiselab2/Y2H-SCORES. An overview of the settings that should be followed to run the software are highlighted as follows:

3.3.1 File Structure

Y2H-SCORES takes the data in files that are distributed in a folder structure separated by bait as "Bait/files." This means that for each bait in the experiment there should be an independent folder that contains the total and fusion counts. It is recommended that all the bait folders are contained in one folder. The paths to each bait folder should be put in the --fofn file, for which absolute paths should be used. The --fofn files should be contained in the same folder as the software files.

3.3.2 Software Testing

Before running the software with the samples of interest, run the test dataset to make sure all the installations are running as expected. Once outputs are generated, the samples of interest can be analyzed.

3.3.3 Count Datasets

Y2H-SCORES uses as inputs the count tables generated by NGPINT. If software other than NGPINT has been used to quantify the preys, the tables should be formatted as shown in the test data and Fig. 2c. For the total count table formatting, keep non-selected counts in columns before the selected counts. In case of non-selected controls that were not included in the design (not recommended), the software is adapted to run. In this scenario, it is important to include counts from a control bait under selection, which should be put as an additional bait dataset. Lastly, the order of the preys (rows) in the total count tables should be consistent across baits, adding zeros as necessary. This step may be required if reads are mapped to different references or gathered from multiple sources.

For the fusion count table formatting, even in designs without a non-selected control, the software is able to use this information and calculate the *in-frame* score, so they should be included it in the dataset, if available. The order of rows in the fusion count table does not have to be consistent across bait screenings.

3.3.4 Normalization

For count normalization, supply raw data as input and the software will implement library-size method for normalization [13, 33]. Methods that are usually used to normalize RNA-Seq for differential expression analysis are not appropriate for Y2H-NGIS counts. Those methods assume that only a low number of genes are differentially expressed, which is usually not the case in Y2H-NGIS experiments. Therefore, the assumptions of those methods do not hold when normalizing Y2H-NGIS counts [13]. If another method of normalization is implemented, the Y2H-SCORES software has the option to indicate that the counts are normalized, so it skips the library size normalization step.

3.3.5 Select Parameters and Run Y2H-SCORES

Set the parameters to run the pipeline as indicated in the GitHub manual, as follows:

--**fofn:** Text file with the full paths to the configuration files.

--**out_dir:** Full path to the output directory. This should be different from the output_directory in the configuration files.

--**spec_groups:** Text file with baits to be analyzed together to calculate the specificity score. Each line should contain each group, separating baits by comma. If no file is provided, the baits will be grouped randomly.

--**spec_p_val:** Threshold for the p-values that are used to calculate the specificity score. Values between [0,1]. Default value of 1. Smaller values indicate more stringent scores.

--**spec_fold_change:** Threshold for the fold-changes that are used to calculate the specificity score. This should be larger or equal to zero. Default of zero. Larger values indicate more stringent scores.

--**enrich_p_val:** The desired threshold for the p-values that are used to calculate the enrichment score. Values between [0,1]. Default value of 1. Smaller values indicate more stringent scores.

--**enrich_fold_change:** Threshold for the fold-changes that are used to calculate the enrichment score. This should be larger or equal to zero. Default of zero. Larger values indicate more stringent scores.

--**normalized:** Boolean value (T or F) indicating if the counts are normalized. Default false (F), then the program will implement library size normalization.

With the toy dataset, the software can be run as:

```
Rscript run_scores.R –fofn "toy_dataset/fofn_for_compute_-
scores.txt" –out_dir "output_toy_dataset/" –spec_p_val 0.5 –
spec_fold_change 2 –enrich_p_val 0.5 –enrich_fold_change 0 –
normalized F
```

3.3.6 Identify the Y2H-SCORES Outputs

Y2H-SCORES generates two main outputs. First, the total scores table which contains a list of all the interactions with the three scores, a summary with their sum, a Borda score and the transcript with the highest *in-frame* score. A second output in recent development, consists of a fasta file with interacting fragments based on the fusion read information. These interaction fragments can be used for cloning purposes and binary tests for validation. It has been observed that they may be differences in binary interaction tests when full-length and fragments of a prey are tested [34]. Therefore, identifying the interaction fragments is necessary for a successful validation of the interactions. If the user is interested in obtaining the prey interacting sequences, the NGPINT V2 outputs that contain the fusion read information (final_report.csv) and full-length fasta sequences (bait_transcriptome_file_for_primer_design.fasta) should be provided. The final_report.csv file is also required to calculate the *in-frame* score. The fasta file should be in the same folder as the other bait-specific files (total counts and final reports). If another pipeline is used instead of NGPINT and there is a final_report.csv file available, the transcriptome can be used as fasta, as long as the coordinates provided in the final report correspond to the sequences. If no fasta or final report file are provided, the interaction fragments will not be generated (*see* **Note 4**).

4 Notes

1. **Biological Properties Measured with Y2H-SCORES**

   The statistical basis of Y2H-SCORES consists of the implementation of two distributions, negative binomial (NB) for total counts and binomial for fusion counts. NB has two parameters termed the mean and the overdispersion, whose estimation is necessary in order to compare samples and/or conditions. In RNA-Seq pipelines, there are several tools that calculate differential enrichment using negative binomial distribution [6, 35]. These have been tested extensively and handle a wide range of datasets [36]. Outputs from these programs facilitate pairwise contrasts that are evaluated with p-values and the fold changes between two samples. Y2H-SCORES uses DESeq2 [6] to model total counts and compare non-selected and selected samples. Once total counts have been modeled, fusion counts can be studied with a binomial distribution. This distribution depends on two variables, the total number of trials and the proportion of successes. Given a total number of fusion counts, among them both *in-frame* and out-of-frame, binomial distribution fits the dataset if we consider *in-frame* reads as successes.

   To contextualize the models that are used in analyzing total and fusion counts it is important to analyze the process that the yeast population goes through during screening. A Y2H-NGIS screen can be seen as an evolving process where a yeast population is subjected to selection, bottlenecks, and genetic drift (Fig. 3a). These forces induce a drastic change in the composition of the Y2H-NGIS samples over time. Experimental procedures using large volumes and large aliquots (see [18] and the accompanying Chapter 19 by Elmore et al.) minimize the effects of bottlenecks and genetic drift. Therefore, we assume that non-selected samples resemble the original prey library

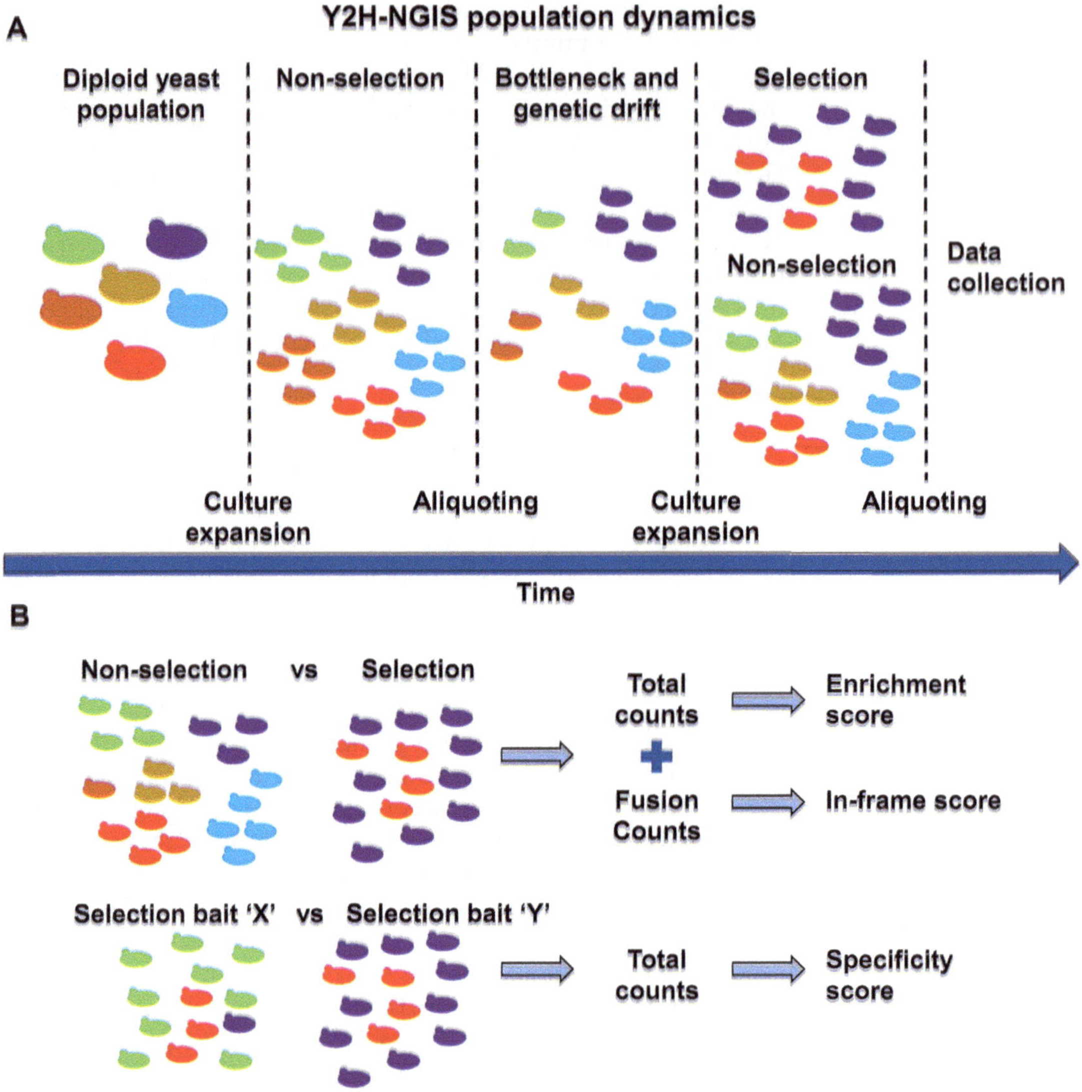

Fig. 3 Y2H-SCORES measures biological properties associated with Y2H-NGIS population dynamics. (**a**) a diploid yeast population expressing different bait-prey combinations goes through different steps during Y2H-NGIS. Culture expansions during non-selection are followed by culture aliquoting which induces bottlenecks in the population. Selection and non-selection also modify the composition of the population before data collection. (**b**) Y2H-SCORES comprise three scores that measure prey enrichment under selection, selection of *in-frame* preys, and specificity across multiple bait screenings. Each score is calculated using total or fusion counts

while selected samples are shaped by the effect of the bait-prey interactions and activation of the selectable marker.

From a biological perspective, a yeast population is composed of cells that express different baits and preys, and if there is a physical interaction, growth under selection occurs. Considering this principle, there should be an *enrichment* in the abundance of those cells when non-selected and selected samples are compared. Another characteristic of true interactions is *specificity,* which can be understood when different baits are compared. It should be expected that unrelated baits have different interactors, therefore, a true interactor should be associated with one bait or a similar subset of baits. Having a measurement of specificity helps to identify auto activators or common interactions across multiple baits. Lastly, considering cDNA prey libraries and the random frame of the fusion between prey fragments and the Y2H vector in the library, a true interactor should be expressed *in-frame* with the activation domain (AD), i.e., the native polypeptide is expressed. These three biological properties inspired the design of the Y2H-SCORES software (Fig. 3b), which contains three ranking scores that measure enrichment, specificity, and *in-frame* information.

2. **End-to-End Case Study: Use of Y2H-NGIS to Identify Interactors of the AVR$_{A13}$ Effector from the Barley Powdery Mildew Fungus, *Blumeria graminis* f. sp. *hordei* [37].**

   The Y2H-NGIS framework described in Chapters 19 (Elmore et al.) and 20 (Velásquez-Zapata et al.) enabled confirmation of fifteen PPI between the barley nucleotide-binding leucine-rich-repeat (NLR) receptor MLA and host proteins [13, 38], as evidenced by an Area Under the Curve (AUC) of 0.96 for the Receiver Operating Characteristic (ROC) curve obtained from the Borda scores, as shown in Fig. 4a. In another example, the interaction between the AVR$_{A13}$ effector from the powdery mildew pathogen, *B. graminis* f. sp. *hordei,* and the barley vesicle-mediated thylakoid membrane biogenesis protein, HvTHF1 was identified. In this scenario, AVR$_{A13}$ was screened as bait using an experimental setting of three biological replicates in the non-selected and selected conditions. Subsequently, the workflow explained in this chapter was used, consisting of NGPINT [29] and Y2H-SCORES [13] software, followed by binary Y2H [3] (Fig. 4b). HvTHF1, corresponding to the barley gene ID HORVU.MOREX.r3.2HG0135590, was identified as a top candidate interactor of AVR$_{A13}$, with score values of 0.97 for enrichment, 0.90 for specificity, and 1 for *in-frame.*

   In the broader context of plant-pathogen interactions, THF1 and its homologs interact with different protein mediators of plant resistance and susceptibility (Fig. 4c). HvTHF1 interacts in yeast and *in planta* with the U-box/armadillo-repeat E3 ligase *HvPUB15* and a partial duplicate, *HvARM1* (for *H. vulgare* Armadillo 1). Neo-functionalization of *HvARM1* increases resistance to powdery mildew, providing a link between plastid function and colonization by biotrophic pathogens [39]. The wheat homolog of HvTHF1, TaToxABP1, interacts with the necrotizing Toxin A effector from

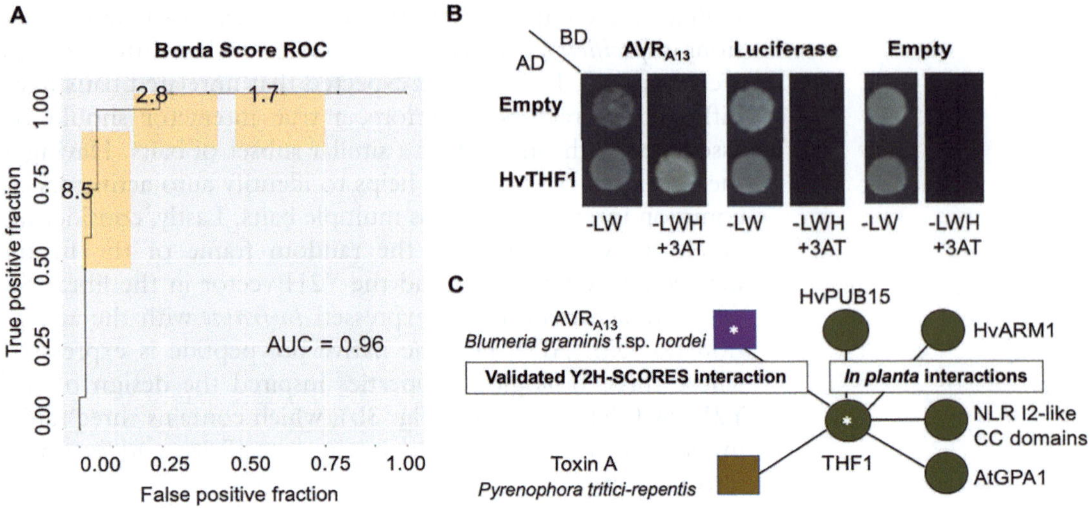

Fig. 4 Experimental performance of the Y2H-NGIS workflow using baits from the interaction between barley and the powdery mildew effector, AVR_{A13}. (**a**) Workflow performance using the published dataset from MLA interactors [5]. True positive and false positive rates are plotted, with a probability of distinguishing between the two of 0.96. (**b**) Binary Y2H test between AVR_{A13} and HvTHF1 showing the diploids control (SC-LW), the stringent interaction (SC-LWH + 3-AT), and tests with luciferase and empty-bait vector to show the specificity of the interaction. (**c**) Validated Y2H-SCORES interaction of the AVR_{A13} effector bait (magenta box with an asterisk) and THF1 prey (green circle with an asterisk). HvPUB15, HvARM1, NLR I2-like CC domains, AtGPA1, and *P. tritici-repentis* Toxin A were shown as interactors of THF1 via bi-molecular fluorescence complementation, co-immunoprecipitation, and/or other in planta methods [37–41]

the tan-spot fungal pathogen, *Pyrenophora tritici-repentis,* which has been associated with reactive oxygen species (ROS) burst [40, 41]. In addition, THF1 also has been found to destabilize several nucleotide-binding leucine-rich-repeat (NLR) resistance proteins by binding their I2-like coiled-coil (CC) domains [42] and the THF1 homolog in *Arabidopsis* interacts with a plasma membrane-localized G-protein, GPA1, involved in a D-glucose signaling pathway [43].

3. Y2H-SCORES can be used to analyze other PPI datasets. As explained above, Y2H-SCORES is based on three fundamental properties of a true interaction using Y2H. Considering this, the software can be extended to identify PPIs from other methods, both in vitro and in vivo, as long as they output count data of any kind. Examples include some in vivo assays based on the same Y2H principles, such as the split-ubiquitin assay [44], Y1H for testing association between protein and DNA [45], and Y3H which tests interactions among proteins and small molecules [46]. Other assays such as co-immunoprecipitation (Co-IP) can also be analyzed with Y2H-SCORES. For example, the software can be implemented with Co-IP data [47] using a design where the samples without antibodies were used as non-selected controls and a

non-specific bait as a specificity control. The results of this analysis are comparable to what was found in the reference software SAINT [48].

4. Y2H-NGIS is a powerful technique to mine protein–protein interaction while increasing the throughput and using the analytical advantages of NGS. General aspects of the technique were presented, which should be considered when designing an experiment. Practical aspects of the analysis of Y2H-NGIS ensure reproducibility and success in the identification of interactions. Two software, NGPINT V2 and Y2H-SCORES, were developed to process Y2H-NGIS reads and rank interactions for binary validation. Recommendations to run both software are presented including input description, pipeline setting, and output check. Also included is a real-life example of the identification of PPI for the barley MLA immune receptor and the AVR_{A13} effector from the powdery mildew pathogen, *B. graminis* f. sp. *hordei*. Lastly, Y2H-SCORES can be extended to work with other proteomic-based datasets that use count data as outputs and that correlate with the biological properties that the software measures.

Acknowledgments

The authors thank Sagnik Banerjee for the original development of NGPINT V1 software. Research supported in part by Fulbright - Minciencias 2015 & Schlumberger Faculty for the Future fellowships to VVZ, USDA-ARS Postdoctoral Research Associateship, and USDA-National Institute of Food and Agriculture (NIFA)-Education and Literacy Initiative (ELI) Postdoctoral Fellowship 2017-67012-26086 to JME, and National Science Foundation - Plant Genome Research Program grant 13-39348, USDA-NIFA grant 2020-67013-31184 and USDA-Agricultural Research Service project 3625-21000-067-00D to RPW. The funders had no role in study design, data collection and analysis, decision to publish, or preparation of the manuscript. Mention of trade names or commercial products in this publication is solely for the purpose of providing specific information and does not imply recommendation or endorsement by the USDA, NIFA, ARS, or the National Science Foundation. USDA is an equal opportunity provider and employer.

References

1. Zitnik M, Sosic R, Feldman MW, Leskovec J (2019) Evolution of resilience in protein interactomes across the tree of life. Proc Natl Acad Sci U S A 116:4426–4433
2. Vidal M, Fields S (2014) The yeast two-hybrid assay: still finding connections after 25 years. Nat Methods 11:1203–1206
3. Dreze M, Monachello D, Lurin C, Cusick ME, Hill DE, Vidal M, Braun P (2010) High-quality binary interactome mapping, 2nd ed. Methods Enzymol https://doi.org/10.1016/S0076-6879(10)70012-4
4. Suter B, Zhang X, Gustavo Pesce C, Mendelsohn AR, Dinesh-Kumar SP, Mao JH (2015)

Next-generation sequencing for binary protein-protein interactions. Front Genet 6: 1–6

5. RCoreTeam (2013) R: a language and environment for statistical computing. Vienna, Austria
6. Love MI, Huber W, Anders S (2014) Moderated estimation of fold change and dispersion for RNA-seq data with DESeq2. Genome Biol 15:550
7. Wickham H (2007) Reshaping data with the reshape package. J Stat, SoftwaresticalSoftware, p 21
8. Wickham H, Averick M, Bryan J et al (2019) Welcome to the {tidyverse}. J Open Source Softw 4:1686
9. Revelle W (2022) psych: procedures for psychological, psychometric, and personality research
10. Venables WN, Ripley BD (2002) Modern applied statistics with S, fourth. Springer, New York
11. Davis TL, Day A, Lianoglou S, Nikelski J, Müller K, Humburg P, FitzJohn R, Choi GJ (2021) Optparse: command line optional argument parser. https://cran.r-project.org/web/packages/optparse/readme/README.html
12. Pagès H, Aboyoun P, Gentleman R, DebRoy S (2020) Biostrings: efficient manipulation of biological strings. In: R Packag. version 2.66.0. https://bioconductor.org/packages/Biostrings
13. Velásquez-Zapata V, Elmore JM, Banerjee S, Dorman KS, Wise RP (2021) Next-generation yeast-two-hybrid analysis with Y2H-SCORES identifies novel interactors of the MLA immune receptor. PLoS Comput Biol 17: e1008890
14. Yachie N, Petsalaki E, Mellor JC et al (2016) Pooled-matrix protein interaction screens using barcode fusion genetics. Mol Syst Biol 12:863
15. Yang JS, Garriga-Canut M, Link N, Carolis C, Broadbent K, Beltran-Sastre V, Serrano L, Maurer SP (2018) Rec-YnH enables simultaneous many-by-many detection of direct protein–protein and protein–RNA interactions. Nat Commun. https://doi.org/10.1038/s41467-018-06128-x
16. Schlecht U, Liu Z, Blundell JR, St Onge RP, Levy SF (2017) A scalable double-barcode sequencing platform for characterization of dynamic protein-protein interactions. Nat Commun 8:1–9
17. Trigg SA, Garza RM, MacWilliams A et al (2017) CrY2H-seq: a massively multiplexed assay for deep-coverage interactome mapping. Nat Methods 14:819–825
18. Pashkova N, Peterson TA, Krishnamani V, Breheny P, Stamnes M, Piper RC (2016) DEEPN as an approach for batch processing of yeast 2-hybrid interactions. Cell Rep 17: 303–315
19. Erffelinck ML, Ribeiro B, Perassolo M, Pauwels L, Pollier J, Storme V, Goossens A (2018) A user-friendly platform for yeast two-hybrid library screening using next generation sequencing. PLoS One 13:1–21
20. Bogdanova EA, Shagina I, Barsova EV, Kelmanson I, Shagin DA, Lukyanov SA (2010) Normalizing cDNA libraries. Curr Protoc Mol Biol. https://doi.org/10.1002/0471142727.mb0512s90
21. Hastie AR, Pruitt SC (2007) Yeast two-hybrid interaction partner screening through *in vivo* Cre-mediated binary interaction tag generation. Nucleic Acids Res. https://doi.org/10.1093/nar/gkm894
22. Nirantar SR, Ghadessy FJ (2011) Compartmentalized linkage of genes encoding interacting protein pairs. Proteomics 11:1335–1339
23. Lewis JD, Wan J, Ford R, Gong Y, Fung P, Nahal H, Wang PW, Desveaux D, Guttman DS (2012) Quantitative interactor screening with next-generation sequencing (QIS-Seq) identifies Arabidopsis thaliana MLO2 as a target of the pseudomonas syringae type III effector HopZ2. BMC Genomics 13:8
24. Weimann M, Grossmann A, Woodsmith J et al (2013) A Y2H-seq approach defines the human protein methyltransferase interactome. Nat Methods 10:339–342
25. Younger D, Berger S, Baker D, Klavins E (2017) High-throughput characterization of protein–protein interactions by reprogramming yeast mating. Proc Natl Acad Sci U S A 114:12166–12171
26. Kessens R, Sorensen N, Kabbage M (2018) An inhibitor of apoptosis (SfIAP) interacts with SQUAMOSA promoter-binding protein (SBP) transcription factors that exhibit pro-cell death characteristics. Plant Direct 2: 1–17
27. Yang F, Lei Y, Zhou M et al (2018) Development and application of a recombination-based library versus library highthroughput yeast two-hybrid (RLL-Y2H) screening system. Nucleic Acids Res 46:1–12
28. Zong T, Yin J, Jin T, Wang L, Luo M, Li K, Zhi H (2020) A DnaJ protein that interacts with soybean mosaic virus coat protein serves as a key susceptibility factor for viral infection. Virus Res 281:197870
29. Banerjee S, Velásquez-Zapata V, Elmore JM, Fuerst G, Wise RP (2021) NGPINT: a next-

generation protein–protein interaction software. Brief Bioinform 22:bbaa351

30. Dobin A, Davis CA, Schlesinger F, Drenkow J, Zaleski C, Jha S, Batut P, Chaisson M, Gingeras TR (2013) STAR: ultrafast universal RNA-seq aligner. Bioinformatics 29:15–21
31. Patro R, Duggal G, Love MI, Irizarry RA, Kingsford C (2017) Salmon provides fast and bias-aware quantification of transcript expression. Nat Methods 14:417
32. Robinson JT, Thorvaldsdóttir H, Winckler W, Guttman M, Lander ES, Getz G, Mesirov JP (2011) Integrative genomics viewer. Nat Biotechnol 29:24–26
33. Dillies MA, Rau A, Aubert J et al (2013) A comprehensive evaluation of normalization methods for Illumina high-throughput RNA sequencing data analysis. Brief Bioinform 14: 671–683
34. Galletta BJ, Rusan NM (2015) Chapter 14 - a yeast two-hybrid approach for probing protein–protein interactions at the centrosome. In: Basto R, Oegema KBT-M in CB (eds) Centrosome & Centriole. Academic Press, pp. 251–277
35. Robinson MD, McCarthy DJ, Smyth GK (2010) edgeR: a Bioconductor package for differential expression analysis of digital gene expression data. Bioinformatics 26:139–140
36. Li X, Brock GN, Rouchka EC, Cooper NGF, Wu D, OToole TE, Gill RS, Eteleeb AM, O'Brien L, Rai SN (2017) A comparison of per sample global scaling and per gene normalization methods for differential expression analysis of RNA-seq data. PLoS One 12:1–22
37. Lu X, Kracher B, Saur IML, Bauer S, Ellwood SR, Wise R, Yaeno T, Maekawa T, Schulze-Lefert P (2016) Allelic barley MLA immune receptors recognize sequence-unrelated avirulence effectors of the powdery mildew pathogen. Proc Natl Acad Sci U S A 113:E6486–E6495
38. Velásquez-Zapata V, Elmore JM, Fuerst G, Wise RP (2022) An interolog-based barley interactome as an integration framework for immune signaling. Genetics 221(2):iyac056. https://doi.org/10.1093/genetics/iyac056
39. Rajaraman J, Douchkov D, Lück S et al (2018) Evolutionarily conserved partial gene duplication in the Triticeae tribe of grasses confers pathogen resistance. Genome Biol 19:1–18
40. Manning VA, Hardison LK, Ciuffetti LM (2007) Ptr ToxA interacts with a chloroplast-localized protein. Mol Plant-Microbe Interact 20:168–177
41. Pandelova I, Figueroa M, Wilhelm LJ, Manning VA, Mankaney AN, Mockler TC, Ciuffetti LM (2012) Host-selective toxins of *Pyrenophora tritici-repentis* induce common responses associated with host susceptibility. PLoS One 7:13–20
42. Hamel LP, Sekine KT, Wallon T, Sugiwaka Y, Kobayashi K, Moffett P (2016) The chloroplastic protein THF1 interacts with the coiled-coil domain of the disease resistance protein N´ and regulates light-dependent cell death. Plant Physiol 171:658–674
43. Huang J, Taylor JP, Chen JG, Uhrig JF, Schnell DJ, Nakagawa T, Korth KL, Jones AM (2006) The plastid protein THYLAKOID FORMATION1 and the plasma membrane G-protein GPA1 interact in a novel sugar-signaling mechanism in *Arabidopsis*. Plant Cell 18:1226–1238
44. Obrdlik P, El-Bakkoury M, Hamacher T et al (2004) K^+ channel interactions detected by a genetic system optimized for systematic studies of membrane protein interactions. Proc Natl Acad Sci 101:12242–12247
45. Reece-Hoyes JS, Marian Walhout AJ (2012) Yeast one-hybrid assays: a historical and technical perspective. Methods 57:441–447
46. Licitra EJ, Liu JO (1996) A three-hybrid system for detecting small ligand-protein receptor interactions. Proc Natl Acad Sci U S A 93: 12817–12821
47. Kanyika BN (2022) Protein-protein interactions in a geminivirus-cassava system. University of the Witwatersrand
48. Teo G, Liu G, Zhang J, Nesvizhskii AI, Gingras A-C, Hyungwon C (2014) SAINTexpress: improvements and additional features in significance analysis of Interactome software Guoci. J Proteome 100:37–43

Chapter 21

Discovering Protein–Protein Interactions using Co-Fractionation-Mass Spectrometry with Label-Free Quantitation

Mopelola O. Akinlaja, R. Greg Stacey, Queenie W. T. Chan, and Leonard J. Foster

Abstract

Proteins generally achieve their functions through interactions with other proteins, so being able to determine which proteins interact with which other proteins underlies much of molecular biology. Co-fractionation (CF) is a mass spectrometry-based method for detecting proteome-wide protein–protein interactions. An attractive feature of CF is that it is not necessary to label or otherwise alter samples. Although we have previously published a widely used protocol for a label-incorporated CF methodology, no published protocols currently exist for the label-free variation. In this chapter, we describe a label-free CF-MS protocol. This protocol takes a minimum of a week, excluding the time for cell/tissue culture. It begins with cell/tissue lysis under non-denaturing conditions, after which intact protein complexes are isolated using size exclusion chromatography (SEC) where they are fractionated according to size. The proteins in each fraction are then prepared for mass spectrometry analysis where the constituent proteins are identified and quantified. Finally, we describe an in-house bioinformatics pipeline, PrInCE, to accurately predict protein complexes. Taken together, co-fractionation methodologies combined with mass spectrometry can identify and quantify thousands of protein–protein interactions in biological systems.

Key words Protein–protein interactions, Co-fractionation, Label-free quantitation, Mass spectrometry, Size exclusion chromatography

1 Introduction

Proteins assemble into cellular machines that underpin nearly all biological processes. Therefore, characterizing these associations, or "protein–protein interactions," is central to understanding biological pathways that are essential to life [1]. There are several high-throughput methods by which protein–protein interactions can be characterized such as affinity-purification mass spectrometry or yeast two-hybrid [2]. However, a more recent method, co-fractionation (CF), has distinct advantages due to its ability to

Shahid Mukhtar (ed.), *Protein-Protein Interactions: Methods and Protocols*, Methods in Molecular Biology, vol. 2690,
https://doi.org/10.1007/978-1-0716-3327-4_21,

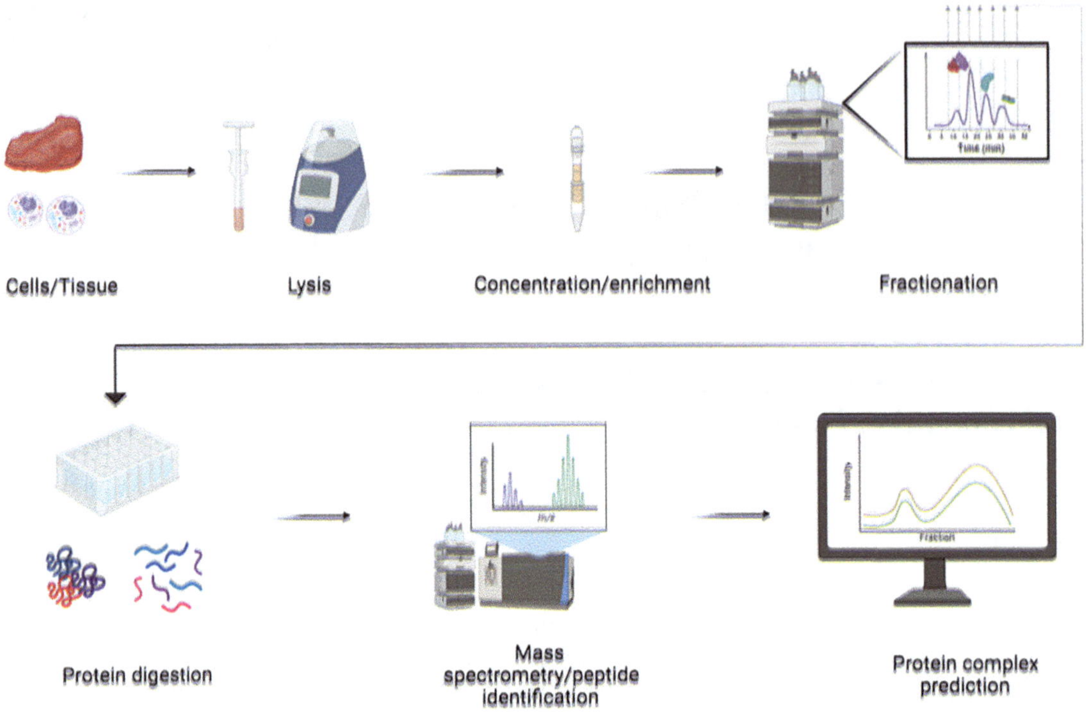

Fig. 1 Schematic of experimental workflow: Cells or tissue are gently lysed and enriched for protein complexes, which are then fractionated using SEC. Fractionated proteins are digested in preparation for mass spectrometry, followed by peptide identification and bioinformatic protein complex prediction. (Created with BioRender.com)

capture all (*see* **Note 1**) protein–protein interactions in a sample, not just interactions for specific targeted proteins (*see* **Note 2**). CF methods also allow the isolation of protein complexes in their native state, avoiding the risk of chemically altering the interactions due to protein tagging or forcing non-physiological expression. The principle of CF relies on two key steps: separating protein complexes in their native form by chromatographic or electrophoretic means and using mass spectrometry to quantify proteins occurring in different fractions post-separation (Fig. 1).

Since 2014, our group has worked on developing and optimizing a sample preparation and data analysis pipeline based on the principles of CF, with a chapter detailing CF methods for labeled proteins [3]. Subsequent iterations of this pipeline have been successfully applied by our group [4, 5], as well as others [6, 7], to study protein complexes and their roles in diseases such as viral pathogenesis and cancer.

In this chapter, we will describe a similar protocol using label-free quantitation (LFQ) as the quantitation method instead of stable isotope labeling by amino acids in cell culture (SILAC). LFQ [8, 9] is an excellent alternative to labeled methods because it can be valuable in situations where SILAC is not amenable [10], for example, in the case of cells that do not easily incorporate isotope labels or tissue samples which can be quite laborious, expensive, or even impossible to label in vivo.

One of the challenges of CF datasets is that they typically require extensive bioinformatic analyses to efficiently find co-fractionating proteins within the dataset. This process is resource-intensive because of the combinatorial explosion of potential solutions: a dataset with 1000 proteins contains more than 10^{57} potential complexes of 30 proteins or less. There are two broad analysis strategies to tackle this problem: complex-centric approaches that measure CF of known complexes, such as is implemented in CCProfiler [11], and machine learning approaches that look for novel interactions such as EPIC and PrInCE [12, 13]. These latter approaches use known protein complexes to train a machine learning classifier to distinguish between interacting and non-interacting pairs of CF profiles and provide a rigorous framework for tackling the complexity of CF data. Given the available software packages (e.g., PrInCE), machine learning methods are an attractive bioinformatic solution for CF data and the one we focus on in this protocol.

Here we describe a detailed method for the isolation of protein complexes from cultured cells or tissue and fractionation by size exclusion chromatography (SEC), followed by the preparation of collected fractions for protein identification by LC-MS/MS. We also describe protein identification and quantitation using MaxQuant followed by protein complex prediction using PrInCE.

2 Materials

2.1 Cell/Tissue Lysis

1. Wash buffer: 50 mM potassium chloride, 50 mM, sodium acetate and 50 mM Tris Base, pH 7.2. Weigh 3.72 g of KCl, 4.1 g of sodium acetate, and 6.06 g Tris base into a flask. Add water to a volume of 1 L. Adjust the pH to 7.2 using 37% HCl. Store at 4 °C.
2. Lysis buffer: Wash buffer (50 mM potassium chloride, 50 mM, sodium acetate, and 50 mM Tris Base, pH 7.2) with HALT protease inhibitor. Dilute 100 μL of 100× HALT protease inhibitor to 1× using the wash buffer for a final volume of 10 mL. Scale up as necessary.
3. Dounce homogenizer for soft tissue and cells. Another homogenizer such as a digital Sonifier may be used to break down tougher tissues prior to cell lysis.
4. Ice bucket.
5. Refrigerated ultracentrifuge capable of >100,000 g.

6. Thick-walled ultracentrifuge tubes.
7. Ultrafiltration spin column with 100 kDa molecular weight cutoff membrane such as (Vivaspin 20, product ID: 28932258). A minimum volume capacity of 2 mL is recommended.
8. Refrigerated centrifuge with a fixed or swinging bucket rotor, as directed by the centrifugal spin filter manufacturer's instructions.

2.2 Protein Complex Isolation by SEC

1. High-performance liquid chromatography (HPLC) system with an isocratic pump, manual injector (it is not necessary to use an autosampler in this case because each sample needs to be prepared fresh), column oven, variable wavelength UV detector, fraction collector, temperature regulator, and degasser (*see* **Note 3**), for example, Agilent 1200 series or newer.
2. High-resolution SEC HPLC column: We use the Biosep 5 μm SEC-s4000 500 Å 300 × 7.8 mm, although columns with similar resolution should suffice (*see* **Note 4**).
3. Mobile Phase: *Same as Wash Buffer.* 50 mM potassium chloride, 50 mM sodium acetate, and 50 mM Tris Base, pH 7.2. Weigh 3.72 g of KCl, 4.1 g of sodium acetate, and 6.06 g of Tris Base into a flask. Add water to a volume of 1 L. Then, adjust the pH to 7.2 using 37% HCl. Store at 4 °C.

   After preparation, pass through a 0.2 μm pore size filter into a new, clean bottle. Store at 4 °C.
4. Deep well plate with seal.
5. Hamilton syringe for sample injection.

   SEC Standards: 7.5 mg/mL carbonic anhydrase, 2.5 mg/mL albumin (2.5 mg/mL), 12.5 mg/mL alcohol dehydrogenase, 7.5 mg/mL B-amylase, 2.5 mg/mL apoferritin, and 3.5 mg/mL thyroglobulin. MW range: 29 kDa–669 kDA. Mix the protein standards in their respective concentrations. Then, aliquot 20 μL into single-use microcentrifuge tubes. Store at −20 °C.

2.3 Sample Preparation for Mass Spectrometry Analysis

1. Chaotropic buffer: 6 M urea, 2 M thiourea in mobile phase.
2. Dithiothreitol (DTT) stock solution: 0.5 μg/μL dithiothreitol.
3. Iodoacetamide (IAA) stock solution: 2.5 μg/μL iodoacetamide.
4. Digestion buffer: 50 mM ammonium bicarbonate.
5. Trypsin (mass spectrometry grade).
6. Endoproteinase Lys-C.
7. 20% Trifluoroacetic acid (TFA).
8. Buffer A: 1% (v/v) TFA in water.

9. Buffer B: 40% (v/v) acetonitrile, 0.1% (v/v) formic acid in water.
10. High capacity C18 desalting columns (e.g., STAGE tips [14]).

2.4 LC-MS/MS

1. Mobile Phase A: 0.1% (v/v) formic acid, 0.5% (v/v) acetonitrile in water (BAKER ANALYZED™ LC-MS Grade).
2. Mobile Phase B: 0.1% (v/v) formic acid, 0.5% (v/v) water (BAKER ANALYZED™ LC-MS Grade in acetonitrile (LC-MS Grade).
3. nanoHPLC column: ionOptiks (25 cm × 75 μm 1.6 μm FSC C18) or similar.
4. A high-resolution mass spectrometer capable of fragmenting many peptide ions per second (for example, Impact II or tim-sTOF instruments from Bruker Daltonics).

2.5 Protein Identification and Quantitation

1. MaxQuant software, downloadable at https://www.maxquant.org/ (hardware requirements are located on the website) [15].

2.6 Protein–Protein Interaction Prediction

1. Computer with Windows or macOS, with at least 12 GB of memory, but preferably 16 GB (*see* **Note 5**).
2. R (version 3.6.0 or greater).
3. PrInCE R package [12].

3 Methods

3.1 Tissue/Cell Lysis and Protein Complex Enrichment

Here we describe a gentle mechanical disruption method for tissue or cell lysis using a Dounce homogenizer. It is important to avoid any harsher lysis methods, such as detergents or chaotropes, to preserve the integrity of protein interactions in the lysate.

All steps should be carried out at 4 °C or less unless otherwise stated.

1. Obtain the biological sample and keep it on ice. For tissues, initial breakdown using liquid nitrogen or Polytron may be required to obtain pieces small enough to be homogenized in a Dounce homogenizer. Collect these into a tube of at least 15 mL capacity. For non-adherent cultured cells, collect them into a tube of at least 15 mL by centrifugation without breaking the cells. For adherent cultured cells, remove the old media.
2. Wash the sample with 10 mL of cold Wash Buffer three times. At the final wash, remove as much Wash Buffer as possible while avoiding sample loss.

3. Add 1–2 mL of lysis buffer to resuspend the sample. For adherent cells, add the volume directly onto the plate or flask and scrape to collect cells. Transfer the same volume to the next plate or flask and repeat until all the cells are pooled on the final plate.
4. Transfer the sample into a Dounce homogenizer. We recommend 3 min or 100 passes with a "tight" pestle on ice for cultured cells for effective lysis, but this is highly dependent on cell type. View under a microscope to track progress.
5. Transfer the lysed cell suspension into thick-walled tubes and centrifuge at 100,000 × *g*, at 4 °C for a minimum of 15 min. Retain the supernatant for the next step.
6. Concentrate the protein lysate, contained in the supernatant from the previous step, to 200 μL or less through a 100 kDa molecular weight cutoff spin filter according to the manufacturer's instructions. Depending on your starting volume, this step can take 1 h or greater.
7. Determine the protein concentration. It is best to aim for a concentration of at least 10 μg/μL to ensure that loading 100 μL of lysate on the HPLC is equivalent to around 1 mg of the lysate (NOTE: We do not recommend exceeding 3 mg of lysate on the column as it can result in clogging the column and fast loss of resolution after multiple runs). The sample should be fractionated by SEC as soon as possible to minimize degradation.

3.2 Protein Complex Isolation by SEC

Here we describe using SEC to separate protein complexes into fractions.

1. Prepare Mobile Phase (same as Wash Buffer). Filter Mobile Phase into a clean bottle using a 0.2 μm filter.
2. Turn on the HPLC instrument, connect the SEC column, and initialize the various compartments (temperature regulator, detector). Then, condition the column with mobile phase for at least 1 h.
3. Create and subsequently initialize an HPLC method as follows:

   Gradient: Isocratic.

   Flow Rate: 0.6 mL/min.

   Pressure: Not exceeding 100 bar.

   Temperature: Between 4 °C and 6 °C.

   Fractions: Collect a minimum of 40 fractions. We have learned that this number of fractions provides sufficient data for subsequent bioinformatic analysis, greatly reducing the number of samples needed to be run on the MS [16]. More fractions can be collected if desired. If choosing

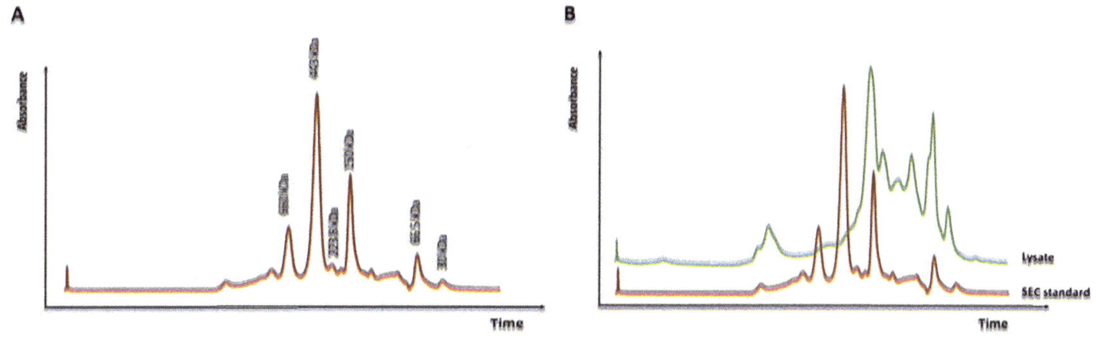

Fig. 2 Sample SEC chromatograms. (**a**) Molecular weights and relative positions of protein standards. (**b**) Overlay of SEC standard and lysate samples

to run a smaller number of fractions, we strongly recommend checking that proteins are present in all fractions collected, i.e., ensure that there is no void volume included in the fraction collection. This is because it is common for later fractions to contain little protein in the case of SEC, reducing the effective number of usable fractions. If already choosing a small number of fractions, this can reduce the effect number below what's needed. We use two Biosep columns that are connected end to end, maximizing resolution. However, with some optimization, it is also possible to use 1 column for the fractionation with good resolution (*see* **Note 6**).

Run Time: 60 min. The first 20 min are typically used to collect void volume and the usable fractions are collected over the next 20 min (between 30 min and 50 min). The remaining time is for post-fraction collection column washing.

4. Dilute SEC standard aliquot to 200 μL using mobile phase.
5. Using a Hamilton syringe, inject the SEC standard mixture into the HPLC, and then run the initialized method. A standard run is necessary because it is a good indicator of what fractions to keep when the samples are run, i.e., protein-containing fractions that exclude void volume (Fig. 2).
6. Inject lysate into the column and collect fractions in a deep-well plate placed in the fraction collector.
7. The fractionated samples can be sealed and stored at −20 °C at this stage. Otherwise, proceed to the next steps.

3.3 Preparation of Fractionated Protein Samples for MS Analysis

Following protein complex fractionation, the protein samples are then digested and desalted in preparation for MS analysis.

1. Based on the volume contained in each fraction and the total number of fractions collected, weigh out urea and thiourea to constitute a final concentration of 6 M and 2 M, respectively, in each well. Solubilize samples in urea and thiourea and stir on a

shaker at room temperature (RT) to ensure complete solubilization. There are alternative in-solution digestion strategies (*see* **Note 7**).

2. Add 1 μg of DTT and incubate at RT for 30 min. Follow this with 5 μg of IAA and incubate at RT for 20 min. There are alternative reagents routinely used in reduction and alkylation steps (*see* **Note 8**).
3. Add 0.1 μg of Lys-C and incubate for 3 h at RT.
4. Dilute sample solution down to ~1 M urea/thiourea using a digestion buffer. Then add 0.5–1 μg of trypsin and incubate for 5 h or overnight at RT.
5. Acidify digested peptides to a pH of ~2.5 or less using TFA (*see* **Note 9**). Confirm pH using pH strips.
6. Desalt peptide solution using high capacity C18 columns [17].
7. Inject peptide samples into a nanoHPLC column coupled to a tandem mass spectrometer and run samples based on a standard shotgun proteomics method [18, 19].

3.4 Protein Identification and Quantitation

There are several existing software for peptide/protein identification (*see* **Note 10**). Here we describe a peptide identification workflow using MaxQuant.

1. Ensure that the latest version of MaxQuant is downloaded.
2. Identify and quantify proteins using the MaxLFQ algorithm in MaxQuant [15]. Be sure to select each fraction as a separate experiment.
3. Export and save *proteinGroups.txt* file for protein complex prediction in R.

3.5 Protein–Protein Interaction Prediction with PrInCE

CF profiles obtained by quantifying proteins across fractions are used as input to the PrInCE workflow. We recommend running this analysis in Rstudio for ease of use. Run times and memory usage are data-dependent (*see* **Note 5**) but a typical PrInCE analysis completes in 1–3 h.

1. Install R, Rstudio, PrInCE package.
2. Read CF profiles into R and store them as a list of *mxn* matrices, where *m* is the number of proteins and *n* is the number of fractions, with rownames equal to protein IDs (Fig. 3, lines 4–14). For MaxQuant and MSFragger tables, this involves selecting table columns with protein IDs and quantifications and removing unnecessary columns, a step that can be performed in R or Excel (*see* **Note 11**).

```r
### 1. load PrInCE package
library(PrInCE, dplyr, ggplot2)

### 2. Read co-fractionation profiles and convert to list of MxN matrices
# replicate 1
profiles = read.csv2("profile_matrix.csv", sep=",") # read profiles
profile_matrix = as.matrix(profiles[,2:ncol(profiles)]) # convert to numeric matrix
rownames(profile_matrix) = profiles[,1] # set rownames to uniprot IDs
# replicate 2
profiles2 = read.csv2("profile_matrix2.csv", sep=",")
profile_matrix2 = as.matrix(profiles2[,2:ncol(profiles2)])
rownames(profile_matrix2) = profiles2[,1]
# combine into list
profile_matrices = list(profile_matrix, profile_matrix2)

### 3. Read gold standard and convert to adjacency matrix
# here we use the CORUM file 'allComplexes.txt'
gold_standard = read.csv2("allComplexes.txt", sep="\t") %>%
  pull(`subunits.UniProt.IDs.`) %>% sapply(., strsplit, ";") %>%
  adjacency_matrix_from_list()

### 4. Perform Gaussian fitting with build_gaussians()
gaussians = lapply(profile_matrices, build_gaussians)

### 5. Calculate pairwise features
features1 = calculate_features(profile_matrices[[1]], gaussians[[1]])
features2 = calculate_features(profile_matrices[[2]], gaussians[[2]])
# combine and impute
features = concatenate_features(list(features1, features2)) %>%
  replace_missing_data()

# 6. sanity check: confirm there are sufficient label=1
label = make_labels(gold_standard, features)
table(label)

### 7. Predict interactions
interactions = predict_interactions(features, gold_standard)

# sanity check: plot precision and top-ranked interactions
plot(interactions$precision)
# plot top interaction
protA = interactions$protein_A[1]
protB = interactions$protein_B[1]
ggplot(data.frame(fractions = rep(1:ncol(profile_matrix),2),
                  value = c(profile_matrix[which(rownames(profile_matrix)==protA),],
                            profile_matrix[which(rownames(profile_matrix)==protB),]),
                  protein = c(rep(protA,ncol(profile_matrix)),
                              rep(protB,ncol(profile_matrix)))),
       aes(x=fractions, y=value, color=protein)) +
  geom_point() + geom_line()
```

Fig. 3 Example R workflow for analyzing a CF dataset with *PrInCE*

3. Read gold standard complexes into R and convert to the appropriate format, either adjacency matrix or list of protein IDs (Fig. 3, lines 16–20) (*see* **Note 12**).
4. Perform Gaussian fitting with *gaussians = build_gaussians(profile_matrices)* (Fig. 3, lines 22–23) (*see* **Note 13**).
5. Calculate pairwise features with *features = calculate_features (gaussians, profile_matrices)* (Fig. 3, lines 25–30).
6. Ensure there are sufficient training examples by running *labels = make_labels(gold_standard, features)* and checking the number of *label = 1* values (Fig. 3, lines 32–34) (*see* **Note 14**).
7. Predict interactions with *interactions = predict_interactions (features, gold_standard)*. Protein–protein interactions are those with an estimated precision value above the chosen threshold (Fig. 3, lines 36–50) (*see* **Note 15**).

4 Notes

1. CF-MS can map the interactome at a high throughput level under native conditions, to a high degree of accuracy and is amenable to more species than most conventional methods of interactome mapping.
2. CF is a faster and more resource-efficient approach to interactome profiling. However, it is important to recognize that the binary interactions identified by these methods do not always imply direct physical interactions in the way that yeast two-hybrid results do. Similar to affinity pull-downs, interactions predicted from CF data represent proteins that are in the same complex.
3. A Sonicator bath for 30 min is sufficient if there is no degasser attached to your HPLC instrument.
4. We recommend attaching a guard column to your column of choice to minimize clogging and preserve the column functionality for longer.
5. Memory requirements are data-dependent. Since PrInCE considers the number of pairs of proteins, which grows with the square of the number of proteins, smaller datasets with <1000 proteins can make do with <12 GB, while datasets with many thousands of proteins could need more than 16 GB.
6. When the flow rate is increased and the run time is shortened, one can get away with using 1 column instead of 2. This will require a bit of optimization.

7. Instead of urea and thiourea, 50% trifluoroacetic acid can be used as a no-salt alternative for the protein digestion steps as well.
8. We sometimes substitute DTT and IAA for tris(2-carboxyethyl)phosphine (TCEP) and chloroacetamide (CAA) for reduction and alkylation, respectively, since the latter are known to be more stable in certain cases.
9. Using a more concentrated TFA solution like 20% minimizes the final volume of peptide solution. This helps the desalting step proceed a bit faster.
10. MS Fragger is a newer peptide identification pipeline that we have explored to identify CF data and we have had success with this tool, as have we when using MaxQuant.
11. MQ and MSFragger return protein groups, i.e., semicolon-separated IDs, rather than single IDs. You can either use the first ID and ignore the rest of the protein group or input the protein groups into PrInCE. Either way, take care that there are IDs in common between the *mxn* matrix rownames and IDs in the gold standard complexes. If the protein groups returned by MQ or MSfragger are not properly parsed, a protein group such as "P06493;E5RIU6" will not be matched to "P06493" in the gold standard.
12. Check there is a reasonable number of IDs in common between your *mxn* CF data matrices and the gold standard. We find that 25 is a rough minimum needed, and ideally there are over 100.
13. Steps 4–7 in Subheading 3.5 can be skipped and replaced with a single *PrInCE()* call, but it can be useful to do these steps for troubleshooting. We find that saving intermediate objects, such as the profile matrix and features, is useful for later analysis.
14. Ideally, there are at least 50 *label* = *1* pairs in your data, as these are the gold standard positive examples used to train the classifier.
15. We find it useful to plot the precision curves of the predicted interactions (Fig. 4). Ideally, the precision curve starts at or near 1 and is relatively smooth. Large jumps mean the precision is estimated from a small number of *label* = *1* training examples.

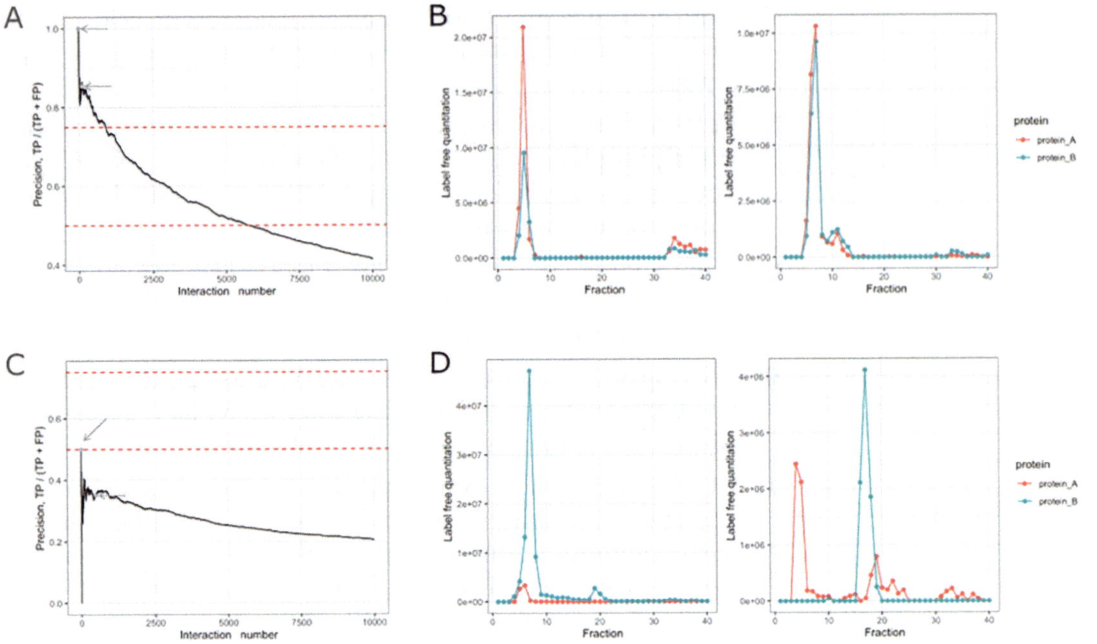

Fig. 4 Example precision and co-fraction profile plots for well-performing (**a** and **b**) and poorly performing (**c** and **d**) CF datasets. Dashed red lines in (**a**) and (**c**) show two representative precision cutoffs (75% and 50%). Note that the dataset in (**c**) and (**d**) has a single interaction that passes precision >50%, either due to data quality, a poorly chosen gold standard, or both

References

1. Poluri KM, Gulati K, Sarkar S (2021) Experimental methods for determination of protein–protein interactions. In: Protein-protein interactions. Springer, Singapore, pp 197–264
2. Kristensen AR, Foster LJ (2013) High throughput strategies for probing the different organizational levels of protein interaction networks. Mol BioSyst 9:2201–2212
3. Kristensen AR, Foster LJ (2014) Protein correlation profiling-SILAC to study protein-protein interactions. Methods Mol Biol 1188: 263–270
4. Scott NE, Rogers LD, Prudova A et al (2017) Interactome disassembly during apoptosis occurs independent of caspase cleavage. Mol Syst Biol 13:906
5. Carlson ML, Greg Stacey R, Young JW et al (2019) Profiling the Escherichia coli membrane protein interactome captured in Peptidisc libraries. eLife 8
6. Kaeser-Pebernard S, Diedrich B, Dengjel J (2019) Identification and regulation of multimeric protein complexes in autophagy via SILAC-based mass spectrometry approaches. Methods Mol Biol 1880:341–357
7. Jakobsen L, Vanselow K, Skogs M et al (2011) Novel asymmetrically localizing components of human centrosomes identified by complementary proteomics methods. EMBO J 30:1520–1535
8. Merl J, Ueffing M, Hauck SM, von Toerne C (2012) Direct comparison of MS-based label-free and SILAC quantitative proteome profiling strategies in primary retinal Müller cells. Proteomics 12:1902–1911
9. Salas D, Stacey RG, Akinlaja M, Foster LJ (2020) Next-generation Interactomics: considerations for the use of co-elution to measure protein interaction networks. Mol Cell Proteomics 19:1–10
10. Zhao L, Cong X, Zhai L et al (2020) Comparative evaluation of label-free quantification strategies. J Proteome 215:103669
11. Bludau I, Heusel M, Frank M et al (2020) Complex-centric proteome profiling by SEC-SWATH-MS for the parallel detection of hundreds of protein complexes. Nat Protoc 15:2341–2386
12. Skinnider MA, Cai C, Stacey RG, Foster LJ (2021) PrInCE: an R/bioconductor package

for protein-protein interaction network inference from co-fractionation mass spectrometry data. Bioinformatics 37:2775. https://doi.org/10.1093/bioinformatics/btab022

13. Hu LZ, Goebels F, Tan JH et al (2019) EPIC: software toolkit for elution profile-based inference of protein complexes. Nat Methods 16: 737–742
14. Rappsilber J, Ishihama Y, Mann M (2003) Stop and go extraction tips for matrix-assisted laser desorption/ionization, nanoelectrospray, and LC/MS sample pretreatment in proteomics. Anal Chem 75:663–670
15. Tyanova S, Temu T, Cox J (2016) The MaxQuant computational platform for mass spectrometry-based shotgun proteomics. Nat Protoc 11:2301–2319
16. Skinnider MA, Foster LJ (2021) Meta-analysis defines principles for the design and analysis of co-fractionation mass spectrometry experiments. Nat Methods 18:806–815
17. Rappsilber J, Mann M, Ishihama Y (2007) Protocol for micro-purification, enrichment, pre-fractionation and storage of peptides for proteomics using StageTips. Nat Protoc 2: 1896–1906
18. Collins MA, An J, Hood BL et al (2015) Label-free LC–MS/MS proteomic analysis of cerebrospinal fluid identifies protein/pathway alterations and candidate biomarkers for amyotrophic lateral sclerosis. J Proteome Res 14: 4486–4501
19. Bauer KM, Lambert PA, Hummon AB (2012) Comparative label-free LC-MS/MS analysis of colorectal adenocarcinoma and metastatic cells treated with 5-fluorouracil. Proteomics 12: 1928–1937

Chapter 22

Protein–Protein Interaction Network Mapping by Affinity Purification Cross-Linking Mass Spectrometry (AP-XL-MS) based Proteomics

Ashima Mehta, Abu Hena Mostafa Kamal, Sharel Cornelius, and Saiful M. Chowdhury

Abstract

Protein–protein interactions (PPIs) are the physical interactions formed among proteins. These interactions are primarily functional, i.e., they arise from specific biomolecular events, and each interaction interface serves a specific purpose. A significant number of methods have been developed for protein interactions in the field of proteomics in the last decade. Advanced mass spectrometry technology significantly contributed to the development of these methods. The rapid advancement of groundbreaking MS technology has greatly aided the mapping of protein interaction from large-data sets comprehensively. This chapter describes the affinity purification (AP) mass spectrometry (MS)-based methods combined with chemical cross-linking (XL) of protein complexes. This chapter includes sample preparation methods involving cell culture, cell treatments with ligands, drugs, and cross-linkers, protein extractions, affinity purification, sodium dodecyl sulfate (SDS) polyacrylamide gel separation, in-solution or in-gel digestion, liquid-chromatography, and mass spectrometry analysis of samples (LC-MS/MS). Application of a cleavable cross-linker, dual cleavable cross-linking technology (DUCCT) in combination with the affinity purification (AP) method has also been described. Methods for data analysis using unmodified and cross-linked peptide analysis are discussed.

Key words Proteomics, Protein, protein interactions, Mass spectrometry, SDS polyacrylamide gel technique, Cross-linking, AP-MS

1 Introduction

A study of protein–protein interactions (PPIs) and protein structures reveals the protein functions in cells. PPI assays can be broadly categorized as in vivo techniques yeast two-hybrid (Y2H) [1] method, in vitro methods (Coimmunoprecipitation [2]), and in silico methods (Computational approaches [3]). Integration of protein domains combined with the localization of proteins reveals a plethora of domains existing within subnetworks and their

Shahid Mukhtar (ed.), *Protein-Protein Interactions: Methods and Protocols*, Methods in Molecular Biology, vol. 2690, https://doi.org/10.1007/978-1-0716-3327-4_22,

correlation with complexes, thereby deciphering the biological roles of proteins with unknown functionalities. Mass spectrometry is now a major instrument for large-scale protein sequencing. Additionally, tandem mass spectrometry helped utilize a fragment-based marker to explore target proteins and their interactor from large-scale cell signaling experiments. Combining chemical labeling techniques with a traditional biochemical approach is capable of pinpointing protein–protein interactions in the native environment of the cells [4].

The affinity purification mass spectrometry (*AP-MS*) approach, a conventional method used for studying protein–protein interactions, utilizes specific binding interactions between the affinity tag or a monoclonal/polyclonal antibody, and macromolecules such as DNA, RNA, lipids, peptides, or proteins (primary target) which act as bait [5]. The inability to distinguish direct interactions from indirect interactions (i.e., between bait and prey proteins) gives rise to high false-positive rates [6, 7]. This approach utilizes mild washing conditions to remove non-specific interactions due to the problem of the detachment of bound proteins from the bait. Thus, weak, and transient interactions are potentially not preserved due to the washing steps. Affinity purification cross-linking mass spectrometry (*AP-XL-MS*) overcomes some limitations by covalently linking different proteins with cross-linkers bearing two or more reactive ends separated by the spacer arm [8]. Cross-linkers can be referred to as molecular rulers as they provide the distance between the cross-linked proteins. This linking also helps apply strong denaturing solvent washing steps during sample preparations and preserves weak and transient interactors. Besides, interactors can be identified by mass spectrometry using non-modified sequences or, more specifically, the specific protein domains by crosslinked peptide analysis.

Different categories of cross-linkers include amine-reactive [9], thiol-reactive [10], carboxyl-reactive [9, 11], MS cleavable [9], and chemical or photocleavable. Cross-linking produces several kinds of peptides after enzymatic digestion, and most are unmodified peptides due to the low reactivity of cross-linkers in an aqueous solvent. Sequencing the inter-crosslinking peptide (two peptides connected with the cross-linkers) is difficult due to the complexity of fragmentation of the two cross-linked peptides during tandem mass spectrometry experiments. Introducing a cleavable group reduces the complexity by cleaving the inter-cross-linked peptides into two parts, and they can be individually sequenced by tandem mass spectrometry. In this chapter, we detailed the identification of interactors by sequencing unmodified peptides. Moreover, we also briefly provided the steps involved in analyzing protein domain using currently available software tools. A novel MS-compatible dual cleavable cross-linking technology (*DUCCT*) has recently been developed by Chowdhury et al. [12]. This advanced cross-linking technology effectively contributes to the mapping of PPIs by reducing the complexities encountered during data analysis. DUCCT functions by generating signature ions in the tandem MS/MS by undergoing sequential collision-induced dissociation (CID) and electron transfer dissociation (ETD) cleavage. CID cleavage results from breaking the labile aspartate-proline (DP) bond, while ETD cleavage occurs by breaking an azo bond in DUCCT [13].

The AP-MS technique reduces data complexity tremendously. The method also requires the transfection of bait proteins with affinity tags. Due to the poor performance of some antibodies, small molecule affinity tags are popular. Due to the advantage, many reagents came out with affinity tags designed in the structure. Ligands or drugs were also designed with affinity tags to enrich the target proteins using this approach. Chemical crosslinkers utilized these affinity tags to enrich cross-linked proteins by affinity chromatography. Affinity-tagged ligands or crosslinkers helped reduce data complexities and pinpoint direct or indirect targets in AP-MS methods. Biotin-avidin affinity chromatography is a typical example of affinity tag-based purification techniques. Besides, the length of spacer chain arms in cross-linkers is important to reach a reasonable distance in the protein complexes. Our lab efficiently conducted AP-MS with spacer-chain controlled cross-linking (SPACC) [14]. This AP-SPACC-MS technique connects more proteins with the bait due to the variable spacer arms of the cross-linkers. This chapter also discusses the procedure to conduct the AP-SPACC-MS experiments using DUCCT (18 Å) and a commercial cross-linker called BS3 (11 Å) [13, 15] (*see* **Note 1**). A general scheme is provided in Fig. 1.

Numerous data analysis software like ASAP [16], pLink [3], X-link [17], and Metamorpheus [18] have been used for the analysis of large amounts of data on cross-linked peptides. However, the software's capabilities are limited due to the increase in search space in cross-linking studies. A few software are available to analyze cleavable cross-linked data, e.g., XlinkX [19] and MeroX [20–22]. However, they still suffer from difficulties in analyzing large-scale data. We have developed a user-friendly software tool Cleave-XL [12] for identifying cross-linked products for DUCCT. This book chapter primarily focuses on identifying unmodified peptides for finding interactions. We will briefly discuss cross-linked peptide analysis using currently availed software tools.

2 Materials

2.1 Cell Culture

1. HEK293-MD2-CD14 cells were transfected with the HA-tagged human TLR2 gene.

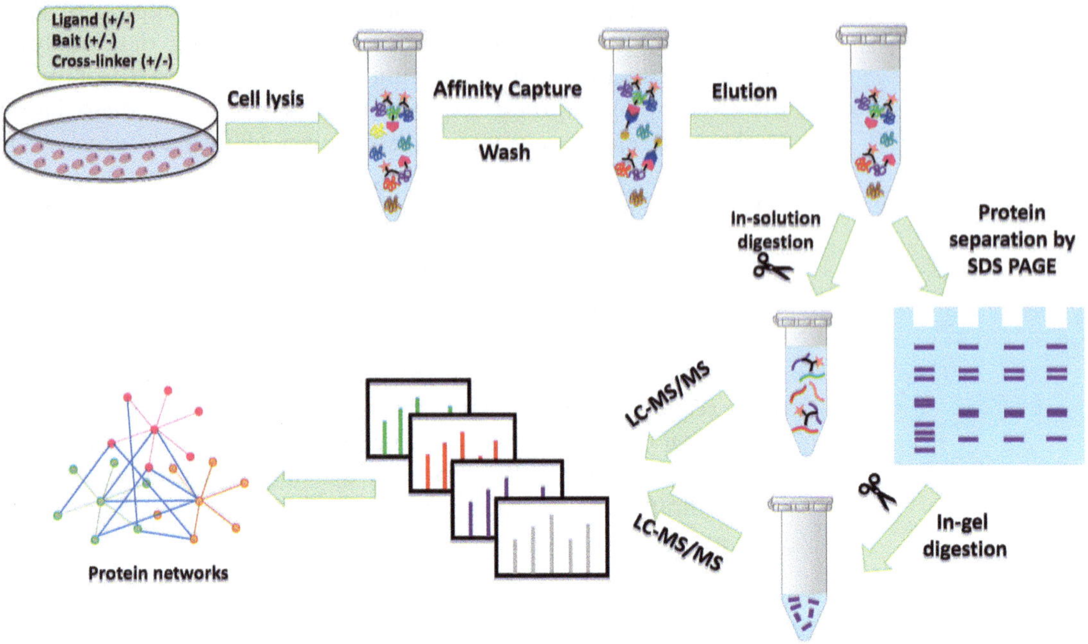

Fig. 1 A general scheme to conduct AP-XL-MS based proteomics. Affinity purification can be achieved using tagged bait proteins, affinity tagged ligands or cross-linkers

2. Growth medium: Complete cell growth medium Dulbecco's Modified Eagle Medium (DMEM) was prepared by adding 10% fetal bovine serum (FBS), 100 U/mL penicillin, 100 μg/mL streptomycin, 50 g/mL hygromycin, and 10 μg/mL blasticidin to prevent the fungal growth.
3. Radio-immunoprecipitation assay buffer (RIPA) buffer: RIPA lysis and extraction buffer and protease inhibitor were mixed in 10:1 ratio.
4. Bicinchoninic acid (BCA) Assay: Pierce BCA Protein Assay Kit.

2.2 Cell Treatments

1. Lipopolysaccharide (LPS) and LPS-biotin were acquired. Prepare the stock solution (5 mg/mL) using endotoxin-free water (*see* **Note 2**).
2. Simvastatin: Prepare 10 μM simvastatin solution in Dimethyl Sulfoxide (DMSO).
3. 1.5 M Tris-HCl buffer (pH 8.0).

2.3 Affinity Purification

1. Tris-buffered saline (TBS).
2. TBST buffer: 0.05% Tween-20 in TBS buffer.
3. Phosphate buffer saline (PBS): pH 7.4.

4. Pierce Monomeric avidin UltraLink Resin. Instead of monomeric avidin, streptavidin can be used.
5. Anti-HA magnetic beads.

2.4 SDS Polyacrylamide Gel

1. Resolving gel buffer: 1.5 M Tris-HCl (dissolve 181.7 g Tris-HCl in 1 liter water, and adjust the pH to 8.8, then filter it, and store it at 4 °C).
2. Stacking gel buffer: 0.5 M Tris-HCl, pH 6.8 (transfer 60.6 g Tris-HCl in 1 liter water, and adjust the pH with concentrated HCl, then filter it, and store at 4 °C).
3. Prepare separation gel by mixing 3.4 mL of water, 2.5 mL of 1.5 M tris buffer, 4 mL 30% Acrylamide, 100 μL 10% sodium dodecyl sulfate (SDS), 10 μL TEMED, 50 μL 10% ammonium persulfate (APS) in a 50 mL falcon tube.
4. Prepare a stacking gel by mixing 2.65 mL of water, 1.25 mL of 0.5 M tris, 1.0 mL of 30% Acrylamide, 50 μL 10% SDS, 5 μL TEMED, 25 μL 10% ammonium persulfate (APS) in a 50 mL falcon tube.
5. Ammonium persulfate (APS): Prepare 10% APS solution by dissolving 100 mg APS in 1 mL ultra-pure water.
6. 30% Acrylamide/bis-acrylamide solution 28.2:0.8. Store at 4 °C.
7. Sodium dodecyl sulfate (SDS): Prepare 10% SDS solution from SDS powder.
8. N,N,N′,N′-Tetramethylenediamine (TEMED) was used.
9. Blue pre-stained protein standard with a broad range of 11–250 kDa.

2.5 Digestion

1. Pierce C18 tips, 100 μL bed.
2. Liquid chromatography-mass spectrometry (LC-MS) grade acetonitrile (ACN) and methanol (MeOH) were obtained.
3. Ammonium bicarbonate, Iodoacetamide (IAA), chloroform, and formic acid were purchased.
4. Dithiothreitol (DTT) was purchased.
5. Proteolytic enzyme trypsin is used to digest the proteins.

2.6 Sample Analysis (Nano ESI-LC-MS/MS)

1. Instrument: LTQ Velos Pro mass spectrometer combined with a Dionex Ultimate 3000 UHPLC system.
2. Acclaim Pepmap C18 column.
3. Solvent system: Mobile Phase A: 0.1% formic acid in water.
4. Mobile Phase B: 0.1% formic acid in 95% acetonitrile, 5% water.

2.7 Dual Cleavable Cross-Linking Technology (DUCCT) and Other Cross-Linkers

1. Fmoc-Gly-wang resin, Fmoc-Asp(O-2-PhiPr), and Fmoc-Asp-OtBu were purchased from EMD Millipore.
2. Fmoc-Pro-OH, 4-formylbenzoic acid, N-hydroxy succinimide (NHS), N,N-dicyclohexylcarbodiimide (DCC), N,N,N′,N′-Tetramethyl-O-(1H-benzotriazol-1-yl)uronium hexafluoro-phosphate, O- (Benzotriazol-1-yl)-N,N,N′,N′- tetramethyluronium hexafluorophosphate (HBTU) were obtained.
3. Dichloromethane (DCM), Dimethylformamide (DMF), Trifluoroacetic acid (TFA) were purchased.
4. Solvent systems for peptide synthesizer PS3, 0.4 M N-methylmorpholine in DMF, 20% piperidine in DMF were purchased from Gyros Protein technologies.
5. 6-Fmoc-hydrazinonicotinic acid was obtained from advanced automated peptide protein technology.

2.8 Data Analysis

1. Proteome Discoverer software (PD), version 2.1 was used for data analysis.

3 Methods

3.1 Cell Culture

Human embryonic kidney (HEK) 293 cells transfected with affinity-tagged Hemagglutinin (HA) immune receptors/or respective cells with affinity-tagged target or antibody-based affinity purification can be utilized.

3.1.1 HA-Tagged HEK293 Cell Culture

1. For warming the medium, keep the growth medium in a water bath at 37 °C for 15 min.
2. Take the cryovial from liquid nitrogen and immediately thaw it at 37 °C for 1 min.
3. The cells were mixed in 5 mL of culture medium into a 15 mL falcon tube.
4. Remove the culture medium after centrifugation at 3000× *g* for 5 min.
5. Collect the cell pellet and gently resuspend it in 1 mL of growth medium.
6. Add 10 mL of pre-made cell growth medium into each 10 cm Petri dish, add the suspended cells drop-wise and incubate at 37 ° C with 5% CO_2.
7. To perform the subculture, remove the medium, and wash with 1X PBS twice.
8. Trypsinize the cells with trypsin-Ethylenediaminetetraacetic Acid (EDTA) and transfer them into a 1.5 mL Eppendorf tube.
9. Collect the cell pellet by centrifugation at 3000 × *g* for 3 min.

10. Resuspend the cell pellet in 1 mL of growth medium.
11. Add 10 mL of growth medium into each petri dish, followed by the addition of suspended cells added based on the experiment, and incubate at 37 °C with 5% CO_2.

3.2 Cell Treatments

3.2.1 Simvastatin Treatment of HA-Tagged TLR2-MD2-CD14 Cells

1. Allow the cells to attain 70% confluency.
2. Treat the cells with simvastatin to a final concentration of 10 μM and incubate them for 24 h followed by stimulation with Lipopolysaccharide (LPS)-biotin.

3.2.2 LPS-Biotin Treatment of HA-Tagged TLR2-MD2-CD14 Cells

1. Allow the growth medium to stand in the 37 °C water bath for 15 min.
2. Prepare the medium with 1 μg/mL of LPS-biotin and incubate at 37 °C for 4–6 h.
3. Wash the cells with PBS twice, then scrape the cells.
4. Centrifuge at 3000× *g* for 3 min and discard the supernatant.
5. Collect the cell pellet for further use.

3.2.3 Cross-Linker

Cross-linker can be either cleavable or non-cleavable. Dual Cleavable Cross-linking Technology (DUCCT~18 Å) [13] and commercially available bissulfosuccinimidyl (BS3 ~ 11 Å) cross-linkers were utilized for this study. Different lengths of spacer arms will provide more coverage for bait to interactors.

1. Prepare 100 mM stock solution of cross-linker in DMSO and treat the cells with cross-linker (DUCCT/BS3) and allow the reaction to proceed for 30 mins.
2. Quench the reaction with 50 mM Tris-HCl buffer (pH 8.0).

3.3 Cell Lysis

1. Prepare RIPA Lysis Buffer cocktail by mixing RIPA buffer with protease inhibitor.
2. Wash the cells twice with PBS and transfer them to a falcon tube. Centrifuge at 3000 rpm for 3 min. Discard the supernatant. During the second wash, the cells should be transferred to a 1.5 mL Eppendorf tube.
3. Add RIPA cocktail and allow proper dispersion into the cell pellet.
4. Pipette and vortex the cells and incubate on ice for 15 min and sonicate for 15 min.
5. Vortex thoroughly and incubate on ice for 15 min.
6. After incubation, centrifuge the cells at 4 °C for 30 min at 20,000× *g* and collect the supernatant for further analysis.
7. Measure the protein concentration using the BCA method.

3.4 Affinity Purification

3.4.1 Biotin Avidin Affinity Purification

1. Allow the resin and reagents to attain room temperature.
2. Centrifuge at 500× *g* for one minute to remove the storage solution.
3. Wash the monomeric avidin resin three times with 1X PBS.
4. Add biotinylated sample to the resin and incubate it at 4 °C overnight. Centrifuge for 1 min at 500× *g* and discard the supernatant.
5. After that, extensively wash the resin with PBS and ultrapure water to get rid of non-specific interactions.
6. Elute the bound proteins in Laemmli buffer. Centrifuge for 1 min at 500× *g* and collect the supernatant.
7. Heat the sample at 95 °C for 5 min and load it onto sodium dodecyl sulfate-polyacrylamide gel electrophoresis (SDS-PAGE) (*see* **Note 3**).

3.4.2 Fusion-Tag Purification

Any fusion tag can be employed for this study such as Myc, and HA. In this work, we have used HA-tagged proteins.

1. HA-tagged proteins can be purified using anti-HA magnetic beads.
2. Wash the anti-HA magnetic beads with 0.05% TBS and vortex gently.
3. Collect the beads against the side of the tube by placing it into a magnetic stand. Discard the supernatant. Repeat this step twice.
4. Add the HA-tagged protein sample to the beads and allow the reaction to proceed overnight at 4 °C with gentle mixing.
5. Place the tube into a magnetic stand and discard the unbound sample.
6. Wash the beads with TBS and ultrapure water twice with gentle mixing.
7. Place the tube into the magnetic stand and discard the supernatant.
8. Elute the beads with Laemmli buffer, heat at 95 °C for 5 min followed by centrifugation, and load onto the gel (*see* **Note 3**).

3.5 Sodium Dodecyl Sulfate-Polyacrylamide Gel Electrophoresis (SDS-PAGE)

1. Pour the contents of separation gel within a 7.25 cm × 10 cm × 1.5 mm gel cassette and overlay a small amount of water on the top of the resolving gel carefully to keep the polymerized gel from drying out.
2. Finally, transfer the stacking gel on top of the resolving gel until it overflows and inserts a 10-well gel comb on top of the stacking gel. Allow the gel to polymerize for about 40 min.

3. Carefully remove the comb from the gel.
4. Place the glass plate sandwich into the electrophoresis core and keep the assembly in the running tank containing the running buffer.
5. Mix the protein sample with Laemmli buffer and heat at 95 °C for 10–20 mins. Load 20 μL of sample into each well using a 200 μL special gel-loading tip.
6. Run the gel with 180 V voltage until the dye reaches the bottom of the gel.
7. Disassemble the running unit, wash the gel with the ultrapure water, and use SYPRO ruby protein Gel stain.

3.6 Digestion

3.6.1 In-Solution Digestion

1. Purified proteins were dissolved in 50 mM ammonium bicarbonate in an Eppendorf tube (*see* **Note 4**).
2. Add 5 μL of 5 mM DTT, vortex, and incubate at 56 °C for 45 min to reduce disulfide bonds.
3. Add 5 μL, 10 mM IAA and incubate at 37 °C for 30 min in the dark (covered in foil) to alkylate cysteines.
4. Bring the volume of the solution up to 150 μL with 50 mM ammonium bicarbonate.
5. Add trypsin solution so that trypsin: protein = 1:50 w/w and incubate at 37 °C for 12–16 h in a thermomixer.
6. Allow the sample to cool to room temperature and quench the digestion by acidifying using 1.5 μL of 10% formic acid.
7. Desalt the sample with C18 columns.
8. Dry the sample in a speed vacuum.
9. Reconstitute the peptides in 0.1% formic acid.
10. Centrifuge the sample at 20,000× *g* for 30 min before transferring the supernatant to the vial for mass spectrometry analysis.

3.6.2 In-Gel Digestion

1. Excise the protein band of interest with a clean scalpel and cut it into smaller pieces.
2. Wash the pieces with double distilled water (DDW) for 10 min.
3. Add 500 μL acetonitrile to shrink the gel pieces and vortex well for 15 min.
4. Discard the supernatant and dry the gel pieces for 10 min.
5. Add 50 μL of 10 mM DTT to the gel pieces. Incubate for 30 min at 56 °C in thermomixer.
6. Spin down the gel pieces for 2 min, allow the tube to stand at room temperature, and carefully remove the supernatant.

7. Add 50 μL of 55 mM IAA and incubate for 30 min at room temperature in the dark.
8. Centrifuge the contents for 2 min, pull off the supernatant, and wash twice with DDW.
9. Add acetonitrile to shrink the gel pieces followed by removal of acetonitrile by speed vac until dry.
10. Add enough trypsin to cover the dry gel pieces then keep it at 37 °C for 30 min.
11. Add 25 μL of 25 mM ammonium bicarbonate and incubate overnight at 37 °C.
12. Centrifuge for 2 min and collect the supernatant.
13. To further extract peptides, add 10 μL of 1% trifluoroacetic acid or 1% formic acid solution to gel pieces and incubate for 10 min, and add 25 μL of 25 mM ammonium bicarbonate.
14. Spin down for 2 min and collect the supernatant.
15. Finally, the sample is ready for LC-MS/MS analysis.

3.7 Sample Analysis (Nano ESI-LC-MS/MS)

LC-MS/MS procedures were detailed using a Thermo velos pro mass spectrometer (*see* **Note 5**).

1. For sample analysis, utilize a Dionex Ultimate 3000 nano-UHPLC system coupled with a nano-ESI-linear ion trap (LTQ) Thermo Velos Pro mass spectrometer.
2. For chromatographic separation, a nano viper analytical C18 column (Acclaim Pepmap 150 mm × 75 μm, 3 μm, 100 Å).
3. Employ 90 min binary gradient system to separate peptides. Mobile phase A: 0.1% formic acid in the water, Mobile phase B: 0.1% formic acid in 95% acetonitrile, 5% water.
4. Set the flow rate of nanocolumn to 300 nL/min and injection volume to 5 μL.
5. Perform positive ion mode acquisition with a fixed source voltage 2.20 kV, and capillary temperature of 275 °C.
6. Perform full MS scans in normal scan mode from 300 to 2000 m/z mass range.
7. Collect MS/MS spectra in Data Dependent Acquisition (DDA) mode to acquire the MS^2 spectra for the 10 most abundant precursor ions. For data acquisition and processing, Xcalibur software can be utilized.
8. In CID activation, set the collisional energy to 35.0%, activation Q of 0.25, and activation time of 30 ms, along with an isolation width of 1.5 Da.

3.8 Data Analysis

1. Search the acquired MS raw files through PD.
2. Download the human protein sequence database from UniProt for the identification of proteins. For all the normal peptides, the search engine node used is Sequest HT.
3. Create processing and a consensus workflow in the workflow editor and set the parameters such as precursor mass tolerance to 2.5 Da and fragment mass tolerance to 0.6 Da. The proteolytic enzyme used for the study was trypsin with at least two missed cleavages. The precursor mass range was set from 350 to 3500 Da.
4. Set a static modification parameter, for example, carbamidomethyl/+57.02146 Da for cysteine and methionine oxidation/+15.00492 Da as a dynamic modification.
5. Decoy searches should be applied to the data sets to deliver false positive estimations. Set the false discovery rates (FDRs) for peptide and protein identifications to 1% with at least two peptides per protein.
6. Filter the results to detect the peptide-spectrum matches (PSMs) with the best score using *XCorr* score $\geq$ 1.5, $\Delta Cn < 0.01$, and q-values of $\leq$0.05 are used for high confident matches (*see* **Note 6**).

4 Notes

1. BS3 is available from Thermo Fisher Scientific. Cross-linkers with specific reactivities, affinity tags (biotin or clickable), and different spacer arms are available from different vendors.
2. Milli-Q ultrapure water was used for all the experiments.
3. Refer to Subheading 3.5 for the procedure of SDS-PAGE Gel electrophoresis.
4. A detailed description of the transfection procedure has been explained in this literature [22]. For in-solution digestion, after washing on bead digestion can be performed.
5. Mass spectrometry analysis can be performed with other mass spectrometers capable of sequencing large-scale samples.
6. For analyzing cross-linked peptides, a few software are available. The software called XlinkX [17] and MeroX [20, 22] can be used in AP-MS for cleavable cross-linkers. For non-cleavable cross-linkers, pLink [3] software is available online. Limited data analysis can be done with Cleave-XL [12], a software for DUCCT.

Acknowledgments

Support from NIGMS, NIH grant number 1UA5GM113216-01 is acknowledged. We recognize funding support for PI Dr. Chowdhury from UT Arlington. UT Systems Shared Proteomics Networks are also acknowledged for mass spectrometry support.

References

1. Ijaz M, Ansari M-u-R, Iqbal M (2018) Introductory chapter: protein-protein interactions and assays. Protein-Protein Interaction Assays. https://doi.org/10.5772/INTECHOPEN.77337
2. MacBeath G, Schreiber SL (2000) Printing proteins as microarrays for high-throughput function determination. Science 289(5485): 1760–1763. https://doi.org/10.1126/SCIENCE.289.5485.1760
3. Purcell S, Neale B, Todd-Brown K et al (2007) PLINK: a tool set for whole-genome association and population-based linkage analyses. Am J Hum Genet 81(3):559–575. https://doi.org/10.1086/519795
4. Ewing RM, Chu P, Elisma F et al (2007) Large-scale mapping of human protein–protein interactions by mass spectrometry. Mol Syst Biol 3(1):89. https://doi.org/10.1038/msb4100134
5. Zacharias AO, Fang Z, Rahman A et al (2021) Affinity and chemical enrichment strategies for mapping low-abundance protein modifications and protein-interaction networks. J Sep Sci 44(1):310–322. https://doi.org/10.1002/JSSC.202000930
6. Fang Z, Baghdady YZ, Schug KA et al (2019) Evaluation of different stationary phases in the separation of inter-cross-linked peptides. J Proteome Res 18(4):1916–1925. https://doi.org/10.1021/ACS.JPROTEOME.9B00114
7. Huttlin EL, Ting L, Bruckner RJ et al (2015) The BioPlex network: a systematic exploration of the human interactome. Cell 162(2): 425–440. https://doi.org/10.1016/j.cell.2015.06.043
8. Steigenberger B, Albanese P, Heck AJR et al (2020) To cleave or not to cleave in XL-MS? J Am Soc Mass Spectrom 31(2):196–206. https://doi.org/10.1021/jasms.9b00085
9. Piersimoni L, Sinz A (2020) Cross-linking/mass spectrometry at the crossroads. Anal Bioanal Chem 412(24):5981–5987. https://doi.org/10.1007/S00216-020-02700-X/FIGURES/4
10. Iacobucci C, Piotrowski C, Rehkamp A et al (2018) The first MS-cleavable, photo-thiol-reactive cross-linker for protein structural studies. J Am Soc Mass Spectrom 30:139–148. https://doi.org/10.1007/s13361-018-1952-8
11. Sriswasdi S (2013) Scholarly commons publicly accessible penn dissertations 2013. In-Depth Analysis of Zero-Length Crosslinking for Structural Mass Spectrometry Spectrometry
12. Chakrabarty JK, Sadananda SC, Bhat A et al (2020) High confidence identification of cross-linked peptides by an enrichment-based dual cleavable cross-linking technology and data analysis tool cleave-XL. J Am Soc Mass Spectrom 31(2):173–182. https://doi.org/10.1021/JASMS.9B00111
13. Chakrabarty JK, Naik AG, Fessler MB et al (2016) Differential tandem mass spectrometry-based cross-linker: a new approach for high confidence in identifying protein cross-linking. Anal Chem 88(20): 10215–10222. https://doi.org/10.1021/ACS.ANALCHEM.6B02886/SUPPL_FILE/AC6B02886_SI_003.XLSX
14. Kamal AHM, Aloor JJ, Fessler MB et al (2019) Cross-linking proteomics indicates effects of simvastatin on the TLR2 interactome and reveals ACTR1A as a novel regulator of the TLR2 signal cascade. Mol Cell Proteomic 18(9):1732–1744. https://doi.org/10.1074/mcp.RA119.001377
15. Chowdhury SM, Du X, Tolić N et al (2009) Identification of cross-linked peptides after click-based enrichment using sequential collision-induced dissociation and electron transfer dissociation tandem mass spectrometry. Anal Chem 81(13):5524–5532. https://doi.org/10.1021/AC900853K
16. Sinz A (2006) Chemical cross-linking and mass spectrometry to map three-dimensional protein structures and protein–protein interactions. Mass Spectrom Rev 25(4):663–682. https://doi.org/10.1002/mas.20082

17. Zheng C, Weisbrod CR, Chavez JD et al (2013) XLink-DB: database and software tools for storing and visualizing protein interaction topology data. J Proteome Res 12(4): 1989–1995. https://doi.org/10.1021/pr301162j
18. Young MM, Tang N, Hempel JC et al (2000) High throughput protein fold identification by using experimental constraints derived from intramolecular cross-links and mass spectrometry. Proc Natl Acad Sci U S A 97(11): 5802–5806. https://doi.org/10.1073/PNAS.090099097
19. Orsburn BC (2021) Proteome discoverer—a community enhanced data processing suite for protein informatics. Proteomes 9(1):15. https://doi.org/10.3390/proteomes9010015
20. Iacobucci C, Götze M, Ihling CH et al (2018) A cross-linking/mass spectrometry workflow based on MS-cleavable cross-linkers and the MeroX software for studying protein structures and protein–protein interactions. Nat Protoc 13(12):2864–2889. https://doi.org/10.1038/s41596-018-0068-8
21. Arlt C, Götze M, Ihling CH et al (2016) Integrated workflow for structural proteomics studies based on cross-linking/mass spectrometry with an MS/MS cleavable cross-linker. Anal Chem 88(16):7930–7937. https://doi.org/10.1021/ACS.ANALCHEM.5B04853
22. Götze M, Pettelkau J, Fritzsche R et al (2015) Automated assignment of MS/MS cleavable cross-links in protein 3D-structure analysis. J Am Soc Mass Spectrom 26(1):83–97. https://doi.org/10.1007/S13361-014-1001-1

Chapter 23

Protein Interaction Screen on a Peptide Matrix (PrISMa)

Daniel Perez-Hernandez, Mattson Jones, and Gunnar Dittmar

Abstract

Protein-protein interactions (PPI) are essential to understanding the cellular function and key mechanisms necessary for life. Although understanding of the interactome and proteome has exploded due to high-throughput methods in the past decade, often limitations in technical methods result in a partial understanding of all PPI. Here we present a protocol dedicated to the Protein Interaction Screen on a peptide Matrix (PrISMa). PrISMa functions as a high-throughput screen unique to targeting weak and transient interactions often missed in other PPI methods. In addition, PrISMa also excels at the mapping of interactions across linear sequences of proteins that are commonly enriched in intrinsically disordered regions (IDRs) which cover 35–40% of the mammalian proteome. This protocol aims to expand the understanding of the targeted proteins by identifying transient interactors.

Key words Interactomics, Mass-spectrometry, Peptide-array, Proteomics

1 Introduction

Protein Interaction Screen on a peptide Matrix (PrISMa) is a high-throughput peptide pull-down assay that has been developed for screening specific interactions based on linear peptide sequences. The novelty of PrISMa is its ability to pinpoint-specific interaction sites within a protein (Table 1). Changes in protein binding due to post-translational modifications (PTMs) or single amino acid exchanges can be easily detected and observed, availing a deeper understanding of weak protein interactions often missed in other PPI methods [1–3].

The screen utilizes a matrix of overlapping peptides that span the amino acid sequence of the protein of interest (bait protein). These matrix peptides are chemically produced by fluorenyl-methoxycarbonyl (FMOC) synthesis allowing the customization and inclusion of PTMs (phosphorylation, acetylation, methylation, etc.) and/or amino acid exchanges [4]. The high local concentration of the synthesized peptides stabilizes low-affinity interactions as well as transient interactions, which are usually lost in

Shahid Mukhtar (ed.), *Protein-Protein Interactions: Methods and Protocols*, Methods in Molecular Biology, vol. 2690, https://doi.org/10.1007/978-1-0716-3327-4_23,

Table 1
Advantages and limitations of the PrISMa method

Advantages	Limitations
High local peptide concentrations can detect low binding affinity and highly specific interactions undetectable with other AP-MS methods	Protein–protein interactions that depend on larger or structured binding sites may not bind (e.g., secondary/tertiary structures)
Rapid experimental design requires no molecular cloning, expression of protein vectors, or purchasing of antibodies	Mild-washing steps result in high background
Mutations or PTMs can be rapidly included in the screen	Large PTMs, like ubiquitin, cannot be synthesized on the membrane
Neighboring matrix peptides can be used for background correction and normalization resulting in higher accuracy	A large sample size requires time for mass spectrometry readout and data analysis
Large protein complexes can be detected	Significant amounts of cell extracts are necessary for the screen
More than 300 different experiments can be completed in less than 2 days, excluding MS and data analysis time	
Can reconfirm previously known interactions	

immunoprecipitation experiments. This enables a larger number of novel and previously known interactors to be identified. In addition, PrISMa is particularly well suited for screening interactions in intrinsically disordered regions (IDRs) that are mediated by short linear motifs (SLiMs).

SLiMs are composed of 3–11 amino acid sequences that bind weakly to protein interactors and reside inside protein IDRs [5]. Proteins bound to these short sequences often enable secondary structure generation of IDRs and modifications to a host of cellular processes including cellular signaling, localization, or regulation mechanisms. Approximately 35–40% of mammalian proteomes are unstructured and can serve as docking sites for protein–protein interactions [6]. Many of the SLiM-mediated interactions are transient or have low binding affinities which are often uncoupled during immunoprecipitation (e.g., affinity-purification (AP)) or genetic (e.g., yeast-two-hybrid interactions) approaches. PrISMa circumvents these limitations by utilizing a matrix of linear peptides combined with a high-throughput screen.

A PrISMa experiment is structured in five steps (Fig. 1). First, the protein or the protein region of interest is selected. PTMs or mutations in the region of interest can be included in the screen utilizing relevant databases (e.g., PTMs: Phosphosite (https://www.Phosphosite.org), Uniprot (https://www.Uniprot.org); mutations: Uniprot or COSMIC (https://cancer.sanger.ac.uk/cosmic)). The second step is the design of the peptide array spanning the region of interest. Peptides are designed with a length of 15 amino acids and continuously shifted by 3 amino acids resulting

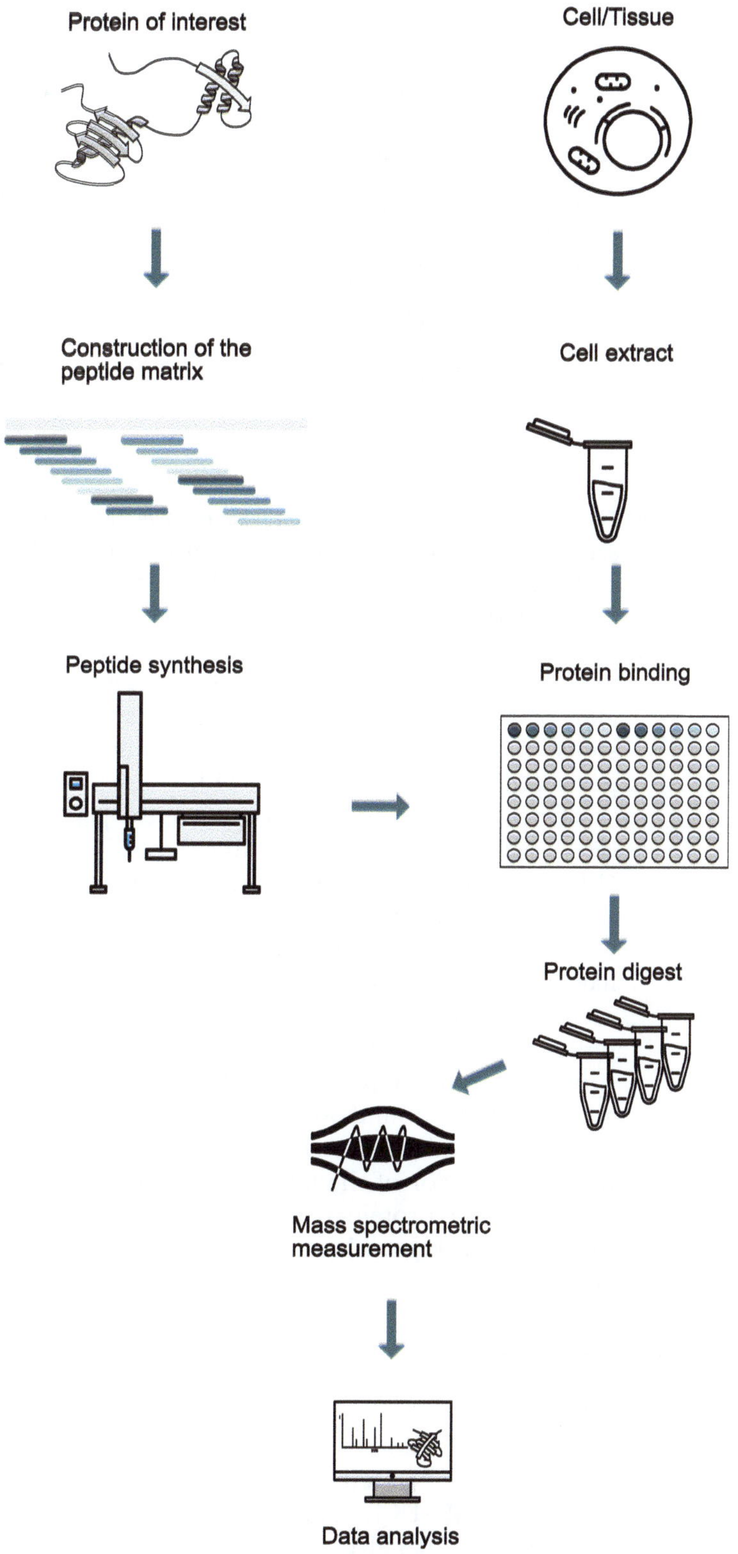

Fig. 1 Workflow of the PrISMa technique. The PrISMa technique starts with two branches, peptide matrix construction, and preparation of the cell extracts. The protein sequence is covered with tilling peptides. The peptides are then chemically synthesized. The membrane is incubated with cell extracts, binding the interacting proteins to the corresponding peptides. The bound proteins are digested into peptides and then analyzed by mass spectrometry

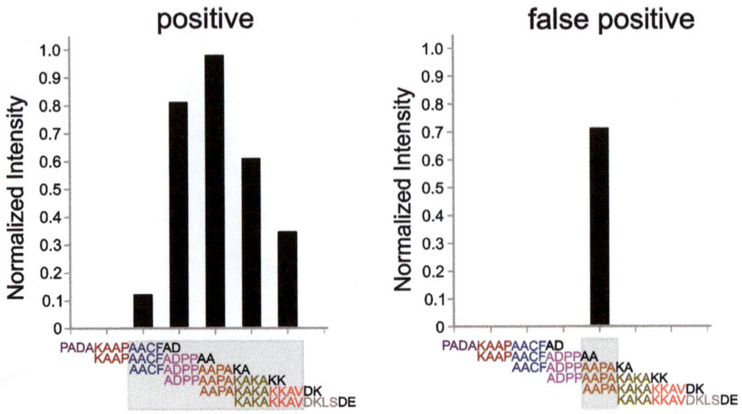

Fig. 2 An example of positive and false positive binding of a protein binding to specific regions of peptide tiles generated from a PrISMa screen. Positive interactors bind over multiple peptide shifts indicating a broad binding region. False positives often lack multiple peaks and indicate nonspecific binding

in peptides tiling the targeted protein region. Peptides with PTMs and/or amino acid mutations are also synthesized onto separate spots on the peptide matrix for a maximum of three modifications per peptide in all combinations [1].

Once the peptide list is compiled, the peptides are synthesized by F-MOC solid-phase synthesis on a polyethylen glycol (PEG)-cellulose membrane (the membrane can be ordered from several manufacturers). Next, the peptide membranes are incubated with any type of protein extract (from cell extracts to subproteomes like nuclear extracts, mitochondria, etc.), which are depleted of membranes and other components which have high unspecific binding capacities. Protein interactors are bound to the peptide spot during this step. After washing, the individual peptide spots are excised, the bound protein denatured, and digested. Finally, the resulting peptides are analyzed with mass spectrometry (MS). The binding behavior of the interactors is compared across all peptide spots. Positive protein–peptide interactions are confirmed statistically by utilizing technical replicates and neighboring peptides (Fig. 2).

2 Materials

2.1 Peptide Array Setup

1. Membrane binding buffer (MBB): 20 mM N-2-hydroxyethylpiperazine-N-2-ethane sulfonic acid (HEPES) (pH = 8), 100 mM KCl, 0.2 mM ethylenediaminetetraacetic acid (EDTA), 20% glycerol (HPLC grade).
2. 0.5 mM dithiothreitol (DTT) (add just before use).

3. Denaturation buffer: 6 M urea, 2 M thiourea, 20 mM HEPES (pH = 7.5).
4. FMOC-synthesized membrane with the peptide library.
5. Protein extract (2–10 mg/mL) can be commercially acquired or produced in-house.
6. Plastic incubation chamber (low protein binding) or in a bag which tightly fits with the peptide membrane.

2.2 Pull-down Assay and Digestion

1. Biopsy punch 2 mm.
2. 96-well PCR plates for digestion (1 well/spot)
3. 10 mM dithiothreitol (DTT)
4. 55 mM 2-chloroacetamide
5. Methanol HPLC grade.
6. Acetonitrile (ACN) hypergrade for MS.
7. Formic acid (FA) for analysis.
8. Endopeptidase Lys-C (Wako, Japan).
9. Modified trypsin, sequencing grade (Promega, WI, USA).
10. Methanol MS grade.
11. C18 Empore disks for stage tips (3 M).
12. StageTip washing buffer: 0.1% trifluoroacetic acid (v/v), 2% acetonitrile (v/v).
13. StageTip elution buffer: 0.1% formic acid (v/v), 80% acetonitrile (v/v).
14. Benchtop centrifuge.
15. Speedvac.
16. 1 mm diameter Polyetheretherketone (PEEK) tubing (Upchurch Scientific).

2.3 Liquid Chromatography Coupled Mass Spectrometry (LC-MS)

1. A nanoflow HPLC system (LC-MS/MS Dionex Ultimate 3000; Tables 2 and 3).
2. A high-resolution mass spectrometer, e.g., TIMS TOFpro (Bruker Daltonics, Germany) or QExactive HF (Tables 2 and 3).
3. Acclaim PepMap RSLC 75 μm × 15 cm nanoViper.
4. Acclaim PepMap RSLC 75 μm × 2 cm nanoViper.
5. MaxQuant software (http://www.maxquant.org/).

2.4 Equipment Setup

Mass spectrometry setup (*see* **Note 1**).

Table 2
Parameters to perform mass spectrometry analysis on a QExactive HF

Mass spectrometer parameters (QExactive HF)	
General	
Runtime	0 to 20 min
Polarity	Positive
Default charge state	2
Full MS	
Microscans	1
Resolution	60,000
AGC target	3e6
Maximum IT	60 ms
Number of scan ranges	1
Scan range	375 to 1700 m/z
Spectrum data type	Profile
dd-MS^2	
Microscans	1
Resolution	17,500
AGC target	1e5
Maximum IT	45 ms
Loop count	12
MSX count	1
Isolation Window	1.2 m/z
Isolation offset	0.0 m/z
Fixed first mass	100.0 m/z
NCE	28
Spectrum data type	Centroid
dd Settings	
Maximum AGC target	4.50e3
Charge exclusion	Unassigned, 1, 6–8, >8
Exclude isotopes	On
Dynamic exclusion	15.0 s

Table 3
Parameters for HPLC (Dionex 3000 Ultimate) coupled to QExactive HF in 20 min gradient

High-performance liquid chromatography parameters (Dionex 3000 ultimate)		
Time (min)	**Flow (μL/min)**	**% MS buffer B**
0	0.500	2
5	0.500	5
16	0.300	45
16.1	0.300	90
18	0.300	90
18.1	0.500	2
20	0.500	2

3 Methods

3.1 Parallel Peptide Pull-Down Assay and Preparation of the Spots for Mass Spectrometric Analysis (See Note 2)

1. Wash the incubation chamber with 70% ethanol followed by phosphate buffered saline (PBS) to remove protein contamination, detergents, salts, chemical polymers, etc.
2. Handle the membrane carefully using forceps without touching the peptide spots. Place the peptide membrane in the incubation chamber and wash with 5 mL of MBB for 10 min.
3. Discard the buffer from the incubation chamber.
4. Incubate the membrane in the incubation chamber with ~5 mL protein extract for 30 min, covering the peptide membrane completely. The protein concentration of the extract should be higher than 2 mg/mL, to ensure the detection of low-affinity interactors [3].
5. Remove the protein extract and wash the peptide membrane with 5–7 mL MBB 3 times for 5 min each.
6. Prepare 96-well plates (use as many plates as needed for all the spots that are contained in the membrane) by putting 20 μL of denaturation buffer in each well.
7. Punch out the spots with a 2 mm biopsy puncher and transfer them to the 96-well plates containing the denaturation buffer (*see* **Note 3**).

3.2 Digestion of the Bound Material (See Note 4)

1. Add 2 μL of 10 mM DTT for 30 min per well.
2. Add 2 μL of 55 mM of Chloroacetamide for 45 min per well.
3. Digest samples with 2 μL of 0.25 μg/μL of sequencing grade endopeptidase LysC (in 20 mM HEPES (pH = 7.5)) for 3 h.

4. Dilute samples with 100 μL of 20 mM HEPES (pH = 7.5) at RT to a final concentration of <1.5 M urea.
5. Digest samples with 2 μL of 0.25 μg/μL of sequencing-grade modified trypsin overnight.
6. Next day, stop the reaction by adding 2 μL of 25% (v/v) trifluoroacetic acid (TFA) (check if the pH is lower than 2 using pH-indicator strips; *see* **Note 5**).
7. Punch out 2 discs of Empore C_{18} material with a flat tip gauge 16 needle.
8. Push the 2 discs into a GPS200 pipette tip using PEEK tubing.
9. Add 20 μL of Methanol to the disc to wet the matrix.
10. Centrifuge the tip at 1000× *g* for 1 min.
11. Add 20 μL of washing buffer to the disc.
12. Centrifuge the tip at 1000× *g* for 1 min.
13. Repeat steps 18 and 19 twice to remove all potential contaminants.
14. Load acidified samples into the tip. Check that pH < 3.
15. Centrifuge the tip at 1000× *g* for 3 min.
16. Wash the sample by repeating steps 18 and 19 three times to remove impurities and contaminants from the sample (*see* **Note 6**).
17. Elute bound peptides with 20 μL of elution buffer.
18. Transfer eluate from each tip in a fresh 96-well plate.
19. Dry the peptides in a speed vacuum concentrator.
20. Store at −20 °C (*see* **Note 7**).

3.3 LC-MS (See Note 8)

1. Resuspend the peptides in 6–8 μL of 0.1% formic acid in HPLC H_2O (sample concentration = 0.1–0.2 μg/μL).
2. Adjust the parameters depending on the mass spectrometer as well as the amount of sample to inject (the PrISMa pull-downs usually produce low peptide amounts, i.e., for a QExactive HF, 100–250 ng should be sufficient).

3.4 Data Analysis

To analyze mass spectrometry data, use MaxQuant (v1.2.2.5 or higher, [8, 9]) with the following search parameters (other search engines might be used):

1. Quantify interactor binding using the LFQ intensity values that are calculated by the MaxQuant software (Table 4).
2. Normalize the signal intensity for each protein to a range of 0 to 1 across the matrix peptides by the following formula:

Table 4
Search engine parameters on MaxQuant software

MaxQuant search parameters	
Time (min)	**Flow (μL/min)**
Fixed modification	Carbamidomethyl on cysteine
Variable modification	Oxidation of methionine, acetylation of the protein N-terminus
A maximum number of missed cleavages	2
Enzyme	Trypsin
Second peptide search	Yes
LFQ analysis	Active
Fasta file	Use the appropriate proteome file for the selected cell lines
FDR	1% for peptides and 1% for proteins
Match between runs	Yes

$$\text{Normalized protein intensity} = \frac{\text{Intensity (spot)} - \text{Min (intensity)}}{\text{Max (intensity)} - \text{Min (intensity)}}$$

3. Filtered out the background interactions for each protein by retaining proteins with a protein binding >0.9 per spot of normalized protein intensity. SLiMs will span several tiling peptides, resulting in partial or full coverage in some peptide sequences. Interacting proteins will show maximal binding signals for peptides containing the full SLiM binding sequence, and adjacent sequences will show decreasing signals (Fig. 2) (*see* **Notes 9** and **10** for "low number of interactors" and "high background," respectively).
4. Use this binding behavior for additional filtering of false positives of the interacting proteins. Exclude proteins that are not interacting with the consecutive peptides (*see* **Note 11** for "removal of false positive interactors."
5. Obtain a list with all identified proteins that bind specifically to one or more spots along the peptide matrix, fulfilling the consecutive binding criteria using this criterion (*see* **Note 12** for expected outcomes).

4 Notes

1. This protocol was used with a QExactive HF (Thermo Scientific), but the workflow is compatible with other high-resolution mass spectrometers. The parameters should be adjusted according to the machine. The recommended parameters for a QExactive HF are described.
2. All the steps should be done in an orbital shaker at 4 °C.
3. The 96 well plates containing the excised peptide spots can be stored in a denaturation buffer at −20 °C for several weeks.
4. All the steps should be done on an orbital shaker at room temperature (RT).
5. Prior to MS analysis, the digested peptides must be desalted. This protocol is based on using the C_{18} StageTips protocol (modified from [7]). However, other desalting methods are applicable. All the steps should be performed at room temperature. Check carefully after each centrifugation that all the liquid passes through the disc.
6. The StageTips can be stored at 4 °C or − 20 °C for several weeks.
7. The plate containing the dried peptides can be stored at 4 °C or − 20 °C for several weeks.
8. We recommend at least three replicates to increase the number of identified significant interactors. We also recommend injecting all replicates consecutively and a blank between different peptide sequences to avoid cross-contamination. We have observed less than 5% carry-over between adjacent samples.
9. A low number of interactors (<500 identified proteins). The membrane binding buffer (MBB) should be used for both incubations and washing of the membrane to avoid protein losses during the washing steps. MBB preserves native structures of protein complexes and allows the binding of complete complexes. Adjustments in the buffer composition could be made, but denaturing buffers will lead to a decreased number of protein complexes identified in the peptide array. Ensure that possible denaturants are used minimally. The concentration of the protein extract is critical to identifying low-abundant or low-affinity binding partners. We recommend using 2 mg/mL as a minimum concentration to perform a reproducible PrISMa experiment. Higher concentrations can also be used but can result in increased unspecific interactors. We also recommend controlling the experimental procedure by including selected

peptides that were previously identified for the protein of interest or validated previously by PrISMa [3].

10. High background. False-positive interactors and high background protein identification often occur by cross-contamination during the punch-out step. We recommend cleaning the tip of the biopsy punch with 70% ethanol after transferring each peptide spot. Reducing the concentration of protein may also help [3].

11. Removal of false-positive interactions. The intensity variation of the peptides is an intrinsic problem in mass spectrometry that will prevent distinguishing true interactors from background identifications. We recommend a minimum of three replicates for label-free approaches to validate the binders on every peptide from the matrix.

12. Expected outcome: We expect to discover proteins and protein complexes that specifically interact with adjacent peptides from the matrix [1–3]. In previous studies, we identified about 500–800 protein interactions per peptide spot. Although PRiSMa excels at PPI discovery, interactions must be confirmed using other methods. A number of molecular, cell biological, or biochemical methods such as proximity-ligations, co-immunoprecipitations, and Bio-ID assays have been successfully used [1–3]. Further confirmation could also be performed using mutagenesis of the detected binding sites to verify the specificity of the amino acid sequence. The accuracy of PrISMa enables the detection of interactome changes caused by the exchange of a single amino acid residues or by the inclusion of a PTM. The exchanged or modified amino acid will modulate the binding affinity, to create either weaker or stronger binding. Those modifications will trigger possible interactors of the specific PTM or mutational binders and also weaker interactors. Classic peptide-based protein pull-downs and their modified counterpart, struggle with high nonspecific binding, which has to be removed for follow-up experiments [10]. The PrISMa screen negates such limitations by including the binding profile and adding additional selection criteria, which allows more stringent filtering. This leads to higher confidence in identified interaction partners and requires minimal confirmation experiments. The peptide–protein interactions of the PrISMa dataset can be analyzed in the next step using a database of protein complexes, like CORUM (https://mips.helmholtz-muenchen.de/corum). The advantage is that the analysis will incorporate the interaction with an entire protein complex instead of a single protein. This can potentially determine several interaction regions between a protein complex and the protein of interest. An example of such an interaction is the interaction of C/EBPß with the Mediator complex [1].

References

1. Dittmar G, Hernandez DP, Kowenz-Leutz E et al (2019) PRISMA: protein interaction screen on peptide matrix reveals interaction footprints and modifications – dependent interactome of intrinsically disordered C/EBPβ. iScience 13:351–370. https://doi.org/10.1016/j.isci.2019.02.026
2. Ramberger E, Sapozhnikova V, Kowenz-Leutz E et al (2021) PRISMA and BioID disclose a motifs-based interactome of the intrinsically disordered transcription factor C/EBPα. iScience 24:102686. https://doi.org/10.1016/j.isci.2021.102686
3. Ramberger E, Suarez-Artiles L, Perez-Hernandez D et al (2021) A universal peptide matrix interactomics approach to disclose motif-dependent protein binding. Mol Cell Proteomics MCP 20:100135. https://doi.org/10.1016/j.mcpro.2021.100135
4. Frank R (2002) The SPOT-synthesis technique: synthetic peptide arrays on membrane supports—principles and applications. J Immunol Methods 267:13–26. https://doi.org/10.1016/S0022-1759(02)00137-0
5. Diella F (2008) Understanding eukaryotic linear motifs and their role in cell signaling and regulation. Front Biosci 13:6580. https://doi.org/10.2741/3175
6. Uversky VN (2019) Intrinsically disordered proteins and their "mysterious" (meta)physics. Front Phys 7:10. https://doi.org/10.3389/fphy.2019.00010
7. Rappsilber J, Mann M, Ishihama Y (2007) Protocol for micro-purification, enrichment, pre-fractionation and storage of peptides for proteomics using StageTips. Nat Protoc 2:1896–1906. https://doi.org/10.1038/nprot.2007.261
8. Cox J, Mann M (2008) MaxQuant enables high peptide identification rates, individualized p.p.b.-range mass accuracies and proteome-wide protein quantification. Nat Biotechnol 26:1367–1372. https://doi.org/10.1038/nbt.1511
9. Cox J, Neuhauser N, Michalski A et al (2011) Andromeda: a peptide search engine integrated into the MaxQuant environment. J Proteome Res 10:1794–1805. https://doi.org/10.1021/pr101065j
10. Meyer K, Kirchner M, Uyar B et al (2018) Mutations in disordered regions can cause disease by creating dileucine motifs. Cell 175:239–253.e17. https://doi.org/10.1016/j.cell.2018.08.019

Chapter 24

Analyzing Protein Interactions by MAC-Tag Approaches

Xiaonan Liu, Kari Salokas, Salla Keskitalo, Patricia Martínez-Botía, and Markku Varjosalo

Abstract

Proteomics methods such as affinity purification (AP) and proximity-dependent labeling (PL) coupled with mass spectrometry (MS) are currently commonly utilized to define interaction landscapes. BioID is one of the PL approaches, and it employs the expression of bait proteins fused to a nonspecific biotin ligase (BirA*), to induce in vivo biotinylation of proximal proteins. We developed the multiple approaches combined (MAC)-tag workflow, which allows for both AP and BioID analysis with a single construct and with almost identical protein purification and MS identification procedures. MAC-tag is a well-established method and has been widely used. Recent developed PL tags such as BioID2 and UltraID are smaller versions of BirA* with faster labeling efficiency. We therefore incorporate these tags into our system to develop MAC2-tag (containing BioID2) and MAC3-tag (containing UltraID) to overcome potential limitations of the original MAC-tag system and broaden the spectrum of applications for MAC-tags. Here, we describe a detailed procedure for the MAC-tag system workflow including cell line generation for the MAC/MAC2/MAC3-tagged protein of interest (POI), sample preparation for AP and PL protein purification, and MS analysis.

Key words Proteomics, Affinity purification, Proximity labeling, MAC-tag, MAC3-tag, Mass spectrometry, Protein interactions, Interactomics

1 Introduction

Protein–protein interactions (PPIs) are involved in almost all biological processes reflecting their essential roles in living cells. Therefore, accurate identification and annotation of PPIs are critical to understanding cellular systems. Affinity purification coupled with mass spectrometry (AP-MS) is one of the most frequently employed methods to discover novel interacting partners in protein complexes [1, 2]. In a typical experiment, a protein of interest (bait protein) is expressed with an epitope tag in cultured cells or tissue samples. The sample is lysed in mild/non-denaturing lysis

Authors Xiaonan Liu and Kari Salokas have equally contributed to this chapter.

Shahid Mukhtar (ed.), *Protein-Protein Interactions: Methods and Protocols*, Methods in Molecular Biology, vol. 2690, https://doi.org/10.1007/978-1-0716-3327-4_24,

conditions and the protein complex is affinity captured from cell lysates by an immobilized antibody/matrix that specifically recognizes the epitope tag. Non-specifically bound proteins are eliminated by extensive washes. This method allows the study of proteins with their associated partners. The functional cross-talk can be illustrated from participating complexes. With improvements in the sensitivity and resolution of MS, thousands of peptides/proteins can be identified in a single analysis. AP-MS has become the method of choice to build the first largest human protein interactome landscape [3].

However, not all protein interactions form a stable complex. Many interactions are transient, and others occur only in certain cellular compartments or at particular times in development [4]. Several alternative and powerful complementary approaches to AP that are based on proximity detection have been developed in recent years [5, 6]. Proximity-labeling (PL)-dependent biotinylation identification (BioID) [5] is the earliest and most widely used one, which is an R118G mutant of wild-type *Escherichia coli* biotin ligase (BirA). This BirA mutant (BirA*) utilizes biotin to form highly reactive biotinoyl-5-AMP that can diffuse to biotinylate nearby proteins within a radius of ~10 nm. BioID can capture weak and transient PPIs that are often overlooked by AP approaches. Indeed, BioID has been applied for PPIs studies of insoluble, inaccessible, or low-abundance structures including chromatin, centrosomes, cell junctions, ER, signaling pathways, and SARS-CoV-2 host cell interactions [7].

Since AP and PL coupled with MS have made substantial contributions to interactomics, integration of both approaches has provided complementary insights into cellular protein interaction networks. We have sought to develop the multiple approaches combined (MAC)-tag system containing the Strep-tag for AP and BirA* for PL into a single construct [8]. Subsequently, cell line generation and protein purification can be carried out with similar protocols [7]. Despite the valuable insights provided by the MAC-tag system [9, 7, 10], it still suffers from some drawbacks. First of all, BirA* of MAC-tag is a 35.4 kDa protein which may hamper the proper localization/conformation of the corresponding bait proteins [11]. Furthermore, 18 h for maximum protein labeling with the MAC-tag approach makes it less than ideal for kinase enrichment or competition experiments [8, 5]. Additionally, BirA* of MAC-tag contains a DNA-binding domain [12], which can complicate the analysis of chromatin-associated bait proteins and lead to artifactual non-specific interacting proteins.

Recently developed PL tags such as BioID2 (26.6 kDa) [6], miniTurbo ID (28 kDa) [13], and UltraID (19.7 kDa) [14] are smaller versions of BirA* with faster labeling efficiency. We therefore incorporated these tags into our system to replace the BirA* of MAC-tag and develop MAC2-tag (Strep&BioID2) and MAC3-tag (strep&UltraID) to overcome potential limitations of the first-generation MAC-tag system. However, biotin supplementation in the culture medium is necessary to prevent a relative biotin deficiency, and alternative versions of BirA*, BioID2, and UltraID are more biotin-sensitive that can utilize the trace amount of endogenous biotin in the culture medium. As a result, they may prevent biotin-dependent cellular processes (e.g., nuclear processes) from utilizing the biotin in culture media. Therefore, we comparatively evaluated the practical application of three generations of MAC-tags on affinity purification workflows.

A previously validated bait, cyclin-dependent kinase 8 (CDK8) [15] that locates in the nucleus was used to better characterize the toolbox of the MAC-tag system. MAC/MAC2/MAC3-tagged CDK8 identified the similar number of high confidence interactors (MAC: 31; MAC2: 40; MAC3:34) (Fig. 1a). The earlier study [8] using affinity purification workflow indicated that MAC-tagged CDK8 offers quantification of Mediator complex composition and allows calculation of stoichiometry. Similar results were obtained this time. All MAC/MAC2/MAC3-tagged CDK8 pull-downs ultimately detected all 31 subunits of the Mediator complex with variation in a few subunits (Fig. 1b). However, a more similar and stronger positive correlation of MS results was shown between MAC-tagged and MAC3-tagged CDK8 (Fig. 1c). While using the MAC2-tag, the overall number of detected subunits of the Mediator disproportionately increases (Fig. 1b). Thus, the abundance of subunits cannot reflect the stoichiometry of the Mediator. It can be reasoned that BioID2 of MAC2-tag may lead to unspecific biotinylation of proximity proteins utilizing endogenous biotin during the cell culture stage.

Furthermore, to determine the performance of transient interaction detection with MAC2/MAC3-tag using proximity labeling workflow, previously well-characterized epidermal growth factor receptor (EGFR) was used as bait protein. EGFR can stimulate several signaling pathways: PLC, JAK, SRC, etc., and is associated with cell proliferation, angiogenesis, invasion, metastasis, and inhibition of apoptosis [16]. MAC-tagged and MAC2-tagged EGFR received conventional biotin supplementation for 24 h and MAC3-tagged EGFR cells were only treated with biotin for 10 min. MAC-tagged EGFR was able to detect interactions with a few PLC isoforms (Fig. 2a). MAC2-tagged EGFR obtains better results, showing additional interaction with JAK. MAC3-tagged EGFR shows the best results (Fig. 2a), as most of the known downstream components, PLC, JAK, and SRC, are detected. Besides, using MAC3-tagged EGFR, interactions with several kinases and phosphatases (Fig. 2a) which are usually weak and transient, could also be captured. These results demonstrate the powerful proximity labeling activities of MAC3-tag for the exploration of transient protein interactions.

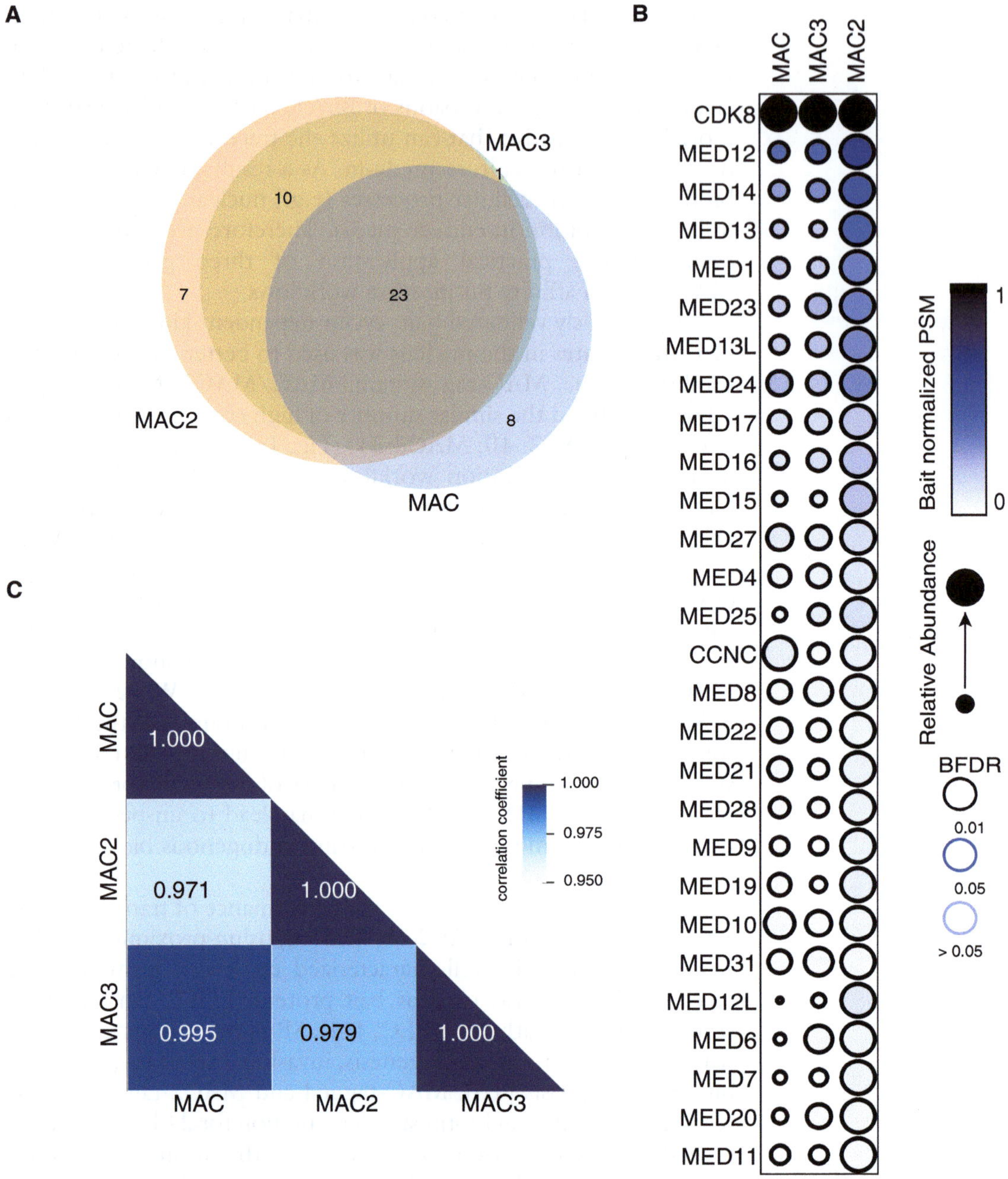

Fig. 1 Application of affinity purification on MAC/MAC2/MAC3-tagged CDK8. (**a**) A Venn diagram compares the number of identified high confidence interactors of MAC/MAC2/MAC3-tagged CDK8 using affinity purification. (**b**) Dot plot visualization of all Mediator complex components obtained by MAC/MAC2/MAC3-tagged CDK8 using affinity purification. Each dot represents the relative quantitative information. (**c**) There is a strong correlation between MAC-tag and MAC3-tag using affinity purification (Pearson's correlation, p-value <0.01, t-test)

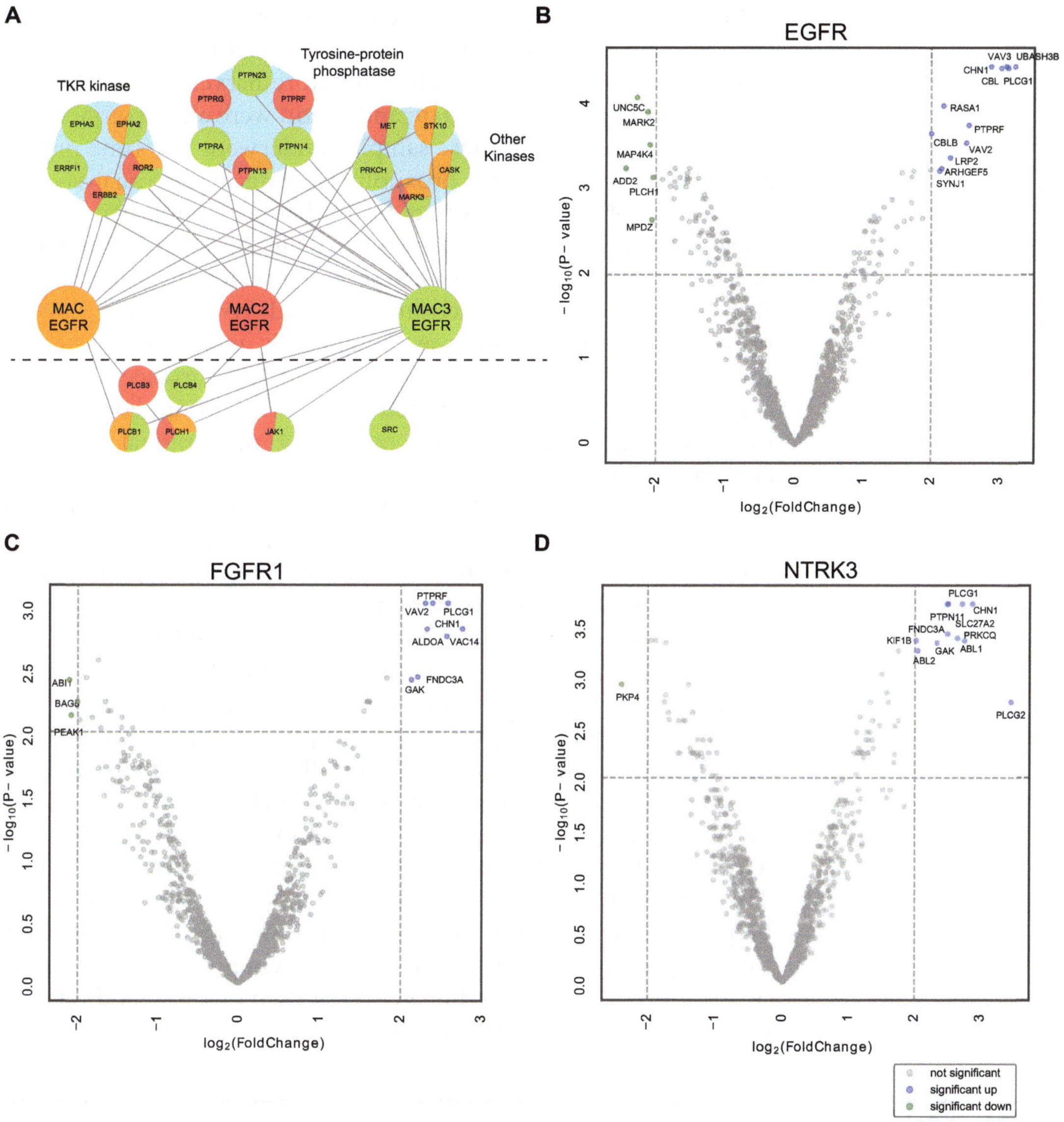

Fig. 2 Differentially identified proteins of MAC/MAC2/MAC3-tagged receptor tyrosine kinases. (**a**) The simplified MAC/MAC2/MAC3-tagged EGFR (MAC-tag: orange; MAC2-tag: pink; MAC3-tag: green) interaction network with a few key high confidence interactors. Interactors are shown and colored according to the specific MAC/MAC2/MAC3-tag. (**b–d**) Volcano plots of differentially identified proteins of MAC3-tagged receptor tyrosine kinases exposed to pervanadate (B: EGFR; C: FGFR1; D: NTRK3). The blue and green dots labeled with protein names represent those considered significantly differentially expressed for the pervanadate treated group compared with the control group (fold change >4 (up or down) and *P*-value <0.01)

Thus, we continued to examine the interaction dynamics on a short timescale of activation (5 min prior to biotin addition). This was achieved by using an extremely potent phosphotyrosine phosphatase inhibitor, pervanadate, to provoke the rapid kinase pathway activation [17]. MAC3-tagged receptor tyrosine kinases (EGFR,

NTRK3, and FGFR1) stable cells were first under tetracycline-mediated protein expression for 18 h. The control samples were only treated with biotin for 10 min before harvesting. For kinases activation samples, pervanadate and biotin were added sequentially at a 5-min interval. All these samples underwent the same protein purification procedures for proximity labeling and MS analysis. The comparison of the interactomics results has been published in detail earlier [9] and is now summarized here (Fig. 2b–d). To summarize, pervanadate treatment reinforced several interactions. For example, PLCG1, which is mainly activated through tyrosine phosphorylation by receptor kinases, was significantly increased. Moreover, MAC3-tagged tyrosine kinases presented less variance and more consistent results between treated and control groups compared to the same experiments carried out using the original MAC-tag approaches [9].

These examples benchmark the utility of our MAC-tag system. MAC2-tag system is similar to MAC-tag in its ability to promiscuously biotinylate proximate proteins but indicates an underperformance for AP workflow. MAC3-tag is fundamentally comparable to the MAC-tag system in the integration of AP and PL workflows, which also expands the application for the detection of subtle changes or transient interactions over a short period of time.

In this chapter, we present a thorough MAC-tag workflow built upon the complementarity of AP and PL with the expansion application of MAC3-tag on kinases activation (Fig. 3). First, we focus on how to prepare the cloning gene of interest (GOI) fused with MAC/MAC2/MAC3-tag for the generation of stable cell lines. Then we move on to the stable cell line selection and sample preparation for different purposes and downstream analysis. Continuously, cell pellets are subjected to protein purification using AP or PL approaches. Purified proteins are then digested and retained peptides are ready for MS analysis.

2 Materials

2.1 Cloning

1. Gateway entry vector of GOI can be requested through Addgene (https://www.addgene.org/) and Genescript (https://www.genscript.com/orf-cdna-clones.html).
2. Gateway destination vector (MAC-tag-C/N; MAC2-tag-C/N; MAC3-tag-C/N). These can be requested through Addgene (https://www.addgene.org/Markku_Varjosalo/).
3. Gateway LR Clonase Enzyme Mix (Life Technologies, Cat # 11791043).
4. Competent cells for bacterial transformation.
5. SOC and LB media.

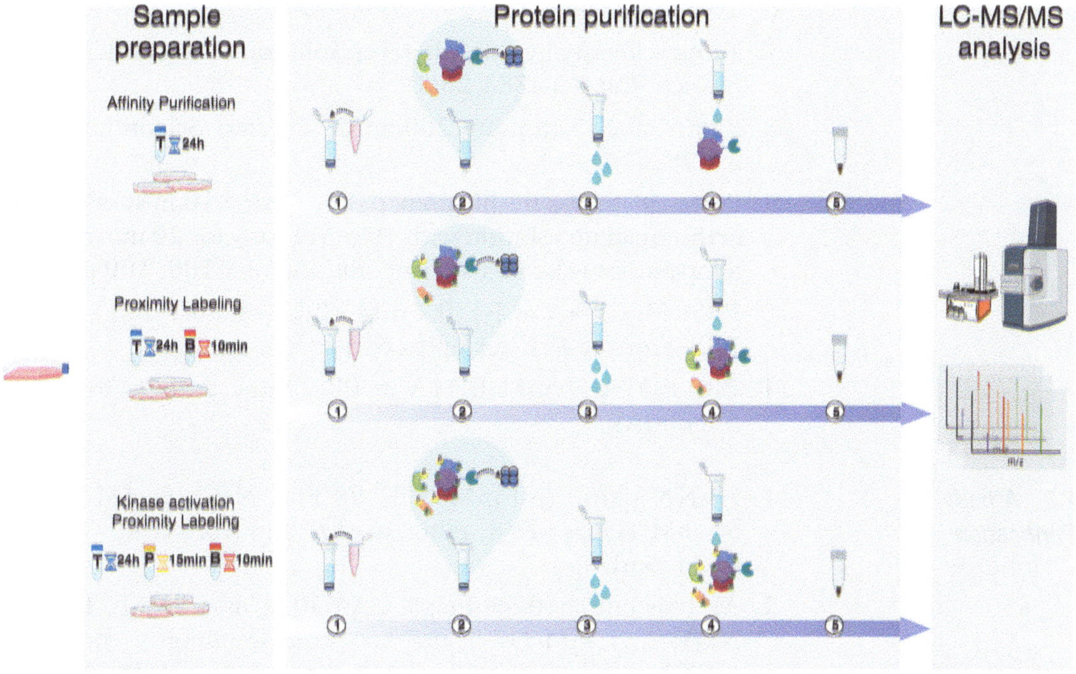

Fig. 3 Schematic overview of the MAC-tag-based workflow. The main procedures for MAC-tag approach include sample preparation, protein pull-down, and LC-MS/MS analysis. Gateway-compatible MAC/MAC2/MAC3-tagged gene of interest vector can be used to generate transgenic stably and inducibly expressing isogenic cell lines. For affinity purification (upper), tetracycline is added for 24 h. For proximity labeling (middle), extra biotin (50 μM) is added besides tetracycline. The incubation time of biotin depends on the generation of the MAC-tag (MAC-tag: ≥ 18 h, MAC2-tag: ≥ 18 h, MAC3-tag: 10 min). For kinase activation, which is only applied to MAC3-tag (lower), tetracycline is added for 24 h. Pervanadate and biotin are added sequentially with a 5 min gap before the harvesting within 15 min. Cells are collected for protein purification, including (1) cell lysis; (2) sample loading; (3) washing; (4) elution; (5) protein sample processing to peptides and desalting. The three pipelines differ in lysis buffer for protein extraction. For AP, mild lysis buffer is used, and for PL and kinase activation, harsh lysis is used. Finally, the reconstituted peptides are loaded for LC-MS/MS analysis

6. LB-ampicillin (100 μg/mL) agar plates.
7. Mini kit for plasmid DNA.
8. pMEI restriction enzyme (NEB, Cat # R0560L).
9. Standard DNA agarose gel system (for clone verification).

2.2 Generation of Stable Cell Lines

1. Flp-In T-REx HEK293 cells (Thermo Fisher, Cat #R78007).
2. pOG44 Flp-Recombinase Expression Vector (Life Technologies, Cat # V600520).
3. Standard growth media: DMEM supplemented with 10% Fetal Bovine Serum, 100 U/mL penicillin/streptomycin.
4. FuGENE® 6 Transfection Reagent (Promega, Cat # E2691).
5. Hygromycin B (Life Technologies, Cat # 10687010).

6. Trypsin-EDTA.
7. Tetracycline hydrochloride stock solution (2 mg/mL; Sigma-Aldrich, Cat # T3383-25G).
8. Biotin stock solution (200 mM; Thermo Scientific, Cat # 29129).
9. Pervanadate was freshly prepared by mixing 10 mM of sodium orthovanadate solution with 10 mM H_2O_2 for 20 min at room temperature (Use at a starting dilution of 1:100; 100 μM).
10. PBS-($MgCl_2$ + $CaCl_2$) (0.1 mM $MgCl_2$ and 0.1 mM $CaCl_2$ in PBS. Store at 4 °C for up to 6 months).
11. PBS-EDTA (1 mM EDTA in PBS. Store at 4 °C for up to 6 months).

2.3 Affinity Purification

1. HENN buffer (50 mM NaF, 90 mM NaCl, 5 mM EDTA, 50 mM Hepes pH 8; Filter-sterilize and store at 4 °C for up to 2 months).
2. AP lysis buffer (0.5% Igepal CA-630, 1 mM PMSF, 1.5 mM Na_3VO_4, 1 mM DTT, and 1 × protease inhibitor cocktail in HENN buffer. Prepare freshly before use) (*see* **Note 1**).
3. Wash buffer 1 (0.5% Igepal CA-630, 1 mM PMSF, 1.5 mM Na_3VO_4, 1 mM DTT, and 1 × protease inhibitor cocktail in HENN buffer; it should be prepared freshly before use).
4. Wash buffer 2 (HENN buffer).
5. Elution buffer (0.5 mM biotin in HENN buffer; it should be prepared freshly before use).
6. Eppendorf LoBind tube (2-mL, Eppendorf, Cat # Z666513).
7. Strep-Tactin® Sepharose® 50% (vol/vol) suspension (IBA life sciences, Cat # 2-1201-010).
8. Bio-Spin® Chromatography Columns (Bio-Rad, Cat # 732-6008).

2.4 Proximity Labeling

1. PL lysis buffer (0.1% SDS, 0.5% Igepal CA-630, 1 mM PMSF, 1.5 mM Na_3VO_4, 1 mM DTT, and 1 × protease inhibitor cocktail in HENN buffer; it should be prepared freshly before use).
2. Wash buffer 1 (same as AP).
3. Wash buffer 2 (same as AP).
4. Elution buffer (same as AP).

2.5 Sample Preparation for Mass Spectrometry

1. Tris (2-carboxyethyl) phosphine hydrochloride (TCEP; Sigma-Aldrich, Cat # C4706-2G).
2. Iodoacetamide (IAA; Sigma-Aldrich, Cat # I1149-25G).

3. Sequencing Grade Trypsin (Promega, Cat # V5113).
4. Peptide Desalting Spin Columns (Choose the right peptide clean-up products for your application).
5. MS sample resuspension buffer (0.1% trifluoroacetic acid and 1% acetonitrile in H_2O. It should be prepared freshly before use. All reagents must be HPLC or MS grade.).

3 Methods

3.1 Generation of MAC/MAC2/MAC3-Tagged Gene of Interest Using Gateway Cloning

Gateway vectors (entry clones) contain modified versions of the att recombination sites and allow rapid transfer of the desired DNA sequence into the expression vector (MAC/MAC2/MAC3-tagged destination clones). The open reading frame (ORF) for the gene/protein of interest can be obtained by PCR or from a public ORFeome collection library. A successful fusion construct can be confirmed using the following protocols:

1. For the Gateway LR reaction, combine the following: (a) Entry clone (for LR reaction) (1 μL of a 300 ng/μL stock, less can be used with compromised LR efficiency); (b) Destination vector (1 μL of a 150 ng/μL stock); (c) TE pH 8.0 up to 5 μL.
2. Thaw LR Clonase on ice. Then add 1 μL of LR Clonase to the reaction (total volume of 5 μL), Vortex for a few seconds to mix and flash spin.
3. Incubate for at least 1 h at 25 °C (*see* **Note 2**).
4. Add 1 μL of Proteinase K solution to the reaction and incubate for 10 min at 37 °C.
5. Add 5 μL of the LR reaction to 25 μL of competent *E. coli* for transformation and plate onto LB-agar plate containing 100 μg/mL ampicillin. Incubate plates at 37 °C overnight (*see* **Note 3**).
6. Pick a single colony and expand in 5 mL LB (ampicillin) overnight in a shaking incubator (220–250 r.p.m.) at 37 °C. Use the overnight culture for plasmid DNA miniprep according to the manufacturer's instructions. A diagnostic restriction digest with pMEI restriction enzyme and DNA-gel electrophoresis helps to identify the correct GOI insert (*see* **Note 4**).

3.2 Generation of Stable Cell Lines Expressing the MAC/MAC2/MAC3-Tagged POI

To avoid cell-to-cell variation of expression levels by transient transfection, we employ the Flp-In™ T-REx™ 293 cell line to rapidly generate an isogenic and inducible stable cell line of MAC/MAC2/MAC3-tagged POI. Selection of integrated cells using Hygromycin B, followed by the establishment of clonal cell lines with a validated fusion construct expression.

1. On Day 0, plate 3×10^5 cells into 6-well plates (*see* **Note 5**).
2. On Day 1, the cells should have 60–80% confluence by visual inspection. Prepare a plasmid mix for each well with a ratio of 9:1 (wt/wt) for pOG44:MAC/MAC2/MAC3-tagged destination vectors. FuGENE 6 (or any other preferred transfection reagents) is used for co-transfection according to the reagent manufacturers' protocols.
3. On Day 2, replace media with fresh complete media.
4. On Day 3, split cells in each well into a separate T175 flask containing a total volume of 35 mL complete cell culture medium.
5. On Day 4, replace media with complete media supplemented with 100 μg/mL hygromycin B to begin the selection of integrated cells.
6. The selection process usually takes approximately 2 weeks, changing selection media every 3–5 days as required to remove dead cells and to allow resistant cell clones to grow (*see* **Note 6**).
7. When the resistant cell clones have grown to a near confluent T175 flask, split the cells with trypsin from each T175 flask into 3–5 × 15 cm Ø dishes (*see* **Note 7**).
8. Expand and grow the cell lines expressing the MAC/MAC2/MAC3-tagged protein of interest in 4 × 15 cm Ø dishes until they reach 80% confluency. Split the 4 × 15 cm Ø dishes into 20 × 15 cm Ø dishes (ratio 1:5), containing a total volume of 30 mL of complete cell culture medium per dish. Cells collected from 4–5 culture plates can be used as one biological replicate. As a minimum requirement, researchers should have two biological replicates for each condition/treatment.

3.3 Generation of Cell Pellets for Protein Purification

Cells are treated for 24 h with tetracycline (2 μg/mL) to induce expression of the MAC-tagged POI. In some cases, fluorescence microscopy can be used to verify the expression and localization of the MAC-tagged bait protein. For proximity labeling, biotin (50 μM) is needed. The biotin treatment time needs to be adjusted depending on the generation of the MAC-tag (MAC-tag: ≥ 18 h, MAC2-tag: ≥ 18 h, MAC3-tag: 10 min). To decipher physiological events regulated by reversible protein phosphorylation, phosphatase inhibitor(s) should be added. Several phosphatase inhibitors are available, and we chose pervanadate, a powerful inhibitor of protein tyrosine phosphatases (PTP) for our research. Pervanadate treatment of cells results in rapid morphological changes within 2 min [18]. Therefore, the cells expressing MAC3-tagged receptor tyrosine kinases are treated with pervanadate for 5 min prior to biotin addition, and cells are then harvested 10 min later. The same approach can be adopted for projects that involve the characterization of specific signaling pathways activation; the exact ligand stimuli can be used before the addition of biotin to capture the dynamics within protein interaction networks.

3.3.1 For Affinity Purification

When the culture plates reach ~80% confluency (normally in ~2–3 d after splitting), induce each plate with 2 μg/mL tetracycline for AP purification.

3.3.2 For Proximity Labeling

When the culture plates reach ~80% confluence induce each plate with 2 μg/mL tetracycline and 50 μM biotin. The incubation time of biotin needs to be adjusted depending on the generation of the MAC-tag (MAC-tag: ≥ 18 h, MAC2-tag: ≥ 18 h, MAC3-tag: 10 min).

3.3.3 For Proximity Labeling with Activation of Kinases Using Pervanadate

When cells expressing MAC3-tagged trypsin kinase reach ~80% confluency on the culture dish, induce each plate with 2 μg/mL tetracycline for protein expression. Pervanadate was added for 15 min before harvesting with a final concentration of 100 μM. After 5 min exposure of cells to pervanadate, 50 μM biotin was added for 10 min prior to sample collection.

1. Remove media of one biological sample (cells from 4–5 culture dishes) into waste. Add 5 mL of PBS-($MgCl_2$ + $CaCl_2$) to the side of each dish, and slowly tilt the dish to gently wash cells. Discard the PBS and place the dishes on ice. The best practice is to collect construct by construct.
2. Add 10 mL ice-cold PBS-EDTA to one of the empty dishes. Pipette up and down several times along the bottom of the plate to detach and harvest the cells. Transfer the cell suspension from the dish to the next dish. Repeat the procedure to harvest cells from 4–5 dishes and collect the suspension into a 15-mL centrifuge tube.
3. Centrifuge cells at a maximum of 1200 g for 5 min at 4 °C. Carefully remove all visible supernatant and snap freeze the pellets using liquid nitrogen, then transfer them to −70 °C to −90 °C freezer for longer-term storage (*see* **Note 8**).

3.4 Protein Purification

The protocol of the MAC-tag approach has been optimized over a few years with different projects [19, 8, 10, 9, 20, 7]. The same protocol should be easily adapted for MAC2-tag and MAC3-tag. Igepal CA-630 as a non-denaturing detergent in lysis buffer is suitable for solubilization and purification of protein complexes in most cases. However, we notice that the n-dodecyl-β-D-maltoside (DDM, 0.5–1% final concentration) performs slightly better for solubilizing membrane and hydrophobic proteins. Researchers can choose the appropriate detergent for their purpose based on the study. It is recommended to perform protein purification in the cold cabinet (4 °C). However, the elution step is preferred to be performed at room temperature.

3.4.1 For Affinity Purification

Resuspend the cell pellet in 3 mL of ice-cold affinity purification-Lysis buffer and incubate it on ice for 15–20 min. During the incubation period, vortex the tube vigorously for 5 sec every 2–5 min. Repeat the vortexing process multiple times to ensure that the entire pellet is resuspended.

3.4.2 For Proximity Labeling

Resuspend the cell pellet in 3 mL of ice-cold proximity labeling-Lysis buffer and incubate it on ice for 5–10 min. Vortex the tube vigorously to ensure that the entire pellet is resuspended. Sonicate the resuspension in an ice bath for 3 cycles, with 3 min of continuous sonication followed by a 5 min pause.

1. Transfer cell lysates into two 2-mL tubes. Centrifuge the samples at 16,000 g for 15 min at 4 °C.
2. Pour the supernatants into a new 2-mL tube and centrifuge again at 16,000 g for 10 min at 4 °C. For quality control, 1% of the supernatants (20 μL) can be reserved for Western blot analysis.
3. During the second round of centrifugation, set up the purification column. Add 400 μL of Strep-Tactin®-bead-slurry to each Bio-Spin® column and equilibrate the Strep-Tactin®-beads once with 1 mL of Lysis buffer.
4. Gently apply cleared supernatants from **step 3** to the rim of the columns without damaging the packed resin. Let liquid escape by gravity and allow the column to drain completely until the entire sample has been loaded.
5. Use 1 mL of ice-cold pull-down wash buffer to wash the resin three times. Let the wash buffer drain completely before moving on to the next step (*see* **Note 9**).
6. Use 1 mL of HENN buffer to wash the resin four times. Allow the wash buffer to flow through the tube's inner rim to ensure that all of the detergents in the lysis buffer and wash buffer have been washed away (*see* **Note 10**).
7. Tightly fit the end cap onto the column tip, and add 200–300 μL of pull-down elution buffer. Vortex the column gently to resuspend the resin in elution buffer and incubate the column at room temperature for 5 min.
8. Place the columns in clean 2-mL LoBind tubes after removing the end caps. Let the elution buffer from the column drain into the LoBind tube completely. To maximize yields, repeat the elution cycle two more times and combine the three elution fractions. For quality control, 1% of the supernatants (10 μL) can be reserved for Western blot analysis.

3.5 Sample Preparation for Mass Spectrometry

Many steps are involved in the strategies used to prepare proteins or more complex proteomic samples for MS analysis. Key steps in bottom-up proteomics approaches include proteolytic cleavage of proteins into short peptide fragments by proteases, and cleanup (desalting) of the final peptide mixture prior to MS analysis. A chemically modified trypsin that selectively cleaves peptide bonds C-terminal to lysine and arginine residues is the most commonly used enzyme for protein digestion. It has the distinct advantage of generating short doubly or triply charged peptides that are water soluble, well separated by liquid chromatography (LC), and well ionized by electrospray ionization (ESI). We usually carry out an extra desalting step on tryptic peptides using pipette tips or spin columns before LC-MS analysis. However, depending on the following LC setup, desalting is not always required before LC-MS analysis.

1. Use pH paper to determine the pH of the elute (around pH 8). Adjust the pH with a 200 mM NH_4HCO_3 solution if necessary.
2. To reduce disulfide bonds, TECP is added to a final concentration of 5 mM. Incubate for 20 min at 37 °C with shaking (*see* **Note 11**).
3. Allow the sample to cool to room temperature before adding freshly prepared IAA to a final concentration of 10 mM for alkylation. Incubate at room temperature for 20 min in the dark.
4. Add 1.5 μg of trypsin (Trypsin:Protein ratio of 1:100–1:20 (wt/wt)) to the solution. Incubate at 37 °C for at least 2 h with shaking (*see* **Note 12**).
5. The following day, use 10% TFA to adjust the pH until the sample is acidic (pH 3). The sample can be directly analyzed by mass spectrometry or extra desalting of the sample can be performed using zip tips or spin columns according to the manufacturers' recommendation.
6. In case of following extra desalting steps, recover the samples in 30 μL of MS sample resuspension buffer and place them in a sonicating water bath at room temperature for 15 min to detach the peptides bound to the plastic. The sample is ready for LC-MS analysis.

3.6 Mass Spectrometry and Data Analysis

The mass spectrometric analysis uses a standard LC-MS/MS setup, which we have also described elsewhere [21]. Recently, with the integration of the Evosep One liquid chromatography system and the Bruker timsTOF Pro mass spectrometer with parallel accumulation serial fragmentation (PASEF) [22], we have been able to maximize the throughput and analytical depth of samples. Purified peptides were separated with the predefined 60 samples per day method (21 min gradient time) on an 8 cm × 150 μm column with 1.5 μm C18 beads (EV1109, Evosep). Mobile phases A and B are 0.1% formic acid in water and 0.1% formic acid in acetonitrile, respectively [23]. The MS analysis is performed in the positive-ion mode using data-dependent acquisition (DDA) in PASEF mode with 10 PASEF scans per topN acquisition cycle. The carry-over is checked by measuring a blank solution, such as the sample resuspension buffer, after each biological sample runs. The tim-sTOF Pro is calibrated regularly according to the manufacturer's guidelines to maintain the best performance. 4 μL of the peptides (400–800 ng) sample is loaded for standard LC-MS (Thermo Scientific Q-Exactive Orbitrap or SCIEX TripleTOF 6600) analysis. For the highly sensitive MS (Bruker Timstof Pro), 1 μL of the peptides (approximately 200 ng) loaded into an EvoTip (Evosep) is enough to achieve good protein identification.

All data processing and analysis can be done with commercial or open-source software, such as Proteome Discoverer [24] and Max-quant [25], which are the two most commonly used. Protein identifications differ slightly between search engines and typically share more than 80% overlap. For the data generated by timsTOF, we used FragPipe (version 13.0) with MSFragger (version 3.0), Philosopher (version 3.2.7), and IonQuant (version 1.4.4) for protein identification and quantification [26]. Within MSFragger, the following parameters are used: trypsin as an enzyme with a maximum of two missed cleavages; precursor mass tolerance of 50 ppm and fragment tolerance of 20 ppm; acetylation of the protein N terminus, oxidation of methionine, and biotinylation of lysine as variable modifications and carbamidomethylation of cysteine residues set as a fixed modification (other parameters were set as default) [26]. The detailed instruction and user manual can be found in the relevant section of the developer's website (https://msfragger.nesvilab.org/).

Large-scale screening of PPI data requires statistical methods for inferring confidence in protein interactions. For such purpose, we routinely use the Significance Analysis of INTeractome (SAINT) software (http://saint-apms.sourceforge.net/Main.html) [27], which assigns confidence scores of PPIs based on protein-level quantitative data. A simple approach to conducting SAINT analysis and other interaction score techniques without installation is to use the MAC-tag platform (http://www.proteomics.fi/). Our platform includes numerous negative controls that can be used for MAC/MAC2/MAC3-tag workflows in the removal of background/nonspecific prey. However, additional control samples are highly recommended for particular experimental setups. Lastly, we developed an application that merges any number of MSFragger output files, so that a single file can be used for the subsequent analysis. It also generates the three files necessary for the SAINT analysis, regardless of the number of input files, and gives a general visualization of the data. The source code is freely available on GitHub (https://github.com/patmartinezb/msfragger2saint).

4 Notes

1. The pH of the cell lysate should be around pH 7.5–8 for proteins' effective binding to the Strep-Tactin beads.
2. Depending on the reaction efficiency (The larger size of the GOI tends to have low efficiency.), the incubation time can be extended up to 24 h.
3. An efficient LR recombination reaction will usually produce >500 colonies. If the LR reaction does not go straight for transformation, the reaction can be stored at −20 °C for a few days until further use.
4. MAC-tag destination vector has two pMEI recognition sites. The correct vector after digestion should appear as two distinct fragments on an agarose gel: One band represents the backbone of MAC-tag about 5000 bp and the other band shows the insert of GOI plus the size of MAC/MAC2/MAC3-tags.
5. Prepare at least 3 wells for each experiment. One well for untransfected cells will be used as a negative control for transfection and selection; One well for MAC/MAC2/MAC3-tagged GFP as a positive control for transfection and selection; One well for MAC/MAC2/MAC3-tagged GOI.
6. One T175 flask typically can obtain ~10–50 colonies (diameter of ~2–5 mm).
7. Cells from one plate can be used for freezing, and cells from the remaining plates can be divided equally for parallel AP and PL workflows.
8. It is important to take into consideration the density and viability of the cells and ensure minimal disruption during harvesting. A rapid cell harvest is very important to preserve protein activity.
9. Detergents (Igepal CA-630/DDM) in wash buffer 1 will aid in the removal of non-specific resin binding while not affecting specific protein binding to the Strep-Tactin® resin.
10. Wash buffer 2 is used to remove any detergents prior to elution. To improve the purity of the final protein products, this washing procedure might be done several times.
11. TCEP or Dithiothreitol (DTT) both are reducing agents that can effectively cleave disulfide bridges and denature proteins.

12. Depending on the manufacturer of trypsin, incubation times may differ. We recommend incubating the protein samples overnight to ensure complete digestion.

Acknowledgments

We thank all members of the Varjosalo laboratory (https://www2.helsinki.fi/en/researchgroups/molecular-systems-biology), especially Tanja Turunen and Antti Tuhkala for optimization of the protocol. This work is funded by grants from the Academy of Finland (nos. 288475 and 294173), the Sigrid Jusélius Foundation, the Finnish Cancer Foundation, Biocentrum Finland, HiLIFE, and POLS (Norway Grants, no. 2020/37 / K / NZ4 / 02761).

References

1. Gingras AC, Gstaiger M, Raught B et al (2007) Analysis of protein complexes using mass spectrometry. Nat Rev Mol Cell Biol 8(8): 645–654. https://doi.org/10.1038/nrm2208
2. Varjosalo M, Sacco R, Stukalov A et al (2013) Interlaboratory reproducibility of large-scale human protein-complex analysis by standardized AP-MS. Nat Methods 10(4):307–314. https://doi.org/10.1038/nmeth.2400
3. Hein Marco Y, Hubner Nina C, Poser I et al (2015) A human interactome in three quantitative dimensions organized by stoichiometries and abundances. Cell 163(3):712–723. https://doi.org/10.1016/j.cell.2015.09.053
4. Bonetta L (2010) Protein-protein interactions: interactome under construction. Nature 468(7325):851–854. https://doi.org/10.1038/468851a
5. Roux KJ, Kim DI, Raida M et al (2012) A promiscuous biotin ligase fusion protein identifies proximal and interacting proteins in mammalian cells. J Cell Biol 196(6):801–810. https://doi.org/10.1083/jcb.201112098
6. Kim DI, Jensen SC, Noble KA et al (2016) An improved smaller biotin ligase for. BioID Proximity Label 27(8):1188–1196. https://doi.org/10.1091/mbc.E15-12-0844
7. Liu X, Huuskonen S, Laitinen T et al (2021) SARS-CoV-2-host proteome interactions for antiviral drug discovery. Mol Syst Biol 17(11): e10396. https://doi.org/10.15252/msb.202110396
8. Liu X, Salokas K, Tamene F et al (2018) An AP-MS- and BioID-compatible MAC-tag enables comprehensive mapping of protein interactions and subcellular localizations. Nat Commun 9(1):1188. https://doi.org/10.1038/s41467-018-03523-2
9. Salokas K, Liu X, Öhman T et al (2022) Physical and functional interactome atlas of human receptor tyrosine kinases. EMBO Report, p e54041. https://doi.org/10.15252/embr.202154041
10. Göös H, Kinnunen M, Salokas K et al (2022) Human transcription factor protein interaction networks. Nat Commun 13(1):766. https://doi.org/10.1038/s41467-022-28341-5
11. Chojnowski A, Sobota RM, Ong PF et al (2018) 2C-BioID: an advanced two component BioID system for precision mapping of protein interactomes. iScience 10:40–52. https://doi.org/10.1016/j.isci.2018.11.023
12. Trinkle-Mulcahy L (2019) Recent advances in proximity-based labeling methods for interactome mapping. F1000Research 8. https://doi.org/10.12688/f1000research.16903.1
13. Branon TC, Bosch JA, Sanchez AD et al (2018) Efficient proximity labeling in living cells and organisms with TurboID. Nat Biotechnol 36(9):880–887. https://doi.org/10.1038/nbt.4201
14. Zhao X, Bitsch S, Kubitz L et al (2021) ultraID: a compact and efficient enzyme for proximity-dependent biotinylation in living cells. J bioRxiv. 2021.2006.2016.448656. https://doi.org/10.1101/2021.06.16.448656
15. Varjosalo M, Keskitalo S, Van Drogen A et al (2013) The protein interaction landscape of

the human CMGC kinase group. Cell Rep 3(4):1306–1320. https://doi.org/10.1016/j.celrep.2013.03.027

16. Wee P, Wang Z (2017) Epidermal growth factor receptor cell proliferation signaling pathways. Cancers (Basel) 9(5):52. https://doi.org/10.3390/cancers9050052
17. Vecchi M, Rudolph-Owen LA, Brown CL et al (1998) Tyrosine phosphorylation and proteolysis. Pervanadate-induced, metalloprotease-dependent cleavage of the ErbB-4 receptor and amphiregulin. J Biol Chem 273(32): 20589–20595. https://doi.org/10.1074/jbc.273.32.20589
18. Bennett PA, Dixon RJ, Kellie S (1993) The phosphotyrosine phosphatase inhibitor vanadyl hydroperoxide induces morphological alterations, cytoskeletal rearrangements and increased adhesiveness in rat neutrophil leucocytes. J Cell Sci 106(Pt 3):891–901. https://doi.org/10.1242/jcs.106.3.891
19. Hietamäki J, Gregory LC, Ayoub S et al (2020) Loss-of-function variants in TBC1D32 underlie syndromic hypopituitarism. J Clin Endocrinol Metabol 105(6):1748–1758. https://doi.org/10.1210/clinem/dgaa078
20. Yellapragada V, Liu X, Lund C et al (2019) MKRN3 interacts with several proteins implicated in puberty timing but does not influence GNRH1 expression. 10. https://doi.org/10.3389/fendo.2019.00048
21. Liu X, Salokas K, Weldatsadik RG et al (2020) Combined proximity labeling and affinity purification−mass spectrometry workflow for mapping and visualizing protein interaction networks. Nat Protoc 15(10):3182–3211. https://doi.org/10.1038/s41596-020-0365-x
22. Meier F, Brunner A-D, Frank M et al (2020) diaPASEF: parallel accumulation–serial fragmentation combined with data-independent acquisition. Nat Methods 17(12):1229–1236. https://doi.org/10.1038/s41592-020-00998-0
23. Skowronek P, Meier F (2022) High-throughput mass spectrometry-based proteomics with dia-PASEF. Methodmol Biol (Clifton, NJ) 2456:15–27. https://doi.org/10.1007/978-1-0716-2124-0_2
24. Orsburn BC (2021) Proteome discoverer – a community enhanced data processing suite for protein informatics. Proteomes 9(1):15. https://doi.org/10.3390/proteomes9010015
25. Tyanova S, Temu T, Cox J (2016) The MaxQuant computational platform for mass spectrometry-based shotgun proteomics. Nat Protoc 11(12):2301–2319. https://doi.org/10.1038/nprot.2016.136
26. Kong AT, Leprevost FV, Avtonomov DM et al (2017) MSFragger: ultrafast and comprehensive peptide identification in mass spectrometry–based proteomics. Nat Methods 14(5):513–520. https://doi.org/10.1038/nmeth.4256
27. Choi H, Larsen B, Lin ZY et al (2011) SAINT: probabilistic scoring of affinity purification-mass spectrometry data. Nat Methods 8(1): 70–73. https://doi.org/10.1038/nmeth.1541

Chapter 25

Identification and Quantification of Affinity-Purified Proteins with MaxQuant, Followed by the Discrimination of Nonspecific Interactions with the CRAPome Interface

Pey Yee Lee and Teck Yew Low

Abstract

Affinity purification coupled to mass spectrometry (AP-MS) is a powerful method to analyze protein–protein interactions (PPIs). The AP-MS approach provides an unbiased analysis of the entire protein complex and is useful to identify indirect interactors. However, reliable protein identification from the complex AP-MS experiments requires appropriate control of false identifications and rigorous statistical analysis. Another challenge that can arise from AP-MS analysis is to distinguish bona fide interacting proteins from the non-specifically bound endogenous proteins or the "background contaminants" that co-purified by the bait experiments. In this chapter, we will first describe the protocol for performing in-solution trypsinization for the samples from the AP experiment followed by LC-MS/MS analysis. We will then detail the MaxQuant workflow for protein identification and quantification for the PPI data derived from the AP-MS experiment. Finally, we describe the CRAPome interface to process the data by filtering against contaminant lists, score the interactions and visualize the protein interaction networks.

Key words Affinity purification, Mass spectrometry, In-solution digestion, MaxQuant, CRAPome

1 Introduction

Affinity purification coupled to mass spectrometry (AP-MS) is often used for capturing, identifying, and quantifying protein–protein interactions in large scale [1, 2]. In the previous chapter, we describe epitope tag-based affinity purification, where both FLAG- and HA-tags are fused to a protein of interest (bait) to facilitate the purification of a host of interacting protein partners called the preys [3–7]. In the past, such affinity-purified protein complexes were usually identified using immunoblotting. With advances in biological mass spectrometry, these co-purified proteins can be readily identified and quantified in one setting, offering much improved sensitivity, speed, and throughput in a non-biased manner [8].

Shahid Mukhtar (ed.), *Protein-Protein Interactions: Methods and Protocols*, Methods in Molecular Biology, vol. 2690, https://doi.org/10.1007/978-1-0716-3327-4_25,

However, the often-adopted MS-based strategy for the identification of peptides and proteins is not without its caveats. Due to the peptide-centric nature of bottom-up MS, the continuity of protein sequence information is lost. In addition to the incomplete daughter ion series generated by tandem MS, this approach faces challenges in the form of false sequence assignment and protein inference. As such, computational proteomics platforms such as Proteome Discoverer or MaxQuant were developed to address these problems by implementing algorithms to estimate the false discovery rates (FDR) at both peptide and protein levels [9]. In addition, these platforms enable the quantification of peptides and proteins using signal intensities from the precursor ions from MS1, or reporter ions from MS2 and MS3.

Another major challenge inherent to AP-MS stems from the co-purification of "background contaminants," referring to high-abundance proteins that are non-specifically bound to the purification system [10]. The identity and quantity of these background contaminants are highly dependent on several experimental parameters such as the cell/tissue types, the epitope tags, subcellular and protein fractionation methods, affinity resin support, and MS instruments. Although tandem affinity purification tags (TAP-tags) allow multiple stringent wash steps, and this physically reduces non-specific binding, weak or transient interactions are often lost. Hence, this calls for more robust methods to discern bona fide interacting partners from background contaminants. To distinguish background contaminants in AP-MS experiments, negative-control purifications are used. These negative controls typically comprise mock purifications using the same support resin and cell lines. However, they do not contain the fused "bait" proteins, but rather, only epitope tags alone are expressed. These universal negative controls help filter the background contaminants from any bait proteins that are subjected to the same purification scheme.

By quantitative comparison of proteins co-purified by the baits against that of the negative controls, one can discern bona fide interacting proteins that are present in a significantly higher amount in the bait experiments. For instance, quantitative MS proteomics data based on label-free quantification or spectra counting can serve as a basis for further statistical analysis using Student's t-tests or ANOVA [11]. In this chapter, we describe the CRAPome interface which incorporates algorithms such as SAINT, FC-A, and FC-B to score and rank each interaction, as well as to build and visualize protein networks [12–14]. Importantly, the CRAPome platform contains a built-in database that contains a diverse range of negative controls derived from many laboratories. These existing negative-controls can be added to help score experimental AP-MS data to improve confidence.

2 Materials

2.1 Trypsin Digestion

1. Lysis buffer: 8 M urea in 50 mM ammonium bicarbonate, pH 7.5.
2. 50 mM ammonium bicarbonate, pH 7.5.
3. 100 mM dithiothreitol DTT in 50 mM ammonium bicarbonate.
4. 100 mM iodoacetamide in 50 mM ammonium bicarbonate.
5. LysC, MS Grade.
6. Trypsin protease, MS Grade.
7. Trifluoroacetic acid.

2.2 Sep-Pak Cleanup

1. Formic acid.
2. Sep-Pak C18 cartridges.
3. Acetonitrile.
4. Wash solvent: 98:2:0.1%—water: ACN: TFA.
5. Elution solvent: 40:60:0.1%—water: ACN: TFA.
6. Vacuum centrifuge.

2.3 UPLC-MS/MS

1. UPLC-MS/MS system.
2. Trapping column (ReproSil-Pur C18-AQ, 3 μm, Dr. Maisch GmbH, Ammerbuch, Germany: 2 cm × 100 μm ID).
3. Analytical column (Agilent Zorbax SB-C18, 1.8 μm, 40 cm × 50 μm).
4. Buffer A: 0.1% formic acid in water.
5. Buffer B: 0.1% formic acid in 98% acetonitrile.

2.4 MaxQuant

1. MaxQuant software.

2.5 CRAPome

1. CPAPome interface.

3 Methods

3.1 Recovery of Immunoprecipitated Proteins and Trypsin Digestion

1. Ensure each affinity purification (AP) experiment consist of at least 3 biological replicates for statistical power.
2. After incubating protein mixture with the affinity resin, bound proteins can be eluted off the affinity beads with 50 μl lysis buffer (8 M urea in 50 mM ammonium bicarbonate, pH 7.5).
3. Repeat **step 2** twice to ensure complete elution. Pool the eluted proteins.

4. Add 100 mM DTT to each sample to a final concentration of 10 mM.
5. Incubate at room temperature for 30 min to reduce disulfide bonds.
6. Add 100 mM iodoacetamide (freshly prepared) to each solution to a final concentration of 20 mM.
7. Incubate at room temperature for 30 min in the dark.
8. Again, add 100 mM DTT to each solution to quench the excess iodoacetamide (*see* **Note 1**).
9. Add 1:50 ratio of LysC: protein concentration and incubate the sample at room temperature for 3 h.
10. Dilute the sample with 50 mM NH_4HCO_3 four-fold to reduce the concentration of urea to less than 2 M.
11. Add trypsin at a 1:50 ratio and incubate the sample at room temperature overnight (16 h).
12. Stop protease activity by adding TFA to a final concentration of 0.5% by volume and incubate at room temperature for 5 min. Remove the precipitates after centrifugation.
13. Perform desalting with Sep-Pak C18 cartridges or other equivalent solid phase extraction (SPE) procedures for sample cleanup.

3.2 Sep-Pak Cleanup

1. Condition the Sep-Pak C18 cartridge twice with 1 mL 100% ACN (Fig. 1).
2. Wash twice with 1 mL wash solvent (98:2:0.1%—water: can: TFA).
3. Load the sample slowly, collect the flowthrough for later analysis.
4. Wash twice with 1 mL wash solvent.
5. Elute twice with 250 μL elution solvent (40:60:0.1%—water:-ACN: TFA) for at least 10 min.
6. Dry down the sample using a vacuum centrifuge at room temperature until most of the elution evaporates. Be careful not to dry the eluant for too long because this can render the pellet difficult to resuspend. These samples can be kept at −20 °C until LC-MS/MS analysis.

3.3 UPLC-MS/MS

1. For UPLC-MS/MS analysis, reconstitute the desalted peptide digest with 10-20 μl of 5% of formic acid.
2. The UPLC was equipped with a double frit trapping column (ReproSil-Pur C18-AQ, 3 μm, Dr. Maisch GmbH, Ammerbuch, Germany: 2 cm × 100 μm ID) and an analytical column

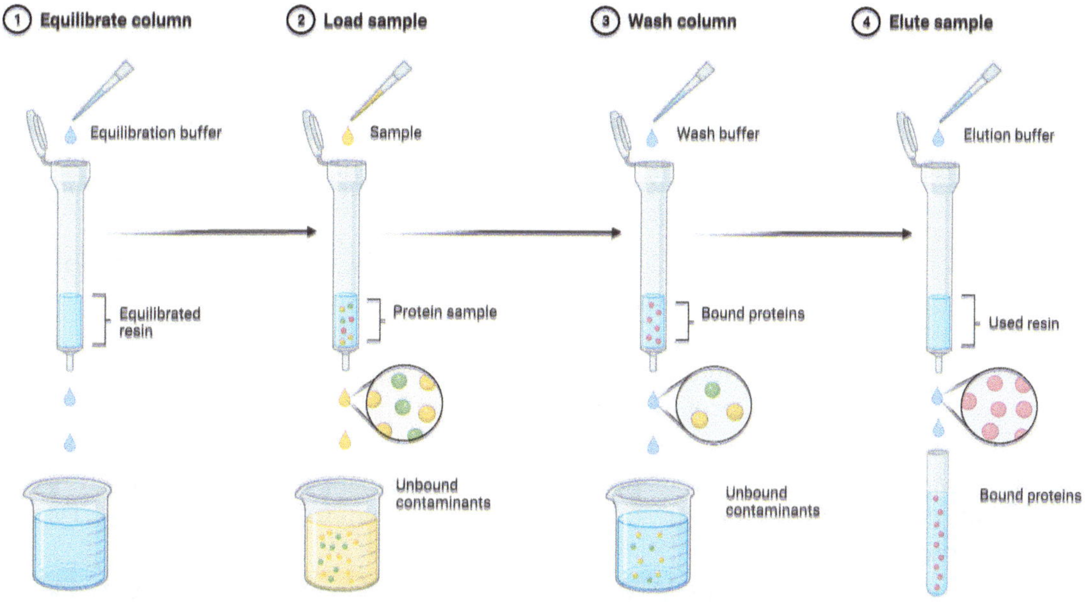

Fig. 1 Schematic representation of Sep-Pak cleanup procedure. The cartridge is first equilibrated to condition the reversed-phase sorbents for sample binding. The sample is then applied to the cartridge, in which proteins are selectively bound and other contaminants will flow through the column. Next, the column is washed to remove all unbound contaminants and finally the captured proteins are eluted from the column using stronger solvent. (Figure was created with BioRender.com)

(Agilent Zorbax SB-C18, 1.8 μm, 40 cm × 50 μm) for online trapping, desalting, and analytical separations.

3. The LC solvents comprise buffer A: 0.1% formic acid in water and buffer B: 0.1% formic acid in 98% acetonitrile. Trapping and desalting were carried out at 5 μL/min for 10 min with 100% buffer A.
4. A 1-hour gradient consisting of: 0-10 min, 0% B at 5.0 μl/min for sample loading; 10.1-40 min, 10% to 40% B at 0.10 μL/min; 40.1-42 min, 40% to 100% B at 0.10 μL/min; 42.1-45 min, 40% to 100% B at 0.10 μL/min; 45.1-59.5 min, 0% B at 0.10 μL/min; 59.6-60.0 min, 0% B at 5.0 μL/min is used for separation of peptides.
5. Eluted peptides are introduced by nano-electrospray into a TripleTOF5600+ System fitted with a nanospray III source (AB SCIEX, Concord, ON) and a coated tip as the emitter (New Objectives, Woburn, MA).
6. Data was acquired using an ion spray voltage of 2.7 kV, curtain gas of 10 psi, nebulizer gas of 10 psi, and an interface heater temperature of 100 °C.
7. The mass spectrometer was operated in information-dependent acquisition mode (IDA), and MS spectra were acquired across the mass range of 350–1250 m/z in high-

resolution mode (>30,000) using 250 ms accumulation time per spectrum.

8. The 20 most abundant precursors ions per cycle at a threshold of 50 counts per second and peptides carrying from 2 up to 5 positive charges were chosen for fragmentation from each MS spectrum with 50 ms minimum accumulation time for each precursor.
9. Dynamic exclusion was set 15 s, and then the precursor was refreshed off the exclusion list.
10. Tandem mass spectra were recorded in high sensitivity mode (resolution >15,000) with rolling collision energy on and with a collision energy spared of 15 V.

3.4 Computational Analysis of LC-MS/MS Data with MaxQuant

1. The analysis of LC-MS/MS data is performed using MaxQuant according to the "minimal workflow" by Tynova et al. [9].
2. Start MaxQuant by double-clicking the MaxQuant.exe file.
3. Go to the "Raw files" tab. Click on the "Load" button to select individual raw data files. Alternatively, click on the "Load folder" button to import an entire folder.
4. The raw files are displayed in a table with columns. You are required to specify the following three parameters for each LC-MS run.
   (a) "Parameter group": Specify the different "Group-specific parameters" for subsets of LC–MS runs.
   (b) "Experiment": Mark samples that belong the same experiment so that the results will be combined in the output table.
   (c) "Fraction": Number the corresponding fractions properly as it is important for the "Matching between runs" feature.
5. Set the type of analysis according to the quantification strategy on the "Type" page. Choose quantification at the MS level (*see* **Note 2**).
   (i) Select "Standard".
   (ii) Set the "Multiplicity" according to the number of labels: Here we select "1" for "Multiplicity" as we do not use any sample labeling.
6. Select the digestion enzyme, i.e., trypsin.
7. Select "Variable modifications" for modifications that may or may not be present on a peptide to be used in the search and transfer them to the right-hand side of the field. Typically, we

select the oxidation of methionine (M) and the acetylation of protein N-terminals.

8. Go to the "Label-free quantification" page and select the "LFQ" option for quantification of unlabeled proteins.
9. Go to the "Sequences" page. Use the "Add file" button to select .fasta files that have been preconfigured with Andromeda and that will be used to generate the peptide search space.
10. Select "Fixed modifications" by moving entries from the box on the left to the box on the right. These are usually sample preparation-specific modifications, which are applied to each occurrence of the specified residue or terminus during database search (e.g., carbamidomethyl (C)).
11. Use the "Number of threads" parameter to set the number of threads to be used by MaxQuant.
12. Press the "Start" button to begin calculations. The "Partial processing" option allows the user to restart the MaxQuant analysis from an intermediate step and therefore to save a large amount of time.
13. Upon the completion of MaxQuant run, open the ProteinGoup.txt result file using a spreadsheet program.
14. Remove the protein identifications comprising common contaminants and decoy hits.
15. Format the list so that it contains protein ID, LFQ quantification, or the spectra count, according to the format required by the CRAPome platform.

3.5 Post-Acquisition Data Analysis with CRAPome

3.5.1 CRAPome 2.0

1. At the CRAPome 2.0 homepage, click on the "Analyze Data" icon, which will lead to the "Analysis Pipeline" page (*see* **Note 3**).

The CRAPome Analysis Pipeline

1. On the "Analysis Pipeline" page, click on the "Start" icon. This leads to the "Upload Data" tab.
2. Before uploading the data file, specify the options for "Organism"; "Experiment Type" and "Quantitation Type" (*see* **Note 4**).
3. Upload the data file, which should be formatted beforehand from the "ProteinGroup.txt" file generated by MaxQuant. Specify the options for "File Type" and "File Format" (*see* **Note 5**).
4. To exclude certain set(s) of data from the analysis, click on "remove" button.

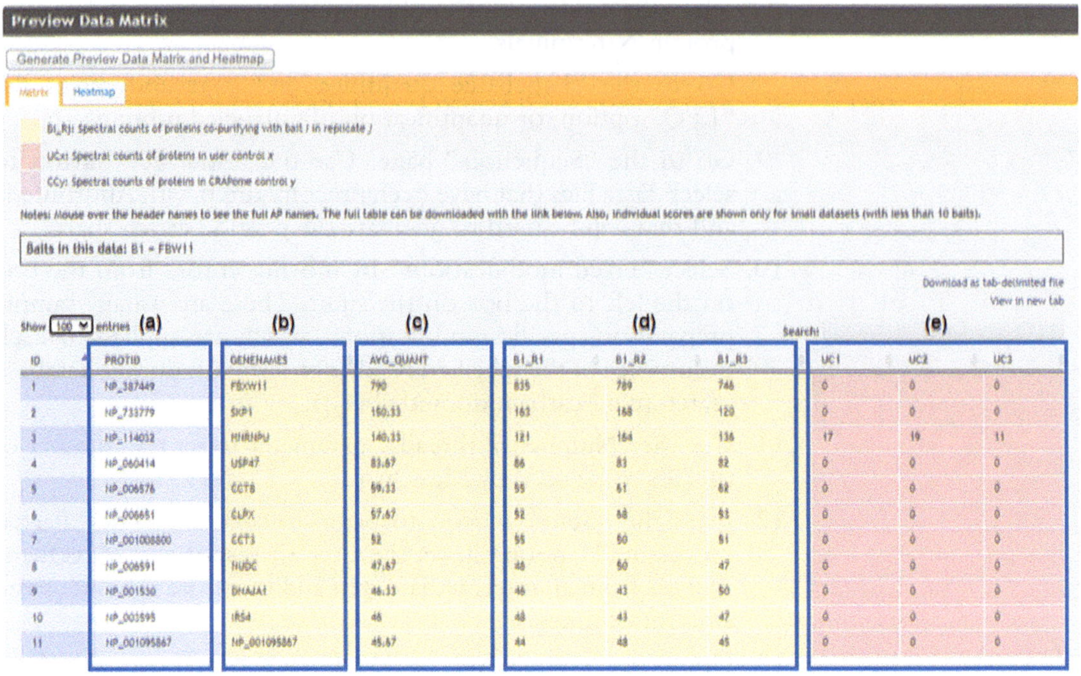

Fig. 2 The "Preview Data Matrix" of CRAPome interface. User can preview the data uploaded in the "Preview Data Matrix." In the example here, the table column shows (**a**) the list of entries in RefSeq protein ID, (**b**) the gene symbols mapped to the entries, (**c**) the averaged spectral counts for each of the entries (**d**) the spectral counts in each of the experiments and (**e**) the spectral counts in each of the negative controls

5. For a quick preview of the data matrix, click on "Preview Data Matrix" (Fig. 2).
6. Proceed to the "Select Controls" tab (Fig. 3). Here, you can select the most appropriate negative controls (deposited in the CRAPome database by other labs) that are most similar to your data using controlled vocabularies (*see* **Note 6**). Selected controls can be saved as a list and reloaded as needed.
7. Next, proceed to the "Score Interactions" tab. Here, Fold Change calculations and SAINT probability scoring can be used to generate ranked lists of bait-prey interactions.
8. Select desired scoring options for Fold Change calculations. Two different Fold Change calculations are generated by default.
9. The first one (FC-A; standard) estimates the background by averaging the spectral counts across the selected controls while the second one (FC-B; stringent) estimates the background by combining the top 3 values for each prey. Combining scores from biological replicates of a bait purification is performed in FC-A by a simple averaging, while FC-B performs a more stringent geometric mean calculation. These parameters are

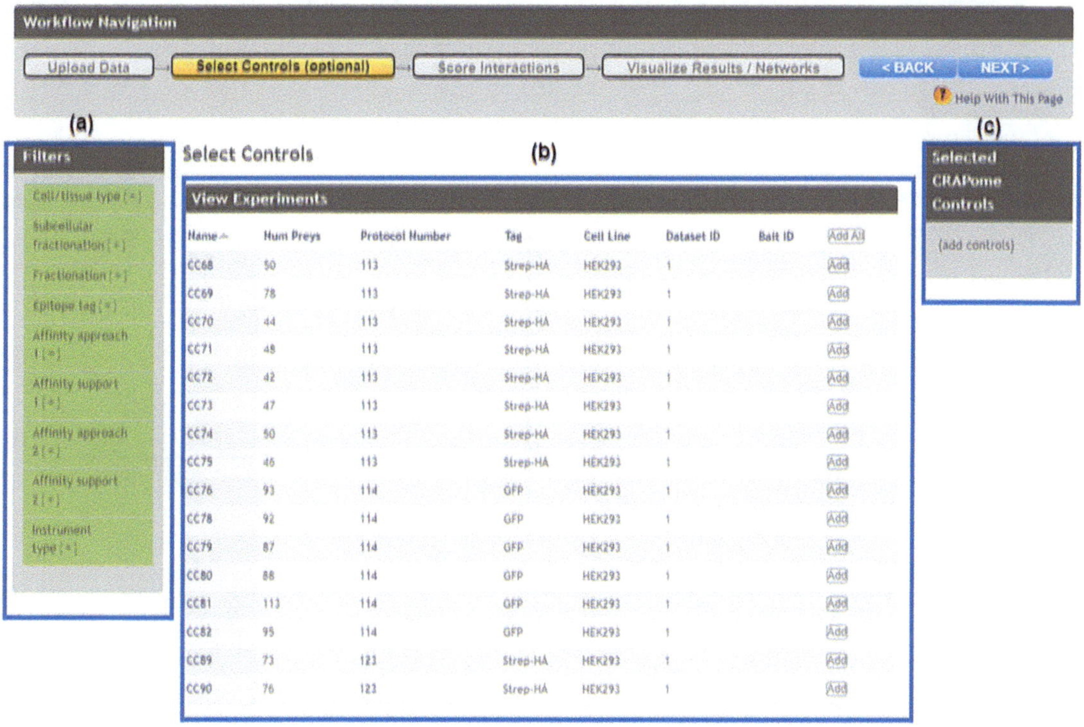

Fig. 3 The "Select Controls" tab of CRAPome interface. (**a**) The "Filters" panel consists of a list of controlled vocabularies to filter the available CRAPome controls. The different options are Cell/tissue type, Subcellular fractionation, Fractionation, Epitope tag, Affinity approach 1, Affinity support 1, Affinity approach 2, Affinity support 2, Instrument type. (**b**) The "View Experiments" panel consists of a table of the controls that passed the selected filters. User can click the control name or the protocol number to display additional information. The controls can be added to the list by clicking the "Add" or "Add all" buttons. (**c**) Added controls will be displayed in the "Selected CRAPome Controls" box

preselected by default but may be modified by the user as required. The user can also specify what set of controls to use (user controls alone or in combination with selected CRAPome controls).

10. You can also select to run SAINT, specify the default options of "lowMode" = 1; "minFold" = 1 and "norm" = 1 (*see* **Note** 7). As with the Fold Change calculations, the user may select which controls to use, and how replicates should be combined. Note that if the number of controls is greater than 10, SAINT generates 10 "virtual controls" by selecting the 10 highest counts for each protein.

11. Once the desired options are selected, press "Run Analysis" (Fig. 4). The new entry will appear at the top of the "Analysis Results" list.

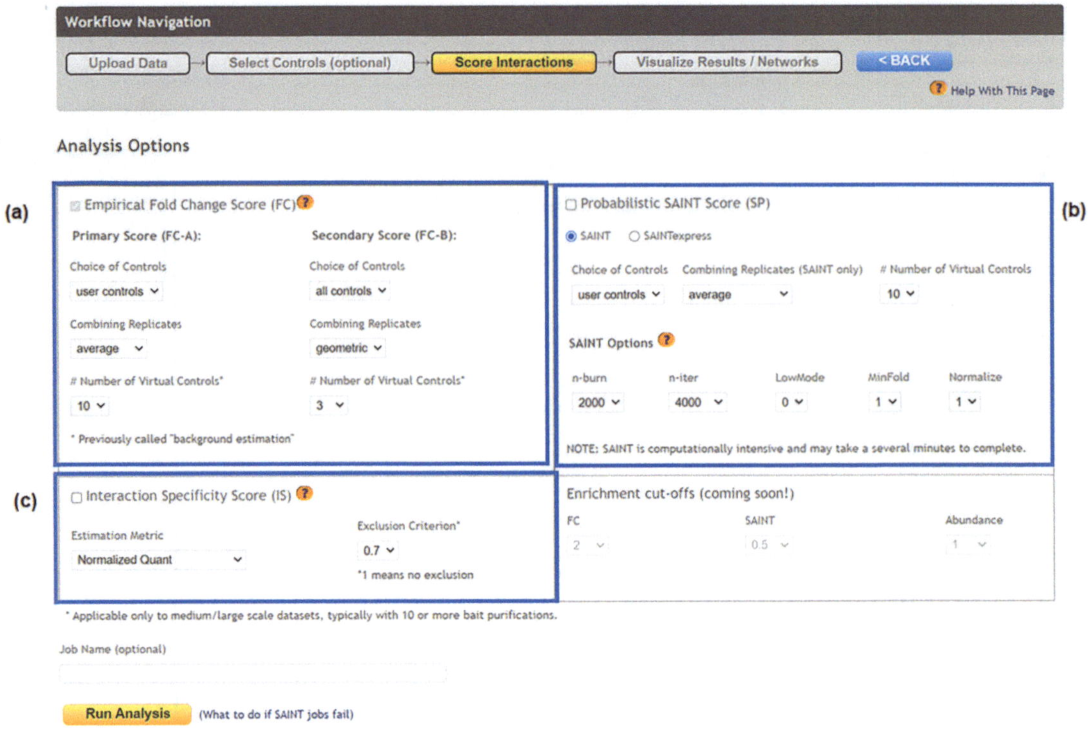

Fig. 4 The "Analysis Options" view of CRAPome interface. (**a**) For the "Empirical Fold Change Score (FC)" option, user can choose either the "Primary Score (FC-A)" or 'Secondary Score (FC-B)'. User can further choose the parameters for "Choice of Controls," "Combining Replicates," and "# Number of Virtual Controls" for the FC score. (**b**) For the "Probabilistic SAINT Score (SP)" option, user can choose to run either "SAINT" or "SAINTexpress" and select the parameters for "Choice of Controls," "Combining Replicates," and "# Number of Virtual Controls." User may employ the preselected default parameters for the "SAINT Options" or modify as required. (**c**) For the "Interaction Specificity Score (IS)" option, user can choose the desired parameter for "Estimation Metric" and "Exclusion Criterion"

12. Since the analysis takes time, you can click on the "refresh" button to check the current status, until "complete" status is shown.
13. Finally, click on "View results" link to view the results. The results can be viewed online in a matrix form or downloaded in a tabular format.

4 Notes

1. DTT is added after alkylation to react with the remaining iodoacetamide to prevent undesirable side reactions that can complicate MS data analysis.
2. There are three main options to distinguish among the MS levels whereby the MS signals generated by peptides can be

quantified. These are the MS1 spectra, as, for instance, with label-free or SILAC-based quantification; in the MS/MS spectra, as with conventional TMT labeling; or in the MS3 spectra, as with the multi-notch approach.

3. This integrated pipeline encompasses uploading of AP-MS data, score interactions, visualize results and build interaction networks. Interactions can be scored in several ways using two complementary criteria: enrichment and specificity. Empirical fold change scores (FC_A/FC_B) and SAINT (SP) use negative controls to compute the enrichment of a prey with respect to the background—real interactions have a high enrichment score. Interaction specificity score (IS) and CompPASS WD-like score (WD) estimate the specificity of a bait-prey interaction in a collection of pull-down experiments. Accordingly, IS and WD scores are applicable only to medium/big data sets, comprising several unrelated baits.
4. The options for data upload:
   (a) Organisms: *H. sapiens*, *M. musculus*, *S. cerevisiae*, *E. coli*, *D. melanogaster*, others.
   (b) Experiment Type: Single step Epitope tag AP-MS, Tandem Epitope tag AP-MS, Proximity Dependent Biotinylation, Endogenous pull-down.
   (c) Quantitation Type: SPC, Intensity.
5. The data should be in either list or matrix format with comma-separated (CSV) or tab-separated columns.
6. The available filters for control selection:
   (a) Cell/tissue type: HeLa, U2OS, PBMC, Jurkat, CEM-T, MRC-5, LS174, BT-549, BJ-5ta fibroblasts.
   (b) Subcellular fractionation: total cell lysate, total lysate +chromatin, nuclear fraction, cytosolic fraction.
   (c) Fractionation: SDS-PAGE, 1D LC-MS, MudPIT, RP-RP, GeLC.
   (d) Epitope tag: FLAG, HA, GFP, TAP, HaloTag, Strep-HA, FLAG-HA, un-transfected.
   (e) Affinity approach 1: M2 anti-FLAG, anti-GFP camel, anti-GFP rabbit, HA-7 anti-HA, HaloLink, IgG, Strep-tactin, SBP, 2xFLAG, anti-GFP mouse, HA.11 anti-HA, Streptavidin, Protein G PhyTip.
   (f) Affinity support 1: agarose, magnetic (dynabead), magnetic (agarose coated), nano-magnetic, microMACS.
   (g) Affinity approach 2: M2 anti-FLAG, anti-GFP camel, calmodulin, HA-7 anti-HA, 2xHA, Streptavidin PhyTip.

(h) Affinity support 2: --, agarose, magnetic bead (dynabead), magnetic beads, agarose coated.

(i) Instrument type: LTQ, LCQ, LTQ-FT, LTQ-Orbitrap, Velos-Orbitrap, 5600 TripleTOF, Q-Exactive, 6600 TripleTOF.

7. SAINT performance and choice of options are described in details by Choi et al. [13].

Acknowledgments

The authors would like to acknowledge the Higher Education Center of Excellence (HICoE) Grant: (JJ-2021-004) awarded by the Ministry of Higher Education of Malaysia to TYL.

References

1. Low TY, Syafruddin SE, Mohtar MA, Vellaichamy A, A Rahman NS, Pung YF et al (2021) Recent progress in mass spectrometry-based strategies for elucidating protein–protein interactions. Cell Mol Life Sci 78:5325-5339
2. Kovanich D, Low TY, Zaccolo M (2023) Using the proteomics toolbox to resolve topology and dynamics of compartmentalized cAMP signaling. Int J Mol Sci 24:4667
3. Low TY, Peng M, Magliozzi R, Mohammed S, Guardavaccaro D, Heck AJ (2014) A systems-wide screen identifies substrates of the SCFβTrCP ubiquitin ligase. Sci Signal 7:rs8
4. Antonova SV, Haffke M, Corradini E, Mikuciunas M, Low TY, Signor L et al (2018) Chaperonin CCT checkpoint function in basal transcription factor TFIID assembly. Nat Struct Mol Biol 25:1119–1127
5. D'Annibale S, Kim J, Magliozzi R, Low TY, Mohammed S, Heck AJ et al (2014) Proteasome-dependent degradation of transcription factor activating enhancer-binding protein 4 (TFAP4) controls mitotic division. J Biol Chem 289:7730–7737
6. Magliozzi R, Low TY, Weijts BG, Cheng T, Spanjaard E, Mohammed S et al (2013) Control of epithelial cell migration and invasion by the IKKβ- and CK1α-mediated degradation of RAPGEF2. Dev Cell 27:574–585
7. Kim J, D'Annibale S, Magliozzi R, Low TY, Jansen P, Shaltiel IA et al (2014) USP17- and SCFβTrCP--regulated degradation of DEC1 controls the DNA damage response. Mol Cell Biol 34:4177–4185
8. Mann M (2008) Can proteomics retire the western blot? J Proteome Res 7:3065
9. Tyanova S, Temu T, Cox J (2016) The MaxQuant computational platform for mass spectrometry-based shotgun proteomics. Nat Protoc 11:2301–2319
10. Dunham WH, Mullin M, Gingras AC (2012) Affinity-purification coupled to mass spectrometry: basic principles and strategies. Proteomics 12:1576–1590
11. Tyanova S, Temu T, Sinitcyn P, Carlson A, Hein MY, Geiger T et al (2016) The Perseus computational platform for comprehensive analysis of (prote)omics data. Nat Methods 13:731–740
12. Mellacheruvu D, Wright Z, Couzens AL, Lambert JP, St-Denis NA, Li T et al (2013) The CRAPome: a contaminant repository for affinity purification-mass spectrometry data. Nat Methods 10:730–736
13. Choi H, Larsen B, Lin ZY, Breitkreutz A, Mellacheruvu D, Fermin D et al (2011) SAINT: Probabilistic scoring of affinity purificationg-mass spectrometry data. Nat Methods 8:70–73
14. Choi H, Glatter T, Gstaiger M, Nesvizhskii AI (2012) SAINT-MS1: Protein-protein interaction scoring using label-free intensity data in affinity purification-mass spectrometry experiments. J Proteome Res 11:2619–2624

Chapter 26

Cataloguing Protein Complexes *In Planta* Using TurboID-Catalyzed Proximity Labeling

Lore Gryffroy, Joren De Ryck, Veronique Jonckheere, Sofie Goormachtig, Alain Goossens, and Petra Van Damme

Abstract

Mapping protein–protein interactions is crucial to understand protein function. Recent advances in proximity-dependent biotinylation (BioID) coupled to mass spectrometry (MS) allow the characterization of protein complexes in diverse plant models. Here, we describe the use of BioID in hairy root cultures of tomato and provide detailed information on how to analyze the data obtained by MS.

Key words Interactomics, Mass spectrometry, Protein, Protein interactions, Proximity-dependent biotinylation, Solanaceae hairy root cultures, TurboID

1 Introduction

Various interactomics technologies have been developed to obtain a comprehensive atlas of the protein–protein interaction (PPI) landscape in plants. The most commonly used high-throughput screening technologies to identify novel plant PPIs are yeast-two-hybrid library (Y2H) screening, eventually coupled to next-generation sequencing, affinity purification (AP), and enzyme-catalyzed proximity labeling (PL) followed by mass spectrometry (MS) [1–3].

In the case of enzyme-catalyzed PL, the protein of interest (POI) or bait is tagged with a (modified) enzyme capable of covalently modifying bait-proximal proteins. PL has the advantage that labeling can be performed in intact cells or tissues, allowing physiologically relevant protein interactions to be captured even when weak or transient. Due to the covalent labeling of proteins, stringent extraction and wash conditions can be applied, which reduces

Authors Lore Gryffroy and Joren De Ryck have contributed equally to this chapter.

Shahid Mukhtar (ed.), *Protein-Protein Interactions: Methods and Protocols*, Methods in Molecular Biology, vol. 2690, https://doi.org/10.1007/978-1-0716-3327-4_26,

false positives and improves solubilization and subsequent identification of membrane proteins [3].

Two main classes of enzymes have been used for proximity-dependent labeling, i.e., abortive promiscuous biotin ligases (e.g., BioID or BirA* [4], BioID2 [5], TurboID and miniTurboID [6, 7]), and peroxidases (e.g., APEX2 [8]). The original proximity-dependent biotin identification (BioID) method makes use of an engineered variant of the *Escherichia coli* biotin ligase BirA (BirA*) and requires exogenous application of biotin for efficient labeling, biotinylated protein purification (by exploiting the high-affinity biotin/avidin interaction), and subsequent MS-based identification of proteins in close vicinity of the POI in living cells or tissue [4]. Promiscuous biotin ligases catalyze the conversion of biotin to biotinoyl-5′-AMP that reacts with amine groups of accessible lysine residues in a radius of ~10 nm [9]. The major disadvantages of BirA* are its slow kinetics (18–24 h) and high optimum working temperature (37 °C). APEX2, an engineered soybean ascorbate peroxidase, has a much faster labeling time (1 min or less) but requires the use of hydrogen peroxide, which is toxic to living samples [8]. More recently, two promiscuous variants of the wild-type biotin ligase BirA were engineered using yeast display-directed evolution, namely TurboID and miniTurboID with 15 and 13 AA mutations, respectively [6, 7]. As compared to BirA*, TurboID (35 kDa) and miniTurboID (28 kDa) have much faster labeling times (circa. 10 min), lower optimum working temperatures (30 ° C), and require non-toxic and easily deliverable biotin to initiate tagging, making it more suitable for use in plants. While TurboID was shown to be most active, permitting shorter labeling times as well as identification of low abundant proteins, the use of mini-TurboID might be preferred when less background and more controlled labeling are required [6, 7].

Biotinylase-catalyzed PL using the promiscuous labeling enzymes TurboID and miniTurboID has successfully been applied in various plant species (e.g., *Arabidopsis thaliana* and *Nicotiana benthamiana*) for the identification of known and novel plant protein interactors and the mapping of plant protein signaling networks [10–13]. The potential applicability of BioID when making use of the promiscuous labeling enzymes BirA*, BioID2, TurboID, and miniTurboID was previously comparatively analyzed in hairy roots of the model plant *Solanum lycopersicum (tomato)* [13]. *For this*, the promiscuous labeling enzymes were genetically fused to an enhanced green fluorescent protein (eGFP) under control of a strong constitutive cauliflower mosaic virus 35S promoter (p*CaMV35S*), and their (auto-)biotinylation activity was assessed through streptavidin blotting. In line with the reported results in various other plant models [10–13], miniTurboID and TurboID were proven to be the most promiscuous variants as manifold faster labeling kinetics was observed compared to other promiscuous biotin ligases tested (i.e., BioID and BioID2) [13].

In this chapter, we focus on TurboID-catalyzed PL in hairy roots of *S. lycopersicum* and describe a detailed step-by-step protocol from the generation of transformed roots to BioID sample preparation, MS, and data analysis. NINJA (Novel Interactor of JAZ)—an adaptor protein involved in plant jasmonate (JA) signaling that connects the co-repressor TOPLESS (TPL) to the JASMONATE-ZIM DOMAIN (JAZ) repressor proteins—was used as representative POI, viewing its well-defined interactome as previously determined by tandem affinity purification (TAP) and Y2H-screening [14]. To overcome false positive proximal partner assignments, relevant controls are highly desirable, and bait controls can be selected dependent on the localization of the POI under investigation. Thanks to the low background levels of biotinylation *in planta*, the addition of biotin allows for induced biotinylation of bait proximal proteins which are subsequently purified by means of streptavidin affinity purification, followed by their MS-based identification (Fig. 1). For the assignment of putative interacting preys, protein abundances between control and bait TurboID setups were compared making use of label-free protein quantification (LFQ). LFQ allows the comparison of a large number of samples and is the preferred method for relative protein quantification in plants due to its low cost and simplicity [15–17]. Finally, the publicly available MaxQuant computational platform combined with the Perseus statistical analysis is described for the rapid, efficient, and user-friendly identification of significantly enriched plant protein and thus putative interactors of the POI [18, 19].

2 Materials

2.1 Bacterial Strains

For cloning purposes, *Escherichia coli* strain DH5α is used. For tomato hairy root transformation, *Agrobacterium rhizogenes* strain ATCC15834 [20] is used.

2.2 Vectors

Baits are cloned into the pDONR221 vector by Gateway BP reaction according to the manufacturer's instructions, thereby creating an entry clone.

1. To generate expression constructs via MultiSite Gateway cloning, entry clones encoding promoter, bait, and TurboID tag are recombined with the multisite Gateway destination vector pKCTAP as described in [21] (Fig. 2). More specifically, the multisite LR Gateway reaction results in translational fusions between the baits and the TurboID labels, of which expression is driven by the XVE promoter (*see* **Note 1**).
2. Generation of the TurboID entry vector has been described in [13]. The pKCTAP destination vector harbors a kanamycin resistance gene (KmR), attR sites, and an eGFP coding sequence (CDS) under the control of the prolD promoter sequences

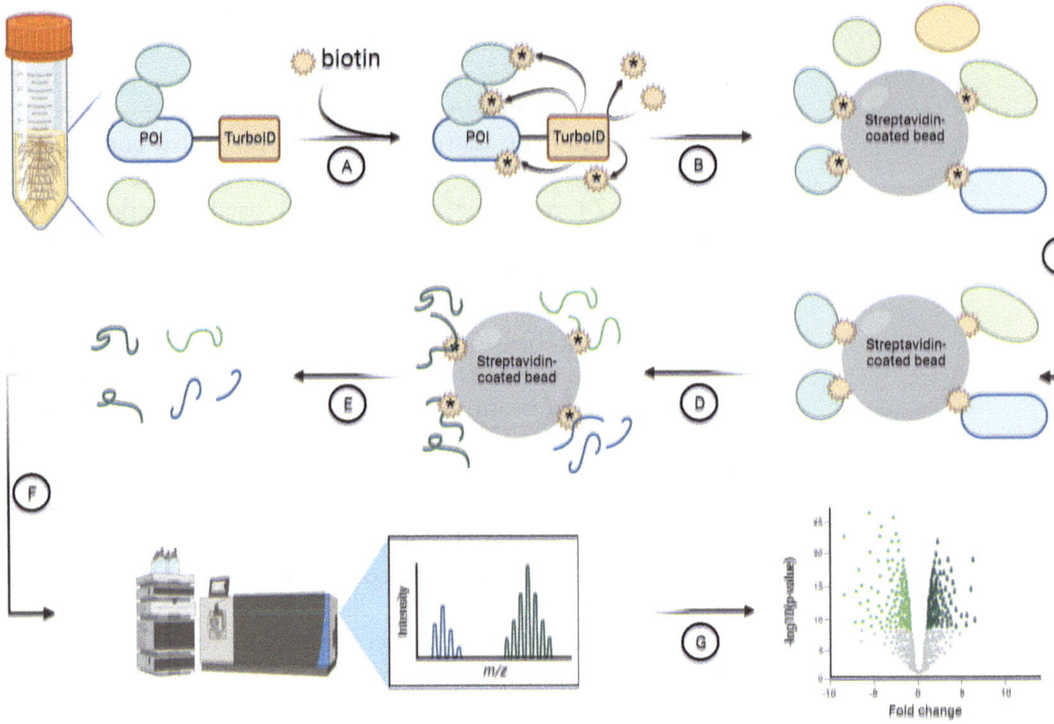

Fig. 1 Principle of proximity-dependent biotin labeling (TurboID) in hairy roots and subsequent sample preparation for mass spectrometry (MS). (**a**) Upon addition of biotin to a hairy root culture expressing the POI-TurboID translational fusion, TurboID will catalyze the activation of biotin by the use of ATP, resulting in reactive biotinyl-5′-AMP (indicated by an asterisk) which will modify proximal proteins by covalent modification of free amines. (**b**) Biotinylated proteins are captured by streptavidin-coated beads and thanks to the strong biotin–streptavidin interaction, (**c**) Denaturing lysis and harsh washing steps can be applied. (**d**) Captured proteins are subjected to on-bead trypsin digestion and, eventually, (**e** and **f**) the supernatant containing the non-biotinylated peptides is collected for LC-MS-analysis. (**g**) Protein identification is performed by MaxQuant database searching, after which a Perseus-guided statistical analysis reveals putative bait-interacting proteins

between the left and right border. The presence of the KmR and expression of eGFP allows for the selection of transformed hairy roots (Fig. 3f). This way, the following expression constructs can be created: XVE-eGFP-TurboID (pKCTAP) and XVE-NINJA-TurboID (pKCTAP). Alternatively, Golden Gibson cloning can be used to generate a binary destination vector encoding two expression cassettes, (1) a translational fusion of the POI with the TurboID proximity label and (2) a fluorescent marker (eGFP) to select for transformed hairy roots (Jacobs and Karimi, unpublished) (Fig. 2, *see* **Note 2**).

2.3 General Materials and Equipment

1. 1.5 mL and 2 mL Eppendorf tubes.
2. Thermomixer for 1.5–2 mL tubes.
3. Vortex mixer.
4. 15 mL centrifuge tubes (sterile).

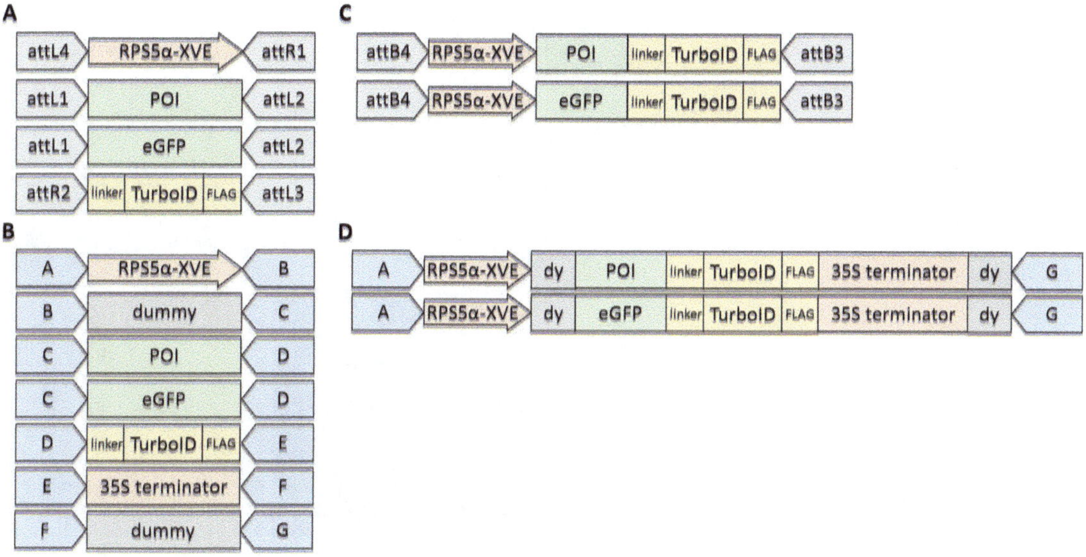

Fig. 2 Overview of available constructs for TurboID-catalyzed PL in plants. Diagrammatic representation of the vectors used in this study and/or available for the community. (**a**) Gateway entry clones and (**b**) Golden Gate building blocks have been used to yield the binary destination vectors and shuttle vectors for Gibson assembly in panel **c** and **d**, respectively. (**c**) Gateway destination vectors encoding a translational fusion of the proteins of interest (POIs) (i.e., NINJA and eGFP) with the TurboID proximity label. (**d**) Shuttle vectors for Gibson assembly, encoding a POI-TurboID fusion with XVE controlled expression and a pCaMV35S driven eGFP fluorescent marker expression cassette which allows for the selection of transformed tomato hairy roots. Shuttle vectors are subsequently recombined in a binary destination vector using Gibson assembly (Jacobs and Karimi, unpublished) (*see* **Note 2**). Gateway att sites are according to Karimi et al. 2007 [21] and GoldenGate overhangs (A-G) according to Lampropoulos et al. 2013 [35]. Linker refers to a 15 bp long sequence (encoding amino acid sequence GGGGS)

5. 50 mL centrifuge tubes (sterile).
6. Centrifuge for 1.5–2 mL tubes.
7. Centrifuge for 15 mL tubes.
8. 70% ethanol.
9. Gloves.
10. Tweezers and scalpel.
11. Micropore^{TM} surgical tape (3M, 1 inch wide, product number 1530).
12. Laminar flow.
13. Growth chamber (24 °C), 16 h/8 h light/darkness photoperiod.
14. Tabletop shaker at 150 rpm.
15. Pipette controller and sterile pipettes (10 mL or 25 mL).

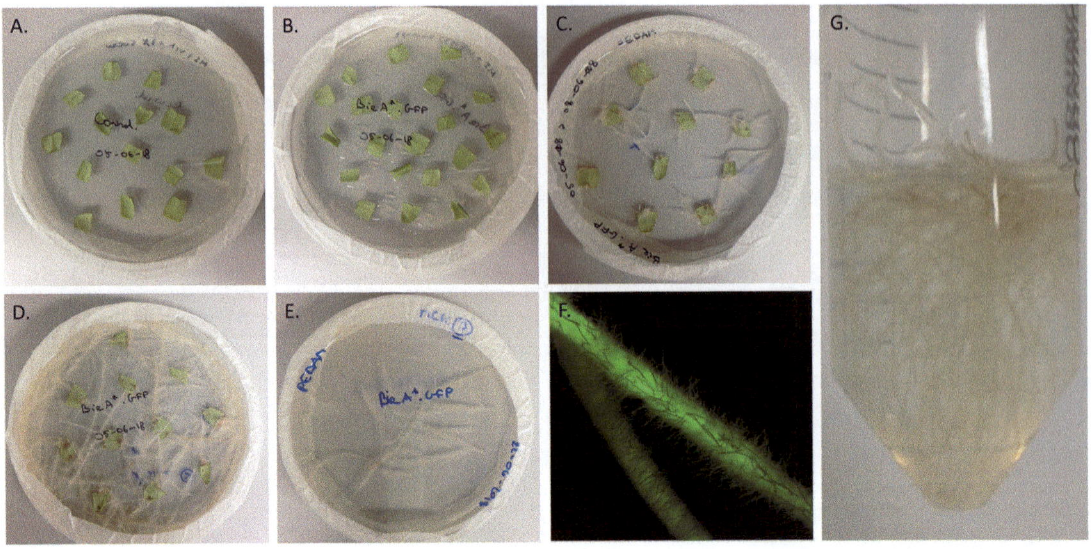

Fig. 3 Tomato hairy roots obtained by means of rhizogenic *Agrobacterium*-mediated transformation. Representative images from a hairy root culture obtained by means of rhizogenic *Agrobacterium*-mediated transformation, transformed with the pCaMV35S::eGFP-BirA* construct (**b–e**) next to a control (non-transformed rhizogenic *Agrobacterium*) setup (**a**) are shown. Images were taken after transformation and 3 days of incubation in the dark (**a** and **b**), as well as 10 days (**c**) and 20 days (**d**) of propagation. Image **e** represents a primary root that was subcloned and propagated for 6 days on a solid MS medium with 3% sucrose added and in the presence of the appropriate antibiotics. (**f**) A transformed hairy root expressing eGFP under the control of prolD (pKCTAP) as observed by fluorescence microscopy. (**g**) A hairy root culture obtained after 3 rounds of sub-selection and growth for 2 weeks in a liquid MS medium with 3% sucrose without the addition of antibiotics

2.4 Hairy Root Transformation

2.4.1 Reagents

1. 3% Sodium hypochlorite (bleach) solution.
2. Magenta box filled with 50 mL of Murashige & Skoog (MS) medium [22] containing 1% sucrose (1% MS) and 1% agar. For 1 L of the medium, add 4.3 g MS (basal salt mixture) (Difco; catalog no. 214530), 0.5 g MES, and 10 g of sucrose. Add 10 g of plant tissue agar after the pH has been set to 5.8 with 1 M KOH. Autoclave the mixture for 20 min. at 121 °C. This medium is used for growing tomato seedlings to harvest the cotyledons for transformation.
3. MS medium containing 3% (w/v) sucrose (3% MS). For 1 L of the medium, add 4.3 g MS (with micronutrients, macronutrients, and vitamins) (Duchefa; product code M0409), 0.5 g MES and 30 g of sucrose. Set the pH to 5.8 with 1 M KOH. For 1 L of the solid medium, add 10 g of plant tissue agar after the pH has been set. Autoclave the mixture for 20 min. at 121 °C. This richer MS medium is used for the selection and growth of hairy root cultures (*see* **Note 3**).

4. Cell culture plate with 40 mL solid MS + 3% sucrose containing 50 μg/mL kanamycin for plant selection and 200 μg/mL cefotaxime for the elimination of *A. rhizogenes* ATCC15834. Allow medium to cool to 65 °C before adding antibiotics.
5. Cell culture plate with 40 mL solid MS + 3% sucrose without antibiotics (*see* **Note 3**).
6. Yeast Extract Beef Broth (YEB) medium for growth of *Agrobacterium rhizogenes* strain ATCC15834: for 1 L of YEB, add 5 g beef extract, 1 g yeast extract, 5 g peptone (bacteriological grade), 5 g sucrose, and 2 mL of 1 M $MgSO_4$. For 1 L of the solid medium, add 15 g of agar (Invitrogen Select Agar; cat. number 30391-023). Autoclave the mixture for 20 min. at 121 °C. After autoclaving, allow to cool to 65 °C and add spectinomycin to a final concentration of 50 μg/mL.
7. Antibiotics (*see* **Note 4**).

2.4.2 Materials and Equipment

1. Tomato cv. Moneymaker seeds.
2. Transparent Magenta boxes.
3. Incubator shaker set at 28 °C and 200 rpm.
4. Spectrophotometer set for reading optical densities (OD) at $\lambda = 600$ nm (OD_{600}).
5. Sterile toothpicks.
6. Cell culture dish (90 × 20 mm).
7. Whatman filter paper.
8. Fluorescence microscope (e.g., Leica stereomicroscope and imaging DFC7000 T Leica microscope camera).

2.5 Induction of Bait-TurboID Expression, Protein Extraction, and Expression Analysis Using Immunoblotting

2.5.1 Reagents

1. Liquid MS added 3% sucrose.
2. Liquid nitrogen.
3. A stock solution of 100 mM 17β-estradiol in dimethyl sulfoxide (DMSO) (Sigma-Aldrich, product code D8418).
4. A stock solution of 50 mM biotin in DMSO.
5. Urea high stringency extraction buffer: 8 M urea (Sigma-Aldrich, product code U1250), 2% sodium dodecyl sulfate (SDS) (Merck, cat. number 822050), 100 mM tris (hydroxymethyl)aminomethane (Tris)-HCl pH 8.0 (Biosolve, cat. number 0020092391BS) and 150 mM NaCl (Chem-lab, cat. number CL00.1429).
6. 4× XT sample buffer (Bio-Rad; cat. n° 1610791).
7. 20× XT reducing agent (Bio-Rad; cat. n° 1610792).
8. 1.0 mm thick 4 to 12% polyacrylamide Criterion Bis-Tris XT-gels (Bio-Rad; cat. n° 3450124).

9. XT MOPS 1D-SDS-PAGE running buffer (Bio-Rad; cat. n° 1610788).
10. 1D-SDS-PAGE electrophoresis and blotting devices.
11. PVDF blotting membrane.
12. Protein standard.
13. Filter paper.
14. Odyssey blocking solution (LI-COR; cat. n° 927-40003).
15. Washing buffer: TBS-T (TBS + 0.1% [v/v] Tween 20).
16. Antibodies and detection reagents (*see* **Note 5**).

2.5.2 Materials and Equipment

1. Liquid nitrogen storage bucket.
2. Eye protection.
3. Aluminum foil.
4. Tissue paper.
5. Dry ice.
6. Liquid nitrogen-cooled pestle and mortar.
7. Micro spoon spatula.
8. Odyssey infrared imaging system (LI-COR) and LICOR Odyssey software for immunoblot image analysis.

2.6 Sample Preparation for Mass Spectrometry

2.6.1 Reagents, Solutions, and Buffers

1. Urea high stringency homogenization buffer (further referred to as lysis buffer): 100 mM tris(hydroxymethyl)aminomethane hydrochloride (Tris-HCl) pH 7.5, 2% (v/v) sodium dodecyl sulfate (SDS), 8 M urea and 150 mM sodium chloride (NaCl), prepared fresh in water (*see* **Note 6**).
2. Streptavidin-agarose bead slurry (Novagen, cat. number 69203).
3. High salt buffer: 1 M NaCl, 100 mM Tris-HCl pH 7.5 in water.
4. Elution buffer: 2% (v/v) SDS, 3 mM biotin and 8 M urea, prepared fresh in phosphate-buffered saline (PBS pH 7.4).
5. 50 mM ammonium bicarbonate prepared fresh in water (pH 7.9).
6. Trypsin/Lys-C Mix, Mass Spec Grade (Promega, cat. number V5073).
7. Hydrogen peroxide (H_2O_2) 30% (v/v) (Sigma-Aldrich, cat. number 1072090250).
8. Acidifying solution (5% trifluoroacetic acid (TFA) LC-grade).
9. LC-grade acetonitrile (ACN).
10. C18 pipette tip conditioning solution: water:ACN 50:50 (v/v).

11. C18 pipette tip washing solution A: 0.1% TFA.
12. C18 pipette tip washing solution B: 0.1% TFA in water:ACN 98:2 (v/v).
13. C18 pipette tip elution solution C: 0.1% TFA in water:ACN 30:70 (v/v).
14. LC-MS peptide resuspension buffer: 2 mM Tris(2-carboxyethyl)phosphine (TCEP) in water:ACN 98:2 (v/v).

2.6.2 Materials and Equipment

1. Protein LoBind tubes (1.5 and 5 mL; Eppendorf).
2. Polypropylene pipette tips.
3. PD-10 Desalting Column with Sephadex G-25 resin, up to 2.5 mL samples (#17085101, GE healthcare Life Sciences).
4. Spinning rotor for 1.5–2 mL tubes.
5. Spinning rotor for 15 mL tubes.
6. SpeedVac concentrator.
7. OMIX C18 pipette tips, 100 μL tip, 10–100 μL elution volume, 1 × 96 tips (#A57003100, Agilent).
8. LC-MS vials.

2.7 LC-MS/MS

1. Various high-resolution LC-MS instruments can be used, e.g., Q Exactive HF Biopharma Quadrupole-Orbitrap mass spectrometer (Thermo Fisher Scientific) coupled to an UltiMate 3000 RSLCnano LC system (Thermo Fisher Scientific). Ionization source: Phoenix PneuNimbus dual-column nanoESI source (MS Wil, Zurich) [23].

2.8 Data Analysis

1. Spreadsheet program (e.g., Microsoft Excel).
2. Galaxy@Belgium (https://usegalaxy.be/).
3. MaxQuant (https://www.maxquant.org/) [18, 19].
4. Software for data handling (e.g., Perseus (https://maxquant.net/perseus/) [19], R Studio, https://www.rstudio.com/).

3 Methods

In this chapter, we describe the detailed step-by-step protocol for TurboID-catalyzed PL in tomato hairy roots from the generation of transformed roots to TurboID sample preparation, MS, and data analysis (Fig. 1). As a proof-of-concept, the adaptor protein NINJA (Solyc05g018320) was selected as bait since NINJA has many known direct and indirect protein interactors [14, 24–27]. NINJA localizes to the plant nucleus where it functions as a negative regulator of JA signaling [14]. eGFP was selected as a relevant bait control based on the subcellular localization of NINJA (i.e., nuclear) and since it has been reported that the eGFP-BioID fusion protein localizes to both the nucleus and cytoplasm [28, 29]. Optimal bait controls are important to avoid false positive proximal partner assignments and should be adjusted according to the POI under investigation.

3.1 Vector Construction for Estradiol-Inducible Bait-TurboID Expression in Rhizogenic Agrobacterium-Transformed Hairy Roots

A translational fusion is generated between the bait and the TurboID proximity label by making use of a standard cloning strategy (e.g., multisite Gateway cloning as described in [21] or Golden Gibson cloning (Jacobs and Karimi, unpublished) (*see* **Note 2**)). Expression of this fusion construct is driven by 17β-estradiol and the estrogen receptor-based transactivator XVE [30]. The high inducibility and tight control of the XVE system in transgenic plants—including hairy roots [31]—was previously demonstrated [30, 31] (*see* **Note 1**). More specifically, the multisite Gateway-compatible binary pKCTAP destination vector enables simultaneous cloning of a promoter sequence (pENTR1), ORF (pENTR2), and a tag of choice (i.e., TurboID) (pENTR3) (*see* **Note 7**), and can thus be used to obtain expression vectors encoding a POI-TurboID fusion of choice with XVE controlled expression. Besides the Gateway gene cassette, pKCTAP contains a KmR cassette and prolD driven eGFP expression cassette between the left (LB) and right (RB) T-DNA border sequences enabling selection of transformed plant cells and visible eGFP+ marker selection of transformed roots (Fig. 2f). Alternatively, Golden Gibson cloning can be used to assemble a binary destination vector encoding a POI-TurboID fusion of choice with XVE controlled expression and a p*CaMV35S* driven eGFP fluorescent maker expression cassette (Jacobs and Karimi, unpublished) (Fig. 2c, d; *see* **Note 2**). Transformation of the binary expression constructs into competent *A. rhizogenes* ATCC15834 was performed by electroporation using standard transformation procedures essentially as described in [32], selection of transformants on YEB selection plates containing the appropriate antibiotics (100 mg/L spectinomycin for the pKCTAP vector), followed by incubation for 3–4 days at 28 °C.

3.2 Hairy Root Transformation, Selection, and Cultivation of Transformed Hairy Roots

All manipulation steps should be performed in a laminar flow and using aseptic conditions.

3.2.1 Seed Sterilization and Germination (7–10 Days)

1. Add at least 20 tomato (*Solanum* spp.) seeds per construct to a sterile 15 mL centrifuge tube. Add 1 mL of commercial 3% (v/v) bleach solution and rotate for 15 min using a test tube vertical rotator.

2. Discard the bleach solution using a pipette controller and a sterile pipette.
3. Rinse with 10 mL of sterile dH_2O. Rotate for 10 min on a test tube rotor. Remove the water using a pipette controller and a serological pipette. Repeat this step twice.
4. Transfer the sterile tomato seeds to Magenta boxes containing 50 mL solid 1% MS. Vernalize the sterile tomato seeds by placing them at 4 °C in the dark for 2 days.
5. After vernalization, transfer the Magenta boxes to a growth chamber under a 16 h/8 h light/darkness photoperiod at 24 °C for 7–10 days until full cotyledon expansion and rapidly following the emergence of the first true leaves.

3.2.2 Growing A. rhizogenes *Strain ATCC15834 (Day − 1)*

1. Pick a transformed *A. rhizogenes* ATCC15834 colony harboring the expression vector, as confirmed by colony PCR, from a Yeast extract beef broth (YEB) selection plate containing spectinomycin (*see* **Note 4**) using a sterile toothpick or sterile pipette tip. Alternatively, pick up *A. rhizogenes* ATCC15834 from a cryopreservation stock.
2. Inoculate 10 mL YEB medium supplemented with spectinomycin (*see* **Note 4**) and grow the culture overnight at 28 °C and 200 rpm.

3.2.3 Hairy Root Transformation (Day 1)

1. Spin down the bacterial culture (5 min at 5000 ×*g*). Discard the supernatant and re-dissolve the bacterial pellet in liquid MS + 3% sucrose to reach a final OD_{600} suspension of ~0.3. A total volume of at least 10 mL per transformant is required for efficient submergence of explants.
2. Fill a sterile cell culture dish with at least 10 mL of liquid MS + 3% sucrose.
3. Gently take the tomato seedling using tweezers from the Magenta boxes and cut the base and top of the cotyledon pair. Place the cut explants immediately into the liquid 3% MS with the abaxial side (bottom) facing up. Do this for at least 30–40 explants per construct (*see* **Note 8**).
4. Discard the liquid MS + 3% sucrose medium and replace it with the bacterial suspension in MS + 3% sucrose using a pipette controller. Soak for 20 min.
5. Transfer the explants to sterile Whatman paper to remove any excess of the bacterial suspension.
6. Transfer the explants from the Whatman filter paper to a cell culture dish containing solid MS + 3% sucrose without antibiotics. About 20–30 explants can be placed onto one cell culture dish with the abaxial side up. Close the cell culture

dish and seal with micropore tape to maintain sterility while permitting oxygen access. Co-cultivate by incubation of the plate containing the explants for 3 days in the dark at 22–25 °C.

3.2.4 Selection and Root Induction (Day 5)

1. After 3 days of co-cultivation, bacterial growth should be observed bordering the cotyledon explants (slimy transparent structures, see Fig. 3b). Transfer the explants to a new MS + 3% sucrose agar dish containing the appropriate antibiotics to select for successful plant T-DNA transformation events and to inhibit further *A. rhizogenes* growth (kanamycin and cefotaxime, respectively) (*see* **Note 4**). Each cell culture dish can fit at least 10 explants.
2. Promote contact of explants with MS-agar by gently pushing the cotyledons onto the MS + 3% sucrose medium using forceps to maximize surface contact of the wounded ends. Close the plate and seal with micropore tape. Incubate the plate containing the explants for 2–3 weeks in the dark at 24 °C until ~2 cm roots emerge from the explants (Fig. 3c, d).

3.2.5 Subculturing of Hairy Roots (Week 4–8)

1. After 2–3 weeks, typically, three to five independent roots (each root reflecting an independent transformation event) have emerged from an explant (Fig. 3c) and eGFP expression (expressed from the T-DNA) can be monitored using a fluorescence microscope (excitation wavelength 488 nm, emission wavelength 509 nm) (Fig. 3f) (*see* **Note 9**).
2. Transfer independent GFP+ roots from ~2 cm in length with similar GFP-fluorescence levels (*see* **Note 10**) to an MS + 3% sucrose cell culture dish containing the appropriate antibiotics (kanamycin for the T-DNA and cefotaxime) (*see* **Note 4**). Gently push the root onto the surface to maximize contact with the medium.
3. Close the plate and seal with micropore tape.
4. Repeat this for at least six to eight GFP-expressing roots per construct (*see* **Note 10**), further referred to as hairy root clones. Grow the hairy root cultures for 1–2 weeks in the dark at 22–28 °C.
5. Using a scalpel and tweezers, cut 2–3 cm of a—preferably branched—root from the hairy root culture and transfer them onto a new MS + 3% sucrose plate containing antibiotics (*see* **Note 4**). Seal the plate with micropore tape and grow it for another 1–2 weeks in the dark at 22–28 °C. Repeat this subculturing step at least two more times (Fig. 3e) (*see* **Note 11**).
6. After three rounds of subculturing, transfer the 2–3 cm of root to a plate containing MS + 3% sucrose medium without antibiotics (*see* **Note 3**).

3.3 Immunoblot Analysis of Bait-TurboID Expression in Transformed Hairy Roots

Since each root reflects an independent transformation event, and since we will make use of three to four hairy root cultures (*see* **Note 12**) corresponding to independent transformations as biological replicates for TurboID followed by MS-analysis (*see* Subheading 3.4), bait expression and (auto-)biotinylation patterns of the eGFP + roots selected for sub-cultivation will be assessed following total protein extraction, 1D-SDS-PAGE, and immunoblot analysis (*see* **Note 13**). For this, the hairy root clones (obtained in Subheading 3.2.5) are grown in liquid MS + 3% sucrose medium without antibiotics, which facilitates uniform root access to β-estradiol and biotin, supplements shown to induce an *in planta* expression and biotinylation [13], respectively. Ideally, per construct, at least four independent hairy root clones with similar bait expression levels and biotinylation patterns are selected for TurboID analysis and further MS sample preparation (*see* **Note 10**).

1. Transfer 2–3 cm of root per clone to 5 mL of liquid MS + 3% sucrose medium without antibiotics in a 50 mL conical tube. Tighten the cap of the tube and then turn it back a quarter turn to permit oxygen access. Seal off the tube with micropore tape. Individual root cultures from the same clone can be grown in separate tubes in parallel and pooled before the grinding of transformed roots in liquid nitrogen (*see* **Note 14**).
2. Grow the liquid cultures for at least 1 week on a tabletop shaker set at 150 rpm at 22–25 °C in the dark (*see* **Note 15**). If the hairy root cultures have not reached the sufficient amount of material required for TurboID analysis, add another 5 mL of liquid 3% MS medium to the tube and continue growing the culture (Fig. 3g). For protein expression analysis by means of immunoblotting, at least 100 mg of freeze-dried ground root material per clone (corresponding to ~400 μg protein) is required (*see* **Note 16**). TurboID sample preparation for MS analysis requires 2.5 g of ground root material per clone corresponding to 10 mg of protein (*see* **Note 16**).
3. Add β-estradiol (for induction of the inducible RPS5α-XVE promoter) to the hairy root liquid culture to obtain a final concentration of 100 μM. Close the tube with micropore tape as indicated before and induce expression for 24 h while incubating the culture on the tabletop shaker at 150 rpm. After 22 h, add biotin to the hairy root culture to obtain a final concentration of 50 μM. Close the tube and grow the hairy root culture for an extra 2 h on the tabletop shaker.
4. After β-estradiol induction and biotin addition, pool the hairy root cultures from the same clone and remove excess medium by blotting the roots onto a tissue.

5. Wrap the collected hairy roots in aluminum foil and immediately transfer them to liquid nitrogen. Frozen samples can be stored at −80 °C until further processing.
6. Using a liquid nitrogen-cooled pestle and mortar, crush the (pooled) hairy roots to a fine powder. Transfer at least 100 mg of the crushed material to a cooled 1.5 mL tube for immunoblot analysis. Transfer the remaining crushed material to a cooled 15 mL centrifuge tube and store at −80 °C until further processing.
7. To 100 mg of ground hairy root material, add 100 μL of lysis buffer (1:1 weight:volume) (*see* **Note 16**).
8. Quickly vortex the sample to ensure uniform re-suspension and homogenization of the crushed material and transfer the tube to liquid nitrogen. Thaw the tube in a water bath at room temperature. Repeat this step twice (*i.e.*, three repetitive freeze-thaw cycles in total to ensure full mechanical disruption and efficient protein extraction when used in combination with tissue grinding).
9. After the third freeze-thaw cycle, centrifuge for 15 min at maximum speed in a tabletop centrifuge cooled to 4 °C. Retrieve the supernatant and repeat the centrifugation step to remove residual debris (*see* **Note 17**).
10. Add SDS-sample buffer and a reducing agent to the protein extract and perform 1D SDS-PAGE and immunoblotting. For immunoblot detection, anti-FLAG (produced in mice) can be used as a primary antibody and as a secondary antibody/conjugate mixture of the anti-mouse IRDye800 CW antibody together with the Streptavidin Alexa 680 conjugate (*see* **Notes 5** and **18**).

3.4 Sample Preparation for MS

3.4.1 Protein Extraction and Streptavidin Beads Enrichment (Day 1)

1. To 2.5 g of ground hairy root material, add 2.5 mL of lysis buffer (1:1 weight:volume) to obtain a volume of ~2500 μL of hairy root lysates, corresponding to a protein concentration of ~4 mg/mL and total protein yield of ~10 mg (*see* **Note 19**).
2. Perform protein extraction as explained in Subheading 3.3, **steps 8** and **9**.
3. Take 50 μL (~200 μg) lysate sample, add 20 μL lysis buffer, 25 μL XT sample buffer (4×), and 5 μL XT reducing agent (20×) for immunoblot analysis (*see* **Note 20**) (**Input sample**, Fig. 4b).
4. Desalt the sample to deplete free biotin using PD-10 Desalting Columns with Sephadex G-25 resin according to the manufacturer's instructions and elute in a 5 mL protein LoBind tube. Note that the sample volume after desalting increases 1.4-fold, up to 3.5 mL.

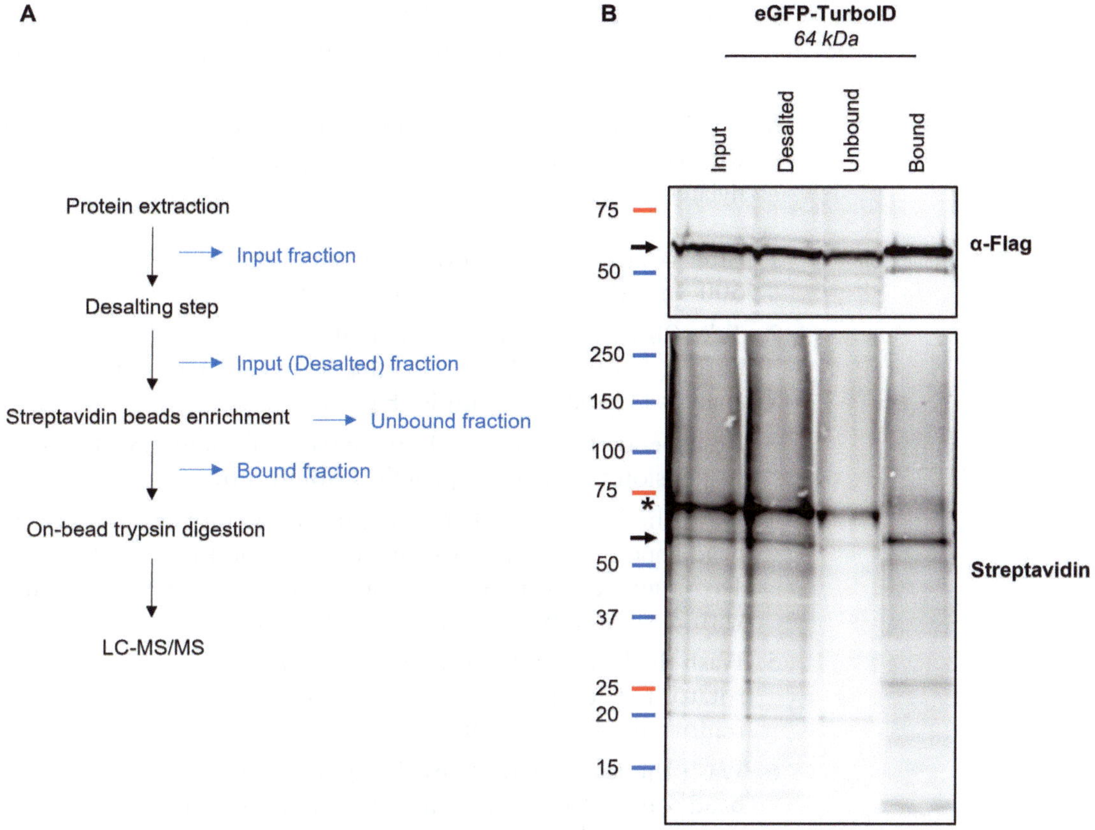

Fig. 4 TurboID-catalyzed proximity labeling in tomato hairy root cultures. XVE::eGFP-TurboID expression construct was used *for rhizogenic Agrobacterium-mediated transformation of S. lycopersicum.* Prior to subjecting protein extracts from transformed hairy roots to TurboID-MS, a quality assessment of the TurboID-MS sample preparation was performed by subjecting (**a**) Input, Input (Desalted), Unbound and Bound fractions collected at various steps of the TurboID procedure to immunoblot analysis. (**b**) Immunoblot results of Input, Input (Desalted), Unbound and Bound fractions (5 times more concentrated). Arrow corresponds to the MW of the eGFP-TurboID fusion protein expressed (i.e., 64 kDa). Asterisk indicates the ~70/75 kDa bands corresponding to the endogenously biotinylated proteins acetyl-CoA carboxylases 1 and 2 (ACC1 and ACC2). A streptavidin/Alexa Fluor™ 680 conjugate was used for the detection of biotinylated proteins, and anti-Flag for the detection of the translational fusion of the promiscuous labeling enzyme TurboID to the bait. Results are representative of the independent root cultures analyzed (i.e., biological TurboID replicate samples)

5. Take 70 μL (~200 μg) of the desalted sample and add 25 μL XT sample buffer (4×) and 5 μL XT reducing agent (20×) for immunoblot analysis (**Desalted Input sample**, Fig. 4b).
6. Pre-wash 400 μL of streptavidin-agarose bead slurry (50% v/v, further referred to as beads) by resuspension in 2 mL of lysis buffer followed by centrifugation for 5 min at 600 ×*g* and removal of the supernatant. Repeat this step twice (*see* **Note 21**).

7. Add 200 μL lysis buffer to 200 μL pre-washed beads to obtain a final volume of 400 μL and add it to the sample (total volume of ~4 mL).
8. Incubate the sample-beads mixture by overnight incubation with overhead rotation at room temperature to capture biotinylated proteins.

3.4.2 Trypsin Digestion (Day 2)

1. After overnight incubation, centrifuge the beads for 2 min at 600 ×*g* and remove the unbound supernatant.
2. Take 70 μL unbound supernatant and add 25 μL XT sample buffer (4×) and 5 μL XT reducing agent (20×) for immunoblot analysis (**Unbound sample**, Fig. 4b).
3. Resuspend the beads in 1 mL lysis buffer and transfer bead suspension to a 1.5 mL protein LoBind tube.
4. Wash the beads with 1 mL lysis buffer by mixing on an overhead rotator followed by centrifugation for 4 min at 600 ×*g* at 18 °C and discard the supernatant. Repeat this step 4 times with 3 × 5 min washes and 1 final 30 min wash.
5. Wash the beads with 1 mL high salt buffer, mix on an overhead rotator for 30 min, centrifuge for 4 min at 600 ×*g* at 18 °C and remove the supernatant.
6. Wash the beads with 1 mL MS-grade water, mix on an overhead rotator for 5 min, centrifuge for 4 min at 600 ×*g* at 18 °C and remove the supernatant.
7. Following the final wash, split the beads 90% (TurboID)/10% (immunoblot analysis), centrifuge both fractions for 4 min at 600 ×*g* and 18 °C and remove the supernatant.
8. Add 70 μL elution buffer to 10% of the beads, incubate for 10 min in a thermomixer at 90 °C and 850 rpm, and centrifuge for 4 min at 600 ×*g*.
9. Transfer the supernatant to a new Eppendorf tube and add 25 μL XT sample buffer (4×) and 5 μL XT reducing agent (20×) to obtain the eluate sample for immunoblot analysis (*see* **Note 20**) (**Bound sample**, Fig. 4b).
10. Wash the aliquot containing 90% of the beads with 1 mL of 50 mM ammonium bicarbonate (pH 7.9), mix on an overhead rotator for 5 min, centrifuge for 4 min at 600 ×*g* at 18 °C, and remove the supernatant. Repeat this step twice.
11. Resuspend the beads in 250 μL of 50 mM ammonium bicarbonate (pH 7.9) (*see* **Note 22**) and add 1 μg of Trypsin/Lys-C Mix resuspended in 250 μL of 50 mM ammonium bicarbonate (pH 7.9).

12. Digest the samples overnight at 37 °C with vigorous mixing (850 rpm) to keep the beads in suspension.
13. After overnight digestion, add an additional 0.5 μg of Trypsin/Lys-C Mix and incubate for 2 h at 37 °C with vigorous mixing (850 rpm).

3.4.3 Peptide Isolation (Day 3)

1. After digestion, centrifuge the beads for 2 min at 600 ×*g* and transfer the supernatant to a fresh protein LoBind tube.
2. Wash the beads twice (2 min at 600 ×*g*) with 250 μL MS-grade water and combine the washes with the original supernatant.
3. Stop digestion by acidification of the peptide solution with 5% TFA to reach a final concentration of 0.1%, centrifuge for 10 min at 16,100 ×*g* (4 °C) to precipitate insoluble material, and transfer the supernatant to fresh protein LoBind tubes.
4. Vacuum dry the samples in the SpeedVac concentrator.

3.4.4 Peptide Purification (Day 4)

1. Redissolve the dried peptides in C18 pipette tip washing solution A, centrifuge for 10 min at 16,100 ×*g* (4 °C) to remove insoluble material and transfer the supernatant to fresh protein LoBind tubes.
2. Perform methionine oxidation by adding H_2O_2 to reach 0.5 % f.c. (v/v) to each sample for 30 min at 30 °C (*see* **Note 23**).
3. Purify the peptides using 100 μL OMIX C18 pipette tips by first conditioning the pipette tip with 100 μL of C18 pipette tip conditioning solution. Discard the solvent.
4. Equilibrate the pipette tip by washing three times with 100 μL of C18 pipette tip washing solution A.
5. Aspirate 100 μL of the acidified samples for 10 cycles to maximize binding efficiency.
6. Wash the pipette tip three times with 100 μL of C18 pipette tip washing solution B.
7. Elute the bound peptides in LC-MS vials with 100 μL of C18 pipette tip elution solution C.
8. Dry the samples in the SpeedVac concentrator and re-dissolve in 15 μL of LC-MS peptide resuspension buffer (*see* **Note 24**).

3.4.5 Quality Assessment of TurboID MS Sample Preparation by Immunoblot Analysis

1. Perform immunoblot analysis on the Input-, Input Desalted-, Unbound- and Bound fractions to assess protein expression, (auto-)biotinylation activity of the bait-TurboID fusion protein, and efficiency of streptavidin-enrichment of biotinylated protein (*see* **Note 20**, Fig. 4b).
2. Use a streptavidin/Alexa Fluor™ 680 conjugate for the detection of biotinylated proteins, and anti-Flag for the detection of the translational fusion of the promiscuous labeling enzyme TurboID to the bait (*see* **Notes 5** and **18**).

3.5 LC-MS/MS Analysis

1. Various high-resolution LC-MS platforms can be used [23].

3.6 MS-Data Analysis

1. Use MaxQuant [18, 19] (optional: "using mqpar.xml") in Galaxy@Belgium (https://usegalaxy.be/) with the integrated Andromeda search engine for peptide and protein identification on the acquired raw files (*see* **Note 25**). Download the FASTA file containing *Solanum lycopersicum* protein entries (downloaded from https://www.uniprot.org/proteomes/UP000004994, taxonomy ID 4081). Generate a FASTA file containing the bait and TurboID protein sequences as separate entries. Upload FASTA files to Galaxy@Belgium. Generate a parameter file in a MaxQuant Version compatible with Galaxy (e.g., 1.6.10.43) and upload this file to Galaxy@Belgium (*see* **Note 26**).
2. In Galaxy, select MaxQuant (optional: "using mqpar.xml"), upload the raw files, the FASTA files (*S. lycopersicum* proteome, bait, and TurboID sequence), and the generated mqpar.xml file, indicate the identifiers (*see* **Note 25**) and select the proteinGroups.txt file as output. Run MaxQuant on the Galaxy server.
3. Further data analysis is performed using the proteinGroups.txt MaxQuant output file in the Perseus software [19].
4. Select the LFQ intensities for all bait/control replicates as main expression values and filter the dataset by removing reversed hits and potential contaminants. Perform a Log_2 transformation of the LFQ intensities. For inspection of the quality and correlation of the data, generate a multiscatter plot with Pearson correlations, a numeric Venn diagram, and histograms. Assign replicate samples to their respective group. Next, filter the protein groups for at least three valid values in at least one group. Impute the data by replacing the missing values from the normal distribution.
5. Use a two- or multiple-sample t-test to identify enriched proteins in the bait samples. For a pairwise comparison with a control condition (in our case, eGFP), visualize the results in a volcano plot (Fig. 5) or Hawaii plot (i.e., multiple volcano plot) by plotting the t-test difference versus the t-test p-value. For a multiple comparison between all protein groups, use an ANOVA test and perform z-scoring to visualize the results in a heatmap or export the output matrix to excel. We recommend the use of a stringent FDR value of 0.01. We typically set the S0 value (artificial within groups variance; this value defines the relative importance of the p-value and difference between means) to 0.1 in the Perseus software. FDR and S0 values can be adjusted accordingly to generate a more relaxed or stringent analysis.

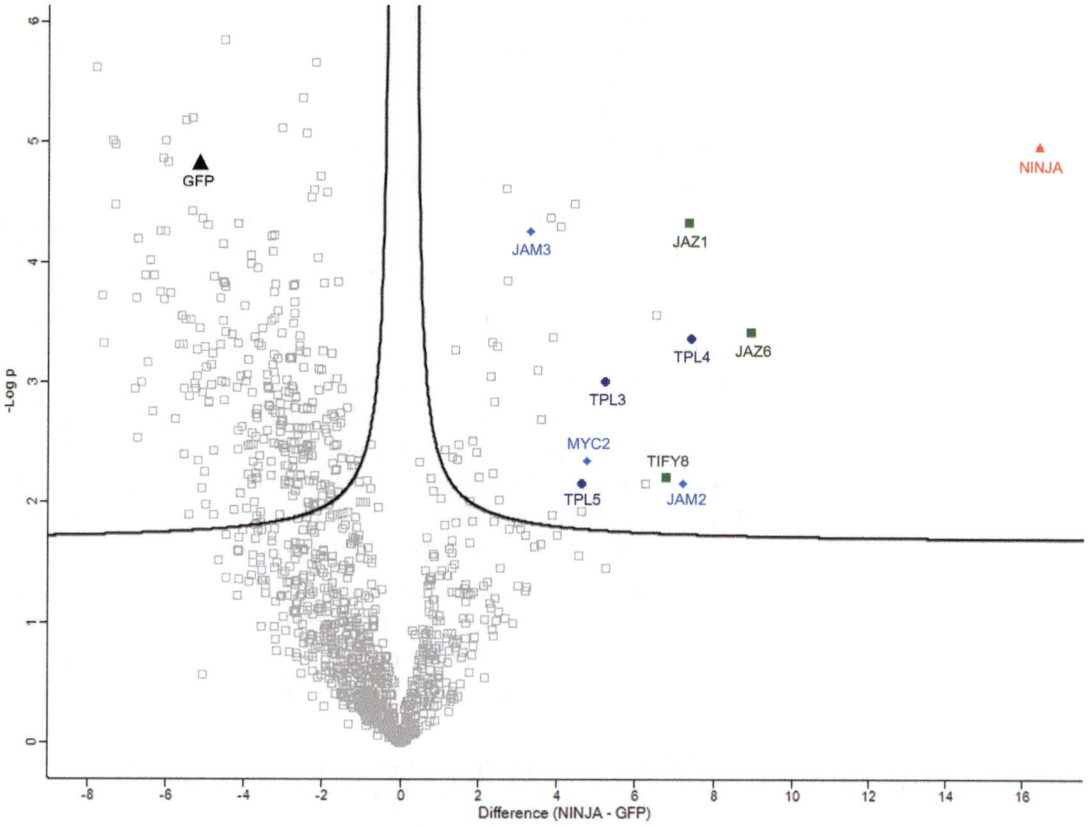

Fig. 5 Volcano plot of pairwise comparison between NINJA and GFP samples. A two-sample t-test was performed to identify enriched proteins in the bait (i.e., NINJA) samples. FDR:0.01. S0:0.1. The t-test difference was plotted versus the *t*-test $-\log(p\text{-value})$. As a proof-of-concept, previously reported direct and indirect interactors of NINJA are highlighted in the volcano plot. Red triangle: NINJA (Solyc05g018320). Green squares: JAZ/TIFY family proteins [24–26], i.e., JAZ1 (Solyc07g042170), JAZ6 (Solyc01g005440), TIFY8 (Solyc06g065650). Dark blue circles: TPL family proteins [14], i.e., TPL3 (Solyc01g100050), TPL4 (Solyc03g116750), TPL5 (Solyc07g008040). Light blue diamonds: bHLH family proteins [27], i.e., MYC2 (Solyc08g076930), JAM2 (Solyc01g096050), JAM3 (Solyc06g083980). Black triangle: eGFP control

6. Functional interpretation of differences in protein interactions can be performed (e.g., using 1D annotation enrichment analysis) (*see* **Note 27**).

4 Notes

1. The inducible XVE promoter was chosen over the constitutive *CaMV35S* promoter since it was often observed that *pCaMV35S*-driven bait expression was lost upon prolonged tomato hairy root cultivation (possibly due to transgene silencing).

2. Golden Gibson cloning can be used as an alternative to Gateway cloning (Jacobs and Karimi, unpublished). More specifically, building blocks (Fig. 2b) are assembled with Golden Gate cloning into the shuttle vectors pGGIB-U1-AG-U2 and pGGIB-U2-AG-U3 to yield pGGIB-U1-XVE-POI-TurboID-35ST-U2 and pGGIB-U2-p*CaMV35S*-eGFP-35ST-U3 (Fig. 2d). In these vectors, unique nucleotide sequences (U-sites, [33]) of 40 bp are flanked by the I-SceI restriction sites. Using Gibson assembly, the fragments were combined in pK-U1-AG-U9 with pGGIB-U3-linker-U9 to yield pK-XVE--POI-TurboID-35ST-p*CaMV35S*-eGFP-35ST. Recombination generates a binary destination vector encoding two expression cassettes: (1) a translational fusion of the POI to the TurboID proximity label and (2) a fluorescent marker (eGFP) for the selection of transformed hairy roots. The structure and sequence of all available entry and destination vectors are accessible online at https://gatewayvectors.vib.be/.
3. After 3 rounds of cultivation, root cultures can be maintained and grown in antibiotics-free half-strength (½) Murashige and Skoog (MS) medium supplemented with 3% sucrose at 22–25 °C. This reduces hairy root growth and reduces the need for sub-cultivation.
4. Antibiotic concentrations: for *E. coli* selection (LB medium), use kanamycin 25 μg/mL (pDONR221) (Duchefa; product code K0126), spectinomycin 50 μg/mL (pKCTAP) (Duchefa; product code S0188); for *A. rhizogenes* selection (YEB medium), use spectinomycin 50 μg/mL (pKCTAP); for 3% MS plates, use 200 μg/mL cefotaxime (Duchefa; catalog no. c0111.0025) in combination with 50 μg/mL kanamycin (pKCTAP). Antibiotics are variable depending on the entry and destination vectors used.
5. Antibodies and detection agents used: Monoclonal ANTI-FLAG® M2 antibody produced in mouse (F3165, Sigma Aldrich) (1:2000), Streptavidin, Alexa Fluor™ 680 Conjugate (S32358, Thermo Scientific) (1:5000), anti-mouse (IRDye® 800CW Goat anti-Mouse IgG Secondary Antibody) (926-32210, LI-COR) (1:10000). Antibodies and detection reagents are diluted in 1/1 [v/v] TBS-T/Odyssey blocking buffer.
6. All solutions are prepared with ultrapure MS-grade water (resistivity of 18.2 MΩ.cm at 25 °C) and analytical-grade or MS-grade reagents.
7. In the case of the 3-site Gateway cloning used here, entry vectors pENTR1, pENTR2, and pENTR3 were produced in a BP Clonase reaction that transferred a PCR amplicon (promoter, ORF without stop codon, and TurboID tag,

respectively) flanked by the appropriate *att* sites (B4 and B1R, B1 and B2, or B2R and B3) in one of the three compatible donor vectors (pDONRP4P1R, pDONR221, or pDONRP2RP3; Invitrogen).

8. For 30–40 explants, at least 15 seedlings are needed, since each seedling contains two cotyledons and each cotyledon, if it is of sufficient size, can be further cut into two parts. The optimal tissue to be selected as explant may depend on the plant model system, as other plant model systems require (young) true leaves. This transformation protocol can thus be expanded to other transformed roots/plant tissue if a transformation protocol is available for the model under study.
9. Chlorophyll interference might be observed in the case of eGFP imaging, particularly in the case of low eGFP expression. It is also possible to use other binary vectors carrying different fluorophores (e.g., mCherry, mRuby) to select for successfully transformed roots.
10. Ideally, six to eight independent roots showing comparable expression of the marker (here eGFP) or separate cultures of single clones should be propagated further per construct. We noticed a good correlation between the expression level of eGFP and bait expression where a low eGFP signal corresponded to lower bait expression.
11. Two rounds of subculturing on the selective medium were frequently found to be insufficient, since in a few cases bacterial growth reappeared. This was not the case for three rounds of subculturing on a selective medium.
12. Repeats of single clones could be opted to minimize differences in bait expression (as quite a huge variation in eGFP—and thus bait—expression can be observed among eGFP+ clones) among replicate samples. Making use of replicate cultures of single clones could therefore increase the reproducibility of the data (and statistical robustness), while nonetheless not considering clonal variation.
13. This analysis is optimally performed once more before final sample collection as silencing of bait-TurboID expression upon prolonged hairy root cultivation has been observed.
14. When cultures are grown in parallel, the required material can be collected in a shorter period. Also noteworthy is that the fold-increase in root material during cultivation over time is reduced upon prolonged cultivation using the conditions here described. To obtain 2.5 g of freeze-dried hairy roots, it is advised to grow replicates of the hairy root clone and pool the replicates before crushing in liquid nitrogen. In our experience, six replicates per clone grown for 2 weeks in liquid 3% MS

medium provides sufficient material for TurboID-MS analysis. More specifically, the amount of hairy root material collected from one replicate sample corresponds to ~500 mg (Fig. 3g).

15. It is important to keep the cultures growing at high rpm (i.e., 150 rpm) to improve aeration. It was noticed that growing them at a lower rpm resulted in browning and reduced growth of the hairy root cultures.
16. The average protein content matching 100 mg of crushed material is ~400 μg. This way, a concentration of ~4 mg/mL is expected requiring 12.5 μL of lysate corresponding to 50 μg of protein to be analyzed on a gel for immunoblotting.
17. Alternatively, a filtration step (e.g., using a polyethylene filter-based spin column [13]) could be performed to remove cellular debris.
18. Alternatively, an anti-BirA antibody can be used to detect protein expression [13]. Biotinylated proteins should be preferable only to be detected in the (Desalted) Input, and Bound fractions but not in the Unbound fraction (Fig. 2b). Otherwise, the Input sample can be reduced accordingly.
19. The protein concentration of the samples can be determined using lysis buffer without SDS (the SDS concentration used is incompatible with the Bradford protein assay). Dependent on the water content (variable dependent on growth conditions of hairy root cultures, e.g., liquid or non-liquid cultures) increase observed upon protein extraction, it can be opted to add the different reagents from the lysis buffer separately in order to obtain a final concentration of 8 M urea in the hairy root lysates. The addition of a separate (solid) component requires quick resuspension to the ground root material to avoid protein degradation.
20. For immunoblot analysis, it is recommended to load 30–50 μg of protein. Lysis buffer should be added to obtain consistent protein concentration during protein extraction to ensure reproducibility. Note that the bound sample is a 5× more concentrated equivalent as compared to the Input fraction, and Input Desalted and Unbound equivalents.
21. Some beads may still be present in the obtained supernatant. To capture the maximal number of beads, do not discard the supernatant after the first centrifugation step. Spin this supernatant down separately and add the recovered beads to the other beads.
22. At this step, samples can be stored in the −20 °C freezer until further processing. At this stage, it is recommended to perform immunoblot analysis to serve as quality control for the TurboID-MS sample processing.

23. Uniform oxidation of methionine to methionine-sulfoxide can be achieved by the use of hydrogen peroxide in acidic conditions. The final concentration and incubation time should be respected since prolonged incubation results in uncontrolled oxidation of methionine to methionine-sulfone, as well as (partial) oxidation of side chains of other amino acids such as cysteine and tryptophan.
24. Samples are dissolved in 2 mM TCEP.HCl (pH of 2.8) to reduce cysteine residues (and disulfide bridges, otherwise hindering MS-identification of disulfide-linked peptides). Since TCEP is active over a wide pH range [34], including lower acidic pH, no pH adjustment of this solution is needed since an acidic environment preserves the reduced status of sulfhydryl functions. TCEP does not interfere with MS detection at the concentration used.
25. The identifier and description parse rule in Galaxy@Belgium need to be adapted to be compatible with the Maxquant parameter file, e.g., identifier parse rule: >(.*) and description parse rule: >([^\s]*) for UniProt protein entries.
26. To generate the parameter file, use the default settings of MaxQuant with slight adaptations: set multiplicity to 1, indicating that no labels were used. Select Acetyl (protein N-term) as a variable modification and methionine oxidation as a fixed modification (*see* **Note 23**). Set the maximal number of modifications per peptide to 5. Indicate Trypsin/P as protease, allow for one missed cleavage, and use only the unique peptides for protein quantification. Save these settings in MaxQuant to use as mqpar.xml file.
27. 1D enrichment tests for every annotation term (such as keyword or molecular function) whether the corresponding numerical values have a preference to be systematically larger or smaller than the global distribution of the values for all proteins.

References

1. Erffelinck ML, Ribeiro B, Perassolo M et al (2018) A user-friendly platform for yeast two-hybrid library screening using next generation sequencing. PLoS One 13:e0201270
2. Bontinck M, Van Leene J, Gadeyne A et al (2018) Recent trends in plant protein complex analysis in a developmental context. Front Plant Sci 9:640
3. Gingras AC, Abe KT, Raught B (2019) Getting to know the neighborhood: using proximity-dependent biotinylation to characterize protein complexes and map organelles. Curr Opin Chem Biol 48:44–54
4. Roux KJ, Kim DI, Raida M, Burke B (2012) A promiscuous biotin ligase fusion protein identifies proximal and interacting proteins in mammalian cells. J Cell Biol 196:801–810
5. Kim DI, Jensen SC, Noble KA et al (2016) An improved smaller biotin ligase for BioID proximity labeling. Mol Biol Cell 27:1188–1196
6. Branon TC, Bosch JA, Sanchez AD et al (2017) Directed evolution of TurboID for efficient proximity labeling in living cells and organisms. bioRxiv:196980
7. Branon TC, Bosch JA, Sanchez AD et al (2018) Efficient proximity labeling in living

cells and organisms with TurboID. Nat Biotechnol 36:880–887
8. Lam SS, Martell JD, Kamer KJ et al (2015) Directed evolution of APEX2 for electron microscopy and proximity labeling. Nat Methods 12:51–54
9. Kim DI, Birendra KC, Zhu W et al (2014) Probing nuclear pore complex architecture with proximity-dependent biotinylation. Proc Natl Acad Sci U S A 111:E2453
10. Kim T-W, Park CH, Hsu C-C et al (2019) Application of TurboID-mediated proximity labeling for mapping a GSK3 kinase signaling network in Arabidopsis. bioRxiv:636324
11. Zhang Y, Song G, Lal NK et al (2019) TurboID-based proximity labeling reveals that UBR7 is a regulator of N NLR immune receptor-mediated immunity. Nat Commun 10
12. Mair A, Xu SL, Branon TC et al (2019) Proximity labeling of protein complexes and cell type specific organellar proteomes in Arabidopsis enabled by TurboID. Elife 8
13. Arora D, Abel NB, Liu C et al (2020) Establishment of proximity-dependent biotinylation approaches in different plant model systems. Plant Cell 32:3388–3407
14. Pauwels L, Barbero GF, Geerinck J et al (2010) NINJA connects the co-repressor TOPLESS to jasmonate signalling. Nature 464:788–791
15. Thelen JJ, Peck SC (2007) Quantitative proteomics in plants: choices in abundance. Plant Cell 19:3339–3346
16. Ramisetty SR, Washburn MP (2011) Unraveling the dynamics of protein interactions with quantitative mass spectrometry. Crit Rev Biochem Mol Biol 46:216–228
17. Cox J, Hein MY, Luber CA et al (2014) Accurate proteome-wide label-free quantification by delayed normalization and maximal peptide ratio extraction, termed MaxLFQ. Mol Cell Proteomics 13:2513–2526
18. Cox J, Mann M (2008) MaxQuant enables high peptide identification rates, individualized p.p.b.-range mass accuracies and proteome-wide protein quantification. Nat Biotechnol 26:1367–1372
19. Tyanova S, Temu T, Cox J (2016) The MaxQuant computational platform for mass spectrometry-based shotgun proteomics. Nat Protoc 11:2301–2319
20. Kajala K, Coil DA, Brady SM (2014) Draft genome sequence of Rhizobium rhizogenes strain ATCC 15834. Genome Announc 2
21. Karimi M, Depicker A, Hilson P (2007) Recombinational cloning with plant gateway vectors. Plant Physiol 145:1144–1154
22. Murashige T, Skoog F (1962) A revised medium for rapid growth and bio assays with tobacco tissue cultures. Undefined 15:473–497
23. Willems P, Fels U, Staes A et al (2021) Use of hybrid data-dependent and -independent acquisition spectral libraries empowers dual-proteome profiling. J Proteome Res 20:1165–1177
24. Chini A, Ben-Romdhane W, Hassairi A, Aboul-Soud MAM (2017) Identification of TIFY/JAZ family genes in Solanum lycopersicum and their regulation in response to abiotic stresses. PLoS One 12:e0177381
25. Chini A, Fonseca S, Fernández G et al (2007) The JAZ family of repressors is the missing link in jasmonate signalling. Nature 448:666–671
26. Cuéllar Pérez A, Nagels Durand A, vanden Bossche R et al (2014) The non-JAZ TIFY protein TIFY8 from Arabidopsis thaliana Is a transcriptional repressor. PLoS One 9:84891
27. Sasaki-Sekimoto Y, Jikumaru Y, Obayashi T et al (2013) Basic helix-loop-helix transcription factors JASMONATE-ASSOCIATED MYC2-LIKE1 (JAM1), JAM2, and JAM3 Are negative regulators of jasmonate responses in Arabidopsis. Plant Physiol 163:291
28. Tanz SK, Castleden I, Small ID, Harvey Millar A (2013) Fluorescent protein tagging as a tool to define the subcellular distribution of proteins in plants. Front Plant Sci 4:214
29. Jonckheere V, van Damme P (2021) N-terminal acetyltransferase Naa40p whereabouts put into N-terminal proteoform perspective. Int J Mol Sci 22
30. Zuo J, Niu QW, Chua NH (2000) An estrogen receptor-based transactivator XVE mediates highly inducible gene expression in transgenic plants. The Plant Journal 24:265–273
31. Rizvi NF, Cornejo M, Stein K et al (2014) An efficient transformation method for estrogen-inducible transgene expression in Catharanthus roseus hairy roots. Plant Cell Tissue Organ Cult (PCTOC) 120:475–487
32. Wen-Jun S, Forde BG (1989) Efficient transformation of *Agrobacterium* spp. by high voltage electroporation. Nucleic Acids Res 17:8385
33. Torella JP, Boehm CR, Lienert F et al (2014) Rapid construction of insulated genetic circuits via synthetic sequence-guided isothermal assembly. Nucleic Acids Res 42:681–689
34. Disulfide reduction using TCEP reaction. https://www.biosyn.com/tew/instruction-of-reduction-reaction-using-tcep.aspx. Accessed 18 Feb 2022
35. Lampropoulos A, Sutikovic Z, Wenzl C et al (2013) GreenGate – a novel, versatile, and efficient cloning system for plant transgenesis. PLoS One 8:e83043

Chapter 27

A Data-Driven Signaling Network Inference Approach for Phosphoproteomics

Imani Madison, Fin Amin, Kuncheng Song, Rosangela Sozzani, and Lisa Van den Broeck

Abstract

Proteins are rapidly and dynamically post-transcriptionally modified as cells respond to changes in their environment. For example, protein phosphorylation is mediated by kinases while dephosphorylation is mediated by phosphatases. Quantifying and predicting interactions between kinases, phosphatases, and target proteins over time will aid the study of signaling cascades under a variety of environmental conditions. Here, we describe methods to statistically analyze label-free phosphoproteomic data and infer posttranscriptional regulatory networks over time. We provide an R-based method that can be used to normalize and analyze label-free phosphoproteomic data using variance stabilizing normalization and a linear mixed model across multiple time points and conditions. We also provide a method to infer regulator-target interactions over time using a discretization scheme followed by dynamic Bayesian modeling computations to validate our conclusions. Overall, this pipeline is designed to perform functional analyses and predictions of phosphoproteomic signaling cascades.

Key words Label-free phosphoproteomics, Kinase regulatory networks, Post-translational modifications, Bayesian modeling

1 Introduction

Plants face many developmental pressures from variations in temperature, water availability, and nutrient availability to newer threats posed by climate change [1]. Posttranslational modifications are major mechanisms regulating rapid, dynamic cellular stress responses so there is a strong need to systemically characterize posttranslational dynamics [2, 3]. Protein phosphorylation is a reversible Posttranslational modification (PTM) that facilitates rapid, dynamic changes to cellular stress responses [4, 5]. Protein activity is rapidly modulated by the addition or subtraction of

Authors Imani Madison and Fin Amin have contributed equally to this chapter.

Shahid Mukhtar (ed.), *Protein-Protein Interactions: Methods and Protocols*, Methods in Molecular Biology, vol. 2690, https://doi.org/10.1007/978-1-0716-3327-4_27,

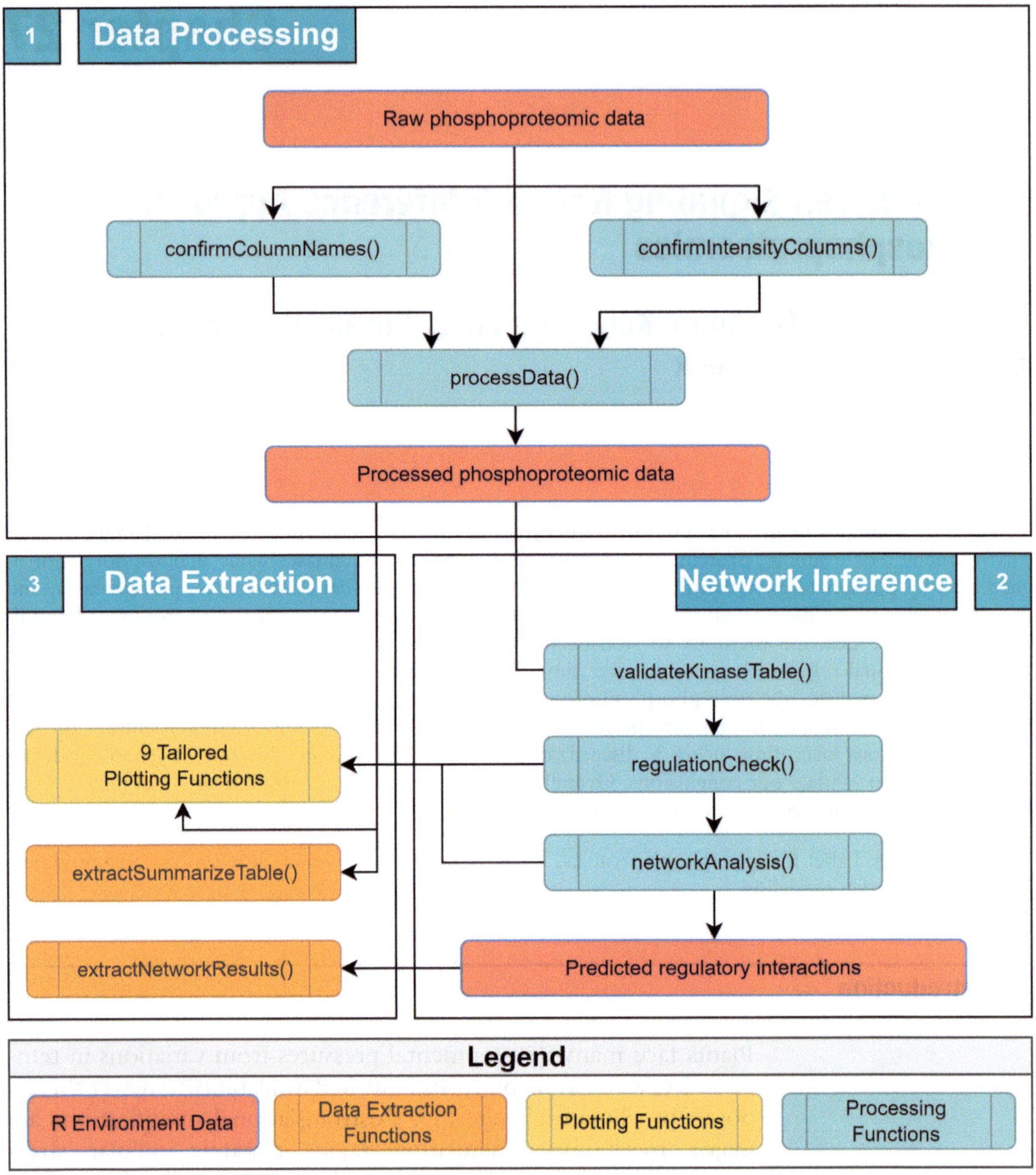

Fig. 1 The NetPhorce R package workflow. The workflow consists of the following three major sections: (1) Data processing: Import and parsing of the raw phosphoproteomic data, followed by data processing, including quality control, normalization, and statistical analysis. (2) Network inference: Identify regulatory interactions between kinases/phosphatases and downstream target phosphosites. (3) Data extraction: generation of structured output tables and a diverse set of visualization graphs

phosphate groups. Kinases phosphorylate target proteins, while phosphatases dephosphorylate them. [4, 5]. Protein phosphorylation or dephosphorylation occurs over a rapid timescale and at a low intracellular abundance [4, 6]. Cellular processes influenced by protein phosphorylation include, but are not limited to, carbon and RNA metabolism, root growth, and defense responses over the course of plant development [4, 7]. Phosphoproteome analysis aims to both identify phosphorylation sites of proteins in an either targeted or label-free manner and to characterize the resulting signaling cascades [5, 8, 9].

There have been recent improvements and increases in phosphoproteome data analysis and signaling network inference methods [4, 7]. Improvements in characterizing phosphorylation events include phosphoproteomic datasets sampled over a fine temporal resolution and improved mass spectrometry (MS) protocols to detect phosphorylation sites, or phosphosites, at a low enrichment [7]. Furthermore, principal component analysis (PCA), protein signaling network inference methods, and machine learning approaches facilitate interpretable descriptions and predictions of protein phosphorylation sites, the resulting protein-protein interactions, and signaling network inferences to develop an overall understanding of signaling networks involved in developmental or stress response mechanisms [6, 7, 10]. Here, we describe a comprehensive R-based pipeline that performs quality controls, statistical analysis, and signaling network inference on time-course label-free phosphoproteomics. A major benefit of this method is that it is "data-driven" and relies on phosphorylation intensities instead of on meta-analyses or aggregated data from the literature on protein phosphorylation sites and consensus sequence motifs, which are limited for non-model species. Moreover, this method partially overcomes the limitations of detecting phosphosites by subsetting the phosphoproteome data set in two subsets. The first subset includes phosphosites with a user-defined number of valid values per replicate for statistical analysis of differential phosphorylation across time and condition. The second subset contains absence/presence phosphosites that are reliably not detected for one or more time points to help infer the timing and triggers of phosphorylation or dephosphorylation. In this chapter, we describe methods to process label-free phosphoproteome time-course data. We also describe a method to infer a signaling network from a time series using dynamic Bayesian principles. To facilitate the use of these methods, we have developed an R package that integrates these analyses, which we will describe further in this chapter. Both of these methods can guide the development of further hypotheses concerning phosphoproteomic dynamics.

2 Materials

1. R 4.1.0 [11].
2. Required R CRAN and *Bioconductor* packages (*see* **Note 1**).

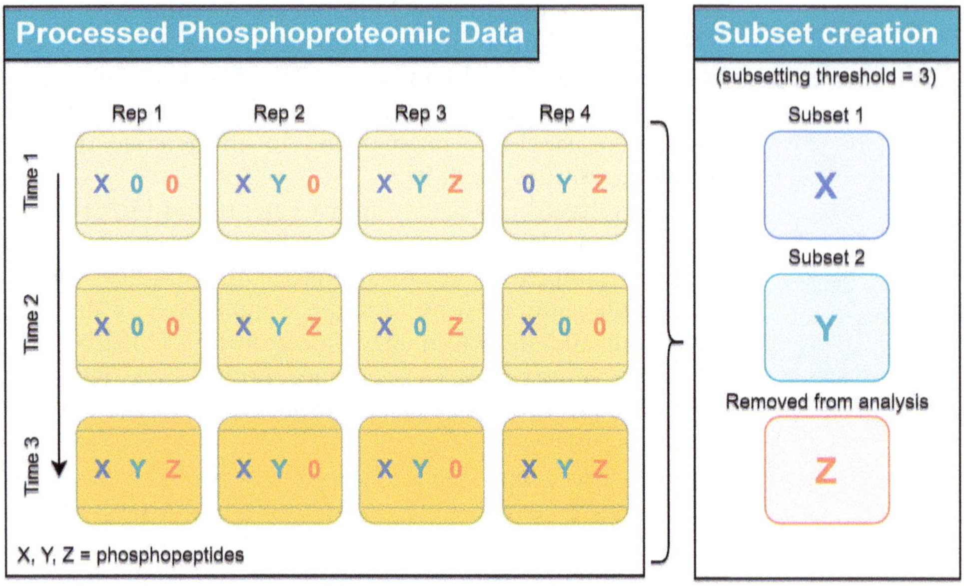

Fig. 2 Diagram depicting how each phosphosite is assigned to either subset 1 or 2 or removed from further analysis. X, Y, and Z represent normalized intensity values of a distinct phosphosite. 0 represents an intensity value of zero. There are 4 technical replicates and 3 time points. Each phosphosite is assigned to a specific subset based on the amount of zero or nonzero values present across each replicate. Phosphosite X is assigned to Subset 1 because its valid values exceed the threshold for each time point. Phosphosite Y is assigned to Subset 2 because its number of zeros or valid values exceed the threshold at each time point. Phosphosite Z is removed from analysis because it did not meet either subset criteria

3. File listing the protein AGI identifier of all kinases and phosphates of the species of interest.
4. MaxQuant output file of a label-free phosphoproteome time-course dataset (*see* **Note 2**).

3 Methods

3.1 R Package for Phosphoproteomic Analysis, Visualization, and Signaling Network Inference

The R package, NetPhorce, contains four main function groups for (i) phosphoproteomic data processing and statistics, (ii) signaling network inference, (iii) the creation of output data frames, and (iv) data visualization (Fig. 1). To use NetPhorce:

1. Install the package from GitHub (https://github.com/ksong4/NetPhorce).
2. Load a MaxQuant search file of a label-free time-course experiment into the R environment.

3. Use the functions confirmColumnNames() and confirmIntensityColumns() to select the necessary data columns, including phosphosite, amino acid, and position, and the intensity columns for further downstream data analysis.
4. Once all the required columns are selected and confirmed, the processData() function should be used to conduct data quality controls and statistical analysis.
5. To retrieve a .csv table containing the processed data, including the normalized values, averages, p-values, and q-values use the function extractSummariseTable(). This processed data can be visualized into heatmaps, dot plots, or line plots, using one or more of the included visualization functions.
6. To perform signaling network inference on the processed data and identify regulatory interactions between kinases/phosphatases and downstream targets, the following three functions should be used sequentially: (i) validateKinaseTable() to identify the kinases and phosphatases within the dataset, (ii) regulationCheck() to set key parameters for signaling network inference that determine stringency and thus accuracy, (iii) networkAnalysis() to score regulatory interactions using Bayesian principles. As with the data processing output, all results are saved in a single R object.
7. The inferred signaling network should be visualized using the plotNetPhorce() function.
8. To retrieve the inferred regulatory interactions and their score in a .csv table, use the function extractNetworkResults().

Detailed tutorials and guides are available on the NetPhorce Website (https://ksong4.github.io/NetPhorce/). In the next section, we describe a detailed protocol to execute the source code underlying the NetPhorce R package encompassing the phosphoproteomic analysis and data-driven signaling network inference.

3.2 Detailed Source Code for Phosphoproteomic Analysis and Signaling Network Inference

3.2.1 Analyzing Label-Free Phosphoproteomics from MaxQuant Searches

1. Open R and install all necessary R packages (*see* **Note 1**).
2. Load the MaxQuant output file (Box 1). The MaxQuant output file should provide the raw intensity quantification data of each identified phosphosite at all time points and replicates.
3. Remove phosphosites that are reverse phosphosites, lack a valid position, are potential contaminants, and are below the set localization probabilization threshold (*see* **Note 2**) (Box 1). By default, the localization probabilization threshold is set to 0.75. However, the user should enter a different value depending on the desired stringency.

4. Assign a unique ID to each phosphosite according to each protein ID, modified amino acid residue, and position of modification site by creating an additional column in your data frame (Box 1).
5. Select the unique ID column and the raw intensity columns within the filtered dataset for further processing. Raw intensity columns can be selected through regular expression (Box 1).

Box 1 Loading and filtering label-free phosphoproteomics data from MaxQuant searches

```
input = read_delim("C:\\Users\\DIRECTORY\\FOLDER\\FILE NAME", delim = "\t",
  quote = '"', col_types = cols(.default = "c")) # Load the MaxQuant output file

Probability = 0.75 # Localization probability threshold value

data.select <- input %>%
  filter(is.na(Reverse)) %>% # Remove reverse phosphosites
  filter(!is.na(`Position`)) %>% # Require a valid position
  filter(is.na(`Potential contaminant`)) %>% # Remove possible contaminants
  filter(as.numeric(`Localization prob`) > as.numeric(Probability)) %>%
  # Remove phosphosites with a localization probability below the set threshold
  mutate(UniqueID = paste(Protein, "_", `Amino acid`, Position, sep = "")) %>%
  # Assign a unique ID and column names for each phosphosite
  dplyr::select(UniqueID, starts_with("Intensity ") & matches("___\\d"))
  # Select the needed intensity columns
```

6. To facilitate further data mining, reorganize the filtered dataset in a long format and split the names of the intensity columns in the separate variables. The column names should include "intensity," "condition," "timepoint," "replicate," and "multiplicity" (*see* **Note 3**) (Box 2). The variable *Order* should be updated depending on the variables included in the column name and the order of those variables.
7. Assign Sample ID to tabulate the experimental factors associated with each phosphosite signal intensity such as condition, time point, and replicate to facilitate the following steps (Box 2).
8. For further statistical analysis, create two data subsets (Fig. 2): (i) the first subset should contain phosphosites with a number of valid values per replicate at one/more time points exceeding the subsetting threshold and will be used for statistical analysis, (ii) the second subset should contain phosphosites with both a number of valid values and zeros exceeding the subsetting threshold per replicate at one/more time points and will be further referred to as the absence/presence subset (Fig. 2).

These phosphosites are considered absent or unphosphorylated at one/more time points. By default, the subsetting threshold value is set to 3; however, the threshold should be adjusted according to the number of replicates and desired stringency (*see* **Note 4**) (Box 2). Any phosphosite that does not meet these criteria should be removed.

Box 2 Subset creation

```
Order = c("intensity","condition", "timepoint", "replicate","Multiplicity")
# Set variables based on the intensity column names

data.t = data.select %>% na_if(0) %>%
  pivot_longer(-UniqueID, values_drop_na=TRUE) %>%
  mutate(value = as.numeric(value)) %>%
  separate(name, into = c(Order), remove=FALSE)
  # Create a long format data frame, containing variables set in the variable Order

data.t = data.t %>%
  unite(SampleID,condition,timepoint,replicate,remove=FALSE) %>%
  unite(UniqueID, UniqueID, Multiplicity) %>% dplyr::select(-name,-intensity)
  # Assign a SampleID and UniqueID to each phosphosite

Threshold = 3 # Subsetting threshold value

data.filtered = data.t %>%
  add_count(UniqueID, timepoint, condition, name = "obs_tp") %>%
  # Count observations that are not zero for each proteinID
  group_by(timepoint) %>% mutate(Reps = (length(unique(replicate)))) %>%
  ungroup() %>% group_by(UniqueID) %>%
  mutate(obs_tr = length(unique(condition))) %>% # Count the number of conditions
  filter(any(obs_tp >= Threshold) &
  all(obs_tp >= Threshold | obs_tp <= (Reps - Threshold))) %>%
  ungroup() %>% filter(obs_tr == max(obs_tr))
  # Keep observations that meet subset requirements
```

3.2.2 Normalization and Statistical Analysis

1. Create a data table, "ExpDesign" to arrange the SampleIDs in a format that is required for the variance stabilization normalization function used in the following normalization steps (Box 3).
2. Transform data to a wide format, to organize data, based on each variable and SampleID, so it will meet the input requirements of the DEP::normalize_vsn() function in the following step (Box 3).
3. Implement variance stabilization normalization using the R/vsn package [13, 14] to normalize the log-transformed dataset to minimize the variation between replicates, *see* **Notes 5–6** (Box 3).

Box 3 Normalizing the data using variance stabilization normalization (vsn)

```
uniqueTimes = data.frame(Origin = unique(data.filtered$timepoint)) %>%
  mutate(Time = as.numeric(str_extract(Origin, "[0-9]+"))) %>%
  arrange(Time) # Sorts the data frame according to the time points

ExpDesign = data.filtered %>% distinct(SampleID) %>%
  # Format variable table that is required for the vsn
  separate(SampleID, into = c("experiment","timepoint","replicate"),
  remove=FALSE) %>%
  mutate(condition = paste(experiment, timepoint, sep = "_")) %>%
  mutate(label = SampleID, ID = SampleID) %>%
  arrange(timepoint) %>%
  column_to_rownames("SampleID")

data.filtered.spread = data.filtered %>%  # Generates a data frame in wide format
  dplyr::select(UniqueID,SampleID,value) %>%
  spread(SampleID,value, fill=NA) %>%
  dplyr::select(UniqueID, ExpDesign %>% rownames) %>%
  column_to_rownames("UniqueID")

row_data = rownames(data.filtered.spread) %>%
  enframe(value = "ID") %>% mutate(name = ID) %>% as.data.frame
  # Creates a data frame with the UniqueIDs as column

se = SummarizedExperiment(assays = log2(as.matrix(data.filtered.spread)), colData =
ExpDesign, rowData = row_data)
# Creates a SummarizedExperiment-class, containing the log transformed intensity
# values and necessary metadata

se_norm = DEP::normalize_vsn(se) # Variance stabilization normalization
data.norm = assays(se_norm)[[1]] %>% as.data.frame() %>%
  rownames_to_column("UniqueID") %>% as_tibble()

data.filtered = data.filtered %>% ungroup() %>%
  complete(UniqueID,timepoint,condition,fill=list(obs_tp = 0)) %>% # Complete
  allows us to consider timepoints with no data
  left_join(data.norm %>% gather(SampleID, normValue, -UniqueID)) %>%
  group_by(UniqueID) %>% mutate(set = if_else(any(obs_tp <
          Threshold),"UniqueSet","StatsSet")) %>% ungroup() %>%
  filter(!is.na(SampleID)) # Labels subset 1, Statset, or 2, UniqueSet
```

4. On subset 1, fit a linear mixed model to statistically analyze the transformed and normalized intensities of phosphosites, *see* **Note 7** (Box 4). Adjust the model (Eq. 1) based on the experimental design. In any model, include a random effect for replicates. In an experiment with one condition and multiple time points, $n = 1$ & $t > 1$, include a fixed effect for each time and replicate variables. In an experiment with multiple conditions but 1 time point, $n > 1$ & $t = 1$, then include a fixed effect for the replicate and condition variables. In an experiment with one condition and several time points $n > 1$ & $t = 1$, then include a main effect for time, condition, replicate variables, and an interaction term of the condition and time.

$$\Upsilon = \begin{cases} \mu + \alpha_i + \gamma_k + \varepsilon & \text{if } n = 1 \quad t > 1 \\ \mu + \beta_j + \gamma_k + \varepsilon & \text{if } n > 1 \quad t = 1 \\ \mu + \alpha_i + \beta_j + \alpha\beta_{ij} + \gamma_k + \varepsilon & \text{if } n > 1 \quad t > 1 \end{cases} \quad (1)$$

Equation 1: Linear mixed model where n is the number of conditions, t is the number of time points, Υ is the phosphorylation intensity. μ is the overall mean, α is the main effect of time, β is the main effect of condition, $\alpha\beta$ is the interaction term of time and condition, γ is the fixed effect of the replicate variable, and ε is the random effect between technical replicates. i is each time point, j is each condition, and k is each phosphosite

5. Perform a p-value correction using the R/qvalue package. If an error is returned, *see* **Note 8** (Box 4).
6. Select the statistically significant phosphosites according to the experimental design (Box 4).

Box 4 Performing statistical analysis by fitting a linear mixed model

```
n = n_distinct(ExpDesign$experiment) # Number of conditions
t = n_distinct(uniqueTimes$Origin) # Number of time points

# Phosphosites for statistical analysis, in Subset 1, will be fitted to a linear
# model based on an experimental design
if(n == 1) { # Experimental design with 1 condition
  data.filtered.aov = data.filtered %>% filter(set == "StatsSet") %>%
    group_by(UniqueID) %>% nest() %>% ungroup() %>%
    mutate(AOV = map(data, ~ aov(normValue ~ replicate + timepoint +
       Error(replicate), data=.x)) ) %>%
    mutate(result = map(AOV,broom::tidy))
  data.filtered.aov.summary = data.filtered.aov %>%
    dplyr::select(UniqueID,result) %>% unnest(cols = c(result)) %>%
    filter(!is.na(p.value)) %>%
    mutate(qvalue = qvalue(p.value, lfdr.out = TRUE)[["qvalues"]])
    # p-value correction step. Rewrite if error according to note 8
} else {
  if (n > 1) { # Experimental design with more than 1 condition
    if (t == 1) { # Experimental design with 1 time point
        data.filtered.aov = data.filtered %>% filter(set == "StatsSet") %>%
            group_by(UniqueID) %>% nest() %>% ungroup() %>%
            mutate(AOV = map(data, ~ aov(normValue ~ replicate + condition +
               Error(replicate), data=.x)) ) %>%
            mutate(result = map(AOV,broom::tidy))
        data.filtered.aov.summary =  data.filtered.aov %>%
            dplyr::select(UniqueID,result) %>% unnest(cols = c(result)) %>%
            filter(!is.na(p.value)) %>%
            filter(term == "condition") %>%
            mutate(qvalue = qvalue(p.value, lfdr.out = TRUE)[["qvalues"]])
            # p-value correction step. Rewrite if error according to note 8
    } else { # Experimental design with more than 1 more time point
        data.filtered.aov = data.filtered %>% filter(set == "StatsSet") %>%
            group_by(UniqueID) %>% nest() %>% ungroup() %>%
            mutate(AOV = map(data, ~ aov(normValue ~ replicate +
               timepoint*condition + Error(replicate), data=.x)) ) %>%
            mutate(result = map(AOV,broom::tidy))
```

(continued)

```
        data.filtered.aov.summary = data.filtered.aov %>%
            dplyr::select(UniqueID,result) %>% unnest(cols = c(result)) %>%
            filter(!is.na(p.value)) %>%
            filter(term == "condition") %>%
            mutate(qvalue = qvalue(p.value, lfdr.out = TRUE)[["qvalues"]])
            # p-value correction step. Rewrite if error according to note 8
      }
    }
}

data.significant.proteins = data.filtered %>%
  # Combine significant differentially phosphorylated phosphosites and Subset 2
  group_by(set, UniqueID,timepoint, condition) %>% filter(obs_tp >= Threshold) %>%
  summarize(m = mean(normValue, na.rm=TRUE)) %>% ungroup() %>%
  complete(UniqueID, condition,fill = list(set = "UniqueSet", m=0),
            timepoint = uniqueTimes$Origin[1]) %>%
  spread(timepoint, m, fill = 0) %>%
  left_join(data.filtered.aov.summary %>% dplyr::select(UniqueID, qvalue)) %>%
  filter(qvalue < 0.05 | set == "UniqueSet") %>%
  gather(time, avgValue, -UniqueID, -qvalue, -condition, -set) %>%
  group_by(condition, UniqueID) %>% add_count(avgValue == 0) %>%
  filter(any(`avgValue == 0` == "FALSE" & n >= 3)) %>%
  dplyr::select(-set, -n, -`avgValue == 0`) %>%
  mutate(tp = as.numeric(str_extract(time,"[:digit:]+"))) %>%
  # Interpret timepoints by digits pattern through regular expression
  mutate(Time = fct_reorder(time,tp))
```

3.2.3 Identify Regulator-Target Pairs for Network Inference

1. Load a file listing all kinases and phosphatases, *see* **Note 9** and Box 5.

Box 5 Identifying kinases/phosphatases within the data

```
Kinases = readxl::read_xlsx("C:\\Users\\DIRECTORY\\FOLDER\\FILE NAME")

data.significant.proteins = data.significant.proteins %>%
    separate(UniqueID, into = c("Model_name", "AA", "multiplicity"), sep = "_",
             remove=FALSE) %>%
    mutate(Model_name = sub("(\\.\\d{1,2})$", "", Model_name)) %>%
    mutate(Model_name = str_to_upper(Model_name)) %>%
    left_join(Kinases %>% mutate(isKinase = TRUE), by = c("Model_name"="ID")) %>%
    tidyr::replace_na(list(isKinase = FALSE))
```

2. Scale the intensity values for each experiment by median centering. Then, discretize the changes in intensities over time of phosphosites to 1 (increasing), 0 (no change), and − 1 (decreasing). Two threshold variables a and b should be set to define the fold change that is considered an increase or decrease in phosphorylation intensity, respectively. A third threshold variable c defines the bottom percentage of all fold changes that should be considered as unchanged (Eq. 2), *see* **Note 10** (Box 6) (Fig. 3).

$$p_\Delta(t) = \begin{cases} 0, & |\,p(t) - p(t-1)\,| < quantile(\forall |p(t) - p(t-1)|,\ c) \\ 1, & p(t) - p(t-1) > a{\cdot}p(t-1) \\ -1, & p(t) - p(t-1) < -b{\cdot}p(t-1) \end{cases} \tag{2}$$

Equation 2: The discretization scheme of the phosphorylation intensities. Where *a, b*, and *c* are user-defined inputs (restricted to >0) that are default set to 0.25, 0.25, and 0.1, respectively

Box 6 Scaling and discretizing the phosphorylation intensities

```
numberOfDiscreteStates = 1
a = 0.25 # Variable a
b = 0.25 # Variable b
c = 0.1 # Variable c

data.significant.proteins = data.significant.proteins %>%
  arrange(condition, time) %>% group_by(condition) %>%
  mutate(conditionMedian = median(avgValue[avgValue > 0], na.rm=TRUE)) %>%
  ungroup() %>% # Scale the intensity values by median centering
  group_by(condition, UniqueID) %>%
  mutate(mvalue = avgValue - conditionMedian) %>%
  mutate(change = mvalue - lag(mvalue)) %>% ungroup() %>% group_by(condition) %>%
  mutate(quantile_threshold = quantile(abs(change), probs = c,
    na.rm=TRUE)) %>% group_by(condition, UniqueID) %>%
  mutate(sigChange = case_when( # Setting p_delta(t) according to equation 4
    change > pmax(abs(a*lag(mvalue)), quantile_threshold ) ~ TRUE,
    change < pmin(-1*abs(b*lag(mvalue)), -quantile_threshold ) ~ TRUE,
    TRUE ~ FALSE
  )) %>%
  mutate(sigChangeSign = sign(change) * sigChange) %>%
  group_by(condition, UniqueID) %>%
  filter(any(sigChangeSign != 0)) %>% ungroup() %>%
  mutate(discreteState = case_when( # Discretize intensity values, step 18
    abs(mvalue) >= abs(conditionMedian) ~ 5,
    mvalue == 0 ~ 1,
    TRUE ~ sign(mvalue)*ntile(abs(mvalue), numberOfDiscreteStates)))
```

3. To limit the potential regulator-target pairs for network inference, *see* **Note 12**, compare the phosphorylation intensity changes between all possible regulator-target pairs across the time course (Eqs. 3a, 3b and 3c). If a pair changes together over 50% of the time, declare that there is a coregulator-target relationship. Additionally, compute this process with the intensity changes of the targets delayed one time step, *see* **Note 13**. Likewise, if the pair changes together over 50% of the time, declare this to be a potential regulator-target relationship. Thus, a protein is a potential regulator of a target protein if and only if it exhibits a change in phosphorylation intensity at the same time or immediately prior to a change in phosphorylation intensity of the target for at least 50% of the time points (Box 7).

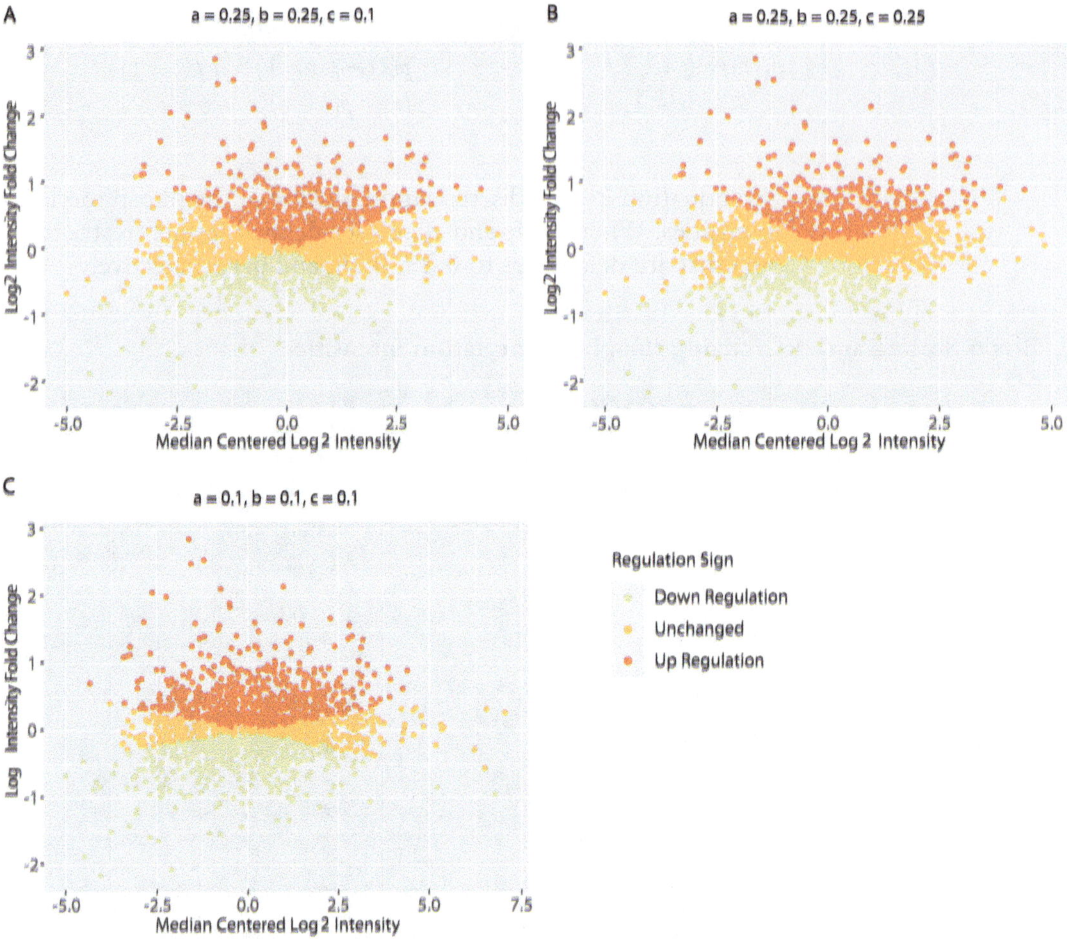

Fig. 3 Selecting the thresholds to determine the directionality (down, unchanged, or up) in any change in phosphorylation status. Dot plots show the $\log_2$ intensity fold change of each phosphosite at each time point against its median centered $\log_2$ intensity. Two threshold variables *a* and *b* are set to define the fold change that is considered an increase or decrease in phosphorylation intensity, respectively. A third threshold variable *c* defines the bottom percentage of all fold changes that should be considered as unchanged

$$\left(\frac{\sum_{t=1}^{n} r_\Delta(t) = p_\Delta(t)}{n}\right) \; or \; \left(\frac{\sum_{t=1}^{n} r_\Delta(t-1) = p_\Delta(t)}{n-1}\right) > 0.5 \quad (3a)$$

$$r_\Delta(t) = r(t) - r(t-1) \quad (3b)$$

$$p_\Delta(t) = p(t) - p(t-1) \quad (3c)$$

4. Equations 3a, 3b and 3c: Identifying candidate regulator-target relationships. Where $r_\Delta(t)$ and $p_\Delta(t)$ represent the discretized changes in candidate regulators and targets, respectively. And the variable t denotes the time step

Box 7 Identifying all possible regulator-target pairs

```
n_levels = data.significant.proteins %>% filter(!is.na(discreteState)) %>%
  distinct(discreteState) %>% nrow
maxConditionMatching = t-1
percentChangeThreshold = 0.5

# Identify all possible target-regulator combinations
data.scoring = data.significant.proteins %>% group_by(condition) %>% nest()

data.scoring = data.scoring %>%
# Expand the possible edges from kinases to genes, then
# filter data where changes correspond
  mutate(potentialEdges = map(data, ~ {
  # Reg -> target are filtered by sigSignChange, not by change in state
    allEdges = .x %>% dplyr::select(UniqueID, isKinase, Time, sigChangeSign) %>%
      filter(!is.na(sigChangeSign)) %>%
      tidyr::expand(nesting(regulator = UniqueID, regulatorCondition = Time,
                    regulatorChange = sigChangeSign, regulatorisKinase = isKinase),
                    nesting(target = UniqueID, targetCondition = Time,
                    targetChange = sigChangeSign)) %>%
      filter(regulatorisKinase) %>% dplyr::select(-regulatorisKinase) %>%
      filter(regulator != target) %>%
      filter(regulatorChange != 0 & targetChange != 0) %>%
      filter(!is.na(levels(targetCondition)[as.numeric(targetCondition)-1])) %>%
      # The regulators proper timepoints and % matching changes
      filter(regulatorCondition == targetCondition | regulatorCondition ==
                    levels(targetCondition)[as.numeric(targetCondition)-1])
    allEdges.filterd = allEdges %>% group_by(regulator, target) %>%
      add_count(coReg = regulatorCondition == targetCondition,
                    wt = (abs(regulatorChange) ==
                    abs(targetChange))/maxConditionMatching,
                    name = "percentChange") %>%
      ungroup() %>% filter(percentChange > percentChangeThreshold)
    return(allEdges.filterd)
  }))
```

3.2.4 Network Inference Through Dynamic Bayesian Scoring

1. To ease computation, discretize the normalized $\log_2$-transformed intensity values. All the intensity values above or below the median with respect to the condition should be discretized as 1 and − 1, respectively. The absent or undetected intensity values should be discretized to 5, *see* **Note 11** (Box 6).
2. To score the identified potential regulator-target relationships, compute the Bayesian Dirichlet equivalent uniform (BDeu), *see* **Note 14** (Box 8), between:
   (a) Each identified regulator-target or coregulator-target.
   (b) Pairs of regulators and their identified target.
   (c) Pairs of regulators and coregulators and their identified target.
   (d) Pairs of coregulators and their identified target

$$BDeu(D,G)=\sum_{i=1}^{n}\sum_{j=1}^{q_i}\left(log\left(\frac{\left(\frac{\alpha}{q_i}\right)}{\left(\sum_{k=1}^{r_i}N_{ijk}+\frac{\alpha}{q_i}\right)}\right)+\sum_{k=1}^{r_i} log\left(\frac{\left(N_{ijk}+\frac{\alpha}{q_i}\right)}{\left(\frac{\alpha}{r_i q_i}\right)}\right)\right) \tag{4}$$

Equation 4: BDeu score where G refers to the Bayesian graph, D refers to the dataset containing the time point observations, N_{ij} indicates the number of data vectors in which target i, has the value k while its parents are in configuration j. α equals $1E^{-15}$, a hyperparameter of the Dirichlet distribution

3. To identify the most probable regulator-target interactions, for each target, select the highest BDeu scoring regulators/coregulators as their respective regulator (Box 8).

Box 8 Computing the regulator-target interaction probabilities with the BDeu score

```
ESS = 1e-15 # Dirichlet distribution parameter

data.scoring.searches = data.scoring %>%
  dplyr::select(condition, potentialEdges) %>% unnest(cols = potentialEdges) %>%
  mutate(regulator = paste(regulator, coReg, sep = "**"), .keep = "unused") %>%
  group_by(condition, target) %>% nest() %>%
  mutate(neighborCombinations = map(data, possibly( ~ {
    allSearchCombinations = bind_rows(
    # Generate combinations of regulators within each searchID to be scored.
      .x %>% distinct(regulator) %>% pull(regulator) %>% as.list %>%
        possibly(~combn(.x, m=1), tibble())() %>% as.data.frame %>%
        gather(searchID,regulator) %>% mutate(searchSizeK = 1),
      .x %>% distinct(regulator) %>% pull(regulator) %>% as.list %>%
        possibly(~combn(.x, m=2), tibble())() %>% as.data.frame %>%
        gather(searchID,regulator) %>% mutate(searchSizeK = 2),
      .x %>% distinct(regulator) %>% pull(regulator) %>% as.list %>%
        possibly(~combn(.x, m=3), tibble())() %>% as.data.frame %>%
        gather(searchID,regulator) %>% mutate(searchSizeK = 3)
    ) %>% as_tibble %>% unnest(cols = c(regulator)) %>%
    tidyr::separate(regulator, c("regulator", "coReg"), sep = "\\*\\*", remove = T)
  },tibble()))) %>% ungroup()

coRegulater = data.scoring.searches %>% dplyr::select(condition, target,
neighborCombinations) %>% unnest(neighborCombinations) %>%
  dplyr::select(condition, target, regulator, coReg) %>% distinct()
  # Regulator is active a time point prior to that of the target;
  # CoReg is active at the same time point of the target

allEdgesCompareState.Nj = data.significant.proteins %>%
  group_by(condition) %>% nest() %>%
  mutate(data = map(data, ~ {.x %>%
      dplyr::select(UniqueID, isKinase, Time, discreteState) %>%
      filter(!is.na(discreteState)) %>%
      tidyr::expand(nesting(regulator = UniqueID, regulatorCondition = Time,
                     regulatorState = discreteState, regulatorisKinase = isKinase),
                     nesting(target = UniqueID, targetCondition = Time, targetState
                     = discreteState)) %>% filter(regulatorisKinase) })) %>%
```

(continued)

```
  unnest(cols = c(data)) %>% right_join(coRegulater) %>%
  group_by(condition) %>% nest() %>%
  mutate(data = map(data, ~ {.x %>% filter((as.numeric(regulatorCondition) ==
                    as.numeric(targetCondition) & coReg == TRUE) |
                    (as.numeric(regulatorCondition) ==
                    as.numeric(targetCondition)-1) & coReg == FALSE) %>%
                    filter(regulator != target)})) %>%
  unnest(cols = c(data))

allEdgesCompareState.Nj.full = data.scoring.searches %>%
  dplyr::select(condition, target, neighborCombinations) %>%
  unnest(neighborCombinations) %>% left_join(allEdgesCompareState.Nj)
data.scoring.genist.1reg = allEdgesCompareState.Nj.full %>%
  filter(searchSizeK == 1) %>% dplyr::select(-regulatorisKinase) %>%
  group_by(condition, target, searchID) %>%
  mutate(Nj = case_when(
    regulatorState == -1 & targetState == -1 ~ "A",
    regulatorState == -1 & targetState == 1 ~ "B",
    regulatorState == -1 & targetState == 5 ~ "C",
    regulatorState == 1 & targetState == -1 ~ "D",
    regulatorState == 1 & targetState == 1 ~ "E",
    regulatorState == 1 & targetState == 5 ~ "F",
    regulatorState == 5 & targetState == 1 ~ "G",
    regulatorState == 5 & targetState == -1 ~ "H",
    regulatorState == 5 & targetState == 5 ~ "I")) %>%
  mutate(qi = case_when(
    regulatorState == -1 ~ "A",
    regulatorState == 1 ~ "B",
    regulatorState == 5 ~ "C")) %>%
  add_count(qi) %>% dplyr::count(condition, target, coReg, searchID, regulator,
searchSizeK, Nj, n, name = "Nj.count") %>%
  mutate(BDei =
           log(gamma(Nj.count+ESS/(n_levels*n_levels^searchSizeK))/gamma(ESS/
           (n_levels*n_levels^searchSizeK)))) %>% ungroup() %>%
  group_by(condition, target, regulator, searchID, searchSizeK, coReg, n) %>%
  summarize(BDei = sum(BDei)) %>%
  mutate_at(vars(BDei), ~ BDei + log(gamma(ESS/n_levels^searchSizeK)/gamma(n+ESS/
           n_levels^searchSizeK))) %>%
  group_by(condition, target, regulator, searchID, searchSizeK, coReg) %>%
  summarize(BDei = sum(BDei)) %>%  ungroup() %>%
  group_by(condition, target) %>% filter(BDei == max(BDei))

tp0 = as.character(uniqueTimes$Origin)[1]

data.scoring.genist.2reg = allEdgesCompareState.Nj.full %>%
  filter(searchSizeK == 2) %>% dplyr::select(-regulatorisKinase) %>%
  group_by(condition, target, searchID) %>%
  filter(regulatorCondition!=tp0 | targetCondition!=tp0 | all(coReg == TRUE)) %>%
  ungroup() %>% group_by(condition, target, searchID, targetCondition) %>%
  mutate(Reg = row_number()) %>%
  pivot_wider(names_from = Reg, values_from = c(regulator, regulatorState,
              coReg, regulatorCondition)) %>%
  ungroup() %>% group_by(condition, target, searchID) %>%
  mutate(Nj = case_when(
    regulatorState_1 == -1 & regulatorState_2 == -1 & targetState == -1 ~ "A",
    regulatorState_1 == -1 & regulatorState_2 == -1 & targetState == 1 ~ "B",
    regulatorState_1 == -1 & regulatorState_2 == -1 & targetState == 5 ~ "C",
    regulatorState_1 == -1 & regulatorState_2 == 1 & targetState == 1 ~ "D",
```

(continued)

```
    regulatorState_1 == -1 & regulatorState_2 == 1 & targetState == -1 ~ "E",
    regulatorState_1 == -1 & regulatorState_2 == 1 & targetState == 5 ~ "F",
    regulatorState_1 == -1 & regulatorState_2 == 5 & targetState == -1 ~ "G",
    regulatorState_1 == -1 & regulatorState_2 == 5 & targetState == 1 ~ "H",
    regulatorState_1 == -1 & regulatorState_2 == 5 & targetState == 5 ~ "I",
    regulatorState_1 == 1 & regulatorState_2 == -1 & targetState == -1 ~ "J",
    regulatorState_1 == 1 & regulatorState_2 == -1 & targetState == 1 ~ "K",
    regulatorState_1 == 1 & regulatorState_2 == -1 & targetState == 5 ~ "L",
    regulatorState_1 == 1 & regulatorState_2 == 1 & targetState == 1 ~ "M",
    regulatorState_1 == 1 & regulatorState_2 == 1 & targetState == -1 ~ "N",
    regulatorState_1 == 1 & regulatorState_2 == 1 & targetState == 5 ~ "O",
    regulatorState_1 == 1 & regulatorState_2 == 5 & targetState == -1 ~ "P",
    regulatorState_1 == 1 & regulatorState_2 == 5 & targetState == 1 ~ "Q",
    regulatorState_1 == 1 & regulatorState_2 == 5 & targetState == 5 ~ "R",
    regulatorState_1 == 5 & regulatorState_2 == -1 & targetState == -1 ~ "S",
    regulatorState_1 == 5 & regulatorState_2 == -1 & targetState == 1 ~ "T",
    regulatorState_1 == 5 & regulatorState_2 == -1 & targetState == 5 ~ "U",
    regulatorState_1 == 5 & regulatorState_2 == 1 & targetState == 1 ~ "V",
    regulatorState_1 == 5 & regulatorState_2 == 1 & targetState == -1 ~ "W",
    regulatorState_1 == 5 & regulatorState_2 == 1 & targetState == 5 ~ "X",
    regulatorState_1 == 5 & regulatorState_2 == 5 & targetState == -1 ~ "Y",
    regulatorState_1 == 5 & regulatorState_2 == 5 & targetState == 1 ~ "Z",
    regulatorState_1 == 5 & regulatorState_2 == 5 & targetState == 5 ~ "AA")) %>%
  mutate(qi = case_when(
    regulatorState_1 == -1 & regulatorState_2 == -1 ~ "A",
    regulatorState_1 == -1 & regulatorState_2 == 1 ~ "B",
    regulatorState_1 == -1 & regulatorState_2 == 5 ~ "C",
    regulatorState_1 == 1 & regulatorState_2 == -1 ~ "D",
    regulatorState_1 == 1 & regulatorState_2 == 1 ~ "E",
    regulatorState_1 == 1 & regulatorState_2 == 5 ~ "F",
    regulatorState_1 == 5 & regulatorState_2 == -1 ~ "G",
    regulatorState_1 == 5 & regulatorState_2 == 1 ~ "H",
    regulatorState_1 == 5 & regulatorState_2 == 5 ~ "I")) %>% add_count(qi) %>%
  dplyr::count(condition, target, coReg_1, coReg_2, searchID, regulator_1,
           regulator_2, searchSizeK, n, Nj, name = "Nj.count") %>%
  mutate(BDei =
         log(gamma(Nj.count+ESS/(n_levels*n_levels^searchSizeK))/gamma(ESS/
         (n_levels*n_levels^searchSizeK)))) %>% ungroup() %>%
  group_by(condition, target, regulator_1, regulator_2, coReg_1, coReg_2,
         searchID, searchSizeK, n) %>%
  summarize(BDei = sum(BDei)) %>%
  mutate_at(vars(BDei), ~ BDei + log(gamma(ESS/n_levels^searchSizeK)/
         gamma(n+ESS/n_levels^searchSizeK))) %>%
  group_by(condition, target, regulator_1, regulator_2, coReg_1, coReg_2,
         searchID, searchSizeK) %>%
  summarize(BDei = sum(BDei)) %>% ungroup() %>% group_by(condition, target) %>%
  filter(BDei == max(BDei))

# We keep only the max scores across entire dataset
data.scoring.genist = bind_rows(data.scoring.genist.1reg %>% mutate(set = "1"),
         data.scoring.genist.2reg %>% pivot_longer(cols = c(ends_with("_1"),
         ends_with("_2")), names_to = c(".value", "set"),
         names_pattern = "(.+)_(.+)")) %>%
  distinct(condition, target, regulator, coReg, BDei) %>%
  group_by(condition, target) %>%
  filter(BDei == max(BDei)) %>% mutate(BDei = exp(BDei), .keep = "unused")
```

4. Determine for each inferred interaction whether it is a phosphorylation or dephosphorylation. If a regulator-target pair showed a similar or opposite trend over time, then the regulation is considered phosphorylation or dephosphorylation, respectively. The phosphorylation dynamics should be evaluated by counting the same and opposite discretized changes over time between the regulator/coregulator and target (calculated in **step 16**) (*see* **Note 10**). If the same changes are larger (smaller) than the opposite changes, then the regulation is accounted as phosphorylation (dephosphorylation). If the same and opposite changes are equal, then the regulation is denoted as ***undetermined*** (Box 9).

Box 9 Identifying the sign of each inferred interaction (phosphorylation or dephosphorylation)

```
data.scoring.sign = data.scoring %>%
# Expand with all identified SigChangeSign,
# count the changes in the same and opposite direction
  dplyr::select(potentialEdges, condition) %>%
  unnest(c(condition, potentialEdges)) %>% dplyr::select(-percentChange) %>%
  mutate(coReg = as.character(coReg), .keep = "unused") %>%
  right_join(data.scoring.genist) %>%
  group_by(condition, target, regulator, coReg) %>%
  dplyr::count(regulatorChange == targetChange) %>%
  pivot_wider(names_from = "regulatorChange == targetChange", values_from = n,
    values_fill = 0) %>%
  mutate(regulation = case_when(
    `FALSE` > `TRUE` ~ "dephosphorylation",
    `TRUE` > `FALSE` ~ "phosphorylation",
    TRUE ~ "undetermined"
  ))
```

4 Notes

1. Install the following R CRAN and bioconductor packages by using install.packages("<package name>"), and load them using library(<package name>): assertthat 0.2.1 [15], DEP 1.14.0 [16], qvalue 2.24.0 [17], SummarizedExperiment 1.22.0 [18], tidyverse 1.3.1 [12], and vsn 3.60.0 [14].
2. MaxQuant compares the hits associated with the forward phosphosites to those of the reverse phosphosites to calculate the amount of false positives. These filtering steps will remove hits associated with the reverse phosphosites as well as those which lack a valid position. During Mass Spectrometry (MS), contaminant peptides may be introduced [19, 20]. Localization probability estimates the likelihood that a residue modification was identified accurately based on the MS ion fragmentation results [21].
3. For the names of the intensity columns in the MaxQuant output file, include a condition, time point, and replicate

variable in the name and separate each variable by the same symbol, preferably an underscore, to facilitate filtering and sample or unique ID creation. For example, WT_0min_A. The condition may refer to different lines, treatments, varieties, etc.

4. By default, the subsetting threshold value is set to 3 to define the number of valid values and number of zeros per replicate across all time points that each phosphosite should have to be assigned to either subset. The subsetting threshold should not be set below 3 to avoid having too few valid values for statistical analysis, but it should be increased to raise stringency. Subset 1 will contain phosphosites whose valid values per replicate at all time points equal or exceed the threshold value (Fig. 2). Phosphosites in subset 1 will be retained for statistical analysis. Subset 2 will contain the phosphosites that equal or exceed the threshold number of valid values or number of zeros per replicate (Fig. 2). Phosphosites in subset 2 will be classified as absent/present phosphosites. For example, a threshold of 3 will result in the following subsets: Subset 1 would contain any phosphosite with 3 or more valid values per replicate. Subset 2 would contain any phosphosite with 3 or more zeros per replicate at one or more time points and the time points not exceeding 3 or more zeros should have 3 or more valid values. Any phosphosites not meeting either of these criteria will not be analyzed further.

5. Normalization makes intensity values between replicates more comparable by reducing variation between technical replicates. Specifically, variance stabilization normalization (vsn) is a useful statistical method to stabilize replicate variation in Mass Spec datasets [13].

6. Variance stabilization normalization is performed using Eq. 5, $h(y)$ such that the variance $Var(h(y)) = v$ is approximately independent of the mean u [14].

$$h(y) = \int^{y} \frac{1}{(v(u))^{1/2}} du \tag{5}$$

7. This linear mixed model assumes that the data of normally distributed residuals with a mean of zero and homoscedasticity of error variance.

8. The p-value correction improves statistical analysis based on multiple hypotheses by estimating q-values from each p-value. When there are too many small p-values, then the script will return an error, specifically, in smooth.spline(lambda, pi0, df = smooth.df). If an error message occurs, then change the

code, marked *#p-value correction step* in the comment in Box 4, to: mutate(qvalue = qvalue(p.value, lfdr.out = TRUE, pi0 = 1)).

9. The Kinases & Phosphatases list [22] will aid in the phosphorylation network construction. This list should have one column of the AGI ID for each protein and a second column of the category, kinase or phosphatase, of each protein.
10. The user-defined thresholds, *a*, *b*, and *c*, control how strictly a change in intensity is considered significant. Very small intensity changes could still be above threshold *a* or *b* as a result of small median-centered intensity values (Fig. 3). To avoid the inclusion of those small changes, a bottom percentage can be set with threshold *c*.
11. The actual values the intensities are discretized to (−1, 1, 5) are arbitrary.
12. This network inference strategy relies on prior knowledge regarding the upstream regulators (kinases and phosphatases), which simplifies inference by reducing the potential interactions that need to be evaluated (only regulators can be upstream). The inference is again simplified by further limiting the potential regulators of each phosphosite, as discussed in **step 17**.
13. The point of having a time delay vs. no time delay is to include both regulators and coregulators. A regulator is a protein that (de)phosphorylates in the time point prior to (de)-phosphorylation of the target; a coregulator is a protein that (de)phosphorylates at the same time point of the target.
14. The Bayesian Dirichlet Equivalent Uniform (BDeu) is used to score the regulator-target relationships. The BDeu maximizes the posterior probability of graph *G* given the dataset *D* over the possible graphs. A potential regulator-target pair is one such Bayesian graph, while the time point observations are the dataset. In other words, the graph (regulator-target pair) which maximizes the BDeu score is the most likely to be a true regulator-target interaction [23].

References

1. Kumar V, Khare T, Sharma M et al (2018) Engineering crops for the future: a phosphoproteomics approach. CPPS 19(4):413–426. https://doi.org/10.2174/1389203718666170209152222
2. Mazzucotelli E, Mastrangelo AM, Crosatti C et al (2008) Abiotic stress response in plants: when post-transcriptional and post-translational regulations control transcription. Plant Sci 174(4):420–431. https://doi.org/10.1016/j.plantsci.2008.02.005
3. Tan H, Yang K, Li Y et al (2017) Integrative proteomics and phosphoproteomics profiling reveals dynamic signaling networks and bioenergetics pathways underlying T cell activation. Immunity 46(3):488–503. https://doi.org/10.1016/j.immuni.2017.02.010
4. Arsova B, Watt M, Usadel B (2018) Monitoring of plant protein post-translational modifications using targeted proteomics. Front Plant Sci 9:1168. https://doi.org/10.3389/fpls.2018.01168

5. Cutillas PR (2017) Targeted in-depth quantification of signaling using label-free mass spectrometry. In: Methods in enzymology. Elsevier, pp 245–268. https://doi.org/10.1016/bs.mie.2016.09.021
6. Subba P, Prasad TSK (2021) Plant phosphoproteomics: known knowns, known unknowns, and unknown unknowns of an emerging systems science frontier. OMICS: J Integrative Biol 25(12):750–769. https://doi.org/10.1089/omi.2021.0192
7. Duan G, Walther D, Schulze WX (2013) Reconstruction and analysis of nutrient-induced phosphorylation networks in Arabidopsis thaliana. Front Plant Sci:4. https://doi.org/10.3389/fpls.2013.00540
8. Dudley E, Bond AE (2014) Phosphoproteomic techniques and applications. In: Advances in protein chemistry and structural biology. Elsevier, pp 25–69
9. Liu Z, Lv J, Liu Y et al (2020) Comprehensive phosphoproteomic analysis of pepper fruit development provides insight into plant signaling transduction. IJMS 21(6):1962. https://doi.org/10.3390/ijms21061962
10. Kanshin E, Giguère S, Jing C et al (2017) Machine learning of global phosphoproteomic profiles enables discrimination of direct versus indirect kinase substrates. Mol Cell Proteomics 16(5):786–798. https://doi.org/10.1074/mcp.M116.066233
11. R Core Team (2021) R: a language and environment for statistical computing. R Foundation for Statistical Computing, Vienna. https://www.R-project.org/
12. Wickham H, Averick M, Bryan J et al (2019) Welcome to the tidyverse. J Open Source Softw 4(43):1686. https://doi.org/10.21105/joss.01686
13. Kanno S, Cuyas L, Javot H et al (2016) Performance and limitations of phosphate quantification: guidelines for plant biologists. Plant Cell Physiol 57(4):690–706. https://doi.org/10.1093/pcp/pcv208
14. Huber W, von Heydebreck A, Sultmann H et al (2002) Variance stabilization applied to microarray data calibration and to the quantification of differential expression. Bioinformatics 18 (Suppl 1):S96–S104. https://doi.org/10.1093/bioinformatics/18.suppl_1.S96
15. Wickham H (2019) Assertthat: easy pre and post assertions. R package version 0.2.1. https://CRAN.R-project.org/package=assertthat
16. Zhang X, Smits AH, van Tilburg GBA et al (2018) Proteome-wide identification of ubiquitin interactions using UbIA-MS. Nat Protoc. https://doi.org/10.1038/nprot.2017.147
17. Storey JD, Bass AJ, Dabney A et al (2021) Qvalue: Q-value estimation for false discovery rate control. R package version 2.24.0. http://github.com/jdstorey/qvalue
18. Morgan M, Obenchain V, Hester J et al (2022) Summarized experiment: summarized experiment container. R package version 1.26.1. https://bioconductor.org/packages/SummarizedExperiment
19. Bittremieux W, Tabb DL, Impens F et al (2017) Quality control in mass spectrometry-based proteomics. Mass Spectrom Rev 37(5): 697–711. https://analyticalsciencejournals.onlinelibrary.wiley.com/doi/10.1002/mas.21544
20. Cox J, Mann M (2008) MaxQuant enables high peptide identification rates, individualized p.p.b.-range mass accuracies and proteome-wide protein quantification. Nat Biotechnol 26(12):1367–1372. https://doi.org/10.1038/nbt.1511
21. Zhou T, Li C, Zhao W et al (2016) MaxReport: an enhanced proteomic result reporting tool for MaxQuant. PLoS One 11(3): e0152067. https://doi.org/10.1371/journal.pone.0152067
22. Lehti-Shiu MD (1602) Shiu S-H (2012) diversity, classification and function of the plant protein kinase superfamily. Phil Trans R Soc B 367: 2619–2639. https://doi.org/10.1098/rstb.2012.0003
23. Heckerman D, Geiger D, Chickering DM (1995) Learning Bayesian networks: the combination of knowledge and statistical data. Mach Learn 20:197–243. https://doi.org/10.1023/A:1022623210503

Chapter 28

Pairwise and Multi-chain Protein Docking Enhanced Using LZerD Web Server

Kannan Harini, Charles Christoffer, M. Michael Gromiha, and Daisuke Kihara

Abstract

Interactions of proteins with other macromolecules have important structural and functional roles in the basic processes of living cells. To understand and elucidate the mechanisms of interactions, it is important to know the 3D structures of the complexes. Proteomes contain numerous protein-protein complexes, for which experimentally determined structures often do not exist. Computational techniques can be a practical alternative to obtain useful complex structure models. Here, we present a web server that provides access to the LZerD and Multi-LZerD protein docking tools, which can perform both pairwise and multi-chain docking. The web server is user-friendly, with options to visualize the distribution and structures of binding poses of top-scoring models. The LZerD web server is available at https://lzerd.kiharalab.org. This chapter dictates the algorithm and step-by-step procedure to model the monomeric structures with AttentiveDist, and also provides the detail of pairwise LZerD docking, and multi-LZerD. This also provided case studies for each of the three modules.

Key words Web server, LZerD, Structure modeling, Protein bioinformatics, Protein-protein docking, Protein structure prediction, Symmetrical docking

1 Introduction

Protein–protein interactions (PPIs) play an important role in the fundamental cellular processes of living cells including cell signaling, antigen–antibody interactions, regulating the function of the proteins, transport, and the formation of structural units. Understanding the structures of complexes is an important step to elucidate the mechanisms of interactions, for example, to prevent or promote the interactions in disease conditions or to design antibodies to prevent the interactions. Experimental determination of complex structures is typically done via methods including X-ray crystallography, cryo-electron microscopy, or nuclear magnetic resonance (NMR). These experimental techniques are expensive and

Shahid Mukhtar (ed.), *Protein-Protein Interactions: Methods and Protocols*, Methods in Molecular Biology, vol. 2690, https://doi.org/10.1007/978-1-0716-3327-4_28,

time-consuming. Hence, computational techniques can be an alternative to model the complexes with available monomeric structures [1–3].

Several protein–protein docking methods have been reported in the literature [4–9]. We have developed protein–protein docking methods, including LZerD [10] for pairwise docking, Multi-LZerD [11] for multiple chain docking, and IDP-LZerD [12, 13] for docking with intrinsically disordered proteins (IDP). LZerD uses the 3D Zernike descriptor (3DZD), a mathematical rotational invariant surface shape representation, which can effectively identify complementary surface shapes from two proteins as docking interface regions [14, 15]. The use of 3DZD facilitates rapid soft evaluation of molecular surface complementarity and is moderately tolerant to conformational changes. The LZerD suite has been ranked at or near the top of all server groups in recent rounds of CAPRI [16–18], the blind community-wide assessment of protein docking methods. Standalone programs of different types of LZerD-based docking are available from the LZerD software suite website (https://kiharalab.org/proteindocking/lzerd.php) for download. LZerD and Multi-LZerD were also implemented as web server, which is available at https://lzerd.kiharalab.org [19, 20]. On the LZerD web server, users can perform protein docking with LZerD and Multi-LZerD from protein tertiary structures provided by users. For cases when users do not have structures to dock, our web server also can model the protein structure from the sequence, using AttentiveDist [21]. Relative to the top existing servers participating in CASP13 [22], AttentiveDist showed competitive performance when evaluated on the CASP13 dataset [21].

This chapter explains the algorithms and usage of three modules of the LZerD web server. Starting with AttentiveDist to model the 3D structure from a protein sequence, we then explore the second module of LZerD for pairwise docking, and finally Multi-LZerD for docking multi-chain proteins with more than two chains. We provide step-by-step procedures and case studies of each of the modules, helping users to effectively utilize different features of the webserver by enforcing user-defined constraints in the docking. The focus here is on multiple chain docking and docking with user-defined constraints specifying residue-residue contacts. The overview of the algorithms is provided below.

1.1 AttentiveDist Protein Structure Prediction Algorithm

For the modeling of individual proteins, users can make use of the AttentiveDist module. The module can take as input a protein sequence with unknown 3D structural information. For a given sequence, four multiple sequence alignments (MSA) with different E-values 0.001, 0.1, 1, and 10 are generated using the DeepMSA [23] method. A neural network predicts the distribution of residue–residue distances and angles, from which the full-atom models are generated using PyRosetta [24]. Finally, models are ranked by a ranksum score obtained from the sum of the model ranks by DFIRE [25], GOAP [26], ITScorePro [27], and Rosetta's REF2015 score [28]. Although using the AttentiveDist module is convenient on the LZerD web server because modeled structures can be seamlessly transferred to the protein docking module, users can also model individual protein structures by other recent structure modeling methods, such as AlphaFold [29], and upload the models to the docking module of LZerD.

1.2 The LZerD Pairwise Docking Algorithm

LZerD uses the 3DZD, a rotation-invariant mathematical moment-based shape descriptor, for representing protein surface shape. It also allows a "soft" representation of the protein surfaces, capturing the shape complementarity of proteins, and providing robustness to a certain degree of conformational changes. Docking poses are scored by a shape-based score, which considers surface shape complementarity that is captured by 3DZD, the directions of the surface normals, interface area, and the atom clashes.

LZerD takes only two protein structures as input. The conformational exploration is performed with a geometric hashing algorithm. The models that are violating the constraints provided by the users like residue-residue distances, interface residues, or symmetry tolerance are rejected. Once the docked models are generated, clustering using the user-defined cutoff, with 4 Å RMSD as default, reduces the numbers to a few tens of thousands by removing the too-similar models. The models are ranked by the ranksum score obtained by the sum of the ranks by DFIRE [25], GOAP [26], and ITScorePro [27]. The ranksum score performed well in the scoring category of CAPRI including the top rank by the top-1 model in the most recently evaluated CAPRI round 50 (joint round with CASP14) [30].

1.3 Multi-LZerD Multiple Chain Docking Algorithm

Multi-LZerD is for docking more than two chains. Users can dock up to six chains simultaneously on the webserver. For docking more than six chains, users can download and run the software locally. In Multi-LZerD, initially, pairwise docking models are generated for all possible pairs of the input chains with LZerD. The generated models are clustered with a user-defined RMSD cutoff, where 10 Å is used as default, removing similar complexes. Then, different multimeric protein complexes are generated by combining pairwise solutions using a genetic algorithm. A physics-based scoring function is used as the fitness function. Models are generated up to a population cutoff (the number of output models) and clustered with the user-defined RMSD cutoff. Final models are ranked in the pairwise scoring functions using ranksum.

2 Materials

To perform protein docking, it is best if users have experimentally determined 3D structures of the individual chains. In case the structure is not available, structures can be modelled from the amino acid sequences of the proteins in question using AttentiveDist.

2.1 Computational Requirements

1. Work station and/or parallel computing specifications: Not required.
2. Operating systems and browser specifications:

   The webserver can run in all operating systems such as Windows, Linux, Mac. It is compatible with the major browsers such as Firefox, Chrome, and Opera.
3. Platform/software loaded: The LZerD webserver presents a convenient interface to the LZerD suite of protein docking tools at https://lzerd.kiharalab.org. Hence manual downloading and installation of any software is not needed.
4. Language used: Python and C++.

3 Method

3.1 Creating an Account in the LZerD Web Server

1. From the front page, access the instruction by clicking "Learn More" or can click "Get Started" to go directly to the job submission page. (https://lzerd.kiharalab.org/upload/), shown in Fig. 1a.

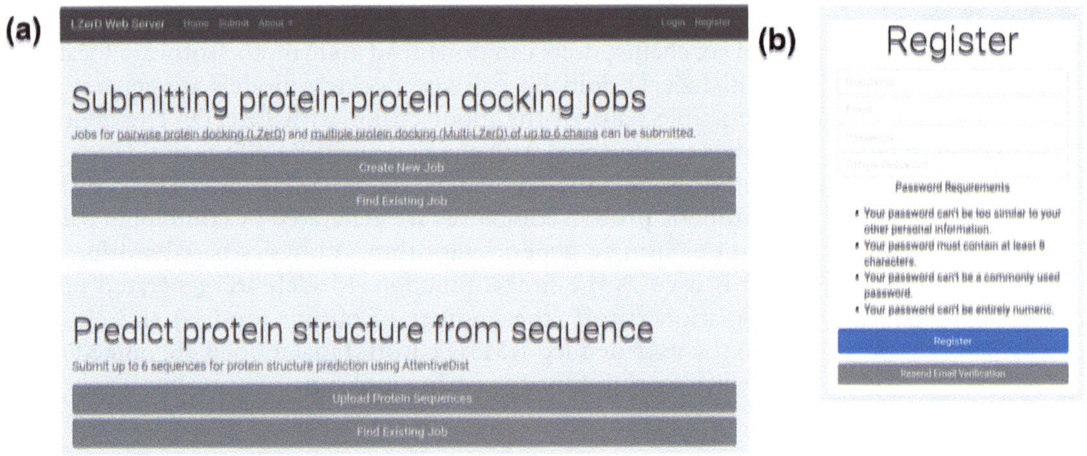

Fig. 1 Screenshots of (**a**) the home page. It contains links to the various job submission pages. (**b**) The registration page

2. Register by clicking "Register" in the top right corner of the main page. Users will be redirected to a page, as shown in Fig. 1b, where they can provide their email address and a unique username and password for their account (*see* **Note 1**).

3.2 Using AttentiveDist

1. For starting from single chain modeling with AttentiveDist, click the "Upload Protein Sequences" under the "Predict protein structure from sequence" category to enter the AttentiveDist submission page.
2. Directly paste or upload the protein sequences in FASTA format can into the textbox. A maximum of six protein sequences with a maximum of 1000 residues each can be submitted at a time. Clicking the "+" button will show additional input panels as needed (Fig. 2a).

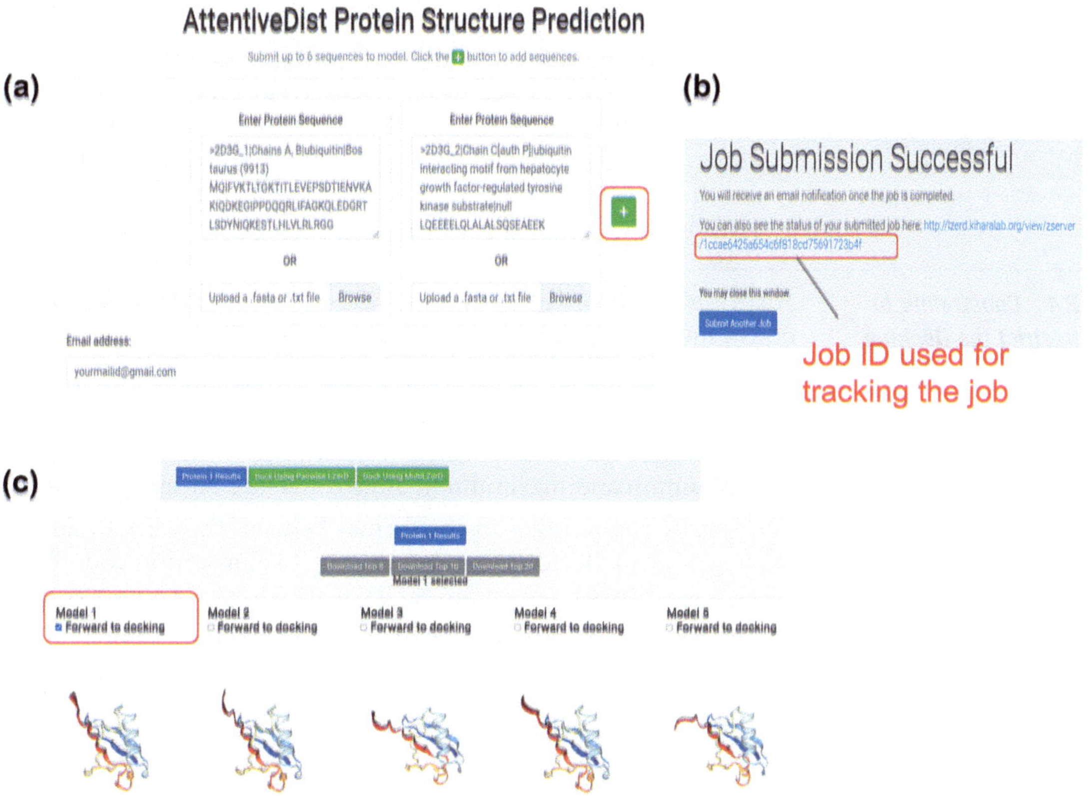

Fig. 2 Screenshots of (**a**) the input page to AttentiveDist and (**b**) the job post-submission page showing the job ID used for tracking. (**c**) Result in summary page of protein structure modeling. The models are shown in the order of ranksum scores. Models can be downloaded in bulk via the Download buttons. The model with the checked "Forward to docking" box will be forwarded to the LZerD docking job submission page

3. Enter an email address to receive a notification when their job completes and submit the job afterward. Once the job is done, applicable users will be notified through email, and the status and results can be viewed using the job ID (Fig. 2b). Once the job is done, users will be redirected to the result summary page (Fig. 2c).
4. Download the models either compressed archives or individually (*see* **Note 2**).

3.3 Using LZerD for Pairwise Docking

1. Start the docking from the main page (Fig. 1a), through "Create New Job," under "Submitting protein-protein docking jobs," which will redirect to the docking submission page (https://lzerd.kiharalab.org/upload/upload/).
2. Upload the subunit structures to dock directly in the PDB format or can be specified with their PDB IDs if they are deposited in the PDB. As mentioned above, if a structure is unknown users can model the structure using the AttentiveDist module.
3. Uploaded proteins 1 and 2 are named receptor protein and ligand proteins, respectively (Fig. 3a). Generated docked models are clustered with an RMSD cutoff of 4 Å as default (Fig. 3b). Changing it to a smaller value will make more models with smaller differences, while making it to a large value will produce fewer models that are more different from each other.

3.4 Constraints to Restrict the Docking

A strong advantage of the LZerD web server is that users can add constraints to restrict the docking. (Fig. 4a).

1. Use the "Add Residue-Residue Constraints'‘area, which restricts the LZerD docking output to models where the distance between the specified residues falls within the specified minimum and maximum in angstroms (see Note 3).
2. Specify the residues in the format "Chain <space> Residue," e.g., "A 113B'‘for chain A, position 113, insertion code B.
3. Set the minimum and maximum distances in the "Min Distance'‘and "Max Distance'‘fields, respectively. For example, if the user wanted to specify that two cysteines should form a disulfide bond, they could set a residue-residue constraint with the minimum distance set to 2.0 Å and the maximum distance set to 3.0 Å.
4. Set the distance minimum to blank and the maximum to 5.0 Å, if the users only know that a residue in the receptor should be in contact with a particular residue in the ligand (*see* **Note 4**).
5. Upload a user-defined constraints as a JSON file by advanced users, and constraints can be previewed in JSON format using

Fig. 3 LZerD pairwise docking submission page. (**a**) Receptor and ligand input fields. (**b**) User-configurable settings for the docking run. The clustering cutoff (default 4.0 Å) controls the redundancy of the docking poses; a lower value will yield more models, while a higher value will yield fewer models. The surface reduction cutoff controls the density of the docking pose search; a lower value results in denser sampling, while a higher value results in sparser sampling. Users can further set their email, whether they want a job start notification, and a title and comment to help organize the job

the "Json Preview"' button. Once the inputs and constraints are provided, users can proceed to click "Submit" to submit the job. An example of constraints is provided in Fig. 4a. For some instances, server may return empty model (*see* **Note 5**).

3.5 Homodimeric Docking

The LZerD web server also has an option to model homodimeric symmetrical complexes.

1. Input the monomeric unit of the protein, after checking the "Switch to C2-symmetric docking," as in Fig. 4b. C2 symmetry of a model can be constructed by transforming the atomic coordinates of the subunit by applying the rigid body transformation from docking twice and calculating the root-mean-square deviation (RMSD) to the original coordinates. The models with RMSD greater than 5 Å are discarded.

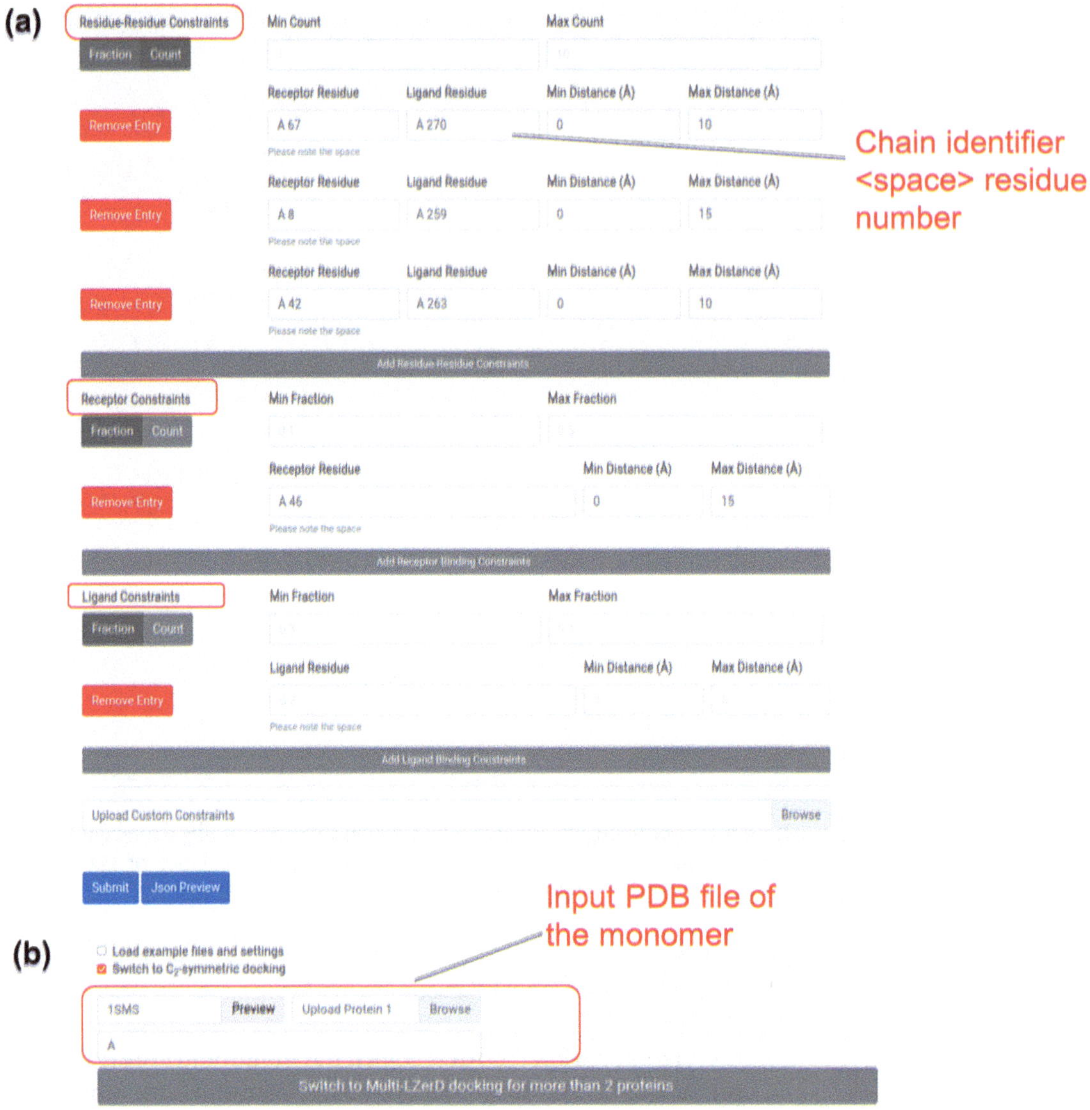

Fig. 4 Constraints users can specify for LZerD docking. (**a**) Here, both residue-residue and receptor binding site constraints have been input. The constraints shown here are discussed in the main text. (**b**) Input area in C2-symmetric homodimer docking mode. Since both receptor and ligand are identical, there is only one structure upload field. It is important to check C2-symmetric docking

2. Follow similar steps as in LZerD pairwise docking: adding constraints, submitting the job, and inspecting the top-ranked models.

3.6 Using Multi-LZerD for Multi-chain Docking

1. Switch to the multiple-chain protein docking by clicking the "Switch to Multi-LZerD docking for more than 2 proteins"-button, highlighted in Fig. 5a, if the complex has more than 2 chains.

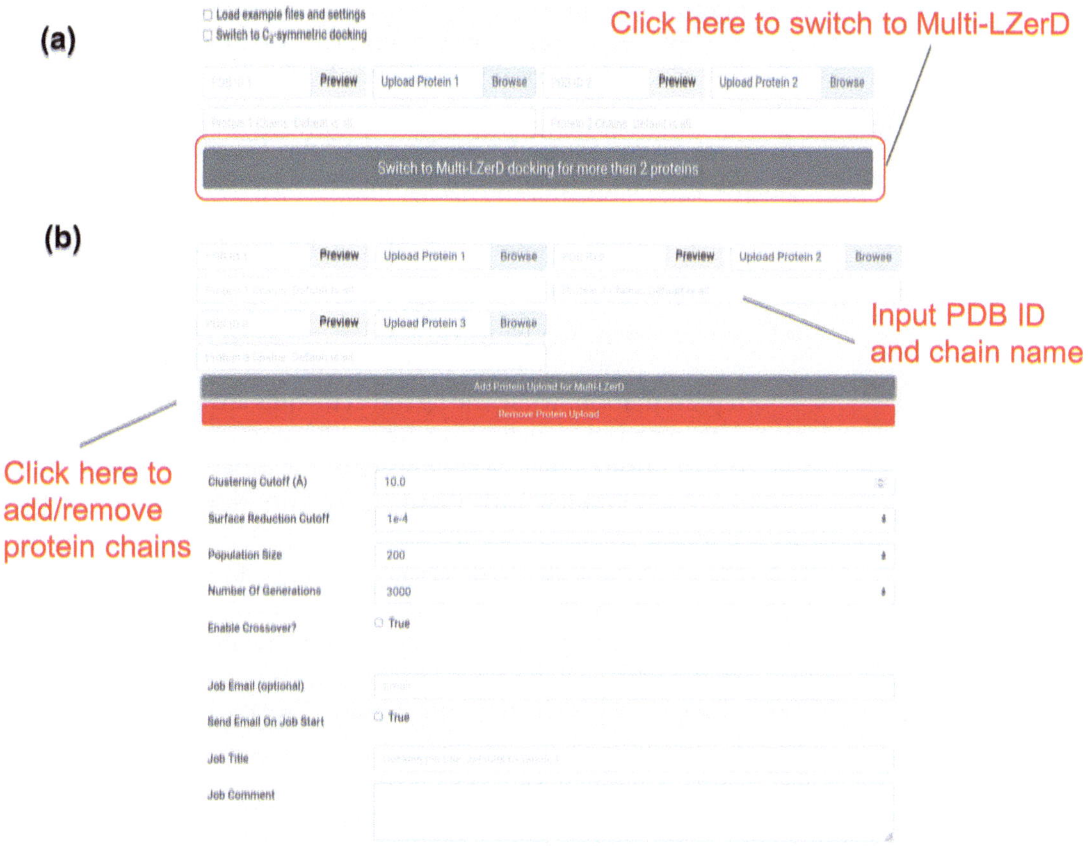

Fig. 5 Multi-LZerD submission. (**a**) Switching to the Multi-LZerD module from the LZerD module. (**b**) The main Multi-LZerD submission page. At the structure input area, clicking the "Add" button will increase the number of input subunits to up to 6. User-configurable settings windows are shown below. The meaning of the clustering cutoff is the same as for pairwise LZerD, but for Multi-LZerD the default value is 10.0 Å. The population size controls the total size of the docking model pool. The number of generations is the maximum number of iterations the genetic algorithm will run for. Crossover controls whether the genetic algorithm will combine models. It is disabled by default

2. Increase the number of input fields by clicking "Add Protein Upload for Multi-LzerD,"until the desired number of chains can be input, as highlighted in Fig. 5b (*see* **Note 6**).
3. Specify constraints as in the LZerD module. In addition, Multi-LZerD has some further settings, Population Size, Number of Generations, and Clustering Cutoff, as shown in Fig. 5c. The number of output models is controlled by the "Population Size" parameter. The "Number of Generations"parameter controls how extensively the docking search space is explored. The "Clustering Cutoff" controls the level of diversity in the output model. Users can typically leave these settings at their default values.

4. Submit the job once any constraints are added. The output page will have options to visualize the top models and users can download them locally to analyze further.

3.7 Output of Docking

After the completion of a docking job, the LZerD web server will provide you with the docked models ranked based on the overall score (ranksum) obtained from the sum of ranks from different scoring functions, namely GOAP, DFIRE, and ITScorePro. The output page has four sections.

1. Starting with the panel on the left, visualize the docked complex, as shown in Fig. 7c, the receptor is shown in a cartoon representation, while the ligand centroids are shown as spheres. The centroid visualization helps the users to understand the distribution of the docking poses of the different models and to select the models with the preferred docking pose for further analysis.
2. Hover the mouse cursor over any centroid to see the rank of the corresponding model and can click to see the full docked structure with both the receptor and that docked ligand pose. In the panel on the right are options to control the number of models for which the centroids are displayed (*see* **Note 7**).
3. In the bottom panel, obtain the ranksum score along with DFIRE, GOAP, and ITScorePro scores of the individual models in a table, as shown in Fig. 7c.
4. Download individual models by clicking the model's name through the download buttons in the rightmost column.
5. Sort this table by any of the component scores (*see* **Note 8**).
6. To demonstrate the utility and flexibility of the LZerD web server, three case studies are shown (*see* **Notes 9**, **10**, **11**, and **12**). In each case study, we illustrate how modules of the web server can be used to effectively model protein structures to near-native quality, in each case evaluating by comparing with experimentally known structures.

4 Notes

1. After verifying their ownership of the email address, newly registered users can log in and start docking. Creating an account provides users the advantage of viewing all the jobs submitted through the account, as well as a three-month job retention time. When a job is done, registered users will be notified by email, which includes a link to the result summary page. Unregistered users will have to keep track of their job IDs, and their jobs may become inaccessible after two weeks.

2. For each of the submitted sequences, the top 20 models can be downloaded locally, while direct 3D visualization is provided on the webpage for the top five models. The scores and rankings of each of the structures are provided in a table. Users can forward the modeled structures to the LZerD or Multi-LZerD submission page by selecting the desired model and selecting "Forward to docking" on the top of the model to perform docking (Fig. 2c).
3. If the users have prior knowledge of the exact distance between residues on different chains, or if the range of distances the interaction between two residues should take place at is known.
4. If either input protein structure is expected to be flexible or is not experimentally determined, then the maximum allowed distance should be increased commensurately with the expected degree of flexibility or modeling error, for example, to 10.0 Å. If the users have prior knowledge that a particular residue should not be interacting with the other chain, then the users can set the minimum to be 5 Å and can leave the maximum blank. In cases where the distances for particular residue pairs are not known, users might want to receive only models having some particular residue either in the ligand or in the receptor at the interface. Then, users can make use of the "Add Receptor/Ligand Binding Constraints" area, where the user can specify the residue of the receptor/ligand to be within a particular distance, say a maximum of 5.0 Å from the other protein. In this case, the nearest atom of the ligand/receptor to the other protein is considered to evaluate the constraint satisfaction condition. In any receptor or ligand binding constraint, only one side of the interaction needs to be specified. Similarly, users can also restrict particular residues to be not in the interface by setting their minimum distance higher, e.g., to 5 Å, while leaving the maximum blank. When multiple conditions are given, by default a model will be selected only if all the constraints are satisfied. If the users want to create soft constraints, then they can specify a minimum number or fraction of the conditions that must be satisfied to select the model. For example, reducing the fraction to 50% could soften the condition by providing the flexibility to select the model even if only 50% of the constraints are satisfied, which can be useful when specifying, e.g., an interaction with either of two distinct putative binding sites.
5. In events if the server returns an empty model set for a given set of constraints, that means that models satisfying the constraints were not obtained during LZerD's conformational sampling. To resolve this, the constraint distance cutoffs should be relaxed, or the fraction of constraints required to be satisfied should be reduced. Users should avoid specifying overly tight

distance ranges, as the smaller the range, the lower the probability that it can appear in the conformational sampling. Additionally, users should avoid specifying constraints that contradict one another since constraints that are impossible to satisfy will always result in empty output.

6. The web server allows users to dock a maximum of 6 chains. For higher-order docking, users can download and run Multi-LZerD locally (https://kiharalab.org/proteindocking/multilzerd.php).
7. By default, the centroids for the top 50 are shown. The shown centroids can be further reduced to the top 30 or 10 models to examine the highest-ranking docked poses, or can instead be increased to show 100, 500, or all poses to examine the wider distribution.
8. The table is sorted by ranksum by default, and a lower ranksum score indicates a stronger consensus between the component scores, a strategy proven effective during CAPRI as previously discussed. Thus, considering the models by the default ordering by ranksum is recommended. Now after understanding the output, from the top panel, users can download the top 10/30/50 or the whole set of docked full-complex models for further analysis. Thus, using the results page, users can view, select, and download the models individually or in bulk based on their needs.
9. Using AttentiveDist to model individual structures and perform pairwise docking; case study 1.

   The first example demonstrates a case where individual protein structures are modeled by AttentiveDist, which are then docked together by LZerD. For this example, we used ubiquitin protein (76 amino acids long) and hepatocyte growth factor-regulated tyrosine kinase substrate (Hrs) protein (21 amino acids long). Hrs protein is important in endosomal protein sorting. It contains a ubiquitin-interacting motif (UIM) which interacts with monoubiquitinated receptors and sorts them to multivesicular bodies for lysosomal degradation. The sequences of the ubiquitin and Hrs proteins are taken from UniProt entries P0CH28 and O14964, respectively.

   FASTA sequences of the proteins were input to AttentiveDist via the web server submission page, as shown in Fig. 6a. The AttentiveDist models for the Hrs UIM are visualized on the results page in ranksum order as shown in Fig. 6b. For evaluation purposes, we note that the top models for ubiquitin and the Hrs UIM had root-mean-square deviations (RMSDs) of 0.95 Å and 0.43 Å, respectively. We note that these RMSDs are not used in model selection, only post-analysis, since they rely on knowledge of the exact native structure. The low

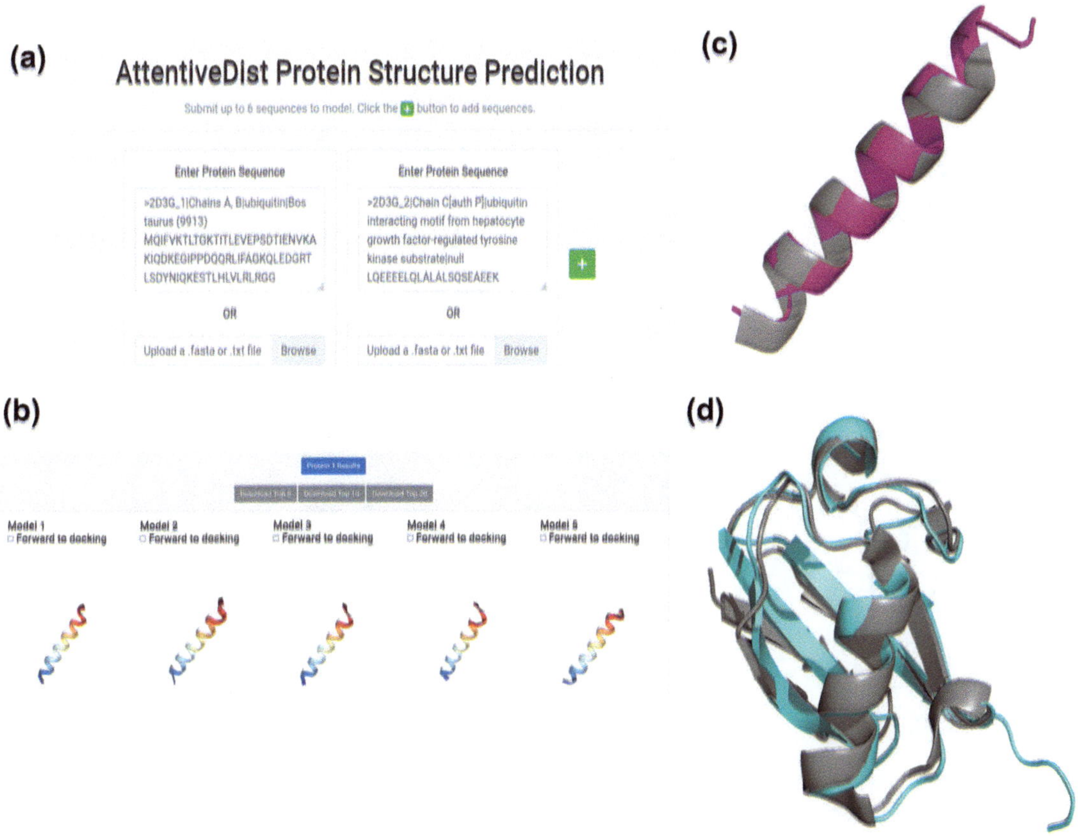

Fig. 6 Case study 1 – monomer modeling with AttentiveDist. In this example, we modeled two proteins, Hrs protein PDB: 2D3G (chain P) and ubiquitin-protein PDB 2D3G (chain B) in preparation for docking in the next case study (**a**) Input Fasta sequences for AttentiveDist based modeling. (**b**) Result summary page with top 5 modeled structures displayed. (**c**) Hrs protein in PDB 2D3G-P in gray aligned with the modeled chain (ranked 1), in magenta. (**d**) ubiquitin (PDB 2D3G-B) represented in gray aligned with the modeled chain, in cyan. The RMSD of Hrs and ubiquitin-protein was 0.95 Å and 0.43 Å, respectively

RMSDs suggest that the modeling of the proteins was reasonably successful. A visual comparison of the top-performing model with the native structure from PDB 2D3G is shown for each chain in (Fig. 6c, d), where it is clear that the models have the correct fold, and indeed nearly identical backbones to their native counterparts.

We selected the top-ranked models from the predictions of AttentiveDist and proceeded to pairwise docking. The ubiquitin and Hrs UIM were provided as the receptor and ligand, respectively (see Fig. 7a), and the docked complexes were analyzed by comparing them with the known experimental structure of the complex available in PDB 2D3G. The top-scored model by ranksum had an RMSD of 1.02 Å when compared with the native structure. The ranksum score for this model was 29. The GOAP and DFIRE scores gave a higher rank of 3, while ITscore gave a rank of 23.

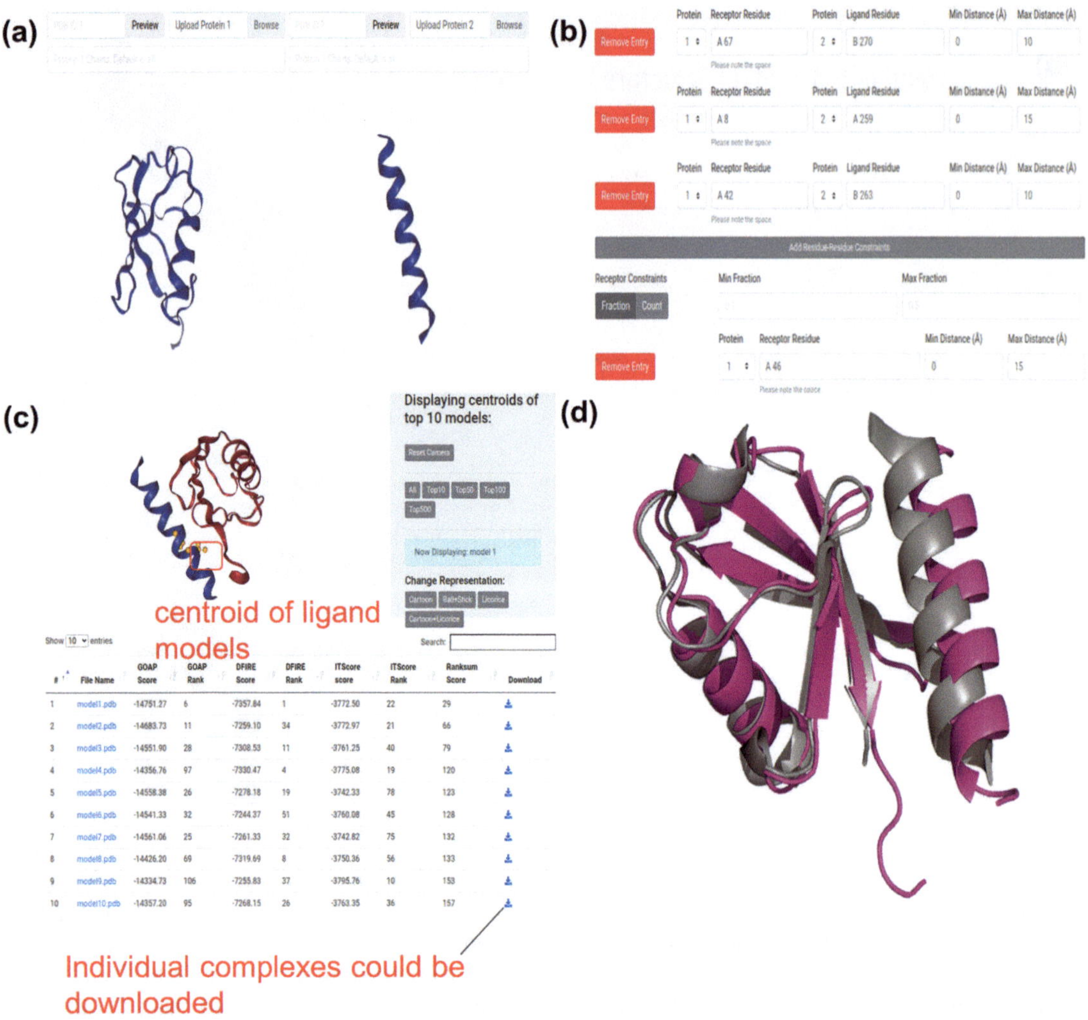

Fig. 7 Case study 1, pairwise protein-protein docking with LZerD. We used Hrs and ubiquitin proteins modeled using AttentiveDist for the docking. (**a**) Input chains obtained from AttentiveDist modeling. (**b**) Constraints used for docking. (**c**) Result summary page. (**d**) Alignment of the experimentally determined complex PDB 2D3G, in gray with the docked complex (ranked 6) from LZerD, in magenta

Although this model has a sufficiently low RMSD for many practical purposes, we then tested how the model would further improve by providing residue distance constraints. We assumed that we had prior knowledge that Ala 46 of chain A is at the protein-protein interface and input the information as "Receptor Binding Constraints." We specified the constraint as "Chain A Ala 46 should have a maximum distance of 15 Å and minimum distance of 0 Å from the other protein." Also, as the second piece of prior knowledge, we input residue-residue contact constraints between chain A Leu 67 and chain B Ser 270, chain A Leu 8 and chain B Glu 259, and chain A Arg 42 and chain B Leu 263. Since these pairs are assumed to be known to contact each other, we input 15 Å as the maximum distance between each pair, providing a large tolerance to the contact distance (Fig. 7b). The results obtained are shown in Fig. 7c. With these constraints, the sixth ranked model with a ranksum score of 128 had an improved RMSD of 0.98 Å. The alignment of the docked complex model with the native is provided in Fig. 7d.

10. Homodimeric protein docking using C2 symmetry; case study 2.

    Next, we show a case of homodimer docking. We used ribonucleotide reductase Rnr4 for this example. Ribonucleotide reductase (RNRs) catalyzes the reduction of ribonucleotides to deoxyribonucleotides. This protein exists both as homodimers and also in complex with other RNRs to perform their functions. To perform homodimer docking, users need to check the box "Switch to C2-symmetric docking," as shown in Fig. 8a. We used a monomer from PDB 1SMS as the input. We provided some residue-residue constraints: Chain A Phe 35 and Chain B Ser 128, Chain A Ser 128 and Chain B Phe 38, Chain A Ala 54 and Chain B Glu 60 were specified as being in contact by setting the minimum and maximum distance parameters to be 0 and 10 Å, respectively, as shown in Fig. 8b in JSON format.

    From the table of complex outputs (Fig. 8c), analyzing the top 10 models from LZerD by comparing with experimentally determined homodimer structure 1SMS, the ninth-ranked model was closest to the native structure, with an RMSD of 4.32 Å and a ranksum value of 217. On the other hand, the top-ranked model was not accurate, with a high RMSD of 22.37 Å. An alignment of the experimental and docked complex is shown in Fig. 8d, showing that our model can capture the correct orientation, having good alignment with the crystal structure.

11. Multi-LZerD docking of the heterotrimeric protein complex; case study 3.

    In this example, we detail the usage of Multi-LZerD to dock three protein chains. We used PDB 1A0R, which is a trimeric complex of the beta (Chain B) and gamma (chain G) subunits of transducin along with a phosducin protein chain (Chain P). After switching to the Multi-LZerD module as shown in Fig. 9a, we provided the PDB IDs and the individual chain ID, as the phosducin binds tightly to the heterotrimeric G-protein transducin, thereby inhibiting the G-protein cycle. We provided the individual chains of the proteins as the Multi-LZerD input as shown in Fig. 9b. We also provided constraints, assuming we know a few contacting residue pairs between the

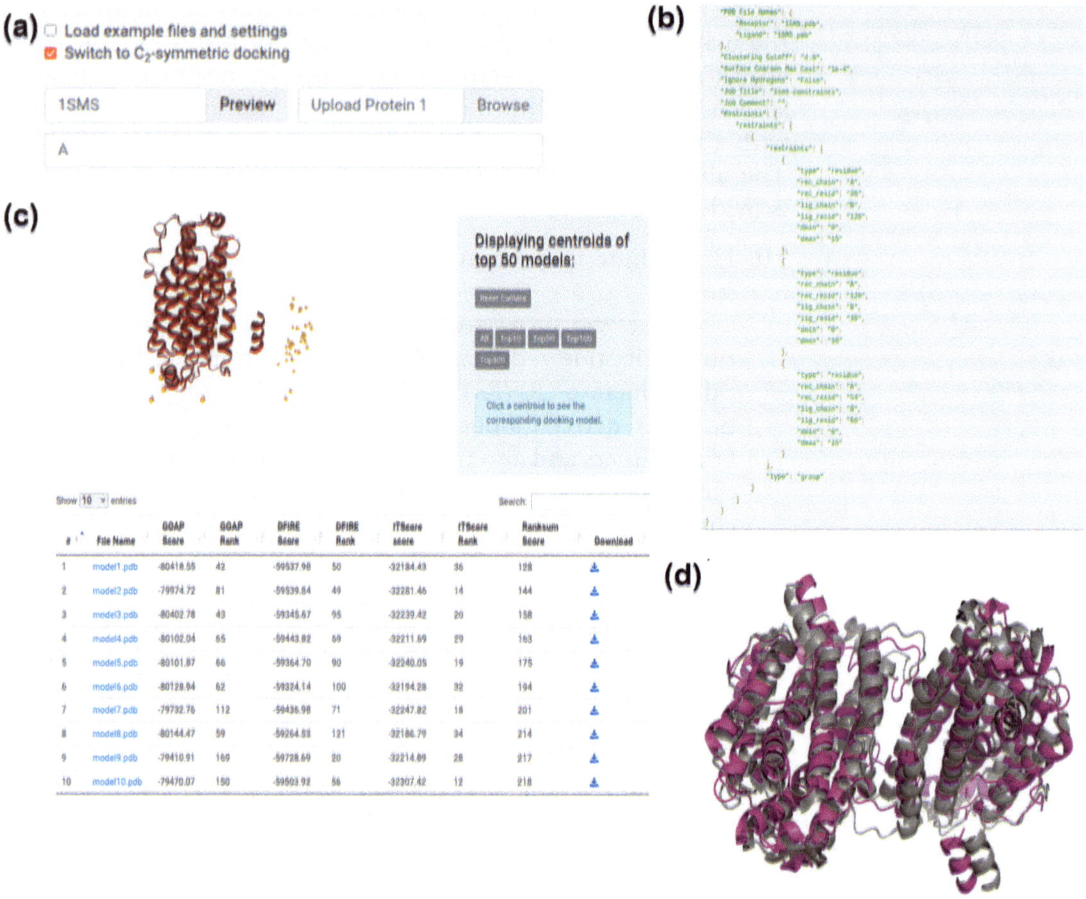

Fig. 8 Case study 2, C2-symmetric homodimer docking with LZerD. In this example, we docked the monomeric protein, ribonucleotide reductase Rnr4 (PDB ID: 1SMS). (**a**) Input protein chain from PDB 1SMS, to perform homodimeric symmetrical docking. (**b**) Input constraints are represented in JSON format. This is the information mentioned in the text. (**c**) Result summary page. (**d**) The structure alignment of the model (ranked 9), generated from C2-symmetric docking from LZerD, in magenta, with the original crystal structure PDB 1SMS, in gray

chains. The residue-residue constraints (Fig. 9c) were provided as "chain B Asp 7 and chain G Leu 15, chain B Arg 283 and chain G Glu 45, chain B Arg 46 and chain P Glu 223, chain B Ser 98 and chain P Arg 94" to be interacting by setting the minimum and maximum distance parameters to be 0 and 5 Å. Also, as we assume we know that Chain G Ser 34 and Chain P Lys 92 do not interact, we set a distance constraint with the minimum set to 10 Å and the maximum to 999 Å. Among the results ordered by the ranksum score (Fig. 9d) after docking, the top-ranked model had a ranksum score of 3, meaning that all the component scoring functions ranked that particular model at the top. Comparing the model with the experimentally determined complex PDB 1A0R, we obtained a RMSD of

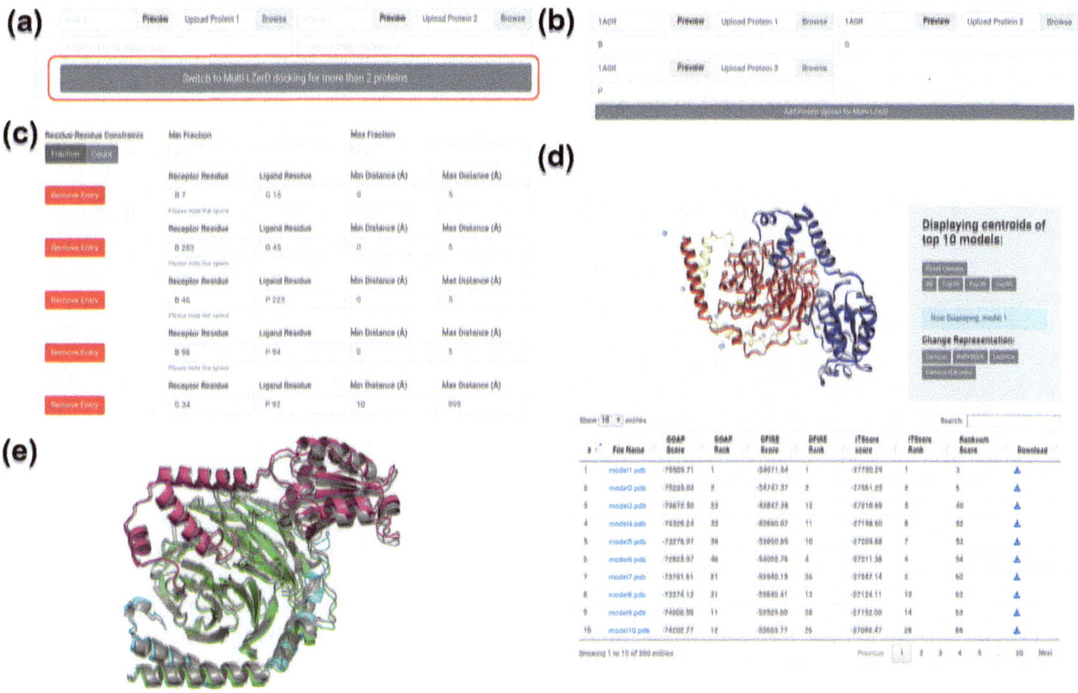

Fig. 9 Case study 3, multimeric protein-protein docking with Multi-LZerD. We docked beta (Chain B, in green) and gamma (chain G, in cyan) subunits of transducin and phosducin protein chain (Chain P, in magenta) (PDB ID: 1AOR). (**a**) Switching to Multi-LZerD via the button. (**b**) Inputting the PDB ID and proteins chains to be docked. (**c**) Residue-residue constraints are provided as described in the text. (**d**) Result summary page. In the 3D structure viewer on the top, the centroids of individual chains for each model are shown as small spheres. By clicking a centroid, the corresponding model is loaded. The centroids are colored to match the modeled chains. In the table below, the scoring details for each model are shown, by default in ranksum order. (**e**) The alignment of the model (ranked 1), generated from Multi-LZerD, is colored based on the chain as mentioned before, with the original crystal structure PDB 1AOR, in gray

1.26 Å. The native complex structure from PDB is shown superimposed with the docked model obtained from Multi-LZerD in Fig. 9e. As shown in our paper [11], Multi-LZerD is able to produce an accurate model for this complex without additional constraints. Here, we added the constraints as a demonstration of how to use the input page of Multi-LZerD.

12. The LZerD webserver provides a convenient and installation-free interface for performing pairwise and multi-chain protein docking. If the monomer protein structures are not known, the web server can predict the structures using AttentiveDist. The web server has numerous options for user-provided biological constraints such as known residue-residue distances or contacts and also the interface residues in protein chains. These constraints enable biologists to input their expert knowledge into the computation, rather than manually searching for models that agree with the experiment after the docking has finished.

Acknowledgments

This work was partly supported by the National Institutes of Health (R01GM133840, R01GM123055, and 3R01GM133840-02S1) and the National Science Foundation (CMMI1825941, MCB1925643, DBI2146026, and DBI2003635). HK and MMG are supported by the Science & Engineering Research Board (SERB), Government of India through the Overseas Visiting Doctoral Fellowship program (OVDF). MMG is partially supported by the Science and Engineering Research Board (SERB), Ministry of Science and Technology, Government of India (No. CRG/2020/000314). CC was supported by a NIGMS-funded predoctoral fellowship (T32 GM132024). The contents of the chapter are solely the responsibility of the authors and do not represent the official views of the NIGMS or NIH.

References

1. Aderinwale T, Christoffer CW, Sarkar D et al (2020) Computational structure modeling for diverse categories of macromolecular interactions. Curr Opin Struct Biol 64:1–8. https://doi.org/10.1016/j.sbi.2020.05.017
2. Gromiha MM, Yugandhar K, Jemimah S (2017) Protein-protein interactions: scoring schemes and binding affinity. Curr Opin Struct Biol 44:31–38. https://doi.org/10.1016/j.sbi.2016.10.016
3. Gromiha MM (2020) Protein interactions: computational methods, analysis and applications. World Scientific, Singapore
4. Pierce BG, Wiehe K, Hwang H et al (2014) ZDOCK server: interactive docking prediction of protein-protein complexes and symmetric multimers. Bioinformatics 30(12): 1771–1773. https://doi.org/10.1093/bioinformatics/btu097
5. van Zundert GCP, Rodrigues J, Trellet M et al (2016) The HADDOCK2.2 web server: user-friendly integrative modeling of biomolecular complexes. J Mol Biol 428(4):720–725. https://doi.org/10.1016/j.jmb.2015.09.014
6. Kozakov D, Hall DR, Xia B et al (2017) The ClusPro web server for protein-protein docking. Nat Protoc 12(2):255–278. https://doi.org/10.1038/nprot.2016.169
7. Lyskov S, Gray JJ (2008) The RosettaDock server for local protein-protein docking. Nucleic Acids Res 36 (Web Server issue), pp W233–W238. https://doi.org/10.1093/nar/gkn216
8. Ritchie DW, Venkatraman V (2010) Ultra-fast FFT protein docking on graphics processors. Bioinformatics 26(19):2398–2405. https://doi.org/10.1093/bioinformatics/btq444
9. Torchala M, Moal IH, Chaleil RA et al (2013) SwarmDock: a server for flexible protein-protein docking. Bioinformatics 29(6): 807–809. https://doi.org/10.1093/bioinformatics/btt038
10. Venkatraman V, Yang YD, Sael L, Kihara D (2009) Protein-protein docking using region-based 3D Zernike descriptors. BMC Bioinformatics 10:407. https://doi.org/10.1186/1471-2105-10-407
11. Esquivel-Rodriguez J, Yang YD, Kihara D (2012) Multi-LZerD: multiple protein docking for asymmetric complexes. Proteins 80(7): 1818–1833. https://doi.org/10.1002/prot.24079
12. Peterson LX, Roy A, Christoffer C et al (2017) Modeling disordered protein interactions from biophysical principles. PLoS Comput Biol 13(4):e1005485. https://doi.org/10.1371/journal.pcbi.1005485
13. Christoffer C, Kihara D (2020) IDP-LZerD: software for modeling disordered protein interactions. Methods Mol Biol 2165:231–244. https://doi.org/10.1007/978-1-0716-0708-4_13
14. Venkatraman V, Sael L, Kihara D (2009) Potential for protein surface shape analysis using spherical harmonics and 3D Zernike descriptors. Cell Biochem Biophys 54(1–3): 23–32. https://doi.org/10.1007/s12013-009-9051-x
15. Kihara D, Sael L, Chikhi R et al (2011) Molecular surface representation using 3D Zernike

descriptors for protein shape comparison and docking. Curr Protein Pept Sci 12(6): 520–530. https://doi.org/10.2174/138920311796957612

16. Lensink MF, Velankar S, Baek M et al (2018) The challenge of modeling protein assemblies: the CASP12-CAPRI experiment. Proteins 86 (Suppl 1):257–273. https://doi.org/10.1002/prot.25419
17. Lensink MF, Brysbaert G, Nadzirin N et al (2019) Blind prediction of homo- and hetero-protein complexes: the CASP13-CAPRI experiment. Proteins 87(12):1200–1221. https://doi.org/10.1002/prot.25838
18. Lensink MF, Nadzirin N, Velankar S et al (2020) Modeling protein-protein, protein-peptide, and protein-oligosaccharide complexes: CAPRI 7th edition. Proteins 88(8): 916–938. https://doi.org/10.1002/prot.25870
19. Christoffer C, Chen S, Bharadwaj V et al (2021) LZerD webserver for pairwise and multiple protein-protein docking. Nucleic Acids Res 49:W359. https://doi.org/10.1093/nar/gkab336
20. Christoffer C, Bharadwaj V, Luu R et al (2021) LZerD protein-protein docking webserver enhanced with de novo structure prediction. Front Mol Biosci 8:724947. https://doi.org/10.3389/fmolb.2021.724947
21. Jain A, Terashi G, Kagaya Y et al (2021) Analyzing effect of quadruple multiple sequence alignments on deep learning based protein inter-residue distance prediction. Sci Rep 11(1):7574. https://doi.org/10.1038/s41598-021-87204-z
22. Kryshtafovych A, Schwede T, Topf M et al (2019) Critical assessment of methods of protein structure prediction (CASP)-round XIII. Proteins 87(12):1011–1020. https://doi.org/10.1002/prot.25823
23. Zhang C, Zheng W, Mortuza SM et al (2020) DeepMSA: constructing deep multiple sequence alignment to improve contact prediction and fold-recognition for distant-homology proteins. Bioinformatics 36(7): 2105–2112. https://doi.org/10.1093/bioinformatics/btz863
24. Chaudhury S, Lyskov S, Gray JJ (2010) PyRosetta: a script-based interface for implementing molecular modeling algorithms using Rosetta. Bioinformatics 26(5):689–691. https://doi.org/10.1093/bioinformatics/btq007
25. Zhou H, Zhou Y (2002) Distance-scaled, finite ideal-gas reference state improves structure-derived potentials of mean force for structure selection and stability prediction. Protein Sci 11(11):2714–2726. https://doi.org/10.1110/ps.0217002
26. Zhou H, Skolnick J (2011) GOAP: a generalized orientation-dependent, all-atom statistical potential for protein structure prediction. Biophys J 101(8):2043–2052. https://doi.org/10.1016/j.bpj.2011.09.012
27. Huang SY, Zou X (2011) Statistical mechanics-based method to extract atomic distance-dependent potentials from protein structures. Proteins 79(9):2648–2661. https://doi.org/10.1002/prot.23086
28. Park H, Bradley P, Greisen P et al (2016) Simultaneous optimization of biomolecular energy functions on features from small molecules and macromolecules. J Chem Theory Comput 12(12):6201–6212. https://doi.org/10.1021/acs.jctc.6b00819
29. Jumper J, Evans R, Pritzel A et al (2021) Highly accurate protein structure prediction with AlphaFold. Nature 596:583. https://doi.org/10.1038/s41586-021-03819-2
30. Lensink MF, Brysbaert G, Mauri T et al (2021) Prediction of protein assemblies, the next frontier: the CASP14-CAPRI experiment. Proteins 89:1800. https://doi.org/10.1002/prot.26222

Chapter 29

Predicting Protein Interaction Sites Using PITHIA

SeyedMohsen Hosseini and Lucian Ilie

Abstract

Several proteins work independently, but the majority work together to maintain the functions of the cell. Thus, it is crucial to know the interaction sites that facilitate protein–protein interactions. The development of effective computational methods is essential because experimental methods are expensive and time-consuming. This chapter is a guide to predicting protein interaction sites using the program "PITHIA." First, some installation guides are presented, followed by descriptions of input file formats. Afterward, PITHIA's commands and options are outlined with examples. Moreover, some notes are provided on how to extend PITHIA's installation and usage.

Key words Protein interactions, Machine learning, Protein interaction residue, Protein interaction site prediction

1 Introduction

In a cell, proteins control many biological systems, and most proteins work together as a team to fulfill their functions. Protein–protein interactions (PPI) are physical contacts between two or more proteins that result from electrostatic forces, hydrogen bonds, and hydrophobic interactions [6]. Interaction sites are the amino acids on the sequence of proteins that facilitate interactions between them.

Researchers will be able to understand various biological processes, diseases, and drug designs more thoroughly by detecting the interaction sites of proteins [23]. A protein's functionality is heavily influenced by its interactions with other proteins, which is why some databases such as PDB [3] and Uniprot [20] provide information about the interactions. In addition to these, there are several databases that focus on protein–protein interactions [7].

It is possible to categorize interaction site prediction into two main categories: experimental [12, 17] and computational [2, 4]. The cost of experimental methods, the amount of time they take, and the labor they require are high. This leads to more

Shahid Mukhtar (ed.), *Protein-Protein Interactions: Methods and Protocols*, Methods in Molecular Biology, vol. 2690, https://doi.org/10.1007/978-1-0716-3327-4_29,

widespread use of computational methods. There are three types of computational models: sequence-based [11, 23, 24], structural-based [15], and hybrid [22]. Many reasons make sequence-based methods more advantageous, including that they provide more information about the sequences, as well as being cheaper and faster. A relatively small number of proteins' structures are available in comparison to the number of sequences.

We recently proposed PITHIA [9]—protein interaction site prediction using multiple sequence alignments and attention—in order to predict the interaction sites. PITHIA employs the MSA-Transformer [16] (MSA stands for multiple sequence alignment) to profile the amino acids of protein sequences. Based on this feature representation, an architecture was developed that utilizes the self-attention model to predict the interaction residues.

2 Materials

2.1 Source Code

This program is written in Keras [5] (Python 3.6.2) with TensorFlow GPU [1] as its backend. PITHIA is available under the GNU General Public License v3.0 at https://github.com/lucian-ilie/PITHIA. A regular source code maintenance is performed on GitHub. On GitHub, users could clone or download PITHIA using the "clone" button or if they are running a UNIX-like system with git installed, then they can git clone https://github.com/lucian-ilie/PITHIA.git. Installation instructions for the git environment can be found at https://git-scm.com/.

2.2 Installation

PITHIA solely employs as input a FASTA file, which contains the names and the sequence of the proteins. The format of a FASTA file presented in Fig. 1:

For each sequence in the FASTA file format, PITHIA creates a file that contains multiple sequence alignments of that protein using HHblits [18] on the UniRef-30 database [19] (dated 2020-03). The Uniref dataset could be installed from http://gwdu111.gwdg.de/~compbiol/uniclust/2022_02/. The size of the dataset before extraction is 58 GB, and after extraction, it requires up to 140 GB. A number of ways exist to install HHblits, such as using Docker [13] or conda, but the one we used was clone

```
>1acbI
TEFGSELKSFPEVVGKTVDQAREYFTLHYPQYDVYFLPEGSPVTLDLRYNRVRVFYNPGTNVVNHVPHVG
>1c1yB
SNTIRVFLPNKQRTVVNVRNGMSLHDCLMKALKVRGLQPECCAVFRLLHEHKGKKARLDWNTDAASLIGEELQVDFL
```

Fig. 1 FASTA format file: each line that starts with the ">" symbol contains the protein ID, followed in the next line by the protein sequence

the code from GitHub and compile it using GCC 4.8 and CMake 2.8.12. A thorough description of how to compile HHblits can be found on their GitHub repository at https://github.com/soedinglab/hh-suite.

In order to obtain the MSA file for each sequence, we have used the following command:

```
hhblits -i $filename -oa3m $pid.a3m -n 4 -d Uni-
Ref30_2020_06
```

In the above command, the option `-i` means input file, the option `-oa3m` tells HHblits to generate an output MSA from the significant hits, `-n` specifies the number of search iterations, and `-d` specifies the dataset. This command will generate a file that contains between one and 500 sequences that can be used to create the embeddings.

There are specific libraries that are necessary for PITHIA such as Numpy, Keras, TensorFlow, and Bio-Transformers. There is a `requirements.txt` file that can be used as follows to install these libraries:

```
pip3 install -r requirement.txt
```

It is strongly suggested to create a virtual environment prior to installing these libraries.

Finally, in order to run PITHIA, the following command should be used:

```
bash run_PITIHA.sh [FASTA file]
```

2.3 Computing Environment

We are able to run our architecture on a UNIX-based server with 64 Gigabytes of RAM, 12 CPU cores, and one GPU (Model T4 with 16 Gigabytes of VRAM). The model takes about 24 h to train for 100 epochs. When embeddings are available, testing takes around 10 s per sequence, and embedding computation takes about 10 to 20 min.

2.4 Input Formatting

PITHIA, like most of the existing classification methods, has two phases. In the training phase, the FASTA file should contain the labels in addition to the names and sequences in order for the learning to happen. Figure 2 demonstrates an acceptable input format.

Input data in the testing phase should only be a FASTA file without the labels. PITHIA predicts, for each position, a number between 0 and 1. A sample output of the code is presented in Fig. 3.

```
>E6Z0R4
MLSEEEIEYRRRDARNALASQRLEGLEPDPQVVAQMERVVVGELETSDVIKDLMERIKREEI
00000100110111111111101111111001000100100000010011001100100101
>P74909
MPRLKVKLVKSPIGYPKDQKAALKALGLRRLQQERVLEDTPAIRGNVEKVAHLVRVEVVE
000000000000000000000000000000000000000000000000000000000000
>Q57839
MRACLKCKYLTNDEICPICHSPTSENWIGLLIVINPEKSEIAKKAGIDIKGKYALSVKE
00000000000000000000000000111111011001111001000001000011000
```

Fig. 2 Illustration of training phase input format: the binary string of labels following the sequence has the same length as the sequence; for each position, 1 indicates a interaction site, whereas 0 means non-binding

```
0    P    0.43742144
1    I    0.48270544
2    D    0.5042083
3    I    0.48167711
4    T    0.8099451
5    A    0.68276757
6    T    0.64708173
7    L    0.6715922
8    R    0.590994
9    C    0.48110902
10   K    0.4784069
11   V    0.31296155
12   A    0.29476246
13   V    0.30727118
```

Fig. 3 Illustration of output format: the first column contains the position number within the protein, the second column is the residue at that position, and the third column contains the prediction of PITHIA

2.5 Feature Extraction

Instead of combining different features such as PSSM, RSA, RAA, and ECO in order to create the feature vector for each amino acid, as most current methods do [11, 23, 24], PITHIA employs an unsupervised approach to extract features by using MSA transformers.

MSA transformers combine multiple sequence alignments (MSA), axial attention [8], and transformers [21]. Axial attention is a specific version of attention that is most suitable for datasets that have two-dimensional structure data. The MSA is an example of two-dimensional structure data. On the row dimension, the attention could help to understand the relation between the sequences, while the columns attention mechanism could help to understand the evolution among the proteins.

The MSA transformer is able to create a context-aware embedding for each amino acid in the sequence. As its side feature, it is able to create a contact map for each protein in an unsupervised approach. This contact map gives us a limited understanding about the three-dimensional structure of proteins. The contact map of a protein indicates the distance between any two residues in its three-dimensional structure. To be more specific, it specifies the likelihood that two residues are situated in contact with each other.

3 Methods

For each target residue, a sliding window centered on its position is employed. In this approach, each target is surrounded by amino acids based on the primary structure of the proteins. Additionally, with respect to the size of the sliding window, those amino acids that do not have the required number of neighbors (first or last few amino acids in the sequence) are padded with zeros.

3.1 MSA Size

In recent years, numerous studies showed that using multiple sequence alignment can improve the performance of the models that predict protein structure [10, 14]. MSA transformers suggest to use first 128 sequences in MSA files to create embeddings; nevertheless, for our specific problem, the results show [9] that 64 sequences have the best results compared to 32 and 128. The process to compute the MSAs is as follows:

1. If the length of the protein is longer than 1024, divide it into chunks with at most 1024 amino acids without overlapping and save them into different FASTA files (*see* **Note 1**).
2. For each FASTA file that contains a sequence, employ HHblits to create the MSA file.
3. Get the first 64 sequences from the MSA file (or all sequences, if less than 64) and feed them to MSA Transformer in order to compute the embeddings (*see* **Note 2**).

3.2 Padding

As stated earlier, the model uses a sliding window, and the size of the input that is fed to the final architecture should be of a fixed size. Therefore, for the resides at the two ends of a protein, padding is required to obtain the same input size. The usual approach in these circumstances is to use zero padding.

3.3 Implementation Limitations

In order to decrease the memory usage, we employed generators to build the input data on the fly. A generator is a special type of function that returns a sequence of values rather than an individual one. Instead of loading the entire dataset, they enable the program to only load the data necessary for the current epoch. Since the input windows for consecutive residues are highly similar to one another, it is necessary to shuffle the data prior to training. Local shuffle and global shuffle are the two options here. Due to the use of generators, we can only use local shuffling.

3.4 Architecture

We developed and compared four different architectures, namely, multiplayer perceptron (MLP), recurrent network (RNN), convolutional network (CNN), and transformer self-attention (TF). Extensive testing [9] showed that a TF model outperforms all the others. The architecture of this model is presented in Fig. 4.

What follows is the overall steps to compute the interaction sites:

1. Compute the MSA files using HHblits.
2. Compute the embeddings using MSA transformers.
3. Create the sliding windows and add zero padding.
4. Compute the interaction sites using the pre-trained model.

3.5 Availability

The trained model, source code, and other related datasets are freely available at https://github.com/lucian-ilie/PITHIA.

The web server is available at http://pithia.csd.uwo.ca/; this is particularly useful for users who do not have the necessary programming background. It has been developed using Python Flask 2.1, Celery 5.2, and Redis 7.0 on Ubuntu 18.04 operating system. The user inputs protein sequences and receives the prediction results via e-mail. The average computation time for a protein of length 500 on the web server is 15 min.

Generally, web servers are synchronous; therefore, whenever a user submits a form, they have to stay on their browsers for a response from the server. On servers that need time and computation to create the appropriate response, developers tend to use asynchronous methods. There are multiple ways to develop asynchronous web servers, for example, some may use cron jobs to repetitively check for submissions and results every 10 min to proceed with the necessary instructions that create unnecessary overhead for the system. We have chosen to use the combination of Celery and Redis as a suitable method for our web server. With this method, the web server would be able to use the processors for the prediction, and the users will get their results faster.

4 Notes

1. The MSA transformer is only able to create embeddings for protein sequences that have less than 1024 amino acids; therefore, the proteins that exceed the required threshold should be divided into 2 or more parts depending on their length. Each chunk of protein should have at most 1024 amino acids in it, and it will be portrayed as a different protein, and the final result should be attached together.

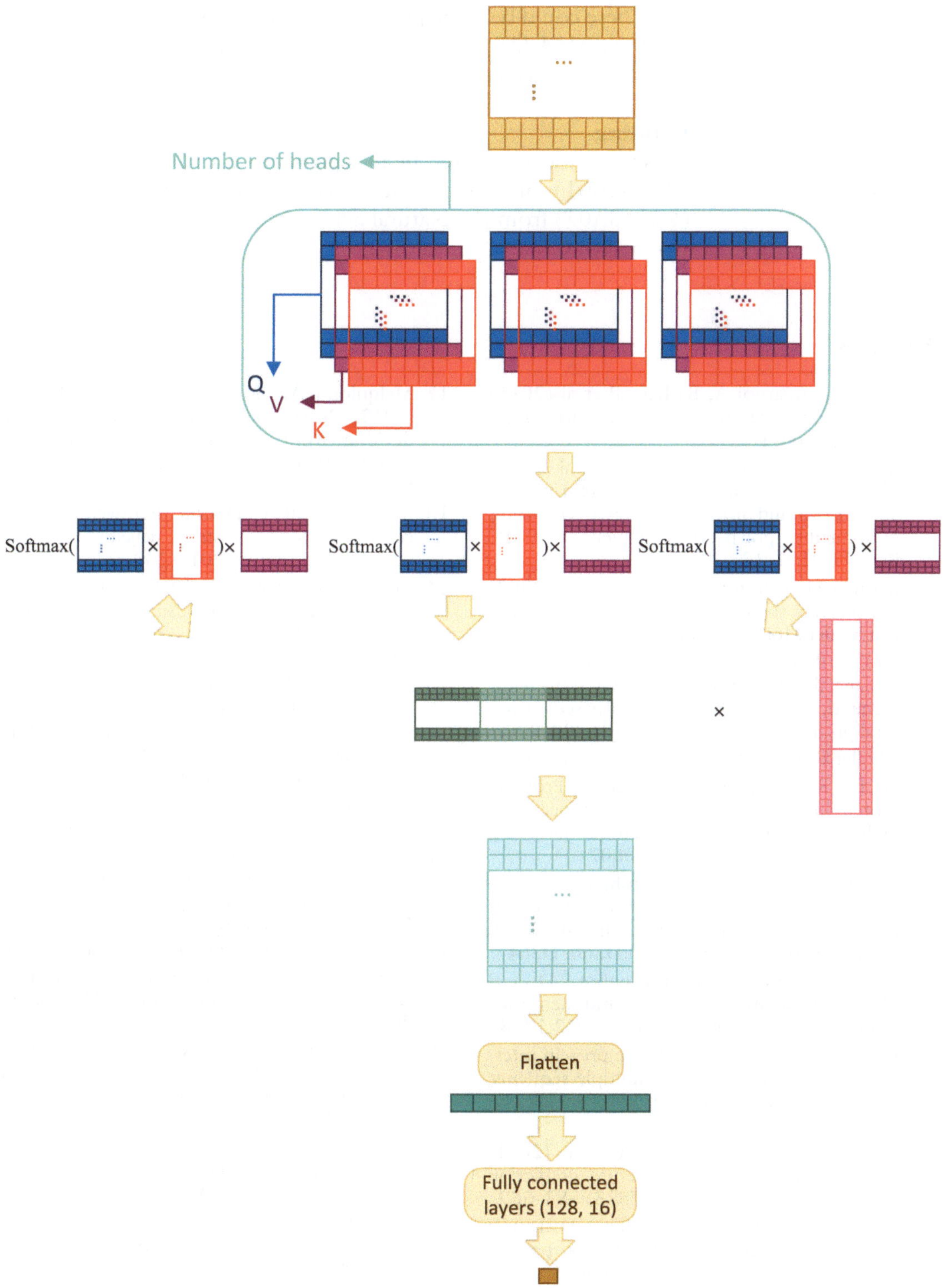

Fig. 4 TF architecture

2. There could be some proteins that do not have 64 sequences in their MSA file; for those proteins, all sequences that are available are selected.

Funding

This work was supported by two grants to L.I.: a Discovery Grant (R3143A01) and a Research Tools and Instruments Grant (R3143A07) from the Natural Sciences and Engineering Research Council of Canada.

References

1. Abadi M, Agarwal A, Barham P et al (2015) TensorFlow: large-scale machine learning on heterogeneous systems
2. Amos-Binks A, Patulea C, Pitre S et al (2011) Binding site prediction for protein-protein interactions and novel motif discovery using re-occurring polypeptide sequences. BMC Bioinf 12(1):1–13
3. Berman H, Henrick K, Nakamura H (2003) Announcing the worldwide protein data bank. Nat Struct Mol Biol 10(12):980–980
4. Cao B, Porollo A, Adamczak R et al (2006) Enhanced recognition of protein transmembrane domains with prediction-based structural profiles. Bioinformatics 22(3):303–309
5. Chollet F et al (2015) Keras. https://keras.io
6. De Las Rivas J, Fontanillo C (2010) Protein–protein interactions essentials: key concepts to building and analyzing interactome networks. PLoS Comput Biol 6(6):e1000807
7. Higurashi M, Ishida T, Kinoshita K (2009) PiSite: a database of protein interaction sites using multiple binding states in the PDB. Nucleic Acids Res 37(suppl_1):D360–D364
8. Ho J, Kalchbrenner N, Weissenborn D et al (2019) Axial attention in multidimensional transformers. arXiv preprint arXiv:1912.12180
9. Hosseini S, Ilie L (2022) Pithia: protein interaction site prediction using multiple sequence alignments and attention. Int J Mol Sci 23(21): 12814
10. Jumper J, Evans R, Pritzel A et al (2021) Highly accurate protein structure prediction with AlphaFold. Nature 596(7873):583–589
11. Li Y, Golding GB, Ilie L (2021) DELPHI: accurate deep ensemble model for protein interaction sites prediction. Bioinformatics 37(7):896–904
12. Melquiond AS, Karaca E, Kastritis PL et al (2012) Next challenges in protein-protein docking: from proteome to interactome and beyond. Wiley Interdiscip Rev Comput Mol Sci 2(4):642–651
13. Merkel D (2014) Docker: lightweight Linux containers for consistent development and deployment. Linux J 2014(239):2
14. Mirabello C, Wallner B (2019) RawMSA: end-to-end deep learning using raw multiple sequence alignments. PloS One 14(8): e0220182
15. Neuvirth H, Raz R, Schreiber G (2004) ProMate: a structure based prediction program to identify the location of protein-protein binding sites. J Mol Biol 338(1):181–199
16. Rao RM, Liu J, Verkuil R et al (2021) MSA transformer. In: International conference on machine learning. PMLR, p 8844–8856
17. Shoemaker BA, Panchenko AR (2007) Deciphering protein-protein interactions. Part I. Experimental techniques and databases. PLoS Comput Biol 3(3):e42
18. Steinegger M, Meier M, Mirdita M et al (2019) HH-suite3 for fast remote homology detection and deep protein annotation. BMC Bioinf 20(1):1–15
19. Suzek BE, Huang H, McGarvey P et al (2007) UniRef: comprehensive and non-redundant UniProt reference clusters. Bioinformatics 23(10):1282–1288
20. The UniProt Consortium (2021) UniProt: the universal protein knowledgebase in 2021. Nucleic Acids Res 49(D1):D480–D489
21. Vaswani A, Shazeer N, Parmar N et al (2017) Attention is all you need. In: Advances in neural information processing systems, vol. 30

22. Zeng M, Zhang F, Wu F-X et al (2020) Protein-protein interaction site prediction through combining local and global features with deep neural networks. Bioinformatics 36(4):1114–1120
23. Zhang B, Li J, Quan L et al (2019) Sequence-based prediction of protein-protein interaction sites by simplified long short-term memory network. Neurocomputing 357:86–100
24. Zhang J, Kurgan L (2019) SCRIBER: accurate and partner type-specific prediction of protein-binding residues from proteins sequences. Bioinformatics 35(14):i343–i353

Chapter 30

Using PlaPPISite to Predict and Analyze Plant Protein–Protein Interaction Sites

Jingyan Zheng, Xiaodi Yang, and Ziding Zhang

Abstract

Proteome-wide characterization of protein–protein interactions (PPIs) is crucial to understand the functional roles of protein machinery within cells systematically. With the accumulation of PPI data in different plants, the interaction details of binary PPIs, such as the three-dimensional (3D) structural contexts of interaction sites/interfaces, are urgently demanded. To meet this requirement, we have developed a comprehensive and easy-to-use database called PlaPPISite (http://zzdlab.com/plappisite/index.php) to present interaction details for 13 plant interactomes. Here, we provide a clear guide on how to search and view protein interaction details through the PlaPPISite database. Firstly, the running environment of our database is introduced. Secondly, the input file format is briefly introduced. Moreover, we discussed which information related to interaction sites can be achieved through several examples. In addition, some notes about PlaPPISite are also provided. More importantly, we would like to emphasize the importance of interaction site information in plant systems biology through this user guide of PlaPPISite. In particular, the easily accessible 3D structures of PPIs in the coming post-AlphaFold2 era will definitely boost the application of plant interactome to decipher the molecular mechanisms of many fundamental biological issues.

Key words Plant, Interactome, Protein, protein interaction site, 3D structures of protein complexes, Database

1 Introduction

Protein–protein interactions (PPIs) are heavily involved in many cellular processes and play crucial roles in maintaining the proper functioning of biological systems [1, 2]. Characterization of proteome-wide PPI network (also termed as interactome) can provide a systematic understanding of protein functions in cells. For instance, the PPI network analysis can provide vital clues for the annotation of functionally unknown proteins by capturing their potential functions through the interaction relationships with functionally known proteins. Therefore, there is an increasing interest in characterizing PPI networks [3]. A variety of high-throughput experimental techniques have been developed to identify PPIs,

Shahid Mukhtar (ed.), *Protein-Protein Interactions: Methods and Protocols*, Methods in Molecular Biology, vol. 2690,
https://doi.org/10.1007/978-1-0716-3327-4_30,

such as yeast two-hybrid assay [4, 5], immunoprecipitation [6–8], and tandem affinity purification [9–11]. However, the experimental methods are generally time-consuming, costly, and laborious, and the available experimental PPI data are still limited.

As an alternative strategy, a plethora of computational methods has been proposed to predict PPIs in the past decades. Most PPI prediction methods rely on the features/patterns of existing PPIs, including homology relationships [12], domain–domain interaction (DDI) [13], co-expression of genes [14, 15], and so on. With the continuous progress of artificial intelligence, many methods based on machine learning have been proposed to facilitate the identification of novel PPIs [16, 17].

Indeed, computational methods have also been widely applied to predict plant PPIs. In general, interolog mapping is the most frequently used PPI inference method. For instance, Gu et al. employed the interolog mapping method to construct a proteome-wide *Oryza sativa* (rice) PPI network called PRIN [18]. They further indirectly validated the reliability of PRIN through other information such as gene ontology (GO) annotation, subcellular localization, and gene expression data. Comparatively, machine learning methods are playing an increasingly important role in predicting plant PPIs when the experimental PPIs in plants have been significantly increased. Zhu et al. employed a support vector machine (SVM) to predict 273,400 PPIs between 10,793 *Zea mays* (maize) proteins, and they further combined the predicted PPIs with experimental data to construct the PPI database for maize (PPIM) [19]. Ding et al. integrated multiple features through machine learning, including functional associations, phylogenetic profiles, protein sequences, and gene co-expression information. Meanwhile, they predicted PPIs of *Arabidopsis thaliana* (Arabidopsis), *Glycine max* (soybean), and maize based on SVM and random forest (RF) [20]. Zhao et al. proposed a computational interactome called AraPPINet, which contains 345,000 predicted PPIs of Arabidopsis. In their method, protein structure, gene expression data, and functional annotation information were used as the input of the RF-based PPI predictor [21]. As a breakthrough technique of machine learning, the application of deep learning in predicting plant PPIs has also been initiated. Very recently, Pan et al. proposed a network embedding-based approach combining deep neural network (DNN) to predict the interactions between different plant proteins, including Arabidopsis, rice, and maize [22].

The predicted PPIs have become a valuable supplement to experimental PPIs and have been incorporated in many plant interactome-related databases, which play an important role in accelerating the functional genomics studies of plants. Li et al. constructed the Arabidopsis protein interactome database AtPID by integrating experimental PPIs and predicted PPIs based on

Table 1
Summary of existing plant PPI prediction resources

Resource	Species	URL	Description
AraPPINet [21]	*Arabidopsis thaliana*	https://netbio.sjtu.edu.cn/arappinet/	345,000 predicted Arabidopsis PPIs based on the RF model
AtPID [23]	*Arabidopsis thaliana*	http://119.3.41.228/atpid/webfile/	45,382 curated PPIs and 118,556 predicted PPIs
AtPIN [24]	*Arabidopsis thaliana*	https://atpin.bioinfoguy.net/cgi-bin/atpin.pl	It is an integration resource for Arabidopsis PPIs, ontology and subcellular localization information
PRIN [18]	*Oryza sativa*	http://bis.zju.edu.cn/prin/	Visualized proteome-wide rice PPI network
GraP [25]	*Gossypium raimondii*	http://structuralbiology.cau.edu.cn/GraP/	Predicted PPI integration and functional genomics analysis platform in which predicted PPI data were integrated
PPIM [19]	*Zea mays*	http://comp-sysbio.org/ppim	273,400 predicted maize PPIs by using the SVM algorithm

various bioinformatics methods, covering 45,382 curated PPIs and 118,556 predicted PPIs [23]. Brandao et al. provided AtPIN, an integration tool that includes information on Arabidopsis PPIs, ontology, and subcellular localization [24]. Zhang et al. constructed a platform of functional genomics analysis in *Gossypium raimondii* (GraP), including the latest functional annotation, gene family classifications, and PPI network information predicted by the interolog mapping method [25]. Available resources for predicting plant PPIs are outlined in Table 1.

To further decipher the interaction detail of a PPI, it is necessary to determine the binding sites/interface between two interacting proteins. Protein interaction sites can be annotated directly from the experimental 3D structures of PPIs. However, the number of 3D experimental structures of plant PPIs is far from enough, so protein–protein docking methods, including template-free [26] and template-based docking [27], are often employed to model the 3D complex structures of interacting proteins from two monomer structures. Compared with template-free docking, template-based docking, such as Homology Modeling of Protein Complex (HMPC) [27] and Protein Interactions by Structural Matching (PRISM) [28], is more reliable in obtaining the complex structure as well as inferring the interaction site information. HMPC is based on the idea that homologous protein complexes share similar binding modes. Once a PPI contains homologous complex templates with known 3D structures, the 3D structure of the query PPI can be easily predicted via HMPC [27]. As an extension of HMPC, the basic principle of PRISM is that two interacting proteins may interact through similar regions if their specific surface regions resemble known interfaces in a template complex [29]. Although protein docking methods have significantly increased the coverage of protein complex structures, accurate protein complex structures for many PPIs are still unavailable. In this context, the identification of interaction regions (i.e., domains and motifs) between two interacting proteins is also necessary and informative. PPIs are often mediated by DDIs or domain–motif interactions (DMIs). Known DDI and DMI data have been accumulated and deposited in some public databases such as iPfam [30] and 3did [31], which significantly facilitate the annotations of DDIs/DMIs in PPIs.

Compared with the available interactome databases, structural-related PPI databases (e.g., the databases recording residue-level interaction sites in the whole interactome) are generally lacking, especially for plant species. In 2016, we constructed a comprehensive database called AraPPISite to provide protein interaction site annotations for the model plant Arabidopsis [32]. AraPPISite allows users to search the 3D structures, interaction sites, DDIs, and DMIs of PPIs. Moreover, it provides rich physicochemical properties of interaction sites. Since AraPPISite only recorded one plant species, we have upgraded it to PlaPPISite which provides interaction details for multiple plant interactomes [33]. Compared with AraPPISite, PlaPPISite incorporates 12 other plant interactomes and significantly increases the coverage of plant PPIs with interaction site annotations, although the PPI data and the corresponding interaction sites are mainly derived from computational methods. Moreover, a convenient prediction platform was incorporated into PlaPPISite, in which users could solely submit two query protein sequences to obtain the protein complex structure and interaction site information. In this chapter, we provide a practical user guide to our PlaPPISite database, which is particularly designed for users inexperienced in protein bioinformatics.

2 Materials

2.1 The Environment of Database and System Requirements

PlaPPISite was constructed based on MySQL 5.5.60 and PHP 5.4.16. The web service runs on Apache 2.4.6 and CentOS 7.4 in Linux. The PPI networks are displayed through a JavaScript graph library called cytoscape.js [34]. In addition, the 3D structures of PPIs were visualized using NGL [35], a web application for molecular visualization based on WebGL (*see* **Note 1** for more details).

2.2 The Input Data

PlaPPISite annotated all experimentally verified and predicted PPIs for 13 plants, including Arabidopsis, *Chlamydomonas reinhardtii*, *Ricinus communis*, soybean, rice, *Selaginella moellendorffii*, *Solanum lycopersicum*, *Solanum tuberosum*, *Vitis vinifera*, maize, *Brachypodium distachyon*, *Populus trichocarpa*, and *Medicago truncatula*. More annotation details are described in **Note 2**. Users can search the database through a single protein ID or a keyword. The PPIs associated with the query protein will be displayed in a table. Otherwise, users can retrieve a PPI by searching for its two protein IDs or keywords. Note that protein IDs accept the following forms: UniProt accession number, Gene ID, or Gene name.

If the query protein or PPI is not in our annotation data, its 3D complex structure can be predicted based on HMPC and the corresponding interaction sites will be further annotated. On the PlaPPISite prediction page, users need to enter the two protein sequences in FASTA format. Proteins represented in FASTA format should have two lines. The first line is the name of the protein, which must start with a ">" sign, followed by the protein ID. The second line is its amino acid sequence, in which each singular letter represents an amino acid. Protein sequences in FASTA format for most species can be downloaded from UniProt (https://www.uniprot.org/) [36].

2.3 The Datasets Deposited in PlaPPISite

To facilitate the research community, there are several datasets available for download in our PlaPPISite database, including: (1) the 3D structures of all protein complexes including experimentally verified and predicted structures; (2) protein interaction sites for the PPIs whose 3D structures are known in our database; (3) all DDIs and DMIs for the PPIs whose 3D structures are unknown in our database; (4) all predicted PPIs and their annotation information, including GO annotation, subcellular localization information, gene expression data, and so on.

3 Methods

As briefly discussed in the subsection of 2.2, PlaPPISite has annotated experimentally verified and predicted PPIs for 13 plants, which can be retrieved by searching for a single protein ID or a pair of protein IDs. If the query PPI cannot be retrieved, relevant annotated information can be obtained through the online prediction platform. Considering that study on plant interactome is still in its infancy, the current version of PlaPPISite may inevitably suffer from some limitations (*see* **Note 3** for more details). The following subsections describe the implementation steps and the resulting data.

3.1 Searching for a Single Protein

1. Open the PlaPPISite website and go to the "Search" page.
2. Input the ID or keyword of a single protein into the search box (the protein "P92978" is taken as a case study (Fig. 1a); *see* **Note 4** for details). Then, click the "Submit" button to obtain the results.

PPI	Gene Name1	Gene Name2	Organism	PPI source	Method	Detail
A4IJ27 - P92978	ROPGEF3	ARAC11	*Arabidopsis thaliana*	Experimentally verified	HMPC	View
F4HRJ4 - P92978	MAPKKK3	ARAC11	*Arabidopsis thaliana*	Predicted	DDI/DMI	View
F4I5N6 - P92978	RIC3	ARAC11	*Arabidopsis thaliana*	Experimentally verified	DDI/DMI	View
F4IVV0 - P92978	RIC1	ARAC11	*Arabidopsis thaliana*	Experimentally verified	DDI/DMI	View
F4J1G1 - P92978	ROPGEF13	ARAC11	*Arabidopsis thaliana*	Experimentally verified	HMPC	View
F4J424 - P92978	RIC5	ARAC11	*Arabidopsis thaliana*	Experimentally verified	DDI/DMI	View
F4JI46 - P92978	ROPGAP2	ARAC11	*Arabidopsis thaliana*	Experimentally verified	DDI/DMI	View
F4K295 - P92978	ROPGEF5	ARAC11	*Arabidopsis thaliana*	Experimentally verified	HMPC	View
P92978 - Q0WNP7	ARAC11	ROPGEF4	*Arabidopsis thaliana*	Experimentally verified	HMPC	View
P92978 - Q1G3Y0	ARAC11	RIC9	*Arabidopsis thaliana*	Experimentally verified	DDI/DMI	View

Fig. 1 An example of searching interaction partners for protein "P92978." (**a**) Search for a single protein in the database. (**b**) PPIs associated with protein "P92978." (**c**) Visualization of the PPI (P92978-A4IJ27) subnetwork.

3. Retrieve all items that appear on the results page, each of which is the interaction information related to the input protein. Each column represents the PPI involved in the input protein, the gene names of two proteins of the PPI, the source species of the PPI, the source of the PPI, the annotation method, and the "View" button to access the detailed information for the PPI (Fig. 1b). In particular, the PPI sources contain two types, i.e., experimentally verified and predicted (*see* **Note 5** for more details about the PPI prediction methods used in PlaPPISite). Regarding the annotation method, the 3D structure of the complex is compiled according to the collection of PDB or prediction of HMPC/PRISM. If the 3D complex structures are unavailable, the corresponding DDI/DMI information is annotated.
4. Click "View" to jump to the detailed information page. As shown in Fig. 1c, a visualization of the PPI subnetwork is provided on the PPI detail page. There are four types of PPI subnetworks in PlaPPISite (*see* **Note 6** for the details of different PPI subnetwork types).
5. Click "Export" to export the network for further analysis. Also, click "Layout" to adjust the layout. Adjust the edge length and node spacing based on the "Edge length" and "Node spacing."
6. In addition, as shown in Fig. 2a–c, the alignment results of each protein in PPI with its template, visualization of the 3D complex structure, and detailed interaction sites are also provided on the details page. It is important to note that the 3D structure of the PPI complex can be viewed from different orientations through rotation. The demonstrations of the 3D complex structures through mouse operations are detailed **Note** 7. In addition to the detailed interaction sites, the corresponding physical and chemical properties, including bond types, conservation score, and the change of Gibbs free energy, are also listed.
7. As shown in Fig. 3a, DDI/DMI annotations are provided for those PPIs whose complex structures cannot be constructed. For PPIs predicted by interolog mapping, source species of interolog templates, GO annotations, and subcellular localizations are provided (Fig. 3b). Similarity measurements of GO annotations, gene expression profiles, and subcellular localizations can be used to further determine the reliability of PPIs.

Fig. 1 (continued) The key implementation steps to search a single protein are: (1) input the protein ID into the search box; (2) click the "Submit" button; (3) click the "View" button to see more information about PPI; (4) query PPI; (5) back to the previous page, that is, query PPI details page; (6) link to the local search page; (7) export the network or layout various network styles; (8) display the network in a random format; (9) adjust the network edge length; (10) adjust the network node spacing

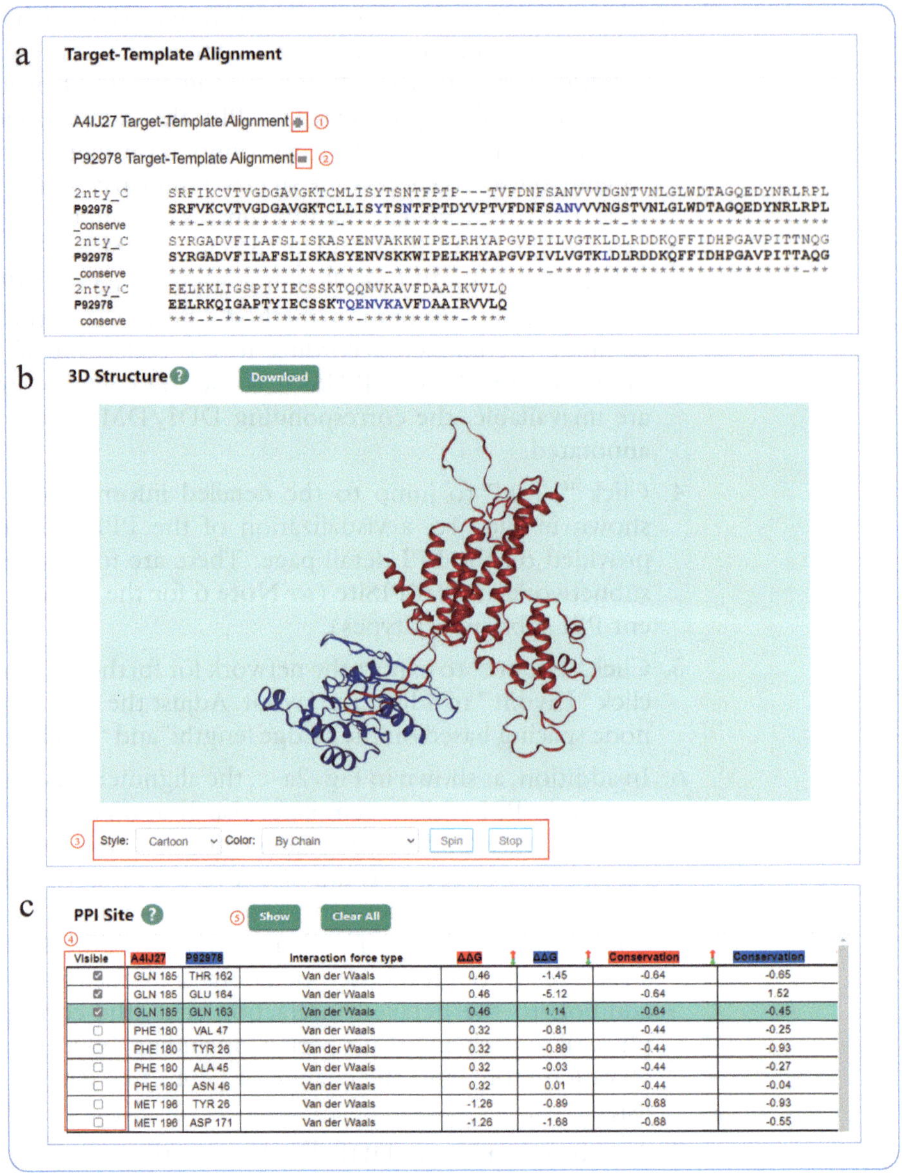

Fig. 2 Detailed annotation information of the PPI (P92978-A4IJ27). (**a**) Results of protein comparison with the template. (**b**) Visualization of the three-dimensional structure of the PPI complex. c. PPI sites and the corresponding physicochemical properties. The implementation details for annotation are: (1) click "+" to display the specific alignment results; (2) click "-" to fold up the specific alignment results; (3) choose alternative colors and styles of the structure. Click the "Spin" button to view the different orientations of the 3D complex structure; (4) users select the residue pairs they are interested in; (5) click the "Show" button to display the chosen residue pairs on the 3D complex structure

Meanwhile, these indirect evidences can be highlighted in the PPI subnetwork, in which nodes are colored by subcellular localizations and edges are colored by gene expression similarity between two nodes (Fig. 3c).

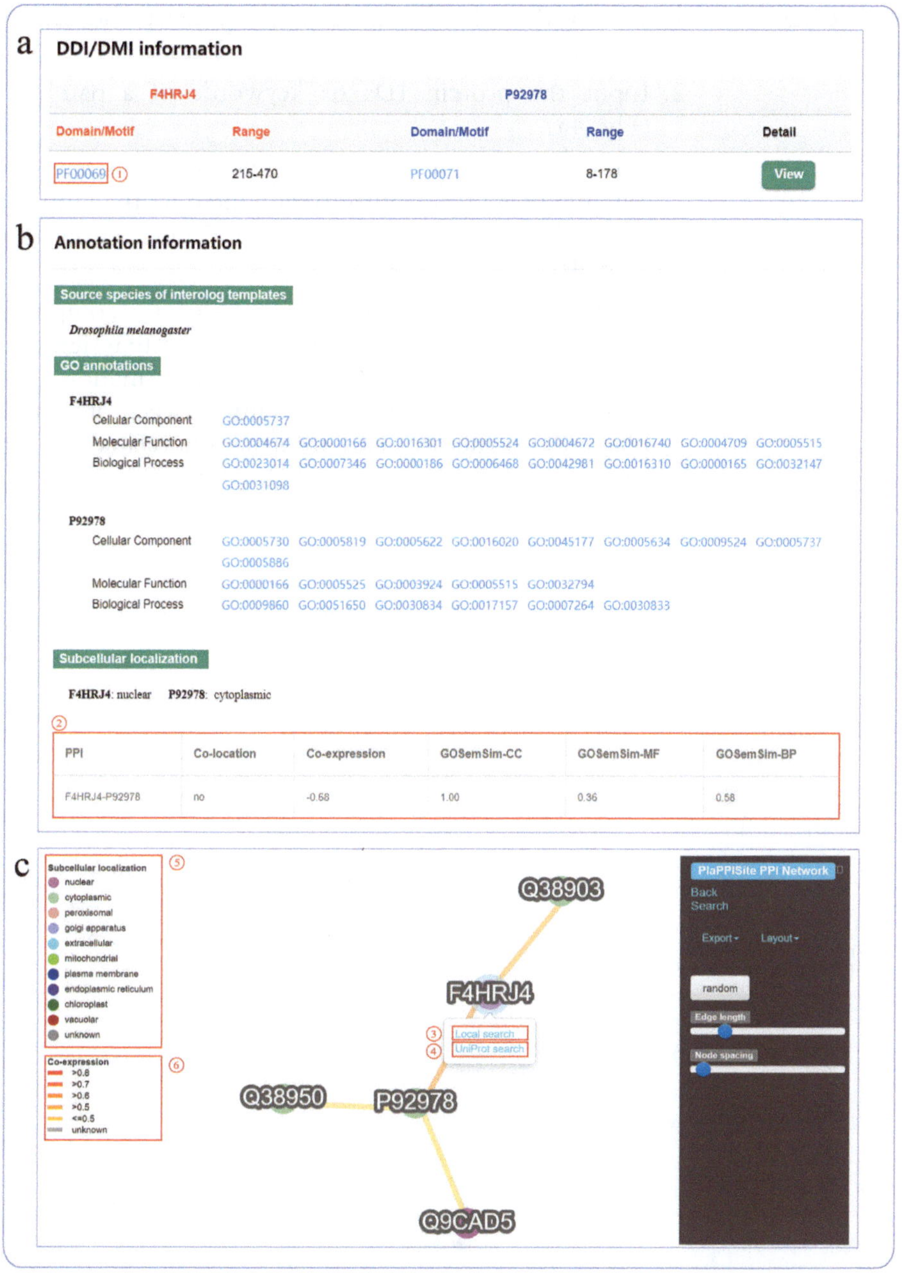

Fig. 3 Detailed annotation information about the PPI (F4HRJ4-P92978). The PPI is predicted and the complex structure cannot be modeled. (**a**) DDI/DMI annotations information. (**b**) Indirect evidence to help users judge the reliability of predicted PPI. (**c**) Visualization of the PPI (F4HRJ4-P92978) subnetwork. The implementation details for annotation are: (1) click the "PF00069" button to jump to the Pfam link to view the annotation; (2) similarity measurements of GO annotations, gene expression profiles, and subcellular localizations; (3) local single protein search; (4) link to UniProt database; (5) predicted subcellular localization; (6) co-expression similarity of predicted PPI

3.2 Search for an Interaction Pair

1. Go to the "PPI" search page to start the retrieval of a protein pair.
2. Input the protein IDs or keywords of a pair of proteins (Fig. 4a).
3. Click the "Submit" button to achieve the search results. The searching result page information includes the UniProt ID and Gene ID of these two proteins, and description information (Fig. 4b).
4. Click the "View" button to further view the complex information. Similar to searching for a single protein described in the subsection of 3.1, the detailed page information includes the PPI subnetwork, alignment results between protein and template, 3D structure, and information related to interaction sites.

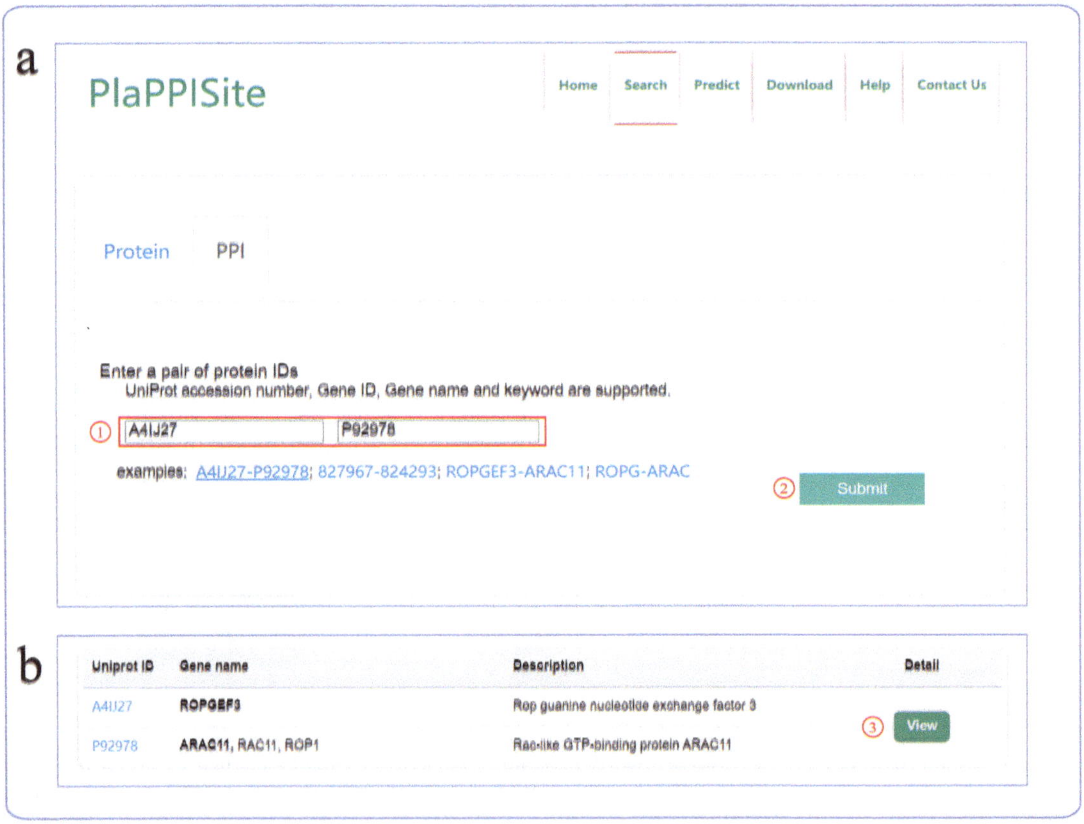

Fig. 4 An example of searching a PPI. (**a**) Search for a pair of PPI (P92978-A4IJ27) in the database. (**b**) Detailed annotation information about the PPI. Note that the key search boxes or buttons (1-3) to implement the PPI searching are labeled

3.3 Prediction of Complex Structure

1. To help construct the complex structures of interacting protein pairs, an online prediction platform was built based on the HMPC method. The process of complex structure construction mainly consists of three steps: template selection, monomer structural modeling, and complex structural modeling. First, enter the "Predict" page to initiate the prediction of complex structure.
2. As shown in Fig. 5a, on the prediction page, input the FASTA sequences of two interacting proteins.
3. Input email address that can receive the prediction result.
4. Click the "SUBMIT" button to start the prediction of complex structures of interacting proteins, which can be predicted based on HMPC. The resulting page displays the two proteins, their respective homologous templates, sequence identity, alignment coverage, and whether the two proteins interact or not (Fig. 5b). At the same time, the 3D model of the complex and the corresponding interaction sites are also provided. Click the corresponding buttons can obtain more information.

4 Notes

1. WebGL is generally used to visualize various forms of protein, DNA, RNA, and other molecules. Our recommended Internet browsers are Firefox 4+, Google Chrome 9+, Opera 12+, Safari 5.1+, Internet Explorer 11+, and Microsoft Edge Build 10,240 + .
2. PlaPPISite provides detailed interaction information on the PPIs of 13 plants. Among them, 121 PPIs have experimental complex structures, and 132,328 PPIs have interaction site information from HMPC and PRISM. PlaPPISite provides DDI or DMI annotations for PPIs whose 3D structures cannot be modeled. On the front page of the PlaPPISite website, we provide the PPI distribution map, i.e., the quantitative statistics of the sources of 3D structures of PPIs in 13 plants.
3. The current version of PlaPPISite still suffers from two limitations. First, the coverage of the PPIs and interaction site information for these plant species is not sufficiently high. In particular, the available complex structures and interaction sites are still far from complete. Second, the PPI and interaction site information deposited in PlaPPISite may inevitably contain false positives, although we have conducted essential procedures to ensure the reliability of the predicted PPIs and protein complex structures. Regarding future perspectives, more powerful bioinformatics methods for PPI prediction, protein–protein interaction site prediction, and protein complex structure

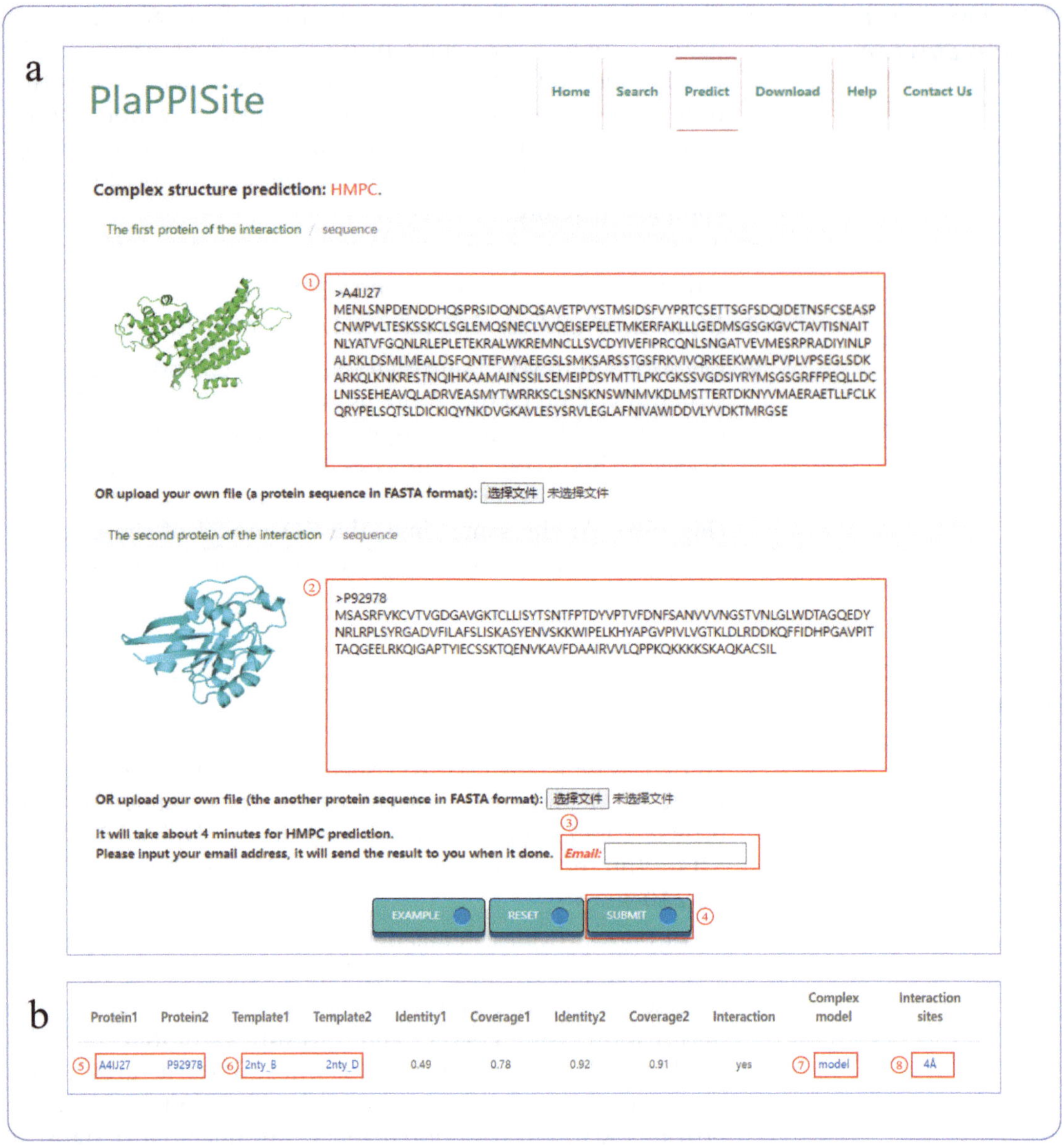

Fig. 5 An example of predicting the complex structure of a PPI. (**a**) Submit the PPI that needs to be predicted. (**b**) The result of complex structure prediction. The implementation steps are: (1) input the sequence of protein "A4IJ27" into the search box; (2) input the sequence of protein "P92978" into the search box; (3) input the email address to receive the result; (4) click the "SUBMIT" button to submit the prediction task; (5) click protein ID to get the sequence; (6) click the "2nty_B" or "2nty_D" button to download the PDB files; (7) click the "model" button to download the complex model; (8) click the "Interaction sites" button for demonstrating interaction details

prediction are under development through community-wide efforts. With the prosperity of deep learning [37] and the huge success of AlphaFold2 [38], it is easily accessible to obtain high-accurate protein 3D models. In this context, more accurate binary PPI predictions will be achieved [39, 40]. Likewise, predicting interaction details from binding regions/residues to 3D conformational dynamics of two interacting proteins will be accelerated [41–43]. Indeed, we are approaching constructing a complete 3D interactome for any plant species, and we anticipate some new plant 3D interactome platforms will be available to the community in the near future.

4. For protein "P92978," we searched for its associated PPIs and the annotation information of these PPIs. "P92978" represents Arabidopsis Rac-like GTP-binding protein ARAC11. This protein may be involved in cell polarity control during the actin-dependent growth at the tip of pollen tubes. A search on PlaPPISite yielded 28 items. Among them, the complex structure of PPI "A4IJ27-P92978" is constructed by HMPC. The protein "A4IJ27" is a guanine-nucleotide exchange factor (GEF) that acts as an activator of Rop (Rho of plants) GTPases by promoting the exchange for GTP. However, PPI "F4HRJ4-P92978" provides DDI/DMI annotation because it cannot construct complex structure. The protein "F4HRJ4" is a mitogen-activated protein kinase kinase kinase 3.
5. We employed the interolog mapping method to predict the PPIs of 13 plants. The basic assumption of this method is that PPIs are conserved in different organisms. In general, if two proteins (A and B) interact, their orthologous proteins in other species (A' and B′) can be predicted to interact.
6. PlaPPISite provides subnetwork visualization for PPIs, which contains four types of PPI networks [1]. PPI network of experimentally verified PPIs (with complex structures from HMPC/PRISM); [2] PPI network of experimentally verified PPIs (with the annotation of DDI/DMI); [3] PPI network of predicted PPIs (with complex structures from HMPC/PRISM); [4] PPI network of predicted PPIs (with the annotation of DDI/DMI). Compared with the experimentally verified PPI network, the main difference of the predicted PPI network is that it provides the predicted subcellular location and the co-expression similarity of PPI to help users to better judge the reliability of the predicted PPIs.
7. Visualization of 3D complex structures is implemented by NGL. And users can interact with 3D structures by mouse operations. If users would like to rotate the view around the center of the canvas, this can be done by left-mouse drag or one-finger drag. Users can translate the view and move the center of rotation by right-mouse drag, two-finger drag, or ctrl + left-mouse drag. Views can be zoomed in and out through the scroll wheel, two-finger pinch, or shift + left mouse drag. Information about atoms or bonds can be displayed by mouse hover or finger tap.

Acknowledgments

We thank our previous colleagues Drs. Hong Li and Shiping Yang for participating in developing PlaPPISite. This work is supported by grants from the National Natural Science Foundation of China (31970645 and 31471249).

References

1. Keskin O, Tuncbag N, Gursoy A (2016) Predicting protein-protein interactions from the molecular to the proteome level. Chem Rev 116(8):4884–4909
2. Berggard T, Linse S, James P (2007) Methods for the detection and analysis of protein-protein interactions. Proteomics 7(16): 2833–2842
3. Braun P, Aubourg S, Van Leene J et al (2013) Plant protein Interactomes. Annu Rev Plant Biol 64:161–187
4. Brückner A, Polge C, Lentze N et al (2009) Yeast two-hybrid, a powerful tool for systems biology. Int J Mol Sci 10(6):2763–2788
5. Paiano A, Margiotta A, De Luca M et al (2019) Yeast two-hybrid assay to identify interacting proteins. Curr Protoc Protein Sci 95(1):e70
6. Bonifacino JS, Dell'Angelica EC, Springer TA (2006) Immunoprecipitation. Curr Protoc Neurosci Chapter 5:Unit 5.24
7. Evans IM, Paliashvili K (2022) Co-Immunoprecipitation Assays. Methods Mol Biol 2475:125–132
8. Lin JS, Lai EM (2017) Protein-protein interactions: co-immunoprecipitation. Methods Mol Biol 1615:211–219
9. Puig O, Caspary F, Rigaut G et al (2001) The tandem affinity purification (tap) method: a general procedure of protein complex purification. Methods 24(3):218–229
10. Li G, Wilson RA (2021) Tandem affinity purification (tap) of low-abundance protein complexes in filamentous fungi demonstrated using Magnaporthe Oryzae. Methods Mol Biol 2356:97–108
11. Elhabashy H, Merino F, Alva V et al (2022) Exploring protein-protein interactions at the proteome level. Structure 30(4):462–475
12. Matthews LR, Vaglio P, Reboul J et al (2001) Identification of potential interaction networks using sequence-based searches for conserved protein-protein interactions or "Interologs". Genome Res 11(12):2120–2126
13. Huang C, Morcos F, Kanaan SP et al (2007) Predicting protein-protein interactions from protein domains using a set cover approach. IEEE/ACM Trans Comput Biol Bioinform 4(1):78–87
14. Lei C, Ruan J (2013) A novel link prediction algorithm for reconstructing protein-protein interaction networks by topological similarity. Bioinformatics 29(3):355–364
15. Sprinzak E, Margalit H (2001) Correlated sequence-signatures as markers of protein-protein interaction. J Mol Biol 311(4): 681–692
16. Guo Y, Yu L, Wen Z et al (2008) Using support vector machine combined with auto covariance to predict protein-protein interactions from protein sequences. Nucleic Acids Res 36(9):3025–3030
17. Chen M, Ju CJ, Zhou G et al (2019) Multifaceted protein-protein interaction prediction based on Siamese residual RCNN. Bioinformatics 35(14):i305–i314
18. Gu H, Zhu P, Jiao Y et al (2011) PRIN: a predicted Rice Interactome network. BMC Bioinf 12:161
19. Zhu G, Wu A, Xu XJ et al (2016) PPIM: a protein-protein interaction database for maize. Plant Physiol 170(2):618–626
20. Ding Z, Kihara D (2019) Computational identification of protein-protein interactions in model plant proteomes. Sci Rep 9(1):8740
21. Zhao J, Lei Y, Hong J et al (2019) AraPPINet: An updated Interactome for the analysis of hormone Signaling crosstalk in Arabidopsis Thaliana. Front Plant Sci 10:870
22. Pan J, You Z-H, Li L-P et al (2022) DWPPI: a deep learning approach for predicting protein–protein interactions in plants based on multi-source information with a large-scale biological network. Front Bioeng Biotechnol 10:807522
23. Li P, Zang W, Li Y et al (2011) AtPID: the overall hierarchical functional protein interaction network Interface and analytic platform for Arabidopsis. Nucleic Acids Res 39(Database issue):D1130–D1133
24. Brandao MM, Dantas LL, Silva-Filho MC (2009) AtPIN: Arabidopsis Thaliana protein interaction network. BMC Bioinf 10:454

25. Zhang L, Guo J, You Q et al (2015) GraP: platform for functional genomics analysis of Gossypium Raimondii. Database (Oxford) 2015:bav047
26. Vakser IA (2014) Protein-protein docking: from interaction to Interactome. Biophys J 107(8):1785–1793
27. Kundrotas PJ, Zhu Z, Janin J et al (2012) Templates are available to model nearly all complexes of structurally characterized proteins. Proc Natl Acad Sci U S A 109(24): 9438–9441
28. Tuncbag N, Gursoy A, Nussinov R et al (2011) Predicting protein-protein interactions on a proteome scale by matching evolutionary and structural similarities at interfaces using PRISM. Nat Protoc 6(9):1341–1354
29. Baspinar A, Cukuroglu E, Nussinov R et al (2014) PRISM: a web server and repository for prediction of protein-protein interactions and Modeling their 3D complexes. Nucleic Acids Res 42(Web Server issue):W285–W289
30. Finn RD, Miller BL, Clements J et al (2014) ipfam: a database of protein family and domain interactions found in the protein data Bank. Nucleic Acids Res 42(Database issue):D364–D373
31. Mosca R, Céol A, Stein A et al (2014) 3did: a Catalog of domain-based interactions of known three-dimensional structure. Nucleic Acids Res 42(Database issue):D374–D379
32. Li H, Yang S, Wang C et al (2016) AraPPISite: a database of fine-grained protein-protein interaction site annotations for Arabidopsis Thaliana. Plant Mol Biol 92(1–2):105–116
33. Yang X, Yang S, Qi H et al (2020) PlaPPISite: a comprehensive resource for plant protein-protein interaction sites. BMC Plant Biol 20(1):61
34. Franz M, Lopes CT, Huck G et al (2016) Cytoscape.Js: a graph theory library for visualisation and analysis. Bioinformatics 32(2): 309–311
35. Rose AS, Hildebrand PW (2015) NGL viewer: a web application for molecular visualization. Nucleic Acids Res 43(W1):W576–W579
36. The UniProt Consortium (2021) UniProt: the universal protein knowledgebase in 2021. Nucleic Acids Res 49(D1):D480–D489
37. Wainberg M, Merico D, Delong A et al (2018) Deep learning in biomedicine. Nat Biotechnol 36(9):829–838
38. Jumper J, Evans R, Pritzel A et al (2021) Highly accurate protein structure prediction with AlphaFold. Nature 596(7873):583–589
39. Hashemifar S, Neyshabur B, Khan AA et al (2018) Predicting protein-protein interactions through sequence-based deep learning. Bioinformatics 34(17):i802–i810
40. Li H, Gong XJ, Yu H et al (2018, 1923) Deep neural network based predictions of protein interactions using primary sequences. Molecules 23(8)
41. Gao M, Nakajima An D, Parks JM et al (2022) AF2Complex predicts direct physical interactions in multimeric proteins with deep learning. Nat Commun 13(1):1744
42. Yuan Q, Chen J, Zhao H et al (2021) Structure-aware protein-protein interaction site prediction using deep graph convolutional network. Bioinformatics 38(1):125–132
43. Tubiana J, Schneidman-Duhovny D, Wolfson HJ (2022) ScanNet: An interpretable geometric deep learning model for structure-based protein binding site prediction. Nat Methods 19(6):730–739

Chapter 31

Machine Learning Methods for Virus–Host Protein–Protein Interaction Prediction

Betül Asiye Karpuzcu, Erdem Türk, Ahmad Hassan Ibrahim, Onur Can Karabulut, and Barış Ethem Süzek

Abstract

The attachment of a virion to a respective cellular receptor on the host organism occurring through the virus–host protein–protein interactions (PPIs) is a decisive step for viral pathogenicity and infectivity. Therefore, a vast number of wet-lab experimental techniques are used to study virus–host PPIs. Taking the great number and enormous variety of virus–host PPIs and the cost as well as labor of laboratory work, however, computational approaches toward analyzing the available interaction data and predicting previously unidentified interactions have been on the rise. Among them, machine-learning-based models are getting increasingly more attention with a great body of resources and tools proposed recently.

In this chapter, we first provide the methodology with major steps toward the development of a virus–host PPI prediction tool. Next, we discuss the challenges involved and evaluate several existing machine-learning-based virus–host PPI prediction tools. Finally, we describe our experience with several ensemble techniques as utilized on available prediction results retrieved from individual PPI prediction tools. Overall, based on our experience, we recognize there is still room for the development of new individual and/or ensemble virus–host PPI prediction tools that leverage existing tools.

Key words Machine learning algorithms, Viral infections, Virus–host protein–protein interactions, Virus bioinformatics, In silico prediction, Ensemble methods

1 Introduction

Viruses are a major cause of infectious diseases, and infectious viral diseases, particularly those leading to global pandemics, have the potential of exerting an abrupt and deep effect both on public health and on the global economy [1]. Even back in 2016, before the emergence of the COVID-19 pandemic, the Institute of Labor Economics had reported that 156 million life-years were lost due to 8 major infectious diseases, the annual economic burden of which was calculated to be 8 trillion USD based on the WHO data [2]. Considering this overwhelming impact, deciphering the inter-

Shahid Mukhtar (ed.), *Protein-Protein Interactions: Methods and Protocols*, Methods in Molecular Biology, vol. 2690,
https://doi.org/10.1007/978-1-0716-3327-4_31,

actions between viruses and their known or possible hosts is a rather crucial task to prevent, if possible, or mitigate the consequences of viral infections [3].

The viral life cycle in a host organism involves the recognition and binding of the virion to the host as a prerequisite step followed by cellular entry, dissemination, and a productive or latent infection that may potentially lead to a viral disease [4]. Throughout this cycle, there is a complex set of interactions between the viruses and their corresponding host cells. The term interaction, on the other hand, has a range of different meanings in molecular biology [5]. For clarification, the protein–protein interaction (PPI) occurring as a specific first physical contact with molecular docking between the viral protein and their respective cellular receptor (s) in vivo is what we herein refer to as "virus–host PPI." To this end, any auxiliary interactions following the initial attachment of a viral protein to a primary host receptor even if they are mediated by proteins, recognitions that employ any non-proteinaceous molecules (e.g., sialic acid), and any other binding between proteins that take place during internalization of virus (e.g., endocytosis or macropinocytosis) and further steps are not covered in PPI context here [6].

To name a few, for instance, current research has revealed that members of the *Coronaviridae* virus family attach through their spike proteins to receptors angiotensin-converting enzyme 2 (ACE2), dipeptidyl peptidase 4 (DPP4), and aminopeptidase N (ANPEP) in human as a host while research on CD147 is ongoing to clarify whether it is a primary or co-receptor [7–9]. The diverse family of *Adenoviridae*, composed of relatively large, non-enveloped, icosahedral viruses, uses their fiber protein and penton protein, in limited cases, to bind primary receptors in host cells. Several adenoviral receptors have been identified and characterized to various extents [10–12]. Accordingly, a current list of receptors manually curated based on the available scientific literature includes Coxsackie and adenovirus receptor (CAR), the most widely studied and well-known receptor, desmoglein-2 (DSG2), integrin subunit alpha-V (ITAV), macrophage scavenger receptor 1 (MSR1), and lung macrophage scavenger receptor SR-A6 (MARCO) in addition to the cluster of differentiation (CD) 46, CD80, and CD86 proteins, which have a variety of other functions [13]. Thus, each one of such a viral attachment protein and host receptor protein pair constitutes a positive PPI of interest for our purposes.

Being the starting point, identification of the prerequisite virus–host PPI is essential in exploring the viral infectivity and pathogenicity as well as in guiding the utilization of viruses in various applications such as drug delivery or cancer vaccines [14, 15]. In this regard, wet-lab research techniques are extensively used to investigate virus–host PPIs. Among them, yeast two-hybrid

(Y2H) and affinity-purification mass spectrometry (AP-MS) are the most widely used in vivo and in vitro techniques, respectively [16]. In addition, other in vitro techniques some specific to viruses such as virus overlay protein binding assays (VOPBAs) [17] as well as the conventional protein microarrays, protease assays, binding assays using flow cytometry, co-immunoprecipitation as well as measurements using surface plasmon resonance (SPR) and Förster Resonance Energy Transfer (FRET) and further genetic or structural validations using, for instance, CRISPR/Cas9 and RNA interference or X-ray crystallography and nuclear magnetic resonance spectroscopy are being used to study virus–host PPIs and to complement each other [16].

Although they are powerful in determining interaction partners or complexes, wet-lab methods are prone to produce false positive (FP) and false negative (FN) results on top of being expensive, time-consuming, and laborious. Furthermore, taking into the enormous diversity of virus and host species and the high number of possible interactions between them, it is not practically plausible to study the entire set of virus–host PPIs even with the help of high-throughput screens [18]. For this reason, a range of computational methods have been developed as a cost-effective and quick alternative and successfully applied both to analyze the existing interaction data from wet-lab experimental approaches and to predict novel PPIs relying on the knowledge gathered from the previously identified interactions. To name a few, PPI prediction by an in silico two-hybrid system based on the assumption of coevolution of interacting proteins [19], a phylogenetic tree approach assuming that interacting proteins should be similar in their evolution history [20], and several domain–domain interaction-based and homology-based prediction methods have been described previously. With the advent of machine learning applications in the field of bioinformatics, on the other hand, gradually increasing attention and effort has been allocated toward making use of machine learning in in silico virus–host PPI prediction and our focus hereinafter will be explaining their usage.

For the newcomers to the field of machine learning (*see* **Notes 1** through **8**), "Notes" section provides an overview of existing machine-learning-based virus–host PPI prediction tools, the challenges, and other considerations involved in the course of developing a binary classification model to generate a virus–host PPI prediction tool. Furthermore, we discuss a Case Study involving the use of ensemble techniques to bring together available virus–host PPI prediction tools.

2 Materials

2.1 Computing Environment

An Intel® Xeon-based server with at least 32 GB system memory.

2.2 Operating Systems and Its Specifications

A server with Linux-based operating system.

2.3 Scripting Language/Software

Python 2.x and 3.x and MATLAB version R2022 (or above) installed on the server.

2.4 Packages/Tools

Standard Python libraries: *csv*, *uuid*, *collections*, *sys*, *warnings*, *os*, *ast*.

Non-standard Python libraries: *matlab.engine*, *Bio.SeqIO*, *pandas*, *numpy*, *Scikit-Learn*.

Virus–host PPI prediction tools: DeNovo, HOPITOR, InterSPPI-HVPPI.

3 Methods

3.1 Set Up the Environment

Ensure the computing environment and software including specified packages, non-standard Python libraries (*see* **Note 1**), and virus–host PPI prediction tools are installed as per the instructions of their respective providers.

3.2 Compile the Dataset

Use a dataset of known virus–host PPIs (positive set) and non-interacting PPIs (negative set). Out of the dataset, generate two tab-separated value (TSV) documents. In the first document, include the protein database accessions for each virus–host protein pair in the list. Additionally, add an indicator as to whether the pair stands for an interacting (positive: 1) or non-interacting (negative: −1) pair, also known as actual class label (i.e., ground truth). In the second TSV file, sort unique accessions and the respective protein sequence at each row. Use these files to feed into the individual PPI tools and thus return their prediction results.

For resources on dataset compilation and associated considerations, see **Note 2**.

3.3 Construct the Feature Vector

Run individual virus–host PPI prediction tools in Python environment using the TSV files from step 3.2 as inputs. Collect DeNovo PPI prediction results that are in 1/−1 designation where 1 denotes interacting and −1 denotes non-interacting. Collect the HOPITOR results that are in the same designation as DeNovo. Next, collect InterSPPI-HVPPI results that are in yes/no designation.

Convert InterSPPI-HVPPI results into same designation as the other tools. In addition to the final prediction results, also collect the posterior probabilities from each tool. As a result of this step, obtain a file with the accession numbers for each virus–host protein pair, three prediction results in 1/−1 designation and posterior probabilities (from DeNovo, HOPITOR, and InterSPPI-HVPPI), and the actual class label combined laterally. Relevant columns of this file stand for a feature vector to be used in the model-generation step.

The feature vectors constructed in this step will be used for ensemble-learning-based models. In case it is intended to develop a novel standalone PPI predictor, select among the described feature-extraction techniques (*see* **Note 3**) to create a feature vector for virus–host protein pairs.

3.4 Apply Ensemble Techniques and Generate the Model

Apply different techniques to compare and select the best performer(s). Here, the use of four ensemble-based approaches (hard voting, soft voting, decision tree, and logistic regression) to combine the results from previously available virus–host PPI prediction tools is described. Implement these ensemble-based models in Python 3.x using Sci-kit learn package.

In hard voting, for each protein pair, take the majority (mode) of the class labels from the three virus–host PPI prediction tools as the consensus prediction. In soft voting, do not take the labels assigned by the tools directly, but rather, compute the average posterior probability of the three virus–host PPI prediction tools. Then, if the average probability is greater than or equal to 0.5, label the PPI as positive, otherwise accept negative is the consensus prediction.

For decision tree and logistic regression, use only the posterior probabilities from the three virus–host PPI prediction tools as features to train a model. Hyperparameter tuning is highly recommended to find the best set of parameters for each model. During the training of decision tree models, for example, use the following set of parameters for hyperparameterization: minimum number of samples required to be at a leaf node (1,2,4), splitting strategy (best, random), a minimum number of samples required to split an internal node (2,5,10), the maximum number of features to consider for the best split (auto, sqrt, log2), and the maximum depth of the tree (a set of evenly spaced 11 numbers over the range of 10–1100, and none). For logistic regression, hyperparameter tuning can be accomplished by experimenting the parameter solvers—standing for the algorithm to use in optimization—(newton-cg, lbfgs, liblinear), penalty (default = l2), and C value—standing for the inverse of regularization strength—(100, 10, 1.0, 0.1, 0.01).

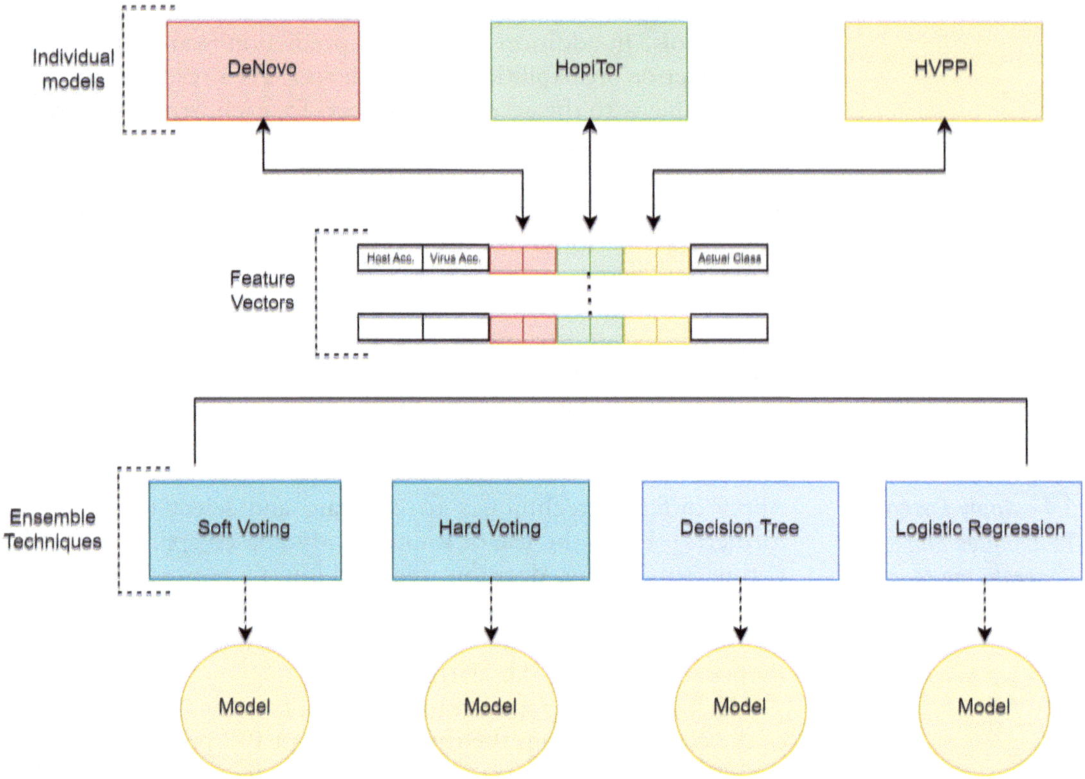

Fig. 1 The overall methodology for applying ensemble techniques to virus–host PPI prediction tools

The user is recommended to see **Note 4** explaining the considerations during a novel model generation, **Note 5** for an overview of the existing tools used herein (DeNovo, HOPITOR, and InterSPPI-HVPPI), and **Note 6** for a brief introduction to ensemble in machine learning.

The overall methodology is illustrated in Fig. 1.

3.5 Evaluate the Performance

Calculate the true positive rate (TPR), true negative rate (TNR), false positive rate (FPR), false negative rate (FNR), positive predictive value (PPV), negative predictive value (NPV), false discovery rate (FDR), accuracy (ACC), F-score (F1), Matthew's correlation coefficient (MCC), and area under the curve (AUC) of your model. A description of these metrics is provided, *see* **Notes 7**.

Following this method and using ensemble techniques, a Case Study was executed (*see* **Note 8**).

4 Notes

1. In order to use non-standard Python packages, user needs to follow their individual installation instructions, e.g., "pip install pandas." In order to use the available virus–host PPI prediction tools (or their corresponding dataset), the following links can be used:

   DeNovo: https://bioinformatics.cs.vt.edu/~alzahraa/denovo_files/Supp_files/ST3ST4ST5.zip

   HOPITOR: https://github.com/foxtrotmike/hopitor

   InterSPPI-HVPPI: http://zzdlab.com/hvppi/download.php

   VirusHostPPI (VHPPI) Dataset: http://bclab.inha.ac.kr/vhppi/Additional%20file%203.zip

2. Publicly available host–pathogen interaction databases (HPIDBs) including viruses (e.g., VirusMentha [21], HPIDB [22]) can be used to extract data for positive interactions. To retrieve the respective protein sequences, please refer to protein databases, e.g., UniProtKB (UniProt Knowledgebase) [23] or RefSeq [24].

   Challenges are involved in the collection of a dataset. Most particularly, retrieving a set of validated negative (non-interacting) virus–host protein pairs is not practically possible. Secondly, considering the number of viral taxa, the size of positive sets is falling short in representing all the viruses, which inevitably causes an imbalance between the positive and negative sets. Finally, the datasets have limited coverage and diversity of host and viral taxa, such that available resources are often predominated by certain organisms (e.g., human or influenza virus and herpes virus strains). The curation of a public benchmark dataset for virus–host PPI prediction would standardize and foster the development of new PPI prediction tools.

   Interactions that have been identified experimentally are utilized as positive examples for model training in PPI prediction tools. In most cases, negative (i.e., non-interacting) instances are created randomly by pairing virus–host proteins. Of note, although called a negative set, there is no experimental setup to verify that these pairs are truly non-interacting. Furthermore, as a general concern in machine-learning-based processes, a large number of interacting and non-interacting pairs are usually required to build effective PPI prediction models. Due to the relatively smaller number of validated pairs in the positive set, in contrast to large number of pairs in negative sets, the datasets tend to have a sharp imbalance that warrants the use of strategies to overcome this class imbalance problem.

3. The feature vectors are required for computational prediction algorithms/models to learn. In recent years, a plethora of feature-extraction techniques has been developed for virus–host PPI prediction. They can be divided into three major groups as structure-based [25–27], sequence-based [28–30], and domain-based [26, 31, 32] techniques. In addition to these techniques, there are feature-extraction methods reported in the literature that consider ontology [33, 34], gene expression [35], and evolutionary profiles [36, 37] of proteins.
4. Create a binary classification model (a virus–host PPI predictor) using machine learning algorithms [26, 34, 38] that can distinguish between positive and negative PPIs. When developing a machine-learning-based classifier, it is critical to think about the complexity of the problem, the available data, and the system's intended usage [39]. This criterion is more important in bioinformatics than in other application domains because of its position as a tool for biological discovery. It is vital to consider the underlying biological consequences while creating a machine learning system and its assessment process in the domain of virus–host interaction prediction. Imbalanced data, large complexity of the feature space, sparseness of known interactions, and probable uncertainty in the labeling of negative training instances are all aspects that influence the learning of a virus–host PPI predictor [40]. The fact that classification examples in virus–host PPI predictors are pairs of proteins introduces a bias that impacts the evaluation of generalization performance. As Park and Marcotte [41] and Hamp and Rost [42] pointed out, machine learning issues with paired inputs provide particular obstacles in assessing the accuracy of such algorithms.
5. Several computational tools to predict virus–host PPIs have been developed that use different machine learning algorithms such as support vector machines (SVMs) [29, 38, 43], random forest (RF) [44], and gradient boosting machine (XGBoost) [28, 45]. Here, to discuss various aspects of in silico methods for virus–host interaction predictions, we made use of some of these previously developed publicly available tools (*see* **Notes 8**, Case Study), and therefore an overview of such tools is provided below to help understand their utilization herein.

   The tool DeNovo [29] collected virus–human PPIs from VirusMentha (accessed June 2014) [21] and protein sequences retrieved from UniProt [23]. After preprocessing and filtering, there remained 5445 unique interactions in the dataset that contains 2340 human proteins and 445 viral proteins, spanning 172 different virus species.

The feature-extraction scheme employed in DeNovo was originally developed by Shen et al. [43]. This feature-extraction scheme briefly encompasses the following: first, 20 amino acids are divided into 7 clusters encoded with numbers from 1 to 7, in the given order {*A*, *V*, *G*}, {*I*, *L*, *F*, *P*}, {*Y*, *M*, *T*, *S*}, {*H*, *N*, *Q*, *W*}, {*R*, *K*}, {*D*, *E*}, and {C}, based on physiochemical similarities involved in PPIs (dipoles and volumes of side chains). Residues in each protein of the PPI pair, one from the host and one from the virus, are mapped to the corresponding cluster numbers. Along each protein sequence, the frequency of each 3-mer is normalized over [0, 1] and used to generate a feature vector ($7^3 = 343$ dimensional). These two normalized feature vectors of a virus–host protein pair (interacting/positive; non-interacting/positive) are concatenated into a single 686-dimensional feature vector.

In its own dataset, the performance metrics average accuracy, sensitivity, and specificity of DeNovo were 97%, 94.5%, and 97.5%, respectively. DeNovo employed a sequence-similarity-based strategy for gathering a negative dataset of virus–host PPIs to be used in training, a cardinal feature distinguishing DeNovo from other SVM-based prediction tools.

DeNovo has inspired researchers with its negative sampling strategy in developing virus–host PPI methods with a similar sampling. HOPITOR is an XGBoost classifier-based host–pathogen predictor that has been developed by Basit and colleagues [28]. HOPITOR also relies on the feature-extraction scheme of Shen et al. Using the same dataset reported by Eid et al. and after training, testing, and comparing different machine-learning-based models, the researchers highlighted the XGBoost-based tool as the most successful predictor with an area under the receiver operating characteristic curve (AUC-ROC) of 0.76 and precision-recall curve (AUC-PR) of 0.53.

In the RF-based classifier InterSPPI-HVPPI developed by Yang and colleagues [44], developers have used the manually curated PPI data from Host–Pathogen Interaction Database (HPIDB; version 3.0) [22] to obtain the positive (interacting) host–pathogen protein pairs wherein 22,653 human–virus PPIs were selected after filtering out the redundant PPIs (based on sequence identity threshold <0.5), non-physical interactions, and any interactions involving a protein size of <30 or > 5000 amino acids. Further, they downloaded protein data available in SwissProt and produced the negative dataset using the Dissimilarity-Based Negative Sampling method, which is the same approach used by Eid et al. Eventually, their dataset has a positive-to-negative ratio of 1:10. These protein sequences were embedded using the doc2vec model. That is, in the feature-extraction process, the authors have created a corpus of sequence information from their protein data to train a doc2vec model that allowed them to yield protein-sequence-specific features. The entire dataset was handled in three random partitions of equal size to reduce sampling bias each of which was further also split into 80% training and 20% test set. Using Barman et al.'s [46] dataset, InterSPPI-HVPPI achieved 79.17% accuracy, 81.85% sensitivity, and 76.45% specificity.

6. The tools, DeNovo, HOPITOR, and InterSPPI-HVPPI are currently trained and tested by their developers using different datasets (for ease of reference, the dataset shared in each tool's publication is herein named after the respective tool) to predict PPIs. However, in machine learning applications, it is quite common that the model developed on a particular dataset does not show similar or consistent performances in other datasets. In order to avoid such fluctuations in prediction results, compensate for the deficiencies of individual tools, and improve the performance of machine learning algorithms ending up with a more robust combinational tool, "ensemble methods" are used [47, 48].

   In general, there are three main types of ensemble techniques called stacking, bagging, and boosting that are used to combine the individual tools, each called an ensemble member. The stacking strategy usually takes in heterogeneous members by learning them in parallel and, to achieve combination, it trains a model with a consensus prediction based on the predictions of weak models. When bagging strategy is used, homogeneous members are often taken where the combinatorial model learns each one independently and in parallel and applies a deterministic averaging process to achieve combination. The boosting strategy also takes in homogeneous members and learns them adaptively in a row.

7. There are several performance metrics used to compare the performance of the resulting binary classification model with other existing tools created with different machine learning algorithms and/or sampling methods. These metrics can be listed as: true positive rate (TPR), true negative rate (TNR), false positive rate (FPR), false negative rate (FNR), positive predictive value (PPV), negative predictive value (NPV), false discovery rate (FDR), accuracy (ACC), F-score (F1), Matthew's correlation coefficient (MCC), and area under the curve (AUC). In the virus–host PPI prediction concept, true positives (TPs) are those host protein and virus protein pairs that are predicted as interacting, and are actually interacting as per the ground truth. True negatives (TNs) are host–virus protein pairs that are predicted to be non-interacting, and are also classified as non-interacting as per the ground truth. False

positive (FP) PPIs are actually non-interacting pairs that are incorrectly predicted to interact. False negatives (FNs) are the PPIs that are predicted as non-interacting but actually do interact. The following set of formulas is used to define/calculate the aforementioned metrics.

$$\mathrm{TPR} = \mathrm{Sensitivity} = \mathit{Recall} = 1 - \mathit{FNR} \quad \frac{TP}{TP + FN}$$

$$\mathrm{TNR} = \mathrm{Specificity} = 1 - \mathrm{FPR} \quad = \frac{TN}{TN + FP}$$

$$\mathrm{FPR} = 1 - TNR = \frac{FP}{TN + FP}$$

$$\mathrm{FNR} = 1 - TPR = \frac{TP}{TP + FN}$$

$$\mathrm{PPV} = \mathrm{Precision} = \frac{TP}{TP + FP}$$

$$\mathrm{NPV} = \frac{TN}{TN + FN}$$

$$\mathrm{FDR} = 1 - \mathrm{PPV} \quad = \frac{FP}{FP + TN}$$

$$\mathrm{Accuracy} = \frac{TP + TN}{TP + TN + FP + FN}$$

$$\mathrm{F - Score} = \frac{2 * \mathrm{Precision} * \mathrm{Recall}}{\mathrm{Precision} + \mathrm{Recall}}$$

$$\mathrm{AUC} = \int_{x=0}^{1} TPR\left(FPR^{-1}(x)\right) dx$$

$$\mathrm{MCC} = \frac{TP * TN - FP * FN}{\sqrt{(TP + FP)(TP + FN)(TN + FP)(TN + FN)}}$$

8. Case Study

Set Up the Environment A server equipped with a processor of Intel® Xeon Silver 4210 CPU (2.20 GHz, 10 cores) and a total 144 GB system memory running Ubuntu version 20.04.3 is used. Python 2.7 and 3.8 and MATLAB version R2022 are installed on the server. Packages described in "Materials" section are installed.

Compile the Dataset and Construct the Feature Vector We used an independent dataset from an SVM-based tool called VirusHostPPI (VHPPI) proposed by Zhou and coworkers [49] to test the individual prediction performances of each tool and then to train and test our models using ensemble techniques. This specific dataset was selected because it was not used by any of the ensemble members, namely DeNovo, HOPITOR, and InterSPPI-HVPPI. The available VHPPI dataset consists of 12,157 positive PPIs and 36,471 negative PPIs, which corresponds to a 1:3 imbalance ratio between positive and negative sets. The challenge of creating a negative set was addressed by, first, obtaining a protein dataset for all the major hosts using UniProt, then removing the protein sequences showing sequence similarity to the protein sequences in the positive dataset over a threshold of 80%, and, finally, pairing the remaining host proteins with the virus proteins, thus constituting the negative (i.e., non-interacting) set. The positive PPI dataset covers 29 unique host organisms and 332 non-redundant viral taxa. A great majority of the virus–host PPIs has human as a host (94.5%) followed by mouse (1.6%) and domestic cow (1.0%).

We downloaded the VHPPI dataset to extract a list of known virus–host PPI interactions (positive set) and a list of non-interacting PPIs (negative set). In the folder Additional_file_3, we took the document Train_POS.xlsx to be used as the positive set and Train_NEG_1vs3.xlsx to be used as negative set. We merged two sets and obtained our input files to run individual virus–host PPI prediction tools.

We took the posterior probabilities computed by each tool to create feature vectors for soft voting, decision tree, and logistic regression ensemble models. Exceptionally, for hard-voting-based ensemble, only class predictions computed by each tool was used to create the feature vector. These feature vectors are next used to train the models using ensemble techniques described next.

Apply Ensemble Techniques and Generate the Model We adapted the stacking ensemble approach to come up with a consensus classification in several ways: hard voting, soft voting, decision tree, and logistic regression.

Evaluate the Performance of Individual Tools In an attempt to make an unbiased comparison among three virus–host PPI prediction tools, we computed the performances of these tools on the independent VHPPI dataset. Of note, we did not train any of the three tools; instead, they were used as is with their default parameters. To compare the success of these tools, the performance evaluation metrics (*see* **Notes 4**) are used. The results are given in Table 1.

Virus–host PPI prediction tools demonstrate varying performances (see Table 1) and oftentimes their performances are sensitive to the dataset used. This is partially attributable to the dataset collection challenges we mentioned earlier. On the other hand, the performances of existing tools call for efforts to develop new prediction tools.

Table 1
Performances of individual virus–host PPI prediction tools on test splits

Tool name	TPR	TNR	PPV	NPV	FPR	FNR	FDR	ACC	F1	MCC	AUC
DeNovo	0.97	0.05	0.25	0.83	0.95	0.03	0.75	0.28	0.40	0.4	0.51
HOPITOR	0.72	0.39	0.28	0.81	0.61	0.28	0.72	0.47	0.40	0.10	0.56
InterSPPI-HVPPI	0.84	0.76	0.54	0.94	0.24	0.16	0.46	0.78	0.66	0.53	0.80

ACC accuracy, *AUC* area under the curve, *F1* F-score, *FDR* false discovery rate, *FNR* false negative rate, *FPR* false positive rate, *MCC* Matthew's correlation coefficient, *NPV* negative predictive value, *PPV* positive predictive value, *TNR* true negative rate, *TPR* true positive rate

Evaluate the Performance of Ensemble Techniques For the cases of decision tree and logistic regression, we adapted two different random sampling—oversampling and undersampling, as both techniques are used for similar bioinformatics problems [50]. After splitting the dataset into two (90% and 10%), we conducted the hyperparameter tuning using five-fold cross-validation in the bigger portion of the dataset. To do this, either oversampling or undersampling strategy is also undertaken to overcome the class imbalance problem in the dataset. Out of the trials, the best performing set of parameters is selected for each of the decision trees and logistic regression techniques. Then we tested each model on the smaller portion of the dataset and thus we computed the performance metrics given in Table 2.

Using ensemble techniques demonstrated slight performance improvements; for instance, the baseline AUC for InterSPPI-HVPPI was 0.80 while the AUC of the decision-tree-based ensemble model was 0.83. Hard voting and soft voting did not help to improve performance. Decision tree and logistic regression resulted better compared to the voting-based techniques. The decision tree had a slightly better performance than logistic regression but more importantly, it performed better than any of the individual tools in terms of AUC. Decision tree models seem to favor sensitivity over specificity, unlike individual tools. In our experiments, oversampling and undersampling strategies did not cause a substantial change. We believe that the availability of additional PPI prediction tools and more PPI data will improve the performance of the ensemble model.

Table 2
Performances of ensemble techniques

Ensemble technique	Hyperparameters	Train-test split	Sampling	TPR	TNR	PPV	NPV	FPR	FNR	FDR	ACC	F1	MCC	AUC
Hard voting	NA	NA	NA	0.87	0.36	0.31	0.90	0.64	0.13	0.69	0.49	0.46	0.22	0.62
Soft voting	NA	NA	NA	0.62	0.81	0.52	0.86	0.19	0.38	0.48	0.76	0.57	0.41	0.72
Decision tree	max_depth = 664, max_feature = auto, min_samples_split = 5, min_samples_leaf = 4, splitter = random	90–10%	Undersampling	0.77	0.89	0.70	0.92	0.11	0.23	0.30	0.86	0.73	0.64	0.83
Logistic regression	C = 100, solver = newton-cg	90–10%	Undersampling	0.74	0.87	0.66	0.91	0.13	0.26	0.34	0.84	0.69	0.59	0.80
Decision tree	max_depth = 10, max_feature = auto, min_samples_split = 2, min_samples_leaf = 1, splitter = best	90–10%	Oversampling	0.74	0.92	0.77	0.92	0.08	0.26	0.23	0.88	0.76	0.68	0.83
Logistic regression	C = 10, solver = newton-cg	90–10%	Oversampling	0.77	0.88	0.68	0.92	0.12	0.23	0.32	0.85	0.72	0.62	0.83

ACC accuracy, *AUC* area under the curve, *F1* F-score, *FDR* false discovery rate, *FNR* false negative rate, *FPR* false positive rate, *MCC* Matthew's correlation coefficient, *NA* not available, *NPV* negative predictive value, *PPV* positive predictive value, *TNR* true negative rate, *TPR* true positive rate

Acknowledgments

This work is funded by the Scientific and Technological Research Council of Turkey (Grant number: 119E664).

References

1. Bloom DE, Cadarette D (2019) Infectious disease threats in the twenty-first century: strengthening the global response. Front Immunol 10:549. https://doi.org/10.3389/fimmu.2019.00549
2. Bloom DE, Kuhn M, Prettner K (2022) Modern infectious diseases: macroeconomic impacts and policy responses. J Econ Lit 60(1):85–131. https://doi.org/10.1257/jel.20201642
3. Jangra RK, Llabres M, Guardado-Calvo P et al (2022) Editorial: influence of Protein-Protein Interactions (PPIs) on the outcome of viral infections. Front Microbiol 13:943379. https://doi.org/10.3389/fmicb.2022.943379
4. Jones JE, Le Sage V, Lakdawala SS (2021) Viral and host heterogeneity and their effects on the viral life cycle. Nat Rev Microbiol 19(4): 272–282. https://doi.org/10.1038/s41579-020-00449-9
5. Wang X, Elston RC, Zhu X (2010) The meaning of interaction. Hum Hered 70(4): 269–277. https://doi.org/10.1159/000321967
6. De Las RJ, Fontanillo C (2010) Protein-protein interactions essentials: key concepts to building and analyzing interactome networks. PLoS Comput Biol 6(6):e1000807. https://doi.org/10.1371/journal.pcbi.1000807
7. Raj VS, Mou H, Smits SL et al (2013) Dipeptidyl peptidase 4 is a functional receptor for the emerging human coronavirus-EMC. Nature 495(7440):251–254. https://doi.org/10.1038/nature12005
8. Shilts J, Crozier TWM, Greenwood EJD et al (2021) No evidence for basigin/CD147 as a direct SARS-CoV-2 spike binding receptor. Sci Rep 11(1):413. https://doi.org/10.1038/s41598-020-80464-1
9. Yeager CL, Ashmun RA, Williams RK et al (1992) Human aminopeptidase N is a receptor for human coronavirus 229E. Nature 357(6377):420–422. https://doi.org/10.1038/357420a0
10. Lasswitz L, Chandra N, Arnberg N et al (2018) Glycomics and proteomics approaches to investigate early adenovirus-host cell interactions. J Mol Biol 430(13):1863–1882. https://doi.org/10.1016/j.jmb.2018.04.039
11. Stasiak AC, Stehle T (2020) Human adenovirus binding to host cell receptors: a structural view. Med Microbiol Immunol 209(3): 325–333. https://doi.org/10.1007/s00430-019-00645-2
12. Zhang Y, Bergelson JM (2005) Adenovirus receptors. J Virol 79(19):12125–12131. https://doi.org/10.1128/JVI.79.19.12125-12131.2005
13. Karabulut OC, Karpuzcu BA, Turk E et al (2021) ML-AdVInfect: a machine-learning based adenoviral infection predictor. Front Mol Biosci 8:647424. https://doi.org/10.3389/fmolb.2021.647424
14. Singh S, Kumar R, Agrawal B (2018) Adenoviral vector-based vaccines and gene therapies: current status and future prospects, pp 1–38
15. Gao J, Zhang W, Ehrhardt A (2020) Expanding the spectrum of adenoviral vectors for cancer therapy. Cancers 12(5):1139
16. Farooq QUA, Shaukat Z, Aiman S et al (2021) Protein-protein interactions: methods, databases, and applications in virus-host study. World J Virol 10(6):288–300. https://doi.org/10.5501/wjv.v10.i6.288
17. Gaggar A, Shayakhmetov DM, Lieber A (2003) CD46 is a cellular receptor for group B adenoviruses. Nat Med 9(11):1408–1412. https://doi.org/10.1038/nm952
18. Brito AF, Pinney JW (2017) Protein-protein interactions in virus-host systems. Front Microbiol 8:1557. https://doi.org/10.3389/fmicb.2017.01557
19. Pazos F, Valencia A (2002) In silico two-hybrid system for the selection of physically interacting protein pairs. Proteins 47(2):219–227. https://doi.org/10.1002/prot.10074
20. Sato T, Yamanishi Y, Kanehisa M et al (2005) The inference of protein-protein interactions by co-evolutionary analysis is improved by excluding the information about the phylogenetic relationships. Bioinformatics 21(17): 3482–3489. https://doi.org/10.1093/bioinformatics/bti564
21. Calderone A, Licata L, Cesareni G (2015) VirusMentha: a new resource for virus-host protein interactions. Nucleic Acids Res 43 (Database issue):D588–D592. https://doi.org/10.1093/nar/gku830

22. Ammari MG, Gresham CR, McCarthy FM, et al (2016) HPIDB 2.0: a curated database for host-pathogen interactions. Database: the journal of biological databases and curation 2016. https://doi.org/10.1093/database/baw103
23. The UPC (2017) UniProt: the universal protein knowledgebase. Nucleic Acids Res 45 (D1):D158–DD69. https://doi.org/10.1093/nar/gkw1099
24. O'Leary NA, Wright MW, Brister JR et al (2016) Reference sequence (RefSeq) database at NCBI: current status, taxonomic expansion, and functional annotation. Nucleic Acids Res 44(D1):D733–D745. https://doi.org/10.1093/nar/gkv1189
25. Bock JR, Gough DA (2001) Predicting protein--protein interactions from primary structure. Bioinformatics 17(5):455–460. https://doi.org/10.1093/bioinformatics/17.5.455
26. Dyer MD, Murali TM, Sobral BW (2007) Computational prediction of host-pathogen protein-protein interactions. Bioinformatics 23(13):i159–i166. https://doi.org/10.1093/bioinformatics/btm208
27. Zhou H, Rezaei J, Hugo W et al (2013) Stringent DDI-based prediction of H. sapiens-M. tuberculosis H37Rv protein-protein interactions. BMC Syst Biol 7(Suppl 6):S6. https://doi.org/10.1186/1752-0509-7-S6-S6
28. Basit AH, Abbasi WA, Asif A et al (2018) Training host-pathogen protein-protein interaction predictors. J Bioinforma Comput Biol 16(4): 1850014. https://doi.org/10.1142/S0219720018500142
29. Eid FE, ElHefnawi M, Heath LS (2016) DeNovo: virus-host sequence-based protein-protein interaction prediction. Bioinformatics 32(8):1144–1150. https://doi.org/10.1093/bioinformatics/btv737
30. Zhou H, Gao S, Nguyen NN et al (2014) Stringent homology-based prediction of H. sapiens-M. tuberculosis H37Rv protein-protein interactions. Biol Direct 9:5. https://doi.org/10.1186/1745-6150-9-5
31. Singhal M, Resat H (2007) A domain-based approach to predict protein-protein interactions. BMC Bioinform 8:199. https://doi.org/10.1186/1471-2105-8-199
32. Zhang A, He L, Wang Y (2017) Prediction of GCRV virus-host protein interactome based on structural motif-domain interactions. BMC Bioinform 18(1):145. https://doi.org/10.1186/s12859-017-1500-8
33. Read TD, Peterson SN, Tourasse N et al (2003) The genome sequence of Bacillus anthracis Ames and comparison to closely related bacteria. Nature 423(6935):81–86. https://doi.org/10.1038/nature01586
34. Tastan O, Qi Y, Carbonell JG et al (2009) Prediction of interactions between HIV-1 and human proteins by information integration. Pac Symp Biocomput:516–527
35. Kshirsagar M, Carbonell J, Klein-Seetharaman J (2013) Multitask learning for host-pathogen protein interactions. Bioinformatics 29(13): i217–i226. https://doi.org/10.1093/bioinformatics/btt245
36. Hamp T, Rost B (2015) Evolutionary profiles improve protein-protein interaction prediction from sequence. Bioinformatics 31(12): 1945–1950. https://doi.org/10.1093/bioinformatics/btv077
37. Zahiri J, Yaghoubi O, Mohammad-Noori M et al (2013) PPIevo: protein-protein interaction prediction from PSSM based evolutionary information. Genomics 102(4):237–242. https://doi.org/10.1016/j.ygeno.2013.05.006
38. Cui G, Fang C, Han K (2012) Prediction of protein-protein interactions between viruses and human by an SVM model. BMC Bioinform 13(Suppl 7):S5. https://doi.org/10.1186/1471-2105-13-S7-S5
39. Wagstaff K (2012) Machine learning that matters. arXiv. https://doi.org/10.48550/arXiv.1206.4656
40. Ben-Hur A, Noble WS (2006) Choosing negative examples for the prediction of protein-protein interactions. BMC Bioinform 7(Suppl 1):S2. https://doi.org/10.1186/1471-2105-7-S1-S2
41. Park Y, Marcotte EM (2012) Flaws in evaluation schemes for pair-input computational predictions. Nat Methods 9(12):1134–1136. https://doi.org/10.1038/nmeth.2259
42. Hamp T, Rost B (2015) More challenges for machine-learning protein interactions. Bioinformatics 31(10):1521–1525. https://doi.org/10.1093/bioinformatics/btu857
43. Shen J, Zhang J, Luo X et al (2007) Predicting protein-protein interactions based only on sequences information. Proc Natl Acad Sci U S A 104(11):4337–4341. https://doi.org/10.1073/pnas.0607879104
44. Yang X, Yang S, Li Q et al (2020) Prediction of human-virus protein-protein interactions through a sequence embedding-based machine learning method. Comput Struct Biotechnol J 18:153–161. https://doi.org/10.1016/j.csbj.2019.12.005
45. Chen C, Zhang Q, Yu B et al (2020) Improving protein-protein interactions prediction accuracy using XGBoost feature selection and

stacked ensemble classifier. Comput Biol Med 123:103899. https://doi.org/10.1016/j.compbiomed.2020.103899

46. Barman RK, Saha S, Das S (2014) Prediction of interactions between viral and host proteins using supervised machine learning methods. PLoS One 9(11):e112034. https://doi.org/10.1371/journal.pone.0112034
47. Polikar R (2006) Ensemble based systems in decision making. IEEE Circuits Syst Mag 6(3): 21–45. https://doi.org/10.1109/mcas.2006.1688199
48. Rokach L (2010) Ensemble-based classifiers. Artif Intell Rev 33(1):1–39. https://doi.org/10.1007/s10462-009-9124-7
49. Zhou X, Park B, Choi D et al (2018) A generalized approach to predicting protein-protein interactions between virus and host. BMC Genomics 19(Suppl 6):568. https://doi.org/10.1186/s12864-018-4924-2
50. Blagus R, Lusa L (2015) Joint use of over- and under-sampling techniques and cross-validation for the development and assessment of prediction models. BMC Bioinform 16:363. https://doi.org/10.1186/s12859-015-0784-9

Chapter 32

Protein–Protein Interaction Network Exploration Using Cytoscape

Aqsa Majeed and Shahid Mukhtar

Abstract

As the protein–protein interaction (PPI) data increase exponentially, the development and usage of computational methods to analyze these datasets have become a new research horizon in systems biology. The PPI network analysis and visualization can help identify functional modules of the network, pathway genes involved in common cellular functions, and functional annotations of novel genes. Currently, a variety of tools are available for network graph visualization and analysis. Cytoscape, an open-source software tool, is one of them. It provides an interactive visualization interface along with other core features to import, navigate, filter, cluster, search, and export networks. It comes with hundreds of in-built Apps in App Manager to resolve research questions related to network visualization and integration. This chapter aims to illustrate the Cytoscape application to visualize and analyze the PPI network using Arabidopsis interactome-1 main (AI-1_{MAIN}) PPI network dataset from Plant Interactome Database.

Key words Cytoscape, Network visualization, Protein–protein interactions, Network biology, Centrality analysis, PPI network modules

1 Introduction

The systems biology approach is used to investigate the attributes of biological systems both in static and dynamic states [1]. Network biology, a branch of systems biology, can enable a detailed understanding of biological components by mapping out the molecular interactions that give rise to a particular phenotype and functional prediction of cellular processes in biological systems [1–10]. Proteins, one of the biological molecules, are extensively involved in local and systemic cellular processes like gene regulation, metabolism, signaling, and transportation. The understanding of protein–protein interactions (PPIs) is essential to unravel the diverse relationships within biological components [11–18].

The protein–protein interactions (PPIs) can be detected using experimental and computational approaches. Some of the key experimental methods are yeast two-hybrid (Y2H, Y3H) [19],

Shahid Mukhtar (ed.), *Protein-Protein Interactions: Methods and Protocols*, Methods in Molecular Biology, vol. 2690, https://doi.org/10.1007/978-1-0716-3327-4_32,

tandem affinity purification, affinity chromatography, and co-immunoprecipitation [20, 21]. The in silico PPI methods include Struct2Net [22], DeepPPI [23], iLoops [24], and PPlevo [25].

Typically, the PPI network graph consists of two main components: the nodes and edges. Network nodes represent molecules such as genes, proteins, or other macromolecules while interactions between molecules or nodes are referred to as edges. The structural analysis of a PPI network graph includes centrality analysis, which gives insights into the node connections in a network and helps to identify functionally important nodes within a protein network [6, 9]. PPI networks can be further divided into sub-networks or modules by selecting densely connected nodes in a PPI graph [26]. Several algorithms have been developed to identify network modules, for example, MCODE [27], ClusterViz [28], and clusterMaker [29]. These clustering algorithms are available as Cytoscape Plugins/Apps.

In centrality analysis, the number of interactions with a given node is termed the degree of centrality. The higher-degree nodes are generally considered hubs and are found at the center of the network. Structural centralities are divided into three groups related to the network topology: (i) path-based centralities measure the number of shortest paths passed by a node, (ii) neighborhood-based centralities measure the impact of individual neighboring nodes, and (iii) iterative refinement centralities calculate closeness of a node to other densely connected nodes in a network [7, 30].

Different standalone software tools and web-based platforms are available to generate, visualize, and analyze PPI networks, for example, programmatic tools such as igraph (for R, Python, and C) [31] and NetworkX (for Python) [32]. Similarly, some of the non-programmatic Java-based tools are Cytoscape [33], Gephi [34], CentiBiN [35], etc. There are various tools available for PPI network visualization and analysis; however, Cytoscape is one of the most commonly used tools to interactively explore networks with a wide variety of Apps previously known as Plugins (active Apps 367 at the time of writing this chapter). These Apps are available from the Cytoscape App Store for a variety of functionalities including graph data import, network layouts, network filtering, functional annotation, module detection, and literature search [36, 37]. The aims of this chapter are to demonstrate how to set up and use Cytoscape to install built-in Apps, create a PPI network graph, use different visual styles and layouts for visualization, find clusters or modules, and centrality analysis.

2 Materials

To use Cytoscape to visualize a PPI network, we will use the AI-1_{MAIN} PPI network dataset from the Arabidopsis Interactome Mapping experiment. Other than that, PPI network datasets can also be retrieved from other protein databases.

2.1 Fetching the AI-1_{MAIN} Dataset from Plant Interactome Database

AI-1_{MAIN} dataset can be downloaded from the following URLs. The downloaded file is in .xls format and can directly open in Excel.

http://interactome.dfci.harvard.edu/A_thaliana/index.php?page=download

http://interactome.dfci.harvard.edu/A_thaliana/doc/AI_interactions.xls

2.2 System Requirements

Cytoscape is a Java-based application developed to run on multiple operating systems (OS): Windows, Linux, and Mac. To execute Cytoscape, the following system requirements should be met.

- **Operating System (OS):** Windows XP, 8, 7, or 10 (64 bit); Mac X 10.11 (El Capitan); and later, Linux Ubuntu 13.x, 12.x, 11.x, or Fedora.
- **Java JDK versions 11** (for Windows and Mac): Cytoscape installer automatically downloads a suitable Java version if none is available on your system.
- **Processor:** 1 GHz to multiple cores.
- **Memory:** 512 MB to 4 GB.

3 Methods

This protocol will illustrate the installation and usage of Cytoscape to visualize and analyze the PPI network graph. This section can be divided into two parts: (i) data acquisition, and (ii) network graph visualization and analysis.

3.1 Data Acquisition

It is important to structure your input data in a Cytoscape-compatible format. The basic PPI input data files can be in .csv or .xls format with two columns. The objects of the first column are usually known as the source and should be connected to the objects of the second or Target column.

3.2 Set Up Cytoscape: Download and Installation

1. Open a Web Browser and Search for the Cytoscape Web Page (http://cytoscape.org). After Loading the Web Page, Select the **Download Cytoscape** Button from the Home Screen

2. Choose the Cytoscape installer according to your operating system (OS) requirements; for the Windows platform, the installer is only available for the 64-bit version.
3. Open the executable file and follow the on-screen instructions to install.
4. After successful installation, launch Cytoscape from your desktop and double-click on the Cytoscape icon. A Cytoscape welcome screen will now appear.
5. To install an App (e.g., ClusterViz), click on the Apps tab in the Menu bar, and from there, open App Manager, then simply type the name of the App in the "search box," and then click on the "Install" button. After installation, the App will appear in the "Currently installed" tab.

3.3 Import Network Data in Cytoscape

1. To import network data from a file, Go to Menu bar → **File → Import → Network from File,** select "AI1_MAIN_PPI.xls" file from the "**PPI_Data**" folder (*see* **Note 1**).
2. Once the network data is imported successfully, you should have a network with 2661 connected nodes and 5664 edges as shown in Fig. 1a.

3.4 Network Customization: Style and Layouts

Initially, the network does not display a particular layout, and Cytoscape will assign a default network layout. To better understand network topology and find out the most important nodes, we can apply different built-in layouts and node styles as shown in Fig. 1b, c (*see* **Note 2**).

3.5 Network Clustering

ClusterViz is one of the Cytoscape Apps that can be used to identify network clusters to identify important functional modules in a network as presented in Fig. 1d. This App can find clusters in a network based on FAG-EC, EAGLE, and MCODE clustering algorithms and provide a comparison of the results of these algorithms to use in downstream analysis (*see* **Note 3**).

3.6 Network Centrality Analysis

In Cytoscape, "Analyze Network" tool computes a comprehensive set of network topological parameters: network centralities such as degree, betweenness centrality, and closeness centrality, along with other network Statistics, such as number of nodes and edges, shortest path lengths, and clustering coefficients (*see* **Note 4**).

3.7 Export Network

Cytoscape provides several options to export networks as image and network data such as nodes, edges, corresponding layouts, and style information in a table format.

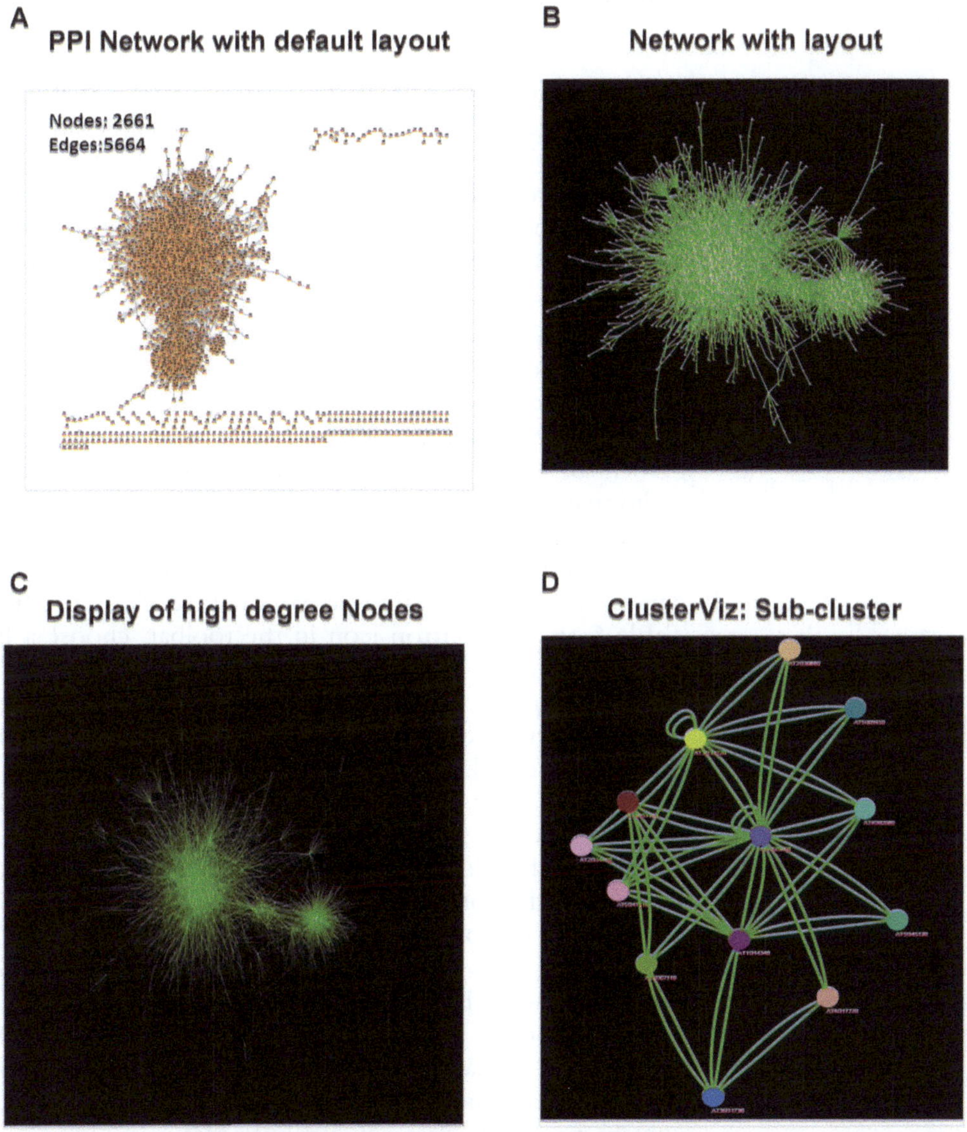

Fig. 1 AI1_MAIN_PPI network graph visualization. (**a**) Display of loaded PPI network with 2661 nodes and 5664 edges. (**b**) Network with customized layout design and styles based on attribute data. (**c**) Display of high-degree nodes in PPI network. (**d**) A cluster of AI1_MAIN_PPI network with 13 densely connected nodes

1. To export the network in a publication-quality image in PNG, JPG, SVG, and PDF formats:

   Click **File → Export → Export as Image...**,

2. To export network data including node, edge, and network tables in. CSV format:

   Click **File → Export → Table to File...**,

3. To save the current session: click the → Save Session icon in the toolbar or click **File → Save As**, give the file name, and save it in .cys format.

4 Notes

1. Create a new folder/directory (with the name of PPI_Data) and place PPI network data in that folder. An alternative way to import network files in Cytoscape is by clicking on **Import Network from File** icon on the toolbar. To navigate the network after it is successfully loaded in Cytoscape, zoom-in and zoom-out icon buttons can be used from the toolbar.
2. From the menu bar, select "Layout → Apply Preferred Layout" to apply a force-directed layout. This step may take a few minutes depending on system requirements.

   Click the Styles tab in the control panel and then choose one of the best-suited node and edge styles from the list of predefined styles. At the bottom of the style tab, there are three additional buttons available to customize the shape, size, color, height, label font, etc. of the nodes, edges, and overall network based on node attribute data (if available). To import attribute data to overlay the existing network data, click on the **Import Table from File** button icon in the toolbar, choose a file to upload, and click OK (at the bottom right corner).

   To change the color and size of network nodes according to node attribute data, click on the Node tab at the bottom of the style pane and it will show all possible mapping properties. Scroll down to show "Fill Color." Click on the arrow on the right to reveal two possible options: Column and Mapping Type. For the Column field, select the column name with attributes data from your node table. In the "Mapping Type" field, choose option "Discrete Mapping" and assign a color for each of the nodes of the given data source by clicking on the tiny button with three dots appearing on the left side. Alternatively, right-click on "Discrete Mapping → Mapping Value Generators → Random Color" to assign each data class a random color as nodes fill only in the case when network data contain many data classes in attribute data.
3. Go to the menu bar "Apps → ClusterViz → Open ClusterViz." The ClusterViz main panel should now be open.

   In the "Find Cluster(s)" tab at the top left, select "in Whole Network," then choose any algorithm from the given three algorithms, i.e., "FAG-EC" in the "Algorithm" tab, then select default parameters or customize these parameters (according to research objective), in "FAG-EC Options" tab; that will affect the results of clustering analysis.

   Click the button "Analyze current network"; now ClusterViz runs the FAG-EC algorithm with a progress bar prompt; this step will take some processing time to complete.

The clustering results can visualize in the result panel in the form of a list, where clusters are sorted by size, score, and modularity. The results of ClusterViz can be exported as a sub-network or to .txt file for further functional enrichment analysis. These clusters can also be expanded or shrunk by adjusting the Size threshold with regard to the "seed" genes.

To export each cluster as a sub-network, select the cluster and click on the "Create Sub-Network" button. This generic pipeline of ClusterViz can be followed for MCODE and EAGLE algorithms as well.

4. Go to Tools → Network Analyzer; then a popup will display to confirm whether the network is directed or undirected. After some processing, results will display in the result panel. Click on the "Node Degree Distribution" button to view the graphical representation of centrality measures. These results can be exported as. PNG files.

   After network analysis, several columns will be added to the Node and Edge Tables including different centrality measures such as degree, betweenness, and closeness centralities as well as clustering coefficient information. One can also visualize these analyses results by drawing histograms and Scatter plots by right-clicking the corresponding column header. From the centrality measure, you can select the nodes with high degree and betweenness centralities and highlight them in the main graph or make a subnetwork.

Acknowledgments

This research was funded by the National Science Foundation (IOS-2038872).

References

1. Garbutt CC, Bangalore PV, Kannar P et al (2014) Getting to the edge: protein dynamical networks as a new frontier in plant-microbe interactions. Front Plant Sci 5:312. https://doi.org/10.3389/fpls.2014.00312
2. Ahmed H, Howton TC, Sun Y et al (2018) Network biology discovers pathogen contact points in host protein-protein interactomes. Nat Commun 9(1):2312. https://doi.org/10.1038/s41467-018-04632-8
3. Arabidopsis Interactome Mapping C (2011) Evidence for network evolution in an Arabidopsis interactome map. Science 333(6042): 601–607. https://doi.org/10.1126/science.1203877
4. Gonzalez-Fuente M, Carrere S, Monachello D et al (2020) EffectorK, a comprehensive resource to mine for Ralstonia, Xanthomonas, and other published effector interactors in the Arabidopsis proteome. Mol Plant Pathol 21(10):1257–1270. https://doi.org/10.1111/mpp.12965
5. Klopffleisch K, Phan N, Augustin K et al (2011) Arabidopsis G-protein interactome reveals connections to cell wall carbohydrates and morphogenesis. Mol Syst Biol 7:532. https://doi.org/10.1038/msb.2011.66
6. Kumar N, Mishra B, Mehmood A et al (2020) Integrative network biology framework elucidates molecular mechanisms of SARS-CoV-

2 pathogenesis. iScience 23(9):101526. https://doi.org/10.1016/j.isci.2020.101526

7. Kumar N, Mishra B, Mukhtar MS (2022) A pipeline of integrating transcriptome and interactome to elucidate central nodes in host-pathogens interactions. STAR Protocols 3(3): 101608. https://doi.org/10.1016/j.xpro.2022.101608
8. Lopez J, Mukhtar MS (2017) Mapping protein-protein interaction using high-throughput yeast 2-hybrid. Methods Mol Biol 1610:217–230. https://doi.org/10.1007/978-1-4939-7003-2_14
9. McCormack ME, Lopez JA, Crocker TH et al (2016) Making the right connections: network biology and plant immune system dynamics. Curr Plant Biol 5:2–12
10. Mishra B, Kumar N, Mukhtar MS (2019) Systems biology and machine learning in plant-pathogen interactions. Mol Plant-Microbe Interact 32(1):45–55. https://doi.org/10.1094/MPMI-08-18-0221-FI
11. Wessling R, Epple P, Altmann S et al (2014) Convergent targeting of a common host protein-network by pathogen effectors from three kingdoms of life. Cell Host Microbe 16(3):364–375. https://doi.org/10.1016/j.chom.2014.08.004
12. Smakowska-Luzan E, Mott GA, Parys K et al (2018) An extracellular network of Arabidopsis leucine-rich repeat receptor kinases. Nature 553(7688):342–346. https://doi.org/10.1038/nature25184
13. Mukhtar MS, Carvunis AR, Dreze M et al (2011) Independently evolved virulence effectors converge onto hubs in a plant immune system network. Science 333(6042):596–601. https://doi.org/10.1126/science.1203659
14. Mott GA, Smakowska-Luzan E, Pasha A et al (2019) Map of physical interactions between extracellular domains of Arabidopsis leucine-rich repeat receptor kinases. Sci Data 6: 190025. https://doi.org/10.1038/sdata.2019.25
15. Mishra B, Sun Y, Howton TC et al (2018) Dynamic modeling of transcriptional gene regulatory network uncovers distinct pathways during the onset of Arabidopsis leaf senescence. NPJ Syst Biol Appl 4:35. https://doi.org/10.1038/s41540-018-0071-2
16. Mishra B, Sun Y, Ahmed H et al (2017) Global temporal dynamic landscape of pathogen-mediated subversion of Arabidopsis innate immunity. Sci Rep 7(1):7849. https://doi.org/10.1038/s41598-017-08073-z
17. Mishra B, Kumar N, Shahid Mukhtar M (2022) A rice protein interaction network reveals high centrality nodes and candidate pathogen effector targets. Comput Struct Biotechnol J 20:2001–2012. https://doi.org/10.1016/j.csbj.2022.04.027
18. Mishra B, Kumar N, Mukhtar MS (2021) Network biology to uncover functional and structural properties of the plant immune system. Curr Opin Plant Biol 62:102057. https://doi.org/10.1016/j.pbi.2021.102057
19. Bruckner A, Polge C, Lentze N et al (2009) Yeast two-hybrid, a powerful tool for systems biology. Int J Mol Sci 10(6):2763–2788. https://doi.org/10.3390/ijms10062763
20. Dunham WH, Mullin M, Gingras AC (2012) Affinity-purification coupled to mass spectrometry: basic principles and strategies. Proteomics 12(10):1576–1590. https://doi.org/10.1002/pmic.201100523
21. Wang S, Wu R, Lu J et al (2022) Protein-protein interaction networks as miners of biological discovery. Proteomics 22(15–16): e2100190. https://doi.org/10.1002/pmic.202100190
22. Singh R, Park D, Xu J et al (2010) Struct2Net: a web service to predict protein-protein interactions using a structure-based approach. Nucleic Acids Res 38(Web Server issue): W508–W515. https://doi.org/10.1093/nar/gkq481
23. Du X, Sun S, Hu C et al (2017) DeepPPI: boosting prediction of protein-protein interactions with deep neural networks. J Chem Inf Model 57(6):1499–1510. https://doi.org/10.1021/acs.jcim.7b00028
24. Planas-Iglesias J, Marin-Lopez MA, Bonet J et al (2013) iLoops: a protein-protein interaction prediction server based on structural features. Bioinformatics 29(18):2360–2362. https://doi.org/10.1093/bioinformatics/btt401
25. Zahiri J, Yaghoubi O, Mohammad-Noori M et al (2013) PPIevo: protein-protein interaction prediction from PSSM based evolutionary information. Genomics 102(4):237–242. https://doi.org/10.1016/j.ygeno.2013.05.006
26. Lin C, Cho Y-r, Hwang WC et al (2007) Clustering methods in protein-protein interaction network. Knowledge discovery in bioinformatics: techniques, methods and application:1–35
27. Bader GD, Hogue CW (2003) An automated method for finding molecular complexes in large protein interaction networks. BMC Bioinf 4(1):1–27
28. Wang J, Zhong J, Chen G et al (2014) ClusterViz: a cytoscape APP for cluster analysis of

biological network. IEEE/ACM Trans Comput Biol Bioinform 12(4):815–822

29. Morris JH, Apeltsin L, Newman AM et al (2011) clusterMaker: a multi-algorithm clustering plugin for Cytoscape. BMC Bioinf 12(1):1–14
30. Mishra B, Kumar N, Mukhtar MS (2022) A rice protein interaction network reveals high centrality nodes and candidate pathogen effector targets. Comput Struct Biotechnol J 20: 2001–2012
31. Csardi G, Nepusz T (2006) The igraph software package for complex network research. InterJournal, Complex Syst 1695(5):1–9
32. Hagberg A, Swart P, S Chult D (2008) Exploring network structure, dynamics, and function using NetworkX. Los Alamos National Lab (LANL), Los Alamos
33. Shannon P, Markiel A, Ozier O et al (2003) Cytoscape: a software environment for integrated models of biomolecular interaction networks. Genome Res 13(11):2498–2504
34. Bastian M, Heymann S (2009) Jacomy M Gephi: an open source software for exploring and manipulating networks. Proc Int AAAI Conf Web Social Media 1:361–362
35. Junker BH, Koschützki D, Schreiber F (2006) Exploration of biological network centralities with CentiBiN. BMC Bioinf 7(1):1–7
36. Su G, Morris JH, Demchak B et al (2014) Biological network exploration with Cytoscape 3. Curr Protoc Bioinformatics 47 (1):8.13: 11-18.13. 24
37. Saito R, Smoot ME, Ono K et al (2012) A travel guide to Cytoscape plugins. Nat Methods 9(11):1069–1076

Chapter 33

Search, Retrieve, Visualize, and Analyze Protein–Protein Interactions from Multiple Databases: A Guide for Experimental Biologists

Vijaykumar Yogesh Muley

Abstract

Functional annotation is lacking for over half of the proteins encoded in genomes and model or representative organisms are not an exception to this trend. One of the popular ways of assigning putative functions to uncharacterized proteins is based on the functions of well-characterized proteins that physically interact with them, i.e., guilt-by-association or functional context approach. In the last two decades, several powerful experimental and computational techniques have been used to determine protein–protein interactions (PPIs) at genome level and are made available through many public databases. The PPI data are often complex and heterogeneously represented across databases posing unique challenges in retrieving, integrating, and analyzing the data even for trained computational biologists, the end users—experimental biologists often struggle to work around the data for the protein of their interests. This chapter provides stepwise protocols to import interaction network of the protein of interest in Cytoscape using PSICQUIC, stringApp, and IntAct App. These are next-generation applications that import PPI from multiple databases/resources and provide seamless functions to study the protein of interest and its functional context directly in Cytoscape.

Key words Protein–protein interactions, Database, PSICQUIC, stringApp, IntAct App, Cytoscape App, Information retrieval, Protein interaction networks

1 Introduction

Proteins are structural and functional components of living systems. They virtually never work alone but form groups or sets by interacting with each other to perform specific tasks. These groups are conventionally referred to as protein complexes and are organized in molecular pathways at a higher level [1]. The physical and functional interactions among proteins that participate in a particular molecular pathway are an important feature of cellular organization and regulation [2, 3]. Because perturbations in the essential or central molecular pathways can collapse the entire

Shahid Mukhtar (ed.), *Protein-Protein Interactions: Methods and Protocols*, Methods in Molecular Biology, vol. 2690, https://doi.org/10.1007/978-1-0716-3327-4_33,

cellular system and can have pathological consequences. Intuitively, protein–protein interactions (PPIs) can be represented in the form of a network or graph in which proteins are nodes (vertices) and are connected by edges (lines) to other proteins they interact with. Many excellent computational tools have been developed for the analysis and visualization of networks including Cytoscape (which was originally developed for biological networks but also works well for complex networks) [4], Gephi (works well for large, complex, and social networks, accessible at https://gephi.org/), and Graphia, which is a recent addition claiming to ease the routine and multivariate analysis of massive networks [5].

In the last two decades, several powerful experimental and computational techniques have been used to determine physical and functional protein–protein interactions (PPIs) [6–8]. These PPIs have been made available through many public databases. The exhaustive list of 690 databases/resources dedicated to PPI and molecular pathways was compiled in 2017 [9], indicating the importance of this field of biological research. Recently, Bajpai et al. compiled, compared, and analyzed 375 databases specialized in PPI data alone, and elaborated their applications from the user's point of view [10]. Evidently, the data and their representation are complex and heterogeneous across databases besides the question of reliability and duplication of reported interactions in them. Retrieving, organizing, and analyzing such data poses a formidable challenge even for well-trained computational biologists [11]. The primary users, experimental biologists, face an additional barrier as they are unable to contemplate the integration of interactions relevant to their research interests across multiple databases.

Significant international collaborative efforts are ongoing to standardize data access from public databases in an integrative and coherent fashion [11–13]. As a first step, the Human Proteome Organization Proteomics Standards Initiative (HUPO-PSI) developed the PSI molecular interaction (PSI-MI) XML format to represent molecular interaction data produced by proteomics experiments [12]. This standardized format was also made available in a simplified tabular form known as the Molecular Interaction Tabular format (MITAB) [14]. PSI-MI formats are widely implemented by many major databases, which have enabled data download in a single format, and are supported by key software tools in the field. However, users would still have to obtain data from multiple resources separately and deal with redundant interactions reported in them. This led to several interesting developments in the field: (1) The International Molecular Exchange Consortium (IMEx) was formed to facilitate the curation of protein–protein interactions from multiple resources to a single non-redundant homogenous set [13] (accessible at https://www.imexconsortium.org/). (2) The Proteomics Standard Initiative Common QUery InterfaCe (PSICQUIC) service was developed to provide a user interface to obtain molecular interaction data from over 30 database and resources (accessible at https://psicquic.github.io/) and comes with implementations for various platforms including R/Bioconductor library [15, 16]. However, the acronym of the application might have hindered its popularity and usability because it's difficult to remember. (3) The Cytoscape stringApp application is a gateway for the STRING database [17], which is popular among experimental biologists (available at https://apps.cytoscape.org/apps/stringApp). (4) Cytoscape IntAct application (IntAct App) provides access to IntAct database [18], which also contains all interactions curated by IMEx consortium [19]. The advantages of using PSICQUIC, stringApp, and IntAct App are that at their backend they collect molecular interactions for query proteins from multiple databases/resources, have some powerful network visualization options, and all can be done from Cytoscape. It is the most used tools in biological network analysis and visualization, with support from a wide variety of plugins and applications for network analysis [17]. Its use has also been demonstrated through several tutorials and documentation available on the Cytoscape website and is within the reach of a research community devoid of computational expertise.

This chapter aims to demonstrate the utility of PSICQUIC, IntAct App, and stringApp in searching and retrieving PPI directly from several databases in Cytoscape for further analysis and visualization. Efforts have been made so that researchers with limited or no computational expertise can systematically retrieve interacting partners of a protein of interest, visualize, analyze, and export (figures and tables) the network data.

2 Materials

1. A query protein name, gene symbol, gene name, or keyword of interest (*see* **Note 1**). Here, I will use the human ATRX gene. In our previous study, it was found massively expressed during the embryonic development of telencephalon, which gives rise to the largest information processing centers of the brain [20] and is a chromatin remodeling protein. Hence, ATRX is expected to interact specifically with transcription factors of nuclear origin and will be demonstrated later in the tutorial.
2. A standard laptop or desktop computer preferably with 4–8 Gb RAM. This protocol should work on almost all laptops and desktops.
3. Internet with reasonable data download speed.

3 Methods

3.1 Cytoscape Installation and Launch

1. Go to https://cytoscape.org/.
2. Click on the *Download* button. It should automatically recognize your operating system (OS) and provide a download link for the Cytoscape installer; if not, look for the "other platforms" option, which will take you to OS-specific installer links. Click on the appropriate link to download Cytoscape.
3. Install the Cytoscape software by double-clicking on the downloaded installer and following the instructions on the installation progress window (*see* **Note 2**). Mostly, click on the next button a few times, accept the terms and conditions of the license (or after reading them), and tick the checkbox, when the installer recommends creating a Desktop icon. It becomes handy for opening Cytoscape for routine use.
4. Launch Cytoscape by clicking its icon from Desktop or Launchpad on macOS, Desktop or Start menu on Windows OS, and Desktop or applications menu (such as GNOME (GNU Network Object Model Environment)) on Linux-based OS. It depends on how you install Cytoscape. It should be noted that Cytoscape can be launched from the command prompt and supports command-line operations for batch analysis.

3.2 Cytoscape Plugin Installation

1. Launch Cytoscape as instructed in step 4, Subheading 3.1.
2. Go to **Apps** and click on **App Manager** from the menu bar of Cytoscape (Fig. 1).
3. As shown in Fig. 1, in the **App Manager**:
   (a) Type PSICQUIC in the search bar.
   (b) Select PSICQUIC Web Service Client from the middle panel.
   (c) Click install.
4. Repeat step 3, Subheading 3.2 by typing stringApp instead of PSICQUIC in the search bar and install it.
5. Repeat step 3, Subheading 3.2. by typing IntAct App instead of PSICQUIC in the search bar and install it.
6. The identifiers (names) of interacting proteins retrieved by PSICQUIC will not be uniform. It is good to map them to routinely used standard names such as gene symbols or gene names for brevity. This functionality is provided by ID mapper and BridgeDb plugins in Cytoscape. They can be installed from App Manager as explained in step 3, Subheading 3.2. For the first plugin, ID should be followed by space, i.e., "ID mapper"

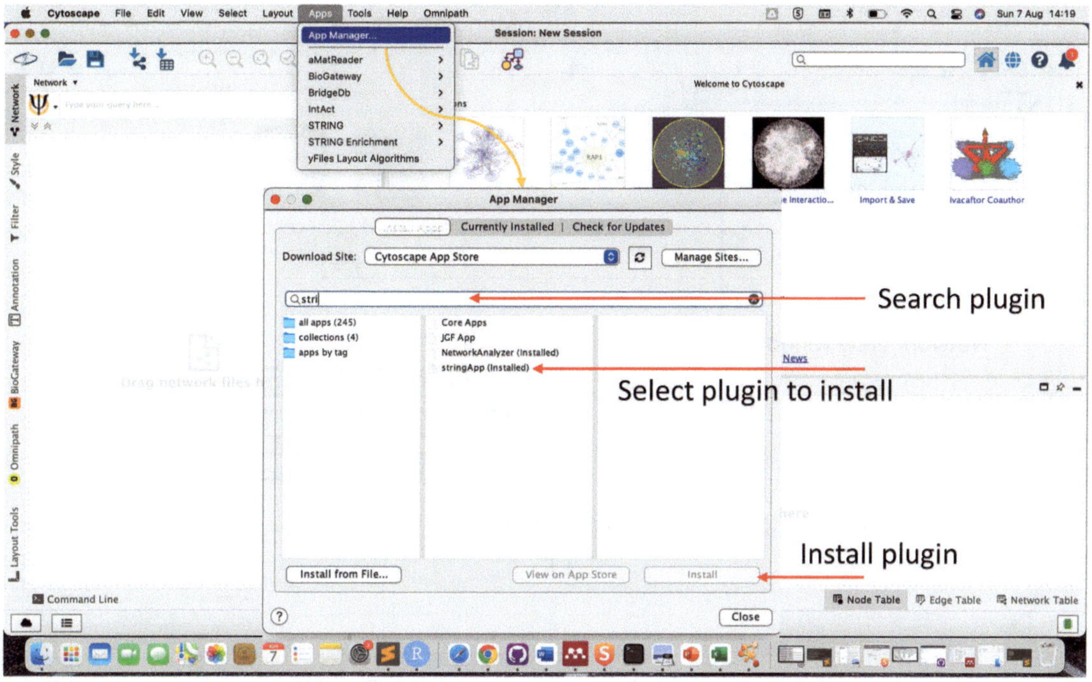

Fig. 1 Plugin installation in the Cytoscape using the application manager from the Apps menu. The plugin can be installed by typing its name in the search bar, and then selecting the plugin of interest among the plugins with a similar name and clicking the install button

[21]. In this protocol, an ID mapper plugin is used, but users should be able to work around BridgeDb as well [22].

7. Assuming all the above steps were carried out successfully, I recommend quitting Cytoscape and then starting before further use. It can be done from the Cytoscape main menu, which should be found in the top left corner.

3.3 Essential Components of Cytoscape Used in This Protocol

1. Launch Cytoscape as explained in step 4, Subheading 3.1. It automatically loads the necessary plugins to run this protocol.
2. Figure 2 shows the typical user interface of Cytoscape and will be used as a reference to highlight panels/options for executing the protocol hereafter.
3. The **Network Search Panel** shown in Fig. 2 and its zoomed details can be used to search and obtain interacting partners of the query protein from various databases registered with PSICQUIC, or from STRING and IntAct databases. The STRING and IntAct databases could be preferred since users do not have to deal with protein identifier mapping to their standard forms.
4. The **Network View Window** displays the obtained interacting proteins in the form of a network or graph.

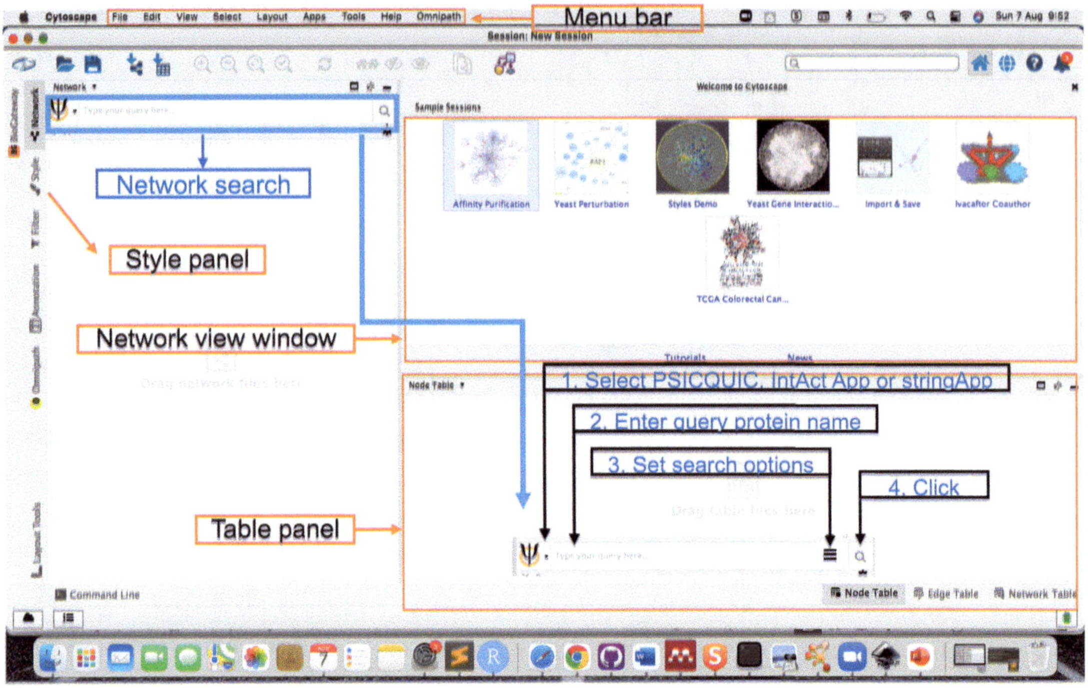

Fig. 2 Cytoscape panels and their components for importing and analysis of protein-protein interactions . The network search box is required for selecting a particular Cytoscape application to retrieve interactions for a query protein, which is shown in detail by zooming it. The style panel is useful for adjusting the visualization of the nodes, edges, and entire network. The table panel shows the attributes associated with the interaction network, nodes, and edges in it. The network view window displays the imported network

5. The **Table Panel** shows information associated with the nodes, edges, and the network. Users can toggle between **Node Table**, **Edge Table**, and an entire network.
6. The **Style Panel** is useful in modifying the default representation of nodes, edges, or the entire network including size, shape, color, labels, and label position.
7. The details of these features are beyond the scope of this protocol, but readers are encouraged to explore https://cytoscape.org/documentation_users.html webpages for excellent tutorials to make the best use of this protocol.

3.4 Import Protein Interactions Using PSICQUIC

1. Interacting partners of the query protein can be downloaded from the **Network Search Panel** in the Cytoscape as follows (please see Fig. 2 for clarity):
   (a) Select PSICQUIC from the **Network Search Panel**.
   (b) Enter the protein name of interest, for example, ATRX here.

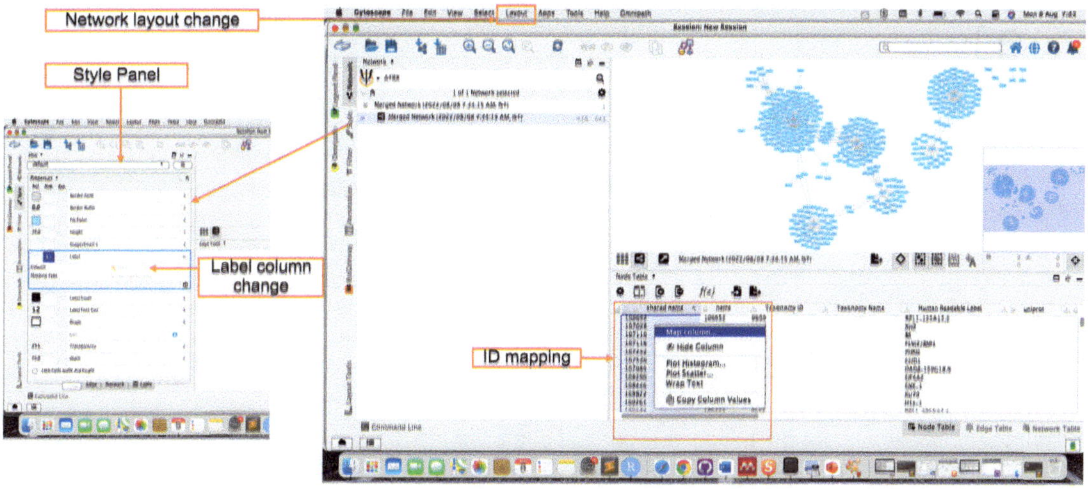

Fig. 3 An example network of the proteins interacting with ATRX imported using the PSICQUIC application. Clusters of networks in the Network View Window are due to different protein identifiers used by the databases registered with the PSICQUIC and also due to interaction data from the different species. The style panel (inset) on the left demonstrates how to change node parameters, i.e., label names in this particular case. In the Table Panel, I opened an ID mapper to convert protein identifiers given in that column (shared name) to standard gene symbols

   (c) Click on the **Search Icon** to launch the PSICQUIC window.

2. The PSICQUIC window shows the number of interactions available in over 30 databases and an option to select some of them or all of them (*see* **Note 3**). Double-check the input query protein or gene name when there is nothing to select in PSICQUIC.
3. Select PPI databases of interest or tick on the **Select All** option in the PSICQUIC window.
4. Set **Automatic Network Merge** operation true by checking the box, so that the interacting partners from multiple databases will be merged and imported as a single network.
5. Click on **Import**.
6. PSICQUIC automatically imports the interacting proteins of the query and displays them in the **Network View Window**, as shown in Fig. 3. Few clusters can be identified in the imported network because PSICQUIC imports interactions based on keywords irrespective of the organisms or the type of protein identifiers. Hence, the interactions with similar identifiers or from the same organisms cluster together.
7. Go to the **View** menu from Cytoscape's **Menu bar** (Fig. 2) and select the **Always Show Graphics Details** option to see the details of the network. Check the difference with and without this option.

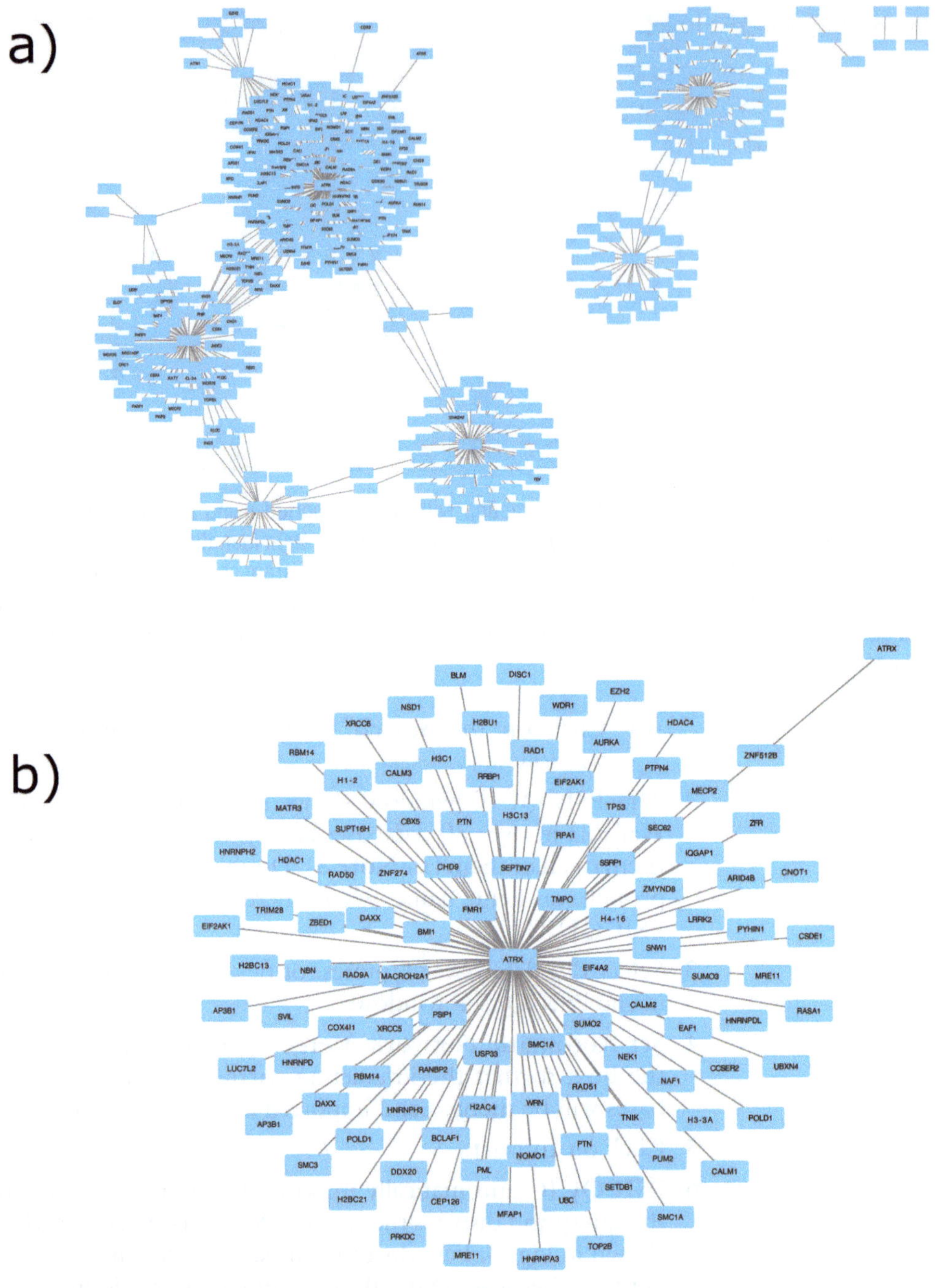

Fig. 4 An example network of the proteins interacting with ATRX imported by the PSICQUIC application after adjustment using Cytoscape functions. (**a**) Gene symbols were mapped to original protein identifiers leaving a lot of nodes without names since many protein names could not be mapped to gene symbols. (**b**) The network after removal of the nodes without gene symbols and performing layout adjustment and scale adjustment as explained in the protocol

8. PSICQUIC does not convert protein synonyms used in different databases to their commonly used standard form. The complete solution to this problem is complex and beyond the scope of this protocol (*see* **Note 4**), and there is no guarantee one can convert all protein names to their gene symbols. However, proteomics data often use UniProt identifiers, which can be mapped to Gene names or symbols using ID mapper, which can be done as follows.
9. Click on the **Node Table** option from **Table Panel** (Fig. 2). Right-click on the column shared name in the **Node Table, and** select the Map Column, as shown in Fig. 3.
10. Select the **Species** of interest, **Map** from *UniProt* to *HGNC*, and click on the **OK** button. It will add a new column to the **Node Table**, which is HGNC (1) in this case. UniProt can be replaced with some other types of identifiers as well. The acronym HGNC stands for the Human Genome Organisation (HUGO) Gene Nomenclature Committee that sets the standards for human gene nomenclature.
11. To display mapped gene symbols on the network nodes, click on the style panel (shown in Fig. 2 and window in Fig. 3) → click on the label → click on the **column** box → select HGNC (1) or any column (of **Node Table**) from which node names should be displayed → click on **Mapping Type** box → select **Passing-through.**
12. As shown in Fig. 4a, the nodes in the network are now labeled with gene names rather than some random numbers (only the network is shown for brevity), and many nodes do not have labels. These nodes could not be mapped to their corresponding gene symbols.
13. Go to **Table Panel**, Sort **Node Table** by clicking on **HGNC (1)** column header (arrow sign). Select empty rows corresponding to unmapped genes in the column and use right-click on the **Select Nodes From Selected Rows** option. This will select all nodes without gene symbols in the **Network View Window** and turn them into yellow color. Then go to the **Edit** menu from the main **Menu bar** and select **Remove Selected Nodes and Edges**. These steps will remove all the nodes in the network for which no gene symbols exist (Fig. 4b).
14. Organize the remaining network nodes and edges from the **Layout** menu by selecting **Apply Preferred Layout** option. There would be some nodes without any connection to other proteins in the network. These were removed from the network for clarity here, but I recommend cross-checking them since they could be potential interacting partners of the query protein, which might have been lost during ID mapping.

15. Finally, the network looks crowded so go to the **Layout** menu from the **Menu bar**, and select **Layout Tools**; a new menu pops up on the left side of the **Table Panel**, which increases the scale until the nodes in the network do not overlap with each other.
16. Export the resulting network from the **File Menu** →Export →Network to Image →input file name in an appropriate format. The exported network should appear as shown in Fig. 4b.

3.5 Import Protein Interactions Using the IntAct App

1. Go to the **Network Search** box (please see Fig. 2 for clarity):
   (a) Select **IntAct Exact Query** or **Fuzzy Search** option (*see* **Note 5**).
   (b) Enter the protein name of interest, for example, ATRX.
   (c) Users can set some **Search options** or keep default settings (see Fig. 2).
   (d) Click on the **Search Icon** to launch the **IntAct App** window.
2. The **IntAct App** window pops up with some choices to be made by the users. In this example, I select protein interactions only from studies in humans.
3. Click on the **Build Network** option to view the network in the **Network View Window.**
4. The network built by IntAct App appears different and Node and Edge Tables in **Table Panel** have a lot more information than the PSICQUIC (Fig. 5).
5. The IntAct App window also pops up next to the **Network View Window** (Fig. 5). There, users can set the color for the edges based on the **Summary score** for the interaction pairs, **Evidence**, or **Mutation**. More about these options can be learned from the IntAct App documentation at https://ebi-intact.github.io/IntActApp/ and https://apps.cytoscape.org/apps/intactapp.
6. The best feature of the IntAct-derived network is that the nodes are labeled with the standard gene symbols; when users select a particular protein in the network, IntAct window also shows information about it and links to external databases to get even more information (Fig. 5).
7. Go to the **View** menu from Cytoscape's **Menu bar** (Fig. 2) and select the **Always Show Graphics Details** option, so that the network will be displayed with all its contents. Check the difference with and without this option.
8. Organize the network nodes and edges from the **Layout** menu by selecting the **Apply Preferred Layout** option. Users can

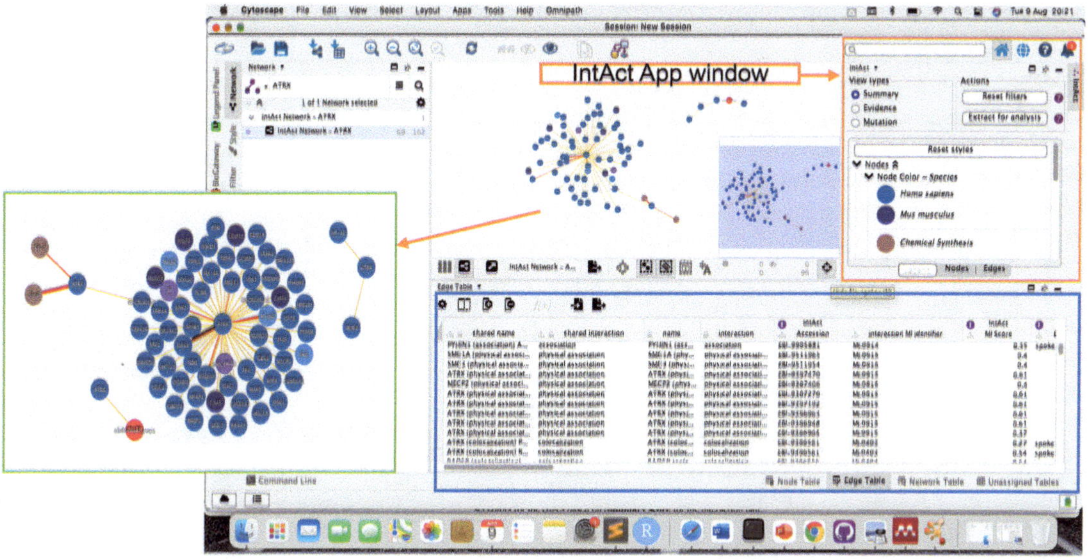

Fig. 5 An example network of the proteins interacting with ATRX imported using the IntAct App. The ATRX gene appears four times in the network representing its various isoforms. Node colors represent species in which interaction was identified. Interactions between nodes are darker in color if they have a strong interaction score, and the width represents the number of evidence that exists for them. The IntAct app menu is highlighted in the figure, and after performing layout and scale adjustments the network appears as it is shown in the inset (left)

play around with the nodes if some of the nodes are placed away from the other nodes manually to place them as desired.

9. Finally, the network looks crowded so go to the **Layout** menu from the **Menu** bar and select **Layout Tools**; a new menu pops up on the left side of the **Table Panel**, which increases the scale until the nodes in the network do not overlap with each other.
10. **Style Panel** can be used to change the node and edge features, and readers are encouraged to read Cytoscape documentation.
11. Export the resulting network from the **File Menu** →Export →Network to Image →input file name in an appropriate format. The exported network should appear as shown in Fig. 5.
12. Export the legends from the **Style Panel** by clicking on the button with three horizontal lines at the top right corner, then click on the **Create legend** and choose the appropriate format and name for the output file.
13. The exported network figures can be combined by image illustrators such as Inkscape (open access) or Adobe Photoshop.

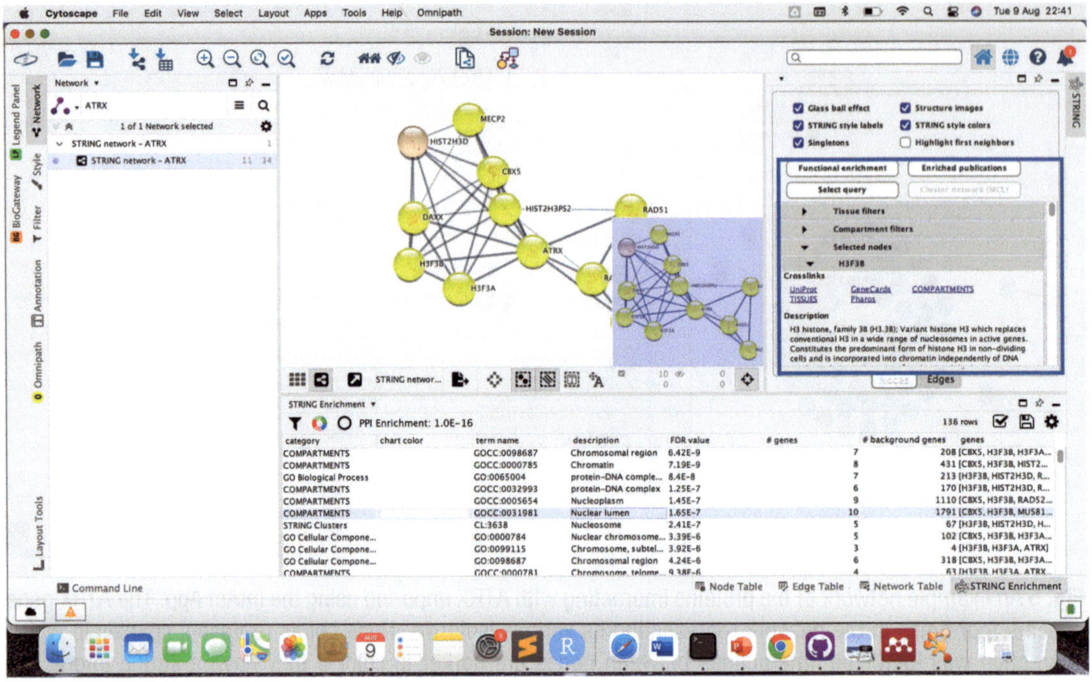

Fig. 6 An example protein-protein interaction network demonstrates the proteins that interact with ATRX, imported from the STRING database. The stringApp showcases several key features for networks analysis, displayed within the box in the figure. One of the features facilatated the functional enrichment analysis for ATRX interacting partners. Moreover, stringApp offers a tool to highlight proteins in the network based on enriched functions by a single mouse click. In this figure, proteins enriched in nuclear lumen compartment are specifically highlighted. Notably, ATRX interacts with 11 proteins that are significantly enriched in nuclear lumen, indicating a probable site for the action of ATRX-associated chromatin remodeling complexes

3.6 Import Protein Interactions Using stringApp

1. Go to the **Network search** box (please see Fig. 2 for clarity):
   (a) Select **STRING protein query.**
   (b) Enter the protein name of interest, for example, ATRX.
   (c) Users can set a few Search options or keep default settings (see Fig. 2).
   (d) Click on the **Search Icon** to import interaction network.
2. By default, the query will be searched in human PPI data, but stringApp allows searching in large numbers of non-human species as well.
3. The network built by the stringApp appears different. The Node and Edge Tables in **Table Panel** have a lot more information than the PSICQUIC or IntAct App (Fig. 6).
4. The stringApp window also pops up next to the **Network View Window**. The stringApp has a lot more functionality than the other applications discussed in this protocol. For example, users can visualize proteins from various subcellular organelles and tissues, and can also perform functional

enrichment analysis (Fig. 6). These are powerful functions and readers should learn about them in more detail from the stringApp documentation to fully leverage their functionality.

5. Some of the best features of the stringApp-derived network are that the nodes are labeled as per their standard gene symbols. When users select a particular protein in the network, the stringApp window shows information about it and links to external databases to get even more information. As mentioned below, users can perform functional enrichment analysis, and highlight proteins in the network that are enriched in specific biological processes.
6. The networks imported from the STRING database through stringApp are usually small, and the default network representation is soothing to the eyes. So, I would not recommend changes to it.
7. Functional enrichment can be done by clicking a button in the stringApp window. The results then appear in the Node Table (Fig. 6). Users can select the enriched term and highlight the nodes that belong to it in the network (Fig. 6). For example, all ATRX interacting proteins belong to the nucleus compartment, which is consistent with its function as a chromatic organizer, and hence it should localize to the nucleus to coordinate gene expression.
8. The enriched publication function can retrieve PubMed citations in which interacting proteins are enriched significantly.
9. The network can be exported as shown above from the File menu →Export →Network to Image →input file name in an appropriate format.

3.7 Merging Networks Imported from PSICQUIC, IntAct App, and stringApp

1. So far we have seen how three different network search applications import interactions in different formats and provide different tools for analysis to the end users.
2. Due to the complexity of the underlying data, there is no one-stop remedy to handling these data in a uniform way, at least as of now. However, once the imported networks are labeled with the gene symbols, they can be merged into a single network for further analysis.
3. This can be done from the Tools menu in the menu bar → Merge Networks → select networks to merge → merge. These series of commands will create a single large network from the networks imported by PSICQUIC, IntAct App, and stringApp. To transform a network, such as the merged one or any other network, into the native format of stringApp, utilize the "**Stringify**" option. First, navigate to the **Apps** menu, select the **STRING** option, and subsequently click on **Stringify**.

This action will prompt you to enter the name of the organism and annotate the network with information from the STRING database.

4. It can be analyzed further using several excellent applications available from Cytoscape. Cytoscape documentation and forum are rich and have solutions for almost every problem a user could encounter.

4 Notes

1. Databases can be searched with exact and fuzzy keywords. However, users should try to provide standard names that are available for a particular protein, such as CBX5 and RutR. This helps in importing relevant PPI for the keyword and avoiding non-specific ones.
2. Cytoscape installation depends on the Java Runtime Environment (JRE). If a Java-related error occurs during installation, Cytoscape documentation has proper installation instructions, and users can easily install Java or JRE using information available on the Java website.
3. Whenever a PPI search application provides an option to select databases, it is worth checking the nature, contents, and reliability of the database to make the best use of the selected option. In PSICQUIC search engine only active databases can be selected. If none of the databases are active to select then it is more likely that users have entered the wrong query, and should double-check spelling or typos.
4. ID mapping is complex especially when different databases represent protein entries with different names. No straightforward solution exists for this problem. The best thing one can do is to first map all UniProt identifiers to their gene symbols, because most databases/proteomics studies use them. Second, attempt to map other types of identifiers to the gene symbols using the ID mapping services in Cytoscape and copy them to the column where UniProt to Gene symbol mapping exists. It is also possible to Export the Node Table and open in Excel to make a column with gene symbols and use them to label the network. However, users are expected to have familiarity with the Cytoscape and ID mapping to do this.
5. IntAct App's fuzzy search option allows users to create a network of proteins associated with specific keywords or biological concepts. I recommend using this option with caution, since many unwanted interactions could be imported. In this case, the users should focus on bigger clusters of the interacting proteins and take precautions for the isolated proteins in the network.

References

1. Muley VY, Acharya V (2013) Genome-wide prediction and analysis of protein-protein functional linkages in bacteria. Springer, New York
2. Muley VY, Ranjan A (2013) Evaluation of physical and functional protein-protein interaction prediction methods for detecting biological pathways. PLoS One 8:e54325. https://doi.org/10.1371/journal.pone.0054325
3. Muley VY, König R (2022) Human transcriptional gene regulatory network compiled from 14 data resources. Biochimie 193:115–125. https://doi.org/10.1016/j.biochi.2021.10.016
4. Otasek D, Morris JH, Bouças J et al (2019) Cytoscape automation: empowering workflow-based network analysis. Genome Biol 20. https://doi.org/10.1186/S13059-019-1758-4
5. Id TCF, Horsewell IS, Id AP et al (2022) Graphia: a platform for the graph-based visualisation and analysis of high dimensional data. PLoS Comput Biol 18:e1010310. https://doi.org/10.1371/journal.pcbi.1010310
6. Muley VY (2012) Improved computational prediction and analysis of protein – protein interaction networks. PhD thesis, Centre for DNA Fingerprinting and Diagnostics, Hyderabad, Manipal University, India. http://hdl.handle.net/10603/5399
7. Snider J, Kotlyar M, Saraon P et al (2015) Fundamentals of protein interaction network mapping. Mol Syst Biol 11:848. https://doi.org/10.15252/msb.20156351
8. Muley VY, Ranjan A (2012) Effect of reference genome selection on the performance of computational methods for genome-wide protein-protein interaction prediction. PLoS One 7:e42057. https://doi.org/10.1371/journal.pone.0042057
9. Bader GD, Cary MP, Sander C (2006) Pathguide: a pathway resource list. Nucleic Acids Res 34:D504. https://doi.org/10.1093/NAR/GKJ126
10. Bajpai AK, Davuluri S, Tiwary K et al (2020) Systematic comparison of the protein-protein interaction databases from a user's perspective. J Biomed Inform 103. https://doi.org/10.1016/J.JBI.2020.103380
11. Koh GCKW, Porras P, Aranda B et al (2012) Analyzing protein–protein interaction networks. J Proteome Res 11(4):2014–2031. https://doi.org/10.1021/pr201211w
12. Hermjakob H, Montecchi-Palazzi L, Bader G et al (2004) The HUPO PSI's molecular interaction format – a community standard for the representation of protein interaction data. Nat Biotechnol 22:177–183. https://doi.org/10.1038/nbt926
13. Orchard S, Kerrien S, Abbani S et al (2012) Protein interaction data curation: the international molecular exchange (IMEx) consortium. Nat Methods 9:345–350. https://doi.org/10.1038/nmeth.1931
14. Kerrien S, Orchard S, Montecchi-Palazzi L et al (2007) Broadening the horizon – level 2.5 of the HUPO-PSI format for molecular interactions. BMC Biol 5:1–11. https://doi.org/10.1186/1741-7007-5-44/figures/5
15. Aranda B, Blankenburg H, Kerrien S et al (2011) PSICQUIC and PSISCORE: accessing and scoring molecular interactions. Nat Methods 8:528–529. https://doi.org/10.1038/nmeth.1637
16. del-Toro N, Dumousseau M, Orchard S et al (2013) A new reference implementation of the PSICQUIC web service. Nucleic Acids Res 41:W601. https://doi.org/10.1093/nar/gkt392
17. Doncheva NT, Morris JH, Gorodkin J, Jensen LJ (2019) Cytoscape StringApp: network analysis and visualization of proteomics data. J Proteome Res 18:623–632. https://doi.org/10.1021/acs.jproteome.8b00702
18. Ragueneau E, Shrivastava A, Morris JH et al (2021) IntAct app: a Cytoscape application for molecular interaction network visualisation and analysis. Bioinformatics 37:3684–3685. https://doi.org/10.1093/bioinformatics/btab319
19. del Toro N, Shrivastava A, Ragueneau E et al (2022) The IntAct database: efficient access to fine-grained molecular interaction data. Nucleic Acids Res 50:D648–D653. https://doi.org/10.1093/NAR/GKAB1006
20. Muley VY, López-Victorio CJ, Ayala-Sumuano JT et al (2020) Conserved and divergent expression dynamics during early patterning of the telencephalon in mouse and chick embryos. Prog Neurobiol 186:101735. https://doi.org/10.1016/j.pneurobio.2019.101735
21. Treister A, Pico AR (2018) Identifier mapping in Cytoscape. F1000Res 7:725. https://doi.org/10.12688/f1000research.14807.2
22. Gao J, Zhang C, van Iersel M et al (2014) BridgeDb app: unifying identifier mapping services for cytoscape. F1000Res 3. https://doi.org/10.12688/f1000research.4521.1

Chapter 34

Centrality Analysis of Protein–Protein Interaction Networks Using R

Vijaykumar Yogesh Muley

Abstract

Proteins are structural and functional components of cells. They interact with each other to drive specific cellular functions. The physical and functional protein interactions are an important feature of cellular organization and regulation. Protein interactions are represented as a network or a graph in which proteins are nodes, and interactions between them are edges. Perturbations in the network affecting essential or central proteins can have pathological consequences. Network or graph theory is a branch of mathematics that provides a conceptual framework to decipher topologically important proteins in the network. These concepts are known as centrality measures. This chapter introduces various centrality metrics and provides a stepwise protocol to quantify protein's strategic positions in the network using an R programming language.

Key words Protein–protein interactions, Protein interaction networks, Network theory, Graph theory, Network centrality, Centrality indices, Betweenness, Network biology

1 Introduction

Living organisms are made up of single or multiple cells. The pool of proteins present within each cell determines its structure and function. Proteins physically or functionally interact with each other to execute their functions [1–3]. These protein–protein interactions (PPI) can be represented as a graph or a network in which nodes (vertices) are proteins, and interactions between them are edges [4]. Networks can have directed or undirected edges. In the directed network, a pair A -> B represents a directed edge from node A to B, while B -> A represents an edge from B to A. In the undirected networks, A -> B and B -> A pairs are equivalent or symmetric. PPI networks are usually undirected.

The network representation is intuitive to understand. For example, in our social networks, let us imagine you have two friends: one is an extrovert and has a big friend circle, and another

Shahid Mukhtar (ed.), *Protein-Protein Interactions: Methods and Protocols*, Methods in Molecular Biology, vol. 2690,
https://doi.org/10.1007/978-1-0716-3327-4_34,

is an introvert. Both know your little secret, and the next day everyone knows about it. Intuitively, a friend with a big social network would be a prime suspect in disclosing your secret to others. Network structure reveals even a bigger picture. In the network, an introverted friend may interact with a few but influential friends with a wider social circle. Hence, though an introvert has few connections, they should also be considered prime suspect or a possible alternative that can spread information as fast as an extrovert but not directly. Network science, a branch of applied mathematics, provides several metrics to quantify such strategically placed nodes in the network. These metrics are commonly known as centrality indices.

Centrality indices assign a rank or numeric value to a node based on its position in the network in relation to remaining nodes [5]. The nodes with high centrality values tend to have central positions in the network, and often have a destructive effect on the network structure upon their removal [6]. Several centrality indices have been proposed in the literature and are mostly developed for social network studies. However, they can be readily applicable to most networks. CentiServer website catalogs known centrality indices, and related web tools and R packages [7]. CentiServer reports a total of 403 centrality indices as of August 2022, indicating the popularity of the subject. However, most centrality indices seem to borrow ideas from the original three indices that were proposed well before the 1970s. These are called degree, betweenness, and closeness centrality [8].

Degree centrality represents the most basic property of any network and is defined as the total number of edges of a given node with other nodes in the network. It assumes linearity, which means nodes with a high number of edges tend to be more important than those having a smaller one. Proteins with a very high number of edges with other proteins are called hubs. Hub proteins are often involved in essential functions. *Betweenness centrality* refers to the number of times a given node is present on the path while traveling from one node to another with the shortest distance or path length [8]. The shortest distance is the minimum number of nodes that need to be crossed to travel between given two nodes in the network. Betweenness centrality is often high for proteins acting as hubs. On the other hand, proteins having a very low degree can also have high betweenness centrality when they are placed between two hubs. These proteins are called bottlenecks or articulation points because they form a bridge between two network components. Malfunctioning of such proteins leads to communication loss between two highly connected components in the network like destroying a bridge on the river separating two parts of the land. *Closeness centrality* measures the number of steps required to access other nodes from a given node [9]. Closeness centrality tells us how quickly and easily information can be spread from a given node to other nodes in the network.

Computing centrality indices is straightforward using several excellent packages that have been developed to use in R. CentiServer website is a good starting point to explore them. igraph and sna are the most used packages, while CentiServer also developed its own package – centiserve [7, 10]. Furthermore, netrankr package is worth mentioning here because it gives users a fresh perspective on network analysis and goes beyond the traditional use of centrality measures [11, 12].

In the subsequent sections, we see how to compute centrality indices for a PPI network using igraph and CINNA packages. The methods described can easily be applied to any network including directed networks by tuning appropriate parameters for the analysis functions. Efforts have been made to allow network analysis with limited or no familiarity with programming.

2 Materials

1. A standard laptop or desktop computer with a minimum of 4 Gb RAM or more. Some centrality algorithms are time-intensive for larger networks since they consume memory as well as processing power and hence should be used cautiously.
2. Internet connection to install R and network analysis packages (libraries).
3. A network file. Here, I use a small toy network from the toyNet.csv file downloaded from https://github.com/muleylab/ppidata. The file format is called comma-separated value (CSV), one of the most frequent file formats used for storing networks. Most network files contain two columns representing pairwise relationships (interaction on each row) between nodes present in the first column and the nodes in the second column. If a file has more than two columns, they usually represent edge attributes such as a column with numerical values indicating the strength of the edge or interaction type (physical or functional), etc. These attributes are often useful in the analysis and visualization of specific properties of the edges. For example, categorical edge attributes can be visualized with different line types or line colors in the network or edge weights can be shown with the relative width of the line or color gradient.

4. A node metadata file containing node attributes. Metadata associated with our network is present in the toyNetNode.csv file, which can be downloaded from https://github.com/muleylab/ppidata.
5. Node or edge attribute files are often optional for analysis.

3 Methods

3.1 Installing R and Network Analysis Packages

1. R can be installed along with its graphical user interface RStudio from https://www.r-project.org and https://www.rstudio.com respectively. Community support is tremendous for R and RStudio and users should be able to install both software by following the installation instructions given on their websites. On Windows operating systems, I also recommend installing RTools, which is also available from the R website.
2. Most analyses described in this chapter require igraph package. In R or Rstudio, igraph can be installed using command install.packages("igraph") at the R console or in the script. Alternatively, open RStudio, then go to the Tools menu → select Install Packages and choose either CRAN or package archive option to install the igraph library directly from the Internet or downloaded source files respectively. Weak Internet connectivity may create some problems during installation; please *see* **Note 1** for troubleshooting this problem. igraph also has excellent documentation at https://igraph.org/ for more information.
3. Now install CINNA package, which provides a wrapper function to compute at least 49 centrality metrics. The installation procedure is the same as above, but here we use a command to install them as follows:

   ```
   install.packages("CINNA")
   ```

   Alternatively, the CINNA package can be installed from the developer's GitHub repository using devtools. First, install devtools by running the following command: install.packages("devtools"). Then, proceed to install CINNA by executing devtools::install_github("https://github.com/jafarilab/CINNA") or download the source file from https://cran.rproject.org/src/contrib/Archive/CINNA/ and follow the installation instructions mentioned in the previous step.
4. To export centrality results to an Excel file, please install the writexl package:

   ```
   install.packages("writexl")
   ```

3.2 Loading Network Analysis Packages in R

1. Open **Rstudio** and go to the **File** menu and click on **New file** and then on an **R script**. It creates a file named **untitled.R** to write R commands. It is always a good idea to save this file with an appropriate name.
2. In the file, users need to type R commands and execute them by clicking on the **Run** option located in the top right corner of the file in RStudio.
3. Type the following commands in the file, select them, and click the Run option to load them in the current R environment or a session:
   (a) library(writexl)
   (b) library(igraph)
   (c) library(CINNA)
4. Now, network analysis and other functions from these libraries are available to use in the current session.

3.3 Reading Network from a File

1. Reading files is easy in R and many functions can be used depending on the file format. Since our PPI network file is saved in CSV format (*see* Subheading 2, "Materials"), we call the read.csv function to import the file and store its contents in a new variable called ppi as shown below:

   ```
   ppi <- read.csv("~/Downloads/toyNet.csv", header = FALSE)
   ```

   The read.csv function is designed to import csv files. Its input is the complete path of the file; here, the name of the file is toyNet.csv, which is present in the Downloads folder, and ~ represents the home directory in Unix/Linux-based system (in Windows, it should be something C:/Users/XXXX/Downloads/toyNet.csv). The header parameter is set to FALSE since our network file does not have column names; otherwise, its value should be set to TRUE. *See* **Note 2** to import files other than csv format. Please note that users often copy commands from the PDF and try to run them in RStudio. In some cases, RStudio may throw an error about the use of double quotes ("). If you encounter such an error, simply replace the copied double quotes with the double quote symbol entered directly from your keyboard.
2. To get more information about the input and output parameters of functions, we can use the help function. For example, get help for the read.csv function as follows:

   ```
   help("read.csv") or ?read.csv
   ```
3. Variables have specific classes in R. Here, the ppi variable is imported as a data.frame since it is a table with two columns.

The first column represents the nodes (proteins) that interact with the nodes of the second column. The head function is handy to look at the first few lines of the PPI variable:

head(ppi)

4. igraph provides a graph_from_data_frame function that turns data.frame into an object of graph class; if data.frame has at least two columns of nodes that will be treated as pairwise interacting pairs, while additional columns convert into edge attributes. Graph class provides an environment specialized to store graphs and their metadata for ease of handling. This class is standard and many other R packages recognize it. Let us call this function and turn our data frame into graphical object G:

   G <- graph_from_data_frame(ppi, directed = FALSE, vertices = NULL)

   The input is ppi data frame containing network, a directed option is set to FALSE since our network is undirected, and the vertices option is kept with its default value, i.e., NULL because we do not want to import node-associated information as of now.

5. igraph has many functions to deal with data stored in G. For a start, let us use a summary function from the igraph package to see what is stored in G.

   summary(G)

   The network summary consists of: a header line that begins with the word IGRAPH; then a randomly generated graph name (not important); followed by a four-character code, where the first character indicates the directionality of the graph by either U or D for undirected or directed network, respectively, the second character is N or hyphen (-), which indicates if the network has a *name* attribute for nodes, the third character is W or hyphen (-), which indicates if the network edges are weighted, and the fourth character is B or hyphen (-), which indicates if nodes have a *type* attribute, which is usually true for bipartite networks, then the number of nodes and edges in the network. The next line shows attributes associated with all the nodes (v) or edges (e) in the network. Our graph shows IGRAPH 325552c UN-- 47 175 – on a header line, which means the G graph is undirected (U in UN--), has a name attribute (N in UN--), and has 47 nodes and 175 edges, and the next line shows + attr: name (v/c), which means only nodes have information associated with them in the name attribute, i.e., name (v/c), which is by default names of all the nodes present in the network.

6. Nodes and edges can be accessed with functions V and E as shown below:

   V(G)

   E(G)

7. The variable G has minimal information required to perform network analysis. Information associated with nodes, i.e., categorical or numerical, can also be added to the graph. For example, we have categorized proteins into different functional categories, which is present in another file called toyNetNode.csv. Let us read this file and have a look at its contents:

   metadata <- read.csv("~/Downloads/toyNetNode.csv", header = TRUE)

   head(metadata)

   In the metadata variable, the first column contains all the protein names present in the ppi data.frame, and the second column contains annotation of them in various functional categories.

8. Let us convert ppi and metadata together into a G object, and check out a summary:

   G <- graph_from_data_frame(ppi, directed = FALSE, vertices = metadata)

   summary(G)

9. There is a new node attribute called Annotation (v/c), which was added from metadata data.frame. We can see its contents with the function vertex_attr:

   vertex_attr(G, name = "Annotation")

   vertex_attr(G)

   The output of the first command was a vector of annotations assigned to nodes because we are particularly interested to see values assigned to the Annotation attribute. The second command shows all attributes. If you observe carefully, they are separated by $Name and $Annotation. R has different classes to store different kinds of data. The output of the vertex_attr(G) command is called a list, which is another fundamental data class in R for storing multiple variables of different classes or different dimensions in one place. In this case, all the attributes associated with nodes are stored in G as a list.

10. Before we perform network analysis, let's remove self-loops and multiple edges between the same nodes.

    G <- simplify(G)

11. Now the graph is ready for analysis.

3.4 Computing Centrality Indices

1. Computing network properties is intuitive and straightforward and can be directly incorporated in the graph object.

   Let us compute degree centrality by the degree function provided in the igraph package:

   ```
   d <- degree(G, mode = "all")
   ```

   As you have noticed, the G variable is often the first input parameter for igraph functions. The mode parameter is set to all because our graph is undirected, hence we do not distinguish incoming edges from the outgoing edges of the nodes. The degree function computes the total number of edges of each node in the network and stores it as a vector in the variable d (*see* **Note 3** for troubleshooting).

2. Let's see the first few entries in variable d.

   ```
   head(d)
   ```

   The number assigned to each node name is its degree.

3. It must be noted that the order of the nodes in the output of the degree function is the same as they are present in graph G or as an attribute. So, the output can be appended in graph G as an additional node attribute using set_vertex_attr function:

   ```
   G <- set_vertex_attr(graph = G, name = "degree_centrality",
     value = d)
   ```

   Let us break down the above command into parts. set_vertex_attr is an igraph function to create or modify node attributes in the graph. It needs three inputs: a graph, i.e., G in the above example; the name for the new node attribute, i.e., degree_centrality; and the value that would be assigned to the newly created degree_centrality attribute, i.e., the variable d. Then the "G <-" updates the original graph with the degree_centrality node attribute in addition to its original contents.

4. The newly added attribute can be verified by command summary(G) and values can be accessed by vertex_attr as aforementioned and shown below:

   ```
   vertex_attr(graph = G, name = "degree_centrality")
   ```

5. R often provides multiple ways to handle operations. For instance, all the above operations can be performed with a single line of code by taking advantage of the graph data structure.

   ```
   V(G)$degree_centrality <- degree(G, mode = "all")
   ```

6. The edge attributes can also be incorporated into the graph in similar fashion (*see* **Note 4**).

7. Many functions work in a similar fashion with little bit of variations in their parameter settings, and their names are intuitive. Let's incorporate some other centrality indices to the graph.

   V(G)$closeness_centrality <- closeness(G, mode = "all")

   V(G)$betweenness_centrality <- betweenness(G, directed = FALSE, normalized = FALSE)

   In the betweenness function, the directedness of the graph must be explicitly mentioned, which I set to FALSE since the network is undirected, and we do not want normalized betweenness scores. Likewise, the clustering coefficient, another centrality metric, can be calculated by using the transitivity function as shown below:

   V(G)$clustering_coefficient <- transitivity(G, type = "local")

   Transitivity function by default computes the clustering coefficient for the entire network, which can be tuned for each node by setting type to local. The clustering coefficient of a node measures the extent of connectivity among its neighboring nodes. Nodes that are part of protein complexes often exhibit high clustering coefficients.

8. Many centrality indices in the igraph return a little bit different data structure of the output. For example, eigen_centrality or hub_score functions return a list of variables in the output as shown below:

   eigen_centrality(G)

9. As explained earlier, the list is a specialized data structure, which can store multiple variables irrespective of their different classes or dimensions. The centrality scores are present in the variable named vector, which can be accessed by:

   eigen_centrality(G)$vector

   The output can be assigned to graph G as a node attribute as shown below:

   V(G)$eigen_centrality <- eigen_centrality(G)$vector

10. Another important centrality function is hub_score, which indicates whether nodes have a propensity to act as hubs in the network.

    V(G)$hub_centrality <- hub_score(G)$vector

11. Recently developed packages have many more centrality indices. Hereafter, we will explore the CINNA package for more centrality metrics, but readers are encouraged to explore some other excellent packages mentioned in the introduction.

12. The CINNA library has a function that finds out the centrality indices applicable to the given graph with the following command:

    proper_centralities(G)

13. With a single function, users can compute all centralities suggested by the above function by feeding them to the calculate_centralities function with the include parameter, as shown below (*see* **Note 5**):

    calc_cent <- calculate_centralities(G, include = proper_centralities(G))

    calc_cent

    The calculate_centralities function returns centrality scores computed for the vector of indices obtained from the proper_centralities function. The calc_cent variable stores computed scores in the form of list.

14. Observe that some centrality indices return NULL values because those cannot be computed for the given network. Let's remove them from the calc_cent variable by the following command:

    calc_cent <- calc_cent[!sapply(calc_cent, is.null)]

15. Now, let us figure out the order of the most important centrality indices based on the structure of a given graph using the pca_centralities function as follows:

    pca_centralities(calc_cent)

    The pca_centralities function applies principal component analysis on the computed centrality values [13] and creates a barplot depicting the order of most important centrality indices for our graph. From the barplot, it appears that the closeness centrality variants and harmonic centrality could be important centrality measurements.

16. It is possible to include only centrality metrics of our interest by using the include parameter. For example:

    calculate_centralities(G, include = c("Kleinberg\'s hub centrality scores", "Harmonic Centrality"))

    The output can be stored in a variable with an appropriate name. Observe that I use the apostrophe punctuation mark with s in the word Kleinberg because apostrophe has a special meaning in R programs; hence, I must explicitly get around the default behavior of R by escaping this metacharacter with backslash (\).

3.5 Exporting Centrality Indices to a File

1. As aforementioned, the output of centrality metrics is stored as node attributes in the graph as a list data class of R. The output of the centrality function from CINNA package is also a list.
2. A list can be transformed into data.frame to export in tabular or Excel format. The following command transforms node attributes stored in the graph object G to data.frame:

   d1 <- as.data.frame(sapply(vertex_attr(graph = G), as.vector))
3. The output of centrality measured by the CINNA package needs a little bit of different syntax to transform the list to data.frame:

   d2 <- as.data.frame(sapply(calc_cent, as.vector))
4. The resulting data frames can be exported to one of the several output formats available in base R or external packages.
5. Let's export d1 data frame to Excel sheet with write_xlsx function from writexl package:

   write_xlsx(d1, path = "centrality_table.xlsx")

   The output file centrality_table.xlsx will be saved to the current working directory (*see* **Note 6**).

   To export to a tabular flat file or text file:

   write.table(d1, file = "centrality_table.txt", col.names = T, row.names = F, quote = F, sep = "\t")
6. The same commands can be adapted to save d2 data.frame

4 Notes

1. Sometimes install.packages function throws an error if there is a problem in downloading the package directly from the Internet due to a weak connection. In such circumstances, the best way is to download the source code of the package from the package repository and install the downloaded file as explained in the Sect. 3.1, **step 2**. If this also does not work, then wget software to automatically download data from the web can be installed and used with the -c option to download the file and then install it locally.
2. R provides several functions to read almost all types of files. Like read.csv, read.table function can be used to read flat files or text files. Microsoft Excel files can be read by using excellent functions from readxl package. Users need to install readxl package in R.
3. Sometimes, the double quote around the "all" input parameter may not be recognized and hence should be typed for

troubleshooting and the same thing must be done for all commands that use input parameters within double quotes.

4. Edge attribute can be added to graph G using function E(graph name). For example, let us say we have a variable called edge_-colors, containing color names for all the edges present in graph G. We can add this attribute to the graph with E(G) $attribute_color <- edge_colors command. This command will create a vector of name attribute_color in graph G, which will have color names corresponding to all the edges in graph G.
5. calculate_centralities function by default computes all types of centrality measures. Some centrality calculations have high computational costs and memory requirements and hence should be used cautiously. It is always best to compute centralities that are relevant to the research problem than computing all of them.
6. Many new users confuse about the current working directory or do not easily find saved files. "getwd()" command displays your current working directory and setwd() function can be used to set the current working directory from the available directories.

References

1. Muley VY, Acharya V (2013) Genome-wide prediction and analysis of protein-protein functional linkages in bacteria. Springer, New York
2. Snider J, Kotlyar M, Saraon P et al (2015) Fundamentals of protein interaction network mapping. Mol Syst Biol 11:848. https://doi.org/10.15252/msb.20156351
3. Muley VY (2012) Improved computational prediction and analysis of protein – protein interaction networks. PhD thesis, Centre for DNA Fingerprinting and Diagnostics, Hyderabad, Manipal University, India. http://hdl.handle.net/10603/5399
4. Barabási AL, Oltvai ZN (2004) Network biology: understanding the cell's functional organization. Nat Rev Genet 5(2):101–113. https://doi.org/10.1038/nrg1272
5. Freeman LC (1978) Centrality in social networks conceptual clarification. Soc Networks 1:215–239. https://doi.org/10.1016/0378-8733(78)90021-7
6. Jalili M, Salehzadeh-Yazdi A, Gupta S et al (2016) Evolution of centrality measurements for the detection of essential proteins in biological networks. Front Physiol 7:375. https://doi.org/10.3389/FPHYS.2016.00375/XML/NLM
7. Jalili M, Salehzadeh-Yazdi A, Asgari Y et al (2015) CentiServer: a comprehensive resource, web-based application and R package for centrality analysis. PLoS One 10:e0143111. https://doi.org/10.1371/JOURNAL.PONE.0143111
8. Freeman LC (1977) A set of measures of centrality based on betweenness. Sociometry 40:35. https://doi.org/10.2307/3033543
9. Sabidussi G (1966) The centrality index of a graph. Psychometrika 31:581–603. https://doi.org/10.1007/BF02289527
10. Butts CT (2008) Social network analysis with sna. J Stat Softw 24:1–51. https://doi.org/10.18637/jss.v024.i06
11. Schoch D (2018) Centrality without indices: partial rankings and rank probabilities in networks. Soc Networks 54:50–60. https://doi.org/10.1016/j.socnet.2017.12.003
12. Schoch D, Brandes U (2016) Re-conceptualizing centrality in social networks. Eur J Appl Math 27:971–985. https://doi.org/10.1017/S0956792516000401
13. Ashtiani M, Mirzaie M, Jafari M (2019) CINNA: an R/CRAN package to decipher Central Informative Nodes in Network Analysis. Bioinformatics 35(8):1436–1437. https://doi.org/10.1093/bioinformatics/bty819

Chapter 35

Protein–Protein Interaction Network Analysis Using NetworkX

Mehadi Hasan, Nilesh Kumar, Aqsa Majeed, Aftab Ahmad, and Shahid Mukhtar

Abstract

In recent years, extracting information from biological data has become a particularly valuable way of gaining knowledge. Molecular interaction networks provide a framework for visualizing cellular processes, but their complexity frequently makes their interpretation difficult. Proteins are one of the primary determinants of biological function. Indeed, most biological activities in the living cells are functionally regulated by protein–protein interactions (PPIs). Thus, studying protein interactions is critical for understanding their roles within the cell. Exploring the PPI networks can open new avenues for future experimental studies and offer interspecies predictions for effective interaction mapping. In this chapter we will demonstrate how to construct, visualize, and analyze a protein–protein interaction network using NetworkX.

Key words Protein–protein interaction, Systems biology, NetworkX, Centrality, Regulatory networks

1 Introduction

The universality of a network's organizational and physical properties across diverse systems provides an opportunity to understand the interaction between a system and its environment [1–11]. Systems Biology is a method that examines intricate biological systems by extensively quantifying a wide range of biomolecules [12–18] at the level of the organism, tissue, and cell [19]. The other purpose of Systems Biology is to identify the interacting systems and their "emergent properties" involved in a biological process. Moreover, a group of proteins and their connections control how biological activities are carried out [20–22]. These interactions are represented as a network called protein–protein interaction (PPI) network (Fig. 1). One of the biological networks that are critical to practically every function in a cell is the network of protein–protein interactions. These interactions, which are controlled by gene

Shahid Mukhtar (ed.), *Protein-Protein Interactions: Methods and Protocols*, Methods in Molecular Biology, vol. 2690, https://doi.org/10.1007/978-1-0716-3327-4_35,

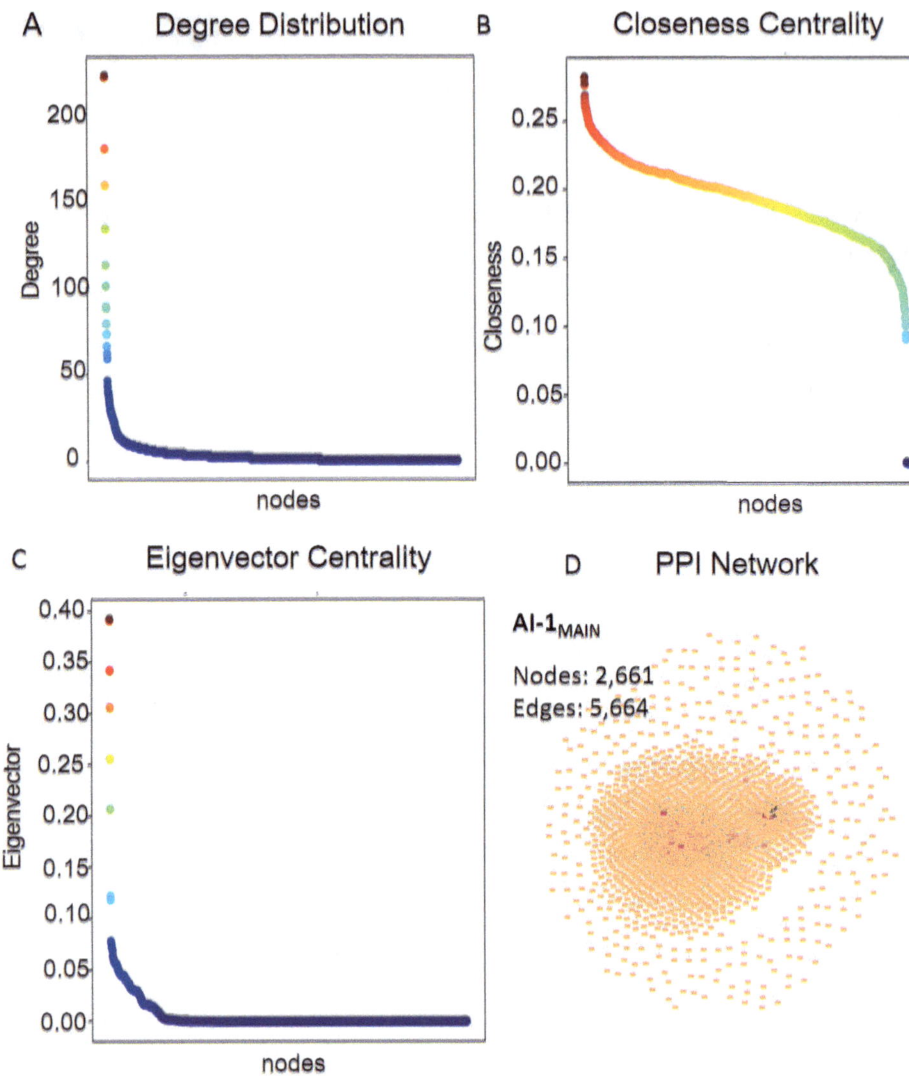

Fig. 1 Protein–protein interaction network and centrality distribution analysis: (**a**) degree distribution; (**b**) closeness centrality; (**c**) eigenvector centrality; and (**d**) PPI network (AI-1_{MAIN} 2661 nodes and 5664 edges) with high-degree nodes in red

expression, involve interactions between proteins and DNA, non-coding RNA and RNA, proteins and proteins, and metabolites and metabolites [8]. Depending on the connections involved, a network's connectedness, complexity, and structure will vary, and edges and nodes (such as proteins or genes in biological networks) may be able to transmit many layers of information. Moreover, protein–protein interaction networks have been utilized to uncover functional modules [23] and conserved interaction patterns and discover novel protein function [24]. Based on network interactions, they can be either directed or undirected [8]. A network is "directed" [8] where the edges have flow direction; that is, they are ordered pairs of vertices (e.g., gene regulatory network), and the direction can be one way (sometimes a two-way direction is also possible). A network is "undirected" [8] where none of the edges has flow direction; they are ordered pairs of vertices (e.g., gene co-expression network). Directed or undirected networks can also have weight, or a measurable value associated with them. This is used to depict concepts such reliability of an interaction.

PPI networks are referred to as "undirected networks" that incorporate actual or anticipated interactions between various proteins to carry out biological functions [8]. PPI networks comprise modules, which are distinct protein–protein interactions that are frequently tailored for a particular phenotype or function [20, 25]. As a result, PPI networks are frequently investigated to find modules—basic protein interactions—that regulate phenotypes and chemical processes.

Furthermore, the use of PPI networks to propose novel protein candidates for molecular functions and phenotypes supported by interactions with known neighbors is another benefit [26, 27]. Although wet lab methods exist for studying novel protein candidates, computational methods for predicting protein function are more rapid, more cost-effective, and less time-consuming than wet lab techniques [28, 29].

How crucial a node or edge is to the connection or information flow of the network is determined by its "centrality" [30]. In general, network topology supports three categories of structural centralities: (i) centralities based on neighborhood (e.g., degree); (ii) centralities based on path (e.g., shortest path length, betweenness, closeness); and (iii) centralities based on iterative refinement (e.g., eigenvector centrality, PageRank). Frequently used centrality properties in PPI networks are degree centrality [30] (complete number of ties a node has), betweenness centrality [30] (number of times a node performs as a link along the shortest path among two other nodes), and closeness centrality [30] (the shortest path distance among a node and other nodes). A sophisticated extension of degree centrality is eigenvector centrality [30]. If a protein's few connections lead to other proteins with many connections, that protein might have a very high eigenvector centrality [30]. A node's importance is measured using eigenvector centrality, which also takes into account the significance of its neighbors [31]. A variation on Eigenvector Centrality [30] is metrics called PageRank [30], which reveals nodes with influence over the network as a whole that goes beyond their immediate connections. Links can only carry influence in one direction and in various degrees since PageRank also takes into account the weight and direction of links. PPI networks make it possible to comprehend how PPIs, submodules, communication between those submodules, and hub proteins are all organized globally [32, 33].

A structural power-law distribution and the fact that just a small number of nodes are connected to virtually all of the network's nodes are revealed by the scale-free network topology found in the majority of biological interactomes previously investigated [34]. Broido and Clauset [35] recently showed in a thorough data-centric analysis that not all networks adhere to scale-free architecture. In contrast to social networks, the study found that the majority of biological and technological networks have the strongest scale freedom.

These densely interconnected proteins, known as "hubs" [14], are therefore frequently significant pharmacological targets in human-disease studies [29] or genomic engineering targets in crop upgrading [36], while nodes that have high betweenness centrality are referred to as "bottlenecks," which are crucial because these nodes act as links between two subnetworks [37] while traversing the entire network.

Furthermore, for protein network analysis multiple R/python packages can be used. Researchers in network science can benefit from practical tools provided by the igraph [38] R package. Cytoscape [39] is an open-source software project that integrates and visualizes biomolecular interaction networks, high-throughput gene-expression data, and other molecular states into an integrated conceptual framework. Gephi [40] is a network analysis and visualization program, which uses a three-dimensional (3D) render engine to speed up research and show massive networks in real time.

Here, we used NetworkX [41] for analyzing protein–protein interaction network. NetworkX [41] is used to create, manipulate, and research the composition of multilayered networks. It's used to investigate large networks that are represented as graphs with nodes and edges. We can also generate a wide variety of random and classic networks, as well as analyze network structure and construct network models.

2 Materials

Unlike many other tools, NetworkX is designed to handle data on a scale relevant to modern problems. We can export protein–protein network data from NetworkX and visualize with other programs, for example, Cytoscape.

2.1 Installation of NetworkX (Windows, Linux)

NetworkX can be set up on a standard UNIX (Linux or macOS) and Windows. Download and run the latest version of Python installer (.exe extension) for Windows. First, you need to install Python (*see* **Note 1**). You can install pip on Windows (*see* **Note 2**), also in Linux (*see* **Note 3**), and then using pip you can install NetworkX on Windows (*see* **Notes 4** and **5**). Additionally, it is recommended to use an integrated development environment (IDE) such as Jupyter Notebook for ease of executing Python scripts (*see* **Note 6**).

2.2 Install NetworkX on Linux

Open command prompt/terminal and execute

```
$ pip install network
```

2.3 Use NetworkX in Python (Jupyter Notebook)

```
import networkx as nx
```

3 Methods

This section discusses the basic steps to use Python package ***NetworkX*** for protein–protein interaction network analysis.

3.1 Use Network Datasets in NetworkX for Protein–Protein Interaction Analysis

A comprehensive understanding of the complex cellular networks that underpin plant-specific functions is still lacking. Here, we employed *Arabidopsis thaliana*'s experimental PPI network **AI-1$_{\text{MAIN}}$** [17] that can be directly downloaded from the web (*see* **Note 7**).

The workflow has four steps: (1) importing the data; (2) creating the network; (3) network centrality analysis; and (4) exporting network to use in other tools.

3.2 Import Required Python Libraries

```
import networkx as nx
import pandas as pd
import matplotlib.colors as colors
import matplotlib.pyplot as plt
import numpy as np
```

3.2.1 Load the Data

```
df = pd.read_csv("AI1_MAIN_PPI.txt",delimi-
ter="\t")
df.head(5)
```

3.2.2 Make the Network Graph from AI-1$_{MAIN}$ Dataset

In a network file, there should be a source column and a target column. In AI-1$_{\text{MAIN}}$ dataset, the source column is "ida" and target column is "idb."

Execute the following code and check nodes and edge numbers (*see* Fig. 1d).

```
G_AI1main = nx.from_pandas_edgelist(df, source='ida', targe-
t='idb')
nx.draw(G_AI1main)


print("Number of nodes =", G_AI1main.number_of_nodes())
print("Number of edges =", G_AI1main.number_of_edges())
```

3.2.3 Network Centrality Analysis

To calculate betweenness centrality, execute the following code.

```
bc = nx.betweenness_centrality(G_AI1main)
nx.set_node_attributes(G_AI1main, bc, "Betweenness
centrality")
```

Execute the following command and get degree centrality.

```
dc = nx.degree_centrality(G_AI1main)
nx.set_node_attributes(G_AI1main, dc, "Degree cen-
trality")
d = dict(nx.degree(G_AI1main))
nx.set_node_attributes(G_AI1main, d, "Degree")
```

Similarly, get eigenvector centrality by executing the following command.

```
ec = nx.eigenvector_centrality(G_AI1main)
nx.set_node_attributes(G_AI1main, ec, "Eigenvector
centrality")
```

Get closeness centrality.

```
cc = nx.closeness_centrality(G_AI1main)
nx.set_node_attributes(G_AI1main, cc, "Closeness
centrality")
```

For example, you can check the centrality information of any specific [AT1G09415] node.

```
print(pd.DataFrame(G_AI1main.nodes["AT1G09415"].
items()))
```

To get centrality attribute to dataframe, execute the following command.

```
AI1main_centrality = pd.DataFrame.from_dict(dict
(G_AI1main.nodes(data=True)), orient='index')
AI1main_centrality.index.name = 'ID'
```

```
AI1main_centrality.to_csv("AI1main_centrality.
csv", sep=",")
AI1main_centrality.head()
```

For plots distribution, use the following Python function "Distribution."

```
def Distribution(df, v=0, dpi=150):
 fig, ax = plt.subplots(1,1, figsize=(6,6))
 Colname = list(df.columns)
 max_v = df.loc[df[Colname[0]].idxmax()]
 max_v = max_v[0]
 if v:
 print(f"max_value:{max_v}")
 #X-axis
 x = np.arange(1,len(df)+1, 1)
 #Y-axis
 y = df[Colname[0]].tolist()
 y.sort(reverse=True)
 if v:
 print(y)
 tag = y
 if v:
 print(tag)
 cmap = plt.cm.jet
 cmaplist = [cmap(i) for i in range(cmap.N)]
 cmap = cmap.from_list('Custom cmap', cmaplist,
cmap.N)
 bounds = np.linspace(1,0.05*len(y),21)
 norm = colors.BoundaryNorm(bounds, cmap.N)
 scat = ax.scatter(x,y,c=tag ,cmap=cmap)#, norm=-
norm)
 n = df[Colname[0]]
 n = n.tolist()
 I = [i for i in range(len(y))]
 for i in I[-5:]:
 txt = n[i]
 if v:
 print(i,txt)
 ax.set_title(Colname[0]+' distrubtion')
 plt.xticks([])
 plt.savefig(Colname[0]+".png", dpi=dpi)
```

For degree distribution (*see* Fig. 1a), execute the following code.

```
Distribution(AI1main_centrality[['Degree']],
dpi=300)
```

Show the degree centrality distribution.

```
Distribution(AI1main_centrality[['Degree central-
ity']], dpi=300)
```

For closeness centrality distribution (*see* Fig. 1b), execute the following code.

```
Distribution(AI1main_centrality[['Closeness cen-
trality']], dpi=300)
```

For eigenvector centrality distribution (*see* Fig. 1c), execute the following code.

```
Distribution(AI1main_centrality[['Eigenvector
centrality']], dpi=300)
```

3.2.4 Exporting the Network

Export network to use in other tools (e.g., Cytoscape)

```
nx.write_edgelist(G_AI1main, "AI1_MAIN_PPI.csv",
delimiter=",", data=False)
nx.write_graphml_lxml(G_AI1main, "AI1_MAIN_PPI.
graphml")
```

4 Notes

1. Install Python on Windows Using Conda

   Execute the in-command line.

```
$ conda install -c conda-forge python
```

2. Install "pip" on Windows

   To manually install "pip" on Windows, you need a copy of **get-pip.py**. Open command prompt and use the following curl command to download python pip script.

```
$ curl https://bootstrap.pypa.io/get-pip.py -o
get-pip.py
```

   Next, run the following command to install pip.

```
$ python get-pip.py
```

Verify that "pip" is installed by running the following command to check the version.

```
$ pip –version
```

3. Install pip on Linux

```
$ python get-pip.py
```

4. Install NetworkX on Windows Using pip

```
$ pip install networkx
```

5. Install NetworkX with All Optional Dependencies (for Linux and Windows)

```
$ pip install networkx[all]
```

6. Install Jupyter Notebook (Linux/Windows)

```
$ pip install notebook
```

Run Jupyter notebook.

```
$ jupyter notebook
```

7. Download AI-1_{MAIN} Network Dataset from the Web

```
Network=pd.read_excel("http://interactome.dfci.harvard.edu/A_thaliana/doc/AI_interactions.xls")

Network = Network[Network.main_screen == 1]
```

Acknowledgments

This research was funded by the National Science Foundation (IOS-2038872).

References

1. Wessling R, Epple P, Altmann S et al (2014) Convergent targeting of a common host protein-network by pathogen effectors from three kingdoms of life. Cell Host Microbe 16(3):364–375. https://doi.org/10.1016/j.chom.2014.08.004
2. Smakowska-Luzan E, Mott GA, Parys K et al (2018) An extracellular network of Arabidopsis leucine-rich repeat receptor kinases. Nature 553(7688):342–346. https://doi.org/10.1038/nature25184

3. Mukhtar MS, Carvunis AR, Dreze M et al (2011) Independently evolved virulence effectors converge onto hubs in a plant immune system network. Science 333(6042):596–601. https://doi.org/10.1126/science.1203659
4. Mott GA, Smakowska-Luzan E, Pasha A et al (2019) Map of physical interactions between extracellular domains of Arabidopsis leucine-rich repeat receptor kinases. Sci Data 6: 190025. https://doi.org/10.1038/sdata.2019.25
5. Mishra B, Sun Y, Howton TC et al (2018) Dynamic modeling of transcriptional gene regulatory network uncovers distinct pathways during the onset of Arabidopsis leaf senescence. NPJ Syst Biol Appl 4:35. https://doi.org/10.1038/s41540-018-0071-2
6. Mishra B, Sun Y, Ahmed H et al (2017) Global temporal dynamic landscape of pathogen-mediated subversion of Arabidopsis innate immunity. Sci Rep 7(1):7849. https://doi.org/10.1038/s41598-017-08073-z
7. Mishra B, Kumar N, Shahid Mukhtar M (2022) A rice protein interaction network reveals high centrality nodes and candidate pathogen effector targets. Comput Struct Biotechnol J 20:2001–2012. https://doi.org/10.1016/j.csbj.2022.04.027
8. Mishra B, Kumar N, Mukhtar MS (2021) Network biology to uncover functional and structural properties of the plant immune system. Curr Opin Plant Biol 62:102057. https://doi.org/10.1016/j.pbi.2021.102057
9. Mishra B, Kumar N, Mukhtar MS (2019) Systems biology and machine learning in plant-pathogen interactions. Mol Plant-Microbe Interact 32(1):45–55. https://doi.org/10.1094/MPMI-08-18-0221-FI
10. McCormack ME, Lopez JA, Crocker TH et al (2016) Making the right connections: network biology and plant immune system dynamics. Curr Plant Biol 5:2–12
11. Lopez J, Mukhtar MS (2017) Mapping protein-protein interaction using high-throughput yeast 2-hybrid. Methods Mol Biol 1610:217–230. https://doi.org/10.1007/978-1-4939-7003-2_14
12. Kumar N, Mishra B, Mukhtar MS (2022) A pipeline of integrating transcriptome and interactome to elucidate central nodes in host-pathogens interactions. STAR Protoc 3(3): 101608. https://doi.org/10.1016/j.xpro.2022.101608
13. Kumar N, Mishra B, Mehmood A et al (2020) Integrative network biology framework elucidates molecular mechanisms of SARS-CoV-2 pathogenesis. iScience 23(9):101526. https://doi.org/10.1016/j.isci.2020.101526
14. Klopffleisch K, Phan N, Augustin K et al (2011) Arabidopsis G-protein interactome reveals connections to cell wall carbohydrates and morphogenesis. Mol Syst Biol 7:532. https://doi.org/10.1038/msb.2011.66
15. Gonzalez-Fuente M, Carrere S, Monachello D et al (2020) EffectorK, a comprehensive resource to mine for Ralstonia, Xanthomonas, and other published effector interactors in the Arabidopsis proteome. Mol Plant Pathol 21(10):1257–1270. https://doi.org/10.1111/mpp.12965
16. Garbutt CC, Bangalore PV, Kannar P et al (2014) Getting to the edge: protein dynamical networks as a new frontier in plant-microbe interactions. Front Plant Sci 5:312. https://doi.org/10.3389/fpls.2014.00312
17. Arabidopsis Interactome Mapping C (2011) Evidence for network evolution in an Arabidopsis interactome map. Science 333(6042): 601–607. https://doi.org/10.1126/science.1203877
18. Ahmed H, Howton TC, Sun Y et al (2018) Network biology discovers pathogen contact points in host protein-protein interactomes. Nat Commun 9(1):2312. https://doi.org/10.1038/s41467-018-04632-8
19. Dhillon BK, Smith M, Baghela A et al (2020) Systems biology approaches to understanding the human immune system. Front Immunol 11:1683. https://doi.org/10.3389/fimmu.2020.01683
20. Shoemaker BA, Panchenko AR (2007) Deciphering protein-protein interactions. Part I. Experimental techniques and databases. PLoS Comput Biol 3(3):e42. https://doi.org/10.1371/journal.pcbi.0030042
21. Spirin V, Mirny LA (2003) Protein complexes and functional modules in molecular networks. Proc Natl Acad Sci U S A 100(21): 12123–12128. https://doi.org/10.1073/pnas.2032324100
22. Wimalagunasekara S, Tirimanne S, Fernando PC (2022) Protein-protein interaction (PPI) network analysis reveals important hub proteins and sub-network modules for root development in rice (*Oryza sativa*). bioRxiv:2022.2006.2006.494990. https://doi.org/10.1101/2022.06.06.494990
23. Fionda V (2011) Biological network analysis and comparison: mining new biological knowledge. Cent Eur J Comput Sci 1(2):185–193. https://doi.org/10.2478/s13537-011-0013-1

24. Fionda V, Palopoli L, Panni S et al (2008) Protein-protein interaction network querying by a "focus and zoom" approach. In: Elloumi M, Küng J, Linial M, Murphy RF, Schneider K, Toma C (eds) Bioinformatics research and development. Springer, Berlin/ Heidelberg, pp 331–346
25. Wei Lan JW, Li M, Peng W, Fangxiang W (2015) Computational approaches for prioritizing candidate disease genes based on PPI networks. Tsinghua Sci Technol 20(5): 500–512. https://doi.org/10.1109/tst.2015.7297749
26. Park S-J, Choi J-S, Kim B-C et al (2009) PutidaNET: Interactome database service and network analysis of *Pseudomonas putida* KT2440. BMC Genomics 10(3):S18. https://doi.org/10.1186/1471-2164-10-S3-S18
27. Sharan R, Ulitsky I, Shamir R (2007) Network-based prediction of protein function. Mol Syst Biol 3:88. https://doi.org/10.1038/msb4100129
28. Kiran M, Nagarajaram HA (2016) Interaction and localization diversities of global and local hubs in human protein–protein interaction networks. Mol BioSyst 12(9):2875–2882. https://doi.org/10.1039/C6MB00104A
29. Apic G, Ignjatovic T, Boyer S et al (2005) Illuminating drug discovery with biological pathways. FEBS Lett 579(8):1872–1877. https://doi.org/10.1016/j.febslet.2005.02.023
30. Hansen D, Shneiderman B, Smith M et al (2020) Social network analysis: measuring, mapping, and modeling collections of connections. In: Analyzing social media networks with NodeXL: insights from a connected world. Elsevier Inc, pp 31–51. https://doi.org/10.1016/B978-0-12-817756-3.00003-0
31. Hansen DL, Shneiderman B, Smith MA et al (2020) Social network analysis: measuring, mapping, and modeling collections of connections. In: Analyzing social media networks with NodeXL, pp 31–51. https://doi.org/10.1016/B978-0-12-817756-3.00003-0
32. Wang X, Gulbahce N, Yu H (2011) Network-based methods for human disease gene prediction. Brief Funct Genomics 10(5):280–293. https://doi.org/10.1093/bfgp/elr024
33. Albert RK (2007) Network inference, analysis, and modeling in systems biology. Plant Cell 19(11):3327–3338. https://doi.org/10.1105/tpc.107.054700
34. Vidal M, Cusick ME, Barabasi AL (2011) Interactome networks and human disease. Cell 144(6):986–998. https://doi.org/10.1016/j.cell.2011.02.016
35. Broido AD, Clauset A (2019) Scale-free networks are rare. Nat Commun 10(1):1017. https://doi.org/10.1038/s41467-019-08746-5
36. Liseron-Monfils C, Ware D (2015) Revealing gene regulation and associations through biological networks. Curr Plant Biol 3. https://doi.org/10.1016/j.cpb.2015.11.001
37. Yu H, Kim PM, Sprecher E et al (2007) The importance of bottlenecks in protein networks: correlation with gene essentiality and expression dynamics. PLoS Comput Biol 3(4):e59. https://doi.org/10.1371/journal.pcbi.0030059
38. Csárdi G, Nepusz T (2006) The igraph software package for complex network research. Int J Complex Syst 1695(5):1–9
39. Shannon P, Markiel A, Ozier O et al (2003) Cytoscape: a software environment for integrated models of biomolecular interaction networks. Genome Res 13(11):2498–2504. https://doi.org/10.1101/gr.1239303
40. Bastian M, Heymann S, Jacomy M (2009) Gephi: an open source software for exploring and manipulating networks. https://doi.org/10.13140/2.1.1341.1520
41. Hagberg A, Swart P, S Chult D (2008) Exploring network structure, dynamics, and function using NetworkX. Los Alamos National Lab (LANL), Los Alamos

Chapter 36

Building Protein–Protein Interaction Graph Database Using Neo4j

Nilesh Kumar and Shahid Mukhtar

Abstract

A cell's various components interact with each other in a coordinated manner to respond to environmental cues and intracellular signals. Compared to the other biological networks, the protein–protein interaction (PPI) is mostly responsible for maintaining signaling pathways. Increasing numbers of experimentally verified and predicted PPIs in plants demand a scalable platform to deal with large and complex datasets. Network/graph data can be organized and analyzed using different tools. This chapter uses Neo4j, a graph database management system, to store and analyze plant PPI networks. To make the graph database and analyze network centrality, we used Arabidopsis interactome-1 main (AI-1_{MAIN}) PPI network.

Key words Graph databases, Protein–protein interactions, Network biology, Centrality analysis, Neo4j, Cypher

1 Introduction

Protein–protein interactions (PPIs) play a vital role in several biological functions and components, including signal transduction, transport, transcription, and adhesion [1, 2]. Using nodes and edges, these networks display how the interactions of proteins occur [3–10]. In a PPI network, the nodes represent proteins, while the interactions between proteins are edges [11–20]. Various approaches can be used to characterize new PPIs, each with its advantages and limitations [2]. Key methods to characterize PPIs include affinity purification-mass spectrometry (AP-MS), yeast two-hybrid system (Y2H), and proximity-dependent biotinylation identification (BioID) [2, 21]. PPlevo [22], Struct2Net [23], ClusPro [24], iLoops [25], BIANA (Biologic Interactions and Network Analysis) Interolog Prediction Server (BIPS) [26], and Consensus Protein–Protein Interaction Site Predictor (cons-PPISP [27] are some of the computational approaches used for PPI inference.

Shahid Mukhtar (ed.), *Protein-Protein Interactions: Methods and Protocols*, Methods in Molecular Biology, vol. 2690,
https://doi.org/10.1007/978-1-0716-3327-4_36,

PPI network data are dispersed across multiple structured and unstructured databases with inconsistent interaction quality [28]. In terms of known PPI (physical) in plants, 81,453 PPIs for *Arabidopsis thaliana* and 346 PPIs for *Oryza sativa* have been included in BioGRID (4.4.211) [29]. Other popular source of plant PPI data is the STRING (11.5) [30] and PlaPPISite [31].

Additionally PPI topological or centrality analysis provides a view of the network as a whole and identifies the key components that make up the network [9–11, 15, 32]. Centrality measures are used to measure how central a given protein is within a PPI network [33]. Three types of centralities can be classified according to the topological features of a network: (i) path-based centralities, (ii) neighborhood-based centralities, and (iii) iterative refinement centralities [11]. The degree centrality, which measures how many proteins interact with each other, is among the most straightforward centralities. The betweenness centrality of a protein is determined by how many shortest paths pass through it. Protein's transitive influence on the PPI can be measured by the eigenvector centrality [34–37].

Different software platforms offer a variety of graph analysis tools. Examples include Python (NetworkX [38], igraph [39], graph-tool [40], etc.), R (igraph [39], tnet [41], etc.), Java (Cytoscape [42], Gephi [43], CentiBiN [44], etc.), C (igraph [39], GraphChi [45], etc.), a Graph database (Neo4j [37] and AllegroGraph [46]), etc. [47, 48]. For the graph analysis, NetworkX [38] (5 k + all citations), igraph [39] (~10 k all citations), and Cytoscape [42] (~30 k all citations) are the most popular choices among systems biologists. However, the amount of data generated by biological networks is increasing rapidly with the advent of high-throughput technology and so is the PPI data. A platform that can analyze graph data and provide a graph database management system (GDMS) is required to store and analyze an enormous amount of data efficiently. In comparison to NetworkX [38], Cytoscape [42], and igraph [39], GDMS like Neo4j [37] and AllegroGraph [46] can be used for graph analysis and management of graph data, such as PPI. Using Neo4j [37], this chapter describes how to create a PPI graph database and perform basic centrality analysis. To create graph databases and centrality analysis using the Neo4j, Arabidopsis interactome-1 main (AI-1_{MAIN}) [19] PPI network is used. When retrieving data from Neo4j's graph, the query language Cypher is used.

2 Materials

Data can be downloaded from databases (BioGRID [29, 49], STRING (Search Tool for Recurring Instances of Neighbouring Genes) [30, 50, 51], IntAct [52–54], etc.), publications (supplementary data), or custom data from any experiment. In this chapter, we will use the AI-1_{MAIN} PPI network [19]. Before building the Neo4j databases, all data needed to be preprocessed (*see* **Note 1**). The preprocessing can be done manually or using a scripting language such as Python, Perl, etc. (*see* **Note 2**). Since PPIs come from different sources, they are likely to have different formats for protein identifiers (such as UniProt-KB, gene names, protein names, STRING IDs, etc.). All these varying protein identifiers needed to be converted into a single common identifier. In addition to PPI files, we need annotation files such as GO annotation (TAIR) or any other desired information sheet. Download the required data from the following URL.

2.1 Plant Interactome Dataset

AI-1MAIN PPI data

http://interactome.dfci.harvard.edu/A_thaliana/index.php?page=download, http://interactome.dfci.harvard.edu/A_thaliana/doc/AI_interactions.xls

2.2 GO Annotation (TAIR)

https://www.arabidopsis.org/download_files/GO_and_PO_Annotations/Gene_Ontology_Annotations/gene_association.tair.gz

2.3 Hardware and Software

Most UNIX and Linux distributions, including Ubuntu 22 and Fedora 36, Red Hat 7, as well as macOS Monterey, are compatible with the Neo4j. This procedure has been tested on a device that meets the following specifications:

1. OS: Windows 11
2. RAM: 16 GB
3. SDD: 256 GB
4. CPU: Intel i7
5. conda 4.12.0
6. Java JDK versions 11
7. Neo4j (1.4.15)
8. APOC (4.4.0.6)
9. Graph Data Science Library (2.0.3)

3 Methods

This section is divided into two parts: (i) preprocessing, and (ii) building a graph using the Neo4j desktop browser application. Neo4j executes database operations using Cypher, a query language. The main objective of this chapter is to describe Neo4j usage with minimal coding. Due to this, data will be imported and exported using the Awesome Procedures On Cypher (APOC) library.

3.1 Preprocessing of PPI and Annotation Data

Preprocess PPI and annotation data to be able to create the PPI graph databases (*see* **Note 2**). Place interaction network (say AI1main_PPI_Interactions_neo.csv) and GO annotation (TAIR [55]) files (say PROTEIN_neo.csv) in a directory (say PPI_Project).

3.2 Neo4j Download and Installation Setup

1. Download Neo4j Desktop from the Neo4j download center for Mac, Linux, and Windows (https://neo4j.com/download-center/#desktop). Open the executable file and follow the on-screen instructions to complete the installation and activation (Windows 11) (*see* **Note 3**).
2. Launch the Neo4j desktop application after the installation and activation have been completed. Click File -> Create project from directory -> <navigate and select PPI_Project folder>.
3. After the successful creation of the project ("PPI_Project"), click on add button > Local DBMS, to create the local databases. Choose a name (say PPI_DBMS) and password (say 123456***) for the database. In the version dropdown menu, select version 4.4.5 and click Create.
4. Once the database (i.e., PPI_DBMS) is created, click on PPI_DBMS (left) to reveal the additional option and information about the databases. In the "Plugins" option, select "APOC" and "Graph Data Science Library" and install them accordingly. Further configuration is required before using the APOC library (*see* **Note 4**).
5. All CSV (comma-separated values) files needed to import in Neo4j should be copied to the import directory (*see* **Note 5**).
6. Open "Neo4j Browser" by clicking start then open (Windows 11). In the Neo4j Browser, execute the following Cypher command to check if there are any data already stored if reusing a previously created database.

```
MATCH (n) return n)
```

In cases the database is not empty, delete all nodes and edges using the following command.

```
MATCH (n) DETACH DELETE n
```

3.3 Import Network and Annotation Data in Neo4j Database

1. Use the following Cypher script to load the network and attribute data into Neo4j.

```
CALL apoc.import.csv(
 [{fileName: 'PORTEIN_neo.csv', labels: ['Pro-
tein']}],
 [{fileName: 'AI1main_PPI_Interactions_neo.
csv', type: 'INTERACTS'}],
 {delimiter: '|', arrayDelimiter: ',', stringIds:
true})
```

2. After the successful execution, 2661 nodes and 5664 relationships/edges should be added to the databases. To print the total number of nodes and edges, use the following Cypher command, respectively.

```
MATCH (n:Protein) RETURN count(n) as Total_Proteins
MATCH ()-[r:INTERACTS]->() RETURN count(r) as To-
tal_Protein_Interactions.
```

3. To display the network or print data in a tabular format, use the following Cypher command.

```
MATCH (n) return n
```

4. Click the download button on the extreme right of the command window to download the snapshot of the network. By default, not all nodes are displayed if the network is large (i.e., 300 nodes only).

3.4 Graph Centrality Analysis

In the Neo4j Desktop application, click on the dropdown menu button on the open button and select the “Graph Data Science” option to launch the “The Graph Data Science Playground.” Keep all the values in the Graph Data Science Playground window, then click “connect” and then “select databases.” In the NEuler window, click select “Run single algorithm” card.

1. Choose “Degree” for the algorithm option in the next window, i.e., “1. Configure.”
2. Change “Any” to Protein for “Label,” “Any” to INTERACTS for “Relationship Type,” and set “Undirected” for “Relationship Orientation.”
3. To save the results in the database, click the “Store results?” radio button and then click “Run Algorithm.”
4. After successful execution in the “2. Result” window, results will be displayed as a table, chart, and visualization (*see* **Note 6**).

5. Using the "New Algorithm" button, select another algorithm (i.e., Betweenness and Eigen centrality) and run the analysis as explained before.

3.5 Export Analyzed Data

To export and save all results as a CSV file, use the following Cypher command in Neo4j browser.

```
CALL apoc.export.cypher.all("AI1_main.cypher")
YIELD file, batches, source, format, nodes, rela-
tionships, properties, time, rows, batchSizeRETURN
file, batches, source, format, nodes, relation-
ships, properties, time, rows, batchSize;
```

4 Notes

1. Among the multiple ways via which data can be imported and exported, it is recommended to use the Awesome Procedures On Cypher (APOC) add-on library for Neo4j since it requires the least Cypher command and is flexible. Other than CSV, APOC also supports other formats such as JSON, GraphMl, XML, HTML, etc. To be able to import a CSV file, header need to be included in the CSV file with a specific delimiter. Below are the top few lines of data used in this chapter (*see* **Note 2**, for preprocessing raw data using python).

```
Protein annotations file:PROTEIN_neo.csv
:ID|name:STRING|GO:STRING[]
AT4G28300|FLOE1|GO:0003674,GO...
AT5G25220|KNAT3|GO:0000978,GO...
PPI file: AI1main_PPI_Interactions_neo.csv
:START_ID|:END_ID
AT1G09415|AT1G64280
AT3G61060|AT5G42190
```

2. Using Python script, preprocess data according to APOC requirements. To install the pandas library in Python, simply use the pip package manager or another alternative method based on the operating system and computing environment.

```
> pip install pandas
Import the required Python libraries.
import pandas as pd
from collections import defaultdict
# Load the gene association data as a pandas data-
frame. Make a dictionary of AGI (Arabidopsis Genome
```

Initiative, unique identifier), gene names, and GO terms using the following script.

```
gene_association = pd.read_csv("https://www.arabi-
dopsis.org/download_files/GO_and_PO_Annotations/
Gene_Ontology_Annotations/gene_association.tair.
gz", sep="\t", header=None, comment="!")
gene_association=gene_association[[1,2,4]].va-
lues.tolist()
AGI_name=dict()
AGI_GO=defaultdict(list)
for i in gene_association:
 AGI, Name, GO = i
 AGI_name[AGI] = Name
 AGI_GO[AGI].append(GO)

# Load interaction data as a pandas dataframe. Keep
only "main_screen" interactions.
Network=pd.read_excel("http://interactome.dfci.
harvard.edu/A_thaliana/doc/AI_interactions.xls")
Network = Network[Network.main_screen == 1]
print(Network.shape)
Network = Network[["ida", 'idb']]
AGI_keep = []
for a,b in Network.values.tolist():
 AGI_keep.append(a)
 AGI_keep.append(b)
AGI_keep = set(AGI_keep)

# Write the protein info to the PROTEIN_neo.csv
file.
fh = open("PROTEIN_neo.csv", "w")
print(":ID|name:STRING|GO:STRING[]", file=fh)
for Protein in AGI_keep:
 ID = Protein
 name = "Unknown"
 GO_Terms = "Unknown"
 if Protein in AGI_name:
 name = AGI_name[Protein]
 if Protein in AGI_GO:
 GO_Terms = ",".join(AGI_GO[Protein])
 print(ID, name, GO_Terms, sep="|", file=fh)
fh.close()

# Write the AI1main_PPI_Interactions_neo.csv File.
fh = open("AI1main_PPI_Interactions_neo.csv", "w")
print(":START_ID|:END_ID", file=fh)
for a,b in Network.values.tolist():
 print(a,b, sep="|", file=fh)
fh.close()
```

3. The Neo4j Desktop needs to be activated. Before downloading the desktop application, fill in the required information on the download window. Once the form is submitted, a key string for activating the product will be displayed. Copy the key string or save it to a text file. After installation, open Neo4j Desktop, click on the "Software Keys" drawer, and then click on "Add activation Key" and paste in the entire contents of the activation key.
4. Configuring the APOC library: Click on the three-dot menu or kebab menu (extreme right of PPI_DBMS panel) and click on "Open folder" -> DBMS -- go to the "conf" folder. Inside the "conf" folder there should be multiple *.conf files (i.e., neo4j.conf, etc.). Create a new "apoc.conf" file inside conf folder. Save the files as ".conf" files under "all files" under the same "conf" folder after typing the following three lines in notepad:

```
apoc.import.file.enabled=true
apoc.import.file.use_neo4j_config=true
apoc.export.file.enabled=true
Then, restart the databases if already running.
```

5. Click on the three-dot menu or kebab menu (extreme right of PPI_DBMS panel) and click on "Open folder" -> Import; it will navigate to the "import" folder. Copy all files (i.e., PROTEIN_neo.csv and AI1main_PPI_Interactions_neo.csv) into the import folder.
6. If the "Store results?" option is selected on the algorithm configure panel, the results of the centrality analysis (such as degree, betweenness, etc.) will be transferred to the network. Click on a single node in the Neo4j Browser to determine whether centrality values are assigned to it or not.

Acknowledgments

This work was supported by the National Science Foundation awards IOS-2038872 to M.S.M.

References

1. Wang S, Wu R, Lu J et al (2022) Protein-protein interaction networks as miners of biological discovery. Proteomics:e2100190. https://doi.org/10.1002/pmic.202100190
2. Walport LJ, Low JKK, Matthews JM et al (2021) The characterization of protein interactions – what, how and how much? Chem Soc Rev 50(22):12292–12307. https://doi.org/10.1039/d1cs00548k
3. Wessling R, Epple P, Altmann S et al (2014) Convergent targeting of a common host protein-network by pathogen effectors from three kingdoms of life. Cell Host Microbe 16(3):364–375. https://doi.org/10.1016/j.chom.2014.08.004

4. Smakowska-Luzan E, Mott GA, Parys K et al (2018) An extracellular network of Arabidopsis leucine-rich repeat receptor kinases. Nature 553(7688):342–346. https://doi.org/10.1038/nature25184
5. Mukhtar MS, Carvunis AR, Dreze M et al (2011) Independently evolved virulence effectors converge onto hubs in a plant immune system network. Science 333(6042):596–601. https://doi.org/10.1126/science.1203659
6. Mott GA, Smakowska-Luzan E, Pasha A et al (2019) Map of physical interactions between extracellular domains of Arabidopsis leucine-rich repeat receptor kinases. Sci Data 6: 190025. https://doi.org/10.1038/sdata.2019.25
7. Mishra B, Sun Y, Howton TC et al (2018) Dynamic modeling of transcriptional gene regulatory network uncovers distinct pathways during the onset of Arabidopsis leaf senescence. NPJ Syst Biol Appl 4:35. https://doi.org/10.1038/s41540-018-0071-2
8. Mishra B, Sun Y, Ahmed H et al (2017) Global temporal dynamic landscape of pathogen-mediated subversion of Arabidopsis innate immunity. Sci Rep 7(1):7849. https://doi.org/10.1038/s41598-017-08073-z
9. Mishra B, Kumar N, Shahid Mukhtar M (2022) A rice protein interaction network reveals high centrality nodes and candidate pathogen effector targets. Comput Struct Biotechnol J 20:2001–2012. https://doi.org/10.1016/j.csbj.2022.04.027
10. Mishra B, Kumar N, Mukhtar MS (2021) Network biology to uncover functional and structural properties of the plant immune system. Curr Opin Plant Biol 62:102057. https://doi.org/10.1016/j.pbi.2021.102057
11. Mishra B, Kumar N, Mukhtar MS (2019) Systems biology and machine learning in plant-pathogen interactions. Mol Plant-Microbe Interact 32(1):45–55. https://doi.org/10.1094/MPMI-08-18-0221-FI
12. McCormack ME, Lopez JA, Crocker TH et al (2016) Making the right connections: network biology and plant immune system dynamics. Curr Plant Biol 5:2–12
13. Lopez J, Mukhtar MS (2017) Mapping protein-protein interaction using high-throughput yeast 2-hybrid. Methods Mol Biol 1610:217–230. https://doi.org/10.1007/978-1-4939-7003-2_14
14. Kumar N, Mishra B, Mukhtar MS (2022) A pipeline of integrating transcriptome and interactome to elucidate central nodes in host-pathogens interactions. STAR Protoc 3(3): 101608. https://doi.org/10.1016/j.xpro.2022.101608
15. Kumar N, Mishra B, Mehmood A et al (2020) Integrative network biology framework elucidates molecular mechanisms of SARS-CoV-2 pathogenesis. iScience 23(9):101526. https://doi.org/10.1016/j.isci.2020.101526
16. Klopffleisch K, Phan N, Augustin K et al (2011) Arabidopsis G-protein interactome reveals connections to cell wall carbohydrates and morphogenesis. Mol Syst Biol 7:532. https://doi.org/10.1038/msb.2011.66
17. Gonzalez-Fuente M, Carrere S, Monachello D et al (2020) EffectorK, a comprehensive resource to mine for Ralstonia, Xanthomonas, and other published effector interactors in the Arabidopsis proteome. Mol Plant Pathol 21(10):1257–1270. https://doi.org/10.1111/mpp.12965
18. Garbutt CC, Bangalore PV, Kannar P et al (2014) Getting to the edge: protein dynamical networks as a new frontier in plant-microbe interactions. Front Plant Sci 5:312. https://doi.org/10.3389/fpls.2014.00312
19. Arabidopsis Interactome Mapping C (2011) Evidence for network evolution in an Arabidopsis interactome map. Science 333(6042): 601–607. https://doi.org/10.1126/science.1203877
20. Ahmed H, Howton TC, Sun Y et al (2018) Network biology discovers pathogen contact points in host protein-protein interactomes. Nat Commun 9(1):2312. https://doi.org/10.1038/s41467-018-04632-8
21. Liu X, Salokas K, Weldatsadik RG et al (2020) Combined proximity labeling and affinity purification-mass spectrometry workflow for mapping and visualizing protein interaction networks. Nat Protoc 15(10):3182–3211. https://doi.org/10.1038/s41596-020-0365-x
22. Zahiri J, Yaghoubi O, Mohammad-Noori M et al (2013) PPIevo: protein-protein interaction prediction from PSSM based evolutionary information. Genomics 102(4):237–242. https://doi.org/10.1016/j.ygeno.2013.05.006
23. Singh R, Park D, Xu J et al (2010) Struct2Net: a web service to predict protein-protein interactions using a structure-based approach. Nucleic Acids Res 38(Web Server issue): W508–W515. https://doi.org/10.1093/nar/gkq481
24. Kozakov D, Hall DR, Xia B et al (2017) The ClusPro web server for protein-protein docking. Nat Protoc 12(2):255–278. https://doi.org/10.1038/nprot.2016.169
25. Planas-Iglesias J, Marin-Lopez MA, Bonet J et al (2013) iLoops: a protein-protein interaction prediction server based on structural features. Bioinformatics 29(18):2360–2362.

https://doi.org/10.1093/bioinformatics/btt401

26. Garcia-Garcia J, Schleker S, Klein-Seetharaman J et al (2012) BIPS: BIANA interolog prediction server. A tool for protein-protein interaction inference. Nucleic Acids Res 40(Web Server issue):W147–W151. https://doi.org/10.1093/nar/gks553
27. Chen H, Zhou HX (2005) Prediction of interface residues in protein-protein complexes by a consensus neural network method: test against NMR data. Proteins 61(1):21–35. https://doi.org/10.1002/prot.20514
28. Alanis-Lobato G, Schaefer MH (2020) Generation and interpretation of context-specific human protein-protein interaction networks with HIPPIE. Methods Mol Biol 2074:135–144. https://doi.org/10.1007/978-1-4939-9873-9_11
29. Oughtred R, Rust J, Chang C et al (2021) The BioGRID database: a comprehensive biomedical resource of curated protein, genetic, and chemical interactions. Protein Sci 30(1):187–200. https://doi.org/10.1002/pro.3978
30. Szklarczyk D, Gable AL, Lyon D et al (2019) STRING v11: protein-protein association networks with increased coverage, supporting functional discovery in genome-wide experimental datasets. Nucleic Acids Res 47(D1):D607–D613. https://doi.org/10.1093/nar/gky1131
31. Yang X, Yang S, Qi H et al (2020) PlaPPISite: a comprehensive resource for plant protein-protein interaction sites. BMC Plant Biol 20(1):61. https://doi.org/10.1186/s12870-020-2254-4
32. Safari-Alighiarloo N, Taghizadeh M, Tabatabaei SM et al (2017) Identification of new key genes for type 1 diabetes through construction and analysis of protein-protein interaction networks based on blood and pancreatic islet transcriptomes. J Diabetes 9(8):764–777. https://doi.org/10.1111/1753-0407.12483
33. Re A, Lecca P (2020) On TD-WGcluster: theoretical foundations and guidelines for the user. Methods Mol Biol 2074:233–262. https://doi.org/10.1007/978-1-4939-9873-9_17
34. Hansen DL, Shneiderman B, Smith MA et al (2020) Social network analysis: measuring, mapping, and modeling collections of connections. In: Analyzing social media networks with NodeXL, pp 31–51. https://doi.org/10.1016/B978-0-12-817756-3.00003-0
35. Golbeck J (2013) Network structure and measures. In: Analyzing the social web, pp 25–44. https://doi.org/10.1016/B978-0-12-405531-5.00003-1
36. Hansen D, Shneiderman B, Smith M et al (2020) Calculating and visualizing network metrics. In: Analyzing social media networks with NodeXL, pp 79–94. https://doi.org/10.1016/B978-0-12-817756-3.00006-6
37. Neo4j graph data platform – the leader in graph databases (2022). https://neo4j.com/
38. Hagberg A, Swart P, S Chult D (2008) Exploring network structure, dynamics, and function using NetworkX. Los Alamos National Lab (LANL), Los Alamos
39. Csardi G, Nepusz T (2006) The igraph software package for complex network research. Int J Complex Syst 1695(5):1–9
40. Peixoto TP (2014) The graph-tool python library. figshare. https://doi.org/10.6084/m9.figshare.1164194.v14
41. tnet: weighted, two-mode, and longitudinal networks analysis (2020). 3.0.16 edn
42. Shannon P, Markiel A, Ozier O et al (2003) Cytoscape: a software environment for integrated models of biomolecular interaction networks. Genome Res 13(11):2498–2504. https://doi.org/10.1101/gr.1239303
43. Bastian M, Heymann S, Jacomy M (2009) Gephi: an open source software for exploring and manipulating networks. In: Proceedings of the international AAAI conference on web and social media, vol 1, pp 361–362
44. Junker BH, Koschutzki D, Schreiber F (2006) Exploration of biological network centralities with CentiBiN. BMC Bioinform 7(1):219. https://doi.org/10.1186/1471-2105-7-219
45. Kyrola A, Blelloch G, Guestrin C GraphChi: large-scale graph computation on just a PC. In: 10th USENIX symposium on operating systems design and implementation (OSDI 12), pp 31–46
46. AllegroGraph. https://allegrograph.com/
47. Jalili M, Salehzadeh-Yazdi A, Asgari Y et al (2015) CentiServer: a comprehensive resource, web-based application and R package for centrality analysis. PLoS One 10(11):e0143111. https://doi.org/10.1371/journal.pone.0143111
48. Jalili M, Salehzadeh-Yazdi A, Gupta S et al (2016) Evolution of centrality measurements for the detection of essential proteins in biological networks. Front Physiol 7:375
49. Stark C, Breitkreutz BJ, Reguly T et al (2006) BioGRID: a general repository for interaction datasets. Nucleic Acids Res 34(Database issue):

D535–D539. https://doi.org/10.1093/nar/gkj109

50. Szklarczyk D, Franceschini A, Wyder S et al (2015) STRING v10: protein-protein interaction networks, integrated over the tree of life. Nucleic Acids Res 43(Database issue):D447–D452. https://doi.org/10.1093/nar/gku1003
51. von Mering C, Jensen LJ, Snel B et al (2005) STRING: known and predicted protein-protein associations, integrated and transferred across organisms. Nucleic Acids Res 33(Database issue):D433–D437. https://doi.org/10.1093/nar/gki005
52. Aranda B, Achuthan P, Alam-Faruque Y et al (2010) The IntAct molecular interaction database in 2010. Nucleic Acids Res 38(Database issue):D525–D531. https://doi.org/10.1093/nar/gkp878
53. Hermjakob H, Montecchi-Palazzi L, Lewington C et al (2004) IntAct: an open source molecular interaction database. Nucleic Acids Res 32(Database issue):D452–D455. https://doi.org/10.1093/nar/gkh052
54. Kerrien S, Aranda B, Breuza L et al (2012) The IntAct molecular interaction database in 2012. Nucleic Acids Res 40(Database issue):D841–D846. https://doi.org/10.1093/nar/gkr1088
55. Swarbreck D, Wilks C, Lamesch P et al (2008) The Arabidopsis information resource (TAIR): gene structure and function annotation. Nucleic Acids Res 36(Database issue):D1009–D1014. https://doi.org/10.1093/nar/gkm965

Index

A

B

C

Shahid Mukhtar (ed.), *Protein-Protein Interactions: Methods and Protocols*, Methods in Molecular Biology, vol. 2690, https://doi.org/10.1007/978-1-0716-3327-4,

D

E

F

G

Q

R

S